ng *Population Genomics Series* deals with the concepts and approaches
genomics and their applications in addressing fundamental and applied
vide variety of organisms. Population genomics is a fast emerging
hich has created a paradigm shift in many fields of life and medical
luding population biology, ecology, evolution, conservation, agricul-
ture, forestry, fisheries, human health and medicine.

n genomics has revolutionized various disciplines of biology including
evolutionary, ecological and conservation genetics, plant and animal
man health, genetic medicine, and pharmacology by allowing to address
ng-standing intractable questions with unprecedented power and accu-
loys large-scale or genome-wide genetic information across individuals
ons and bioinformatics, and provides a comprehensive genome-wide
nd new insights that were not possible before.

n genomics has provided novel conceptual approaches, and is tremen-
ncing our understanding the roles of evolutionary processes, such as
enetic drift, gene flow, and natural selection, in shaping up genetic
individual loci and across the genome and populations, disentangling
ecific effects from the genome-wide effects, detecting and localizing the
enomic elements, improving the assessment of population genetic
r processes such as adaptive evolution, effective population size, gene
ture, inbreeding and outbreeding depression, demography and biogeog-
esolving evolutionary histories and phylogenetic relationships of extant
pecies. Population genomics research is also providing key insights into
basis of fitness, local adaptation, ecological and climate acclimation and
speciation, complex ecologically and economically important traits, and
insect resistance in plants, animals and/or humans. In fact, population
esearch has enabled the identification of genes and genetic variants
vith many disease conditions in humans, and it is facilitating genetic
nd pharmacology. Furthermore, application of population genomics
d approaches can facilitate plant and animal breeding, forensics, delin-
nservation genetic units, understanding evolutionary and genetic impacts
management practices and climate and environmental change, and
n and sustainable management of plant and animal genetic resources.

me editors in this Series have been carefully selected and topics written
scholars from around the world.

Population Genomics

This pioneer
of populatio
topics in a
discipline, w
sciences, inc
ture, horticu
Populatio
population,
breeding, hu
novel and lo
racy. It emp
and populat
perspective
Populatio
dously adva
mutation, g
variation at
the locus-sp
functional
parameters
flow, admix
raphy, and
and extinct
the genomic
adaptation,
disease and
genomics
associated
medicine a
concepts an
eation of co
of resource
conservatio
The vol
by leading

Charlotte Lindqvist • Om P. Rajora
Editors

Paleogenomics

Genome-Scale Analysis of Ancient DNA

Springer

Editors
Charlotte Lindqvist
Department of Biological Sciences
University at Buffalo (SUNY)
Buffalo, NY, USA

School of Biological Sciences
Nanyang Technological University
Singapore, Singapore

Om P. Rajora
Faculty of Forestry and Environmental
Management
University of New Brunswick
Fredericton, NB, Canada

ISSN 2364-6764 ISSN 2364-6772 (electronic)
Population Genomics
ISBN 978-3-030-04752-8 ISBN 978-3-030-04753-5 (eBook)
https://doi.org/10.1007/978-3-030-04753-5

Library of Congress Control Number: 2018966339

This Springer imprint is published by the registered company Springer Nature Switzerland AG
The registered company address is: Gewerbestrasse 11, 6330 Cham, Switzerland

To my children, Torben and Siri.

Charlotte Lindqvist

Respectfully dedicated to my late parents, Shri Than Singh and Shrimati Bhagwati Rajora.

Om P. Rajora

Preface

Paleogenomics centers on recovering the DNA of historic or long-dead organisms to reconstruct and analyze their genomes. Ancient DNA (aDNA) refers to the preserved but often highly degraded genetic material recovered from remains found among paleontological and archaeological settings and in museum and other archival collections, ranging in age from hundreds of thousands of years to less than 100 years old. In the last 30 years, aDNA analysis has grown into a compelling research tool that has radically transformed many scientific fields. Providing a direct window into past organisms, environments, and events, analysis of aDNA has been applied to fields as diverse as evolutionary biology, ecology, anthropology, agriculture, and medicine. Over the last decade, advances in high-throughput DNA sequencing technologies have revolutionized aDNA research and engendered a substantial growth in the volume of aDNA sequence data and publications. Entire nuclear genomes of ancient individuals and extinct species have become approachable targets for in-depth research. New insights into such ancient genomes, or paleogenomes, and their links to modern ones, hold incredible promise to capture, on the fly, organismal evolution as it happens, and the responses of organisms and their genomes to a changing world.

This book presents a review of recent technological developments, methodological advances, and applications in paleogenomics and provides research examples from pathogens to primates. Paleogenomics is a rapidly evolving field, and the goal of this book is to present a snapshot view of its history, current status, and future prospects, taking a broad viewpoint covering a range of topics. The contributors of this book include researchers who pioneered studies in paleogenomics and who are still at the forefront of the most recent developments and applications in aDNA research. The front-running topic in paleogenomics research has clearly been the study of human evolution, wherein many exciting new findings have been made regarding our own evolutionary history and that of our extinct relatives. With this book, however, we strove to not only provide an overview of advances and challenges in methodology, which are specifically discussed in the first part of the book ("Concepts, Technical Advances and Challenges"), but also provide a taste of the wealth and breadth of paleogenomics applications. Hence, in the second part of the book ("Paleogenomics Case Studies: From Ancient Pathogens to Primates"), topics are covered that include ancient human pathogens and viruses, reconstruction of past environments, domestication of crop plants, use of museum collections, such as herbaria, animal domestication (with specific chapters on dog, cat, and horse evolution), challenges of studying nonhuman primates, genomic structural variants in human and other ancient genomes, and insights into the extinction process itself. We anticipate that this diversity of topics will be of interest to a variety of readers, including undergraduate and graduate students, professionals and experts in the field, as well as anyone excited by the extraordinary scientific insights that paleogenomics can offer. We thank the many contributors to the volume and are grateful to the numerous peer reviewers of all chapters.

Buffalo, NY, USA — Charlotte Lindqvist
Fredericton, NB, Canada — Om P. Rajora

Contents

Contributors

Robin G. Allaby School of Life Sciences, University of Warwick, Coventry, UK

Freek T. Bakker Biosystematics Group, Wageningen University & Research, Wageningen, The Netherlands

Justin R. Bradley Department of Biological Sciences, University at Buffalo, The State University of New York (SUNY), Buffalo, NY, USA

Michael Buckley School of Earth and Environmental Sciences, Manchester Institute of Biotechnology, University of Manchester, Manchester, UK

Love Dalén Department of Bioinformatics and Genetics, Swedish Museum of Natural History, Stockholm, Sweden

Evangelos A. Dimopoulos The Palaeogenomics and Bio-Archaeology Research Network, Research Laboratory for Archaeology and History of Art, University of Oxford, Oxford, UK

Laurent A. F. Frantz The Palaeogenomics and Bio-Archaeology Research Network, Research Laboratory for Archaeology and History of Art, University of Oxford, Oxford, UK

School of Biological and Chemical Sciences, Queen Mary University of London, London, UK

Eva-Maria Geigl Institut Jacques Monod, CNRS, UMR 7592, University Paris Diderot, Paris, France

M. Thomas P. Gilbert Centre for GeoGenetics, Natural History Museum of Denmark, University of Copenhagen, Copenhagen, Denmark

Norwegian University of Science and Technology, University Museum, Trondheim, Norway

Jamie Wood Long-Term Ecology Laboratory, Landcare Research, Lincoln, Canterbury, New Zealand

Duo Xu Department of Biological Sciences, University at Buffalo, The State University of New York (SUNY), Buffalo, NY, USA

Part I
Concepts, Technical Advances and Challenges

Technical Advances and Challenges in Genome-Scale Analysis of Ancient DNA

Tianying Lan and Charlotte Lindqvist

Abstract Through improvements in methods to recover ancient DNA (aDNA) and applications of high-throughput DNA sequencing technology, aDNA research has entered into a new era of paleogenomics. Technical advances, including improved knowledge of DNA degradation patterns, identification of variation in endogenous DNA yield in particular skeletal elements, and aDNA-specific modifications to DNA extraction and library construction methods, significantly improved aDNA recovery rates and conversion efficiency of aDNA into sequencing libraries. Thus, in recent years, this has enabled whole-genome sequencing of numerous ancient individuals and extinct species, as well as substantial growth in the volume of ancient DNA studies and sequence data. These ever-growing datasets as well as the degraded nature of ancient DNA have also brought in challenges and innovations to bioinformatic methods for data processing and analysis. Applications of bioinformatic tools should be proceeded with caution by evaluating their feasibility and accuracy on aDNA data, and more standardized analytical pipelines are in high demand. In this chapter, we provide an overview of some of the major technical advances and challenges in aDNA and paleogenomic research.

Keywords Bioinformatics · Contamination · Library preparation · Next-generation sequencing · Paleogenomics · Postmortem DNA damage · Targeted enrichment

---

T. Lan · C. Lindqvist (✉)
Department of Biological Sciences, University at Buffalo (SUNY), Buffalo, NY, USA

School of Biological Sciences, Nanyang Technological University, Singapore, Singapore
e-mail: CL243@buffalo.edu

Charlotte Lindqvist and Om P. Rajora (eds.), *Paleogenomics*,
Population Genomics [Om P. Rajora (Editor-in-Chief)],
https://doi.org/10.1007/13836_2018_54,

patterns from high-throughput sequencing, today most aDNA studies follow certain standards, including working in a dedicated, physically isolated aDNA lab area, which often follows clean room requirements. However, even when criteria for authenticity are stringently followed, one can never guarantee that contamination is entirely avoided (Gilbert et al. 2005). Thus, it is crucial to assess the risk of contamination throughout any aDNA study and, more importantly, to examine potential contamination at each experimental stage, as well as among the obtained results.

3 Ancient DNA Extraction Methods

3.1 Ancient DNA Extraction from Animal Remains

The ultrashort length and low abundance of endogenous DNA fragments, together with the potential large amount of exogenous contamination, have brought many challenges to the recovery of DNA from degraded and ancient specimens. Hence, it has been vital to develop extraction methods that not only recover DNA molecules efficiently but also remove contaminating compounds that may inhibit downstream enzymatic reactions. Generally, the teeth and bone are excellent sources for aDNA, but systematic comparisons of different portions of skeletal remains have demonstrated that endogenous DNA yield differs across different parts of the bones or teeth. For example, although the inner dentine part of the tooth has commonly been used when extracting DNA from teeth, significantly higher yield of endogenous DNA has recently been observed in the tooth cementum layer (Adler et al. 2011; Higgins et al. 2013; Damgaard et al. 2015). Also, the denser part of the petrous bone was shown to contain higher DNA content than less dense skeletal elements (Gamba et al. 2014; Pinhasi et al. 2015) and even tooth cementum (Hansen et al. 2017). Thus, tooth cementum and petrous bones are both preferred substrates in paleogenomic studies. However, often high sample-to-sample variation is present, and limited material prevents access to optimal substrates.

During the last three decades, a range of aDNA extraction methods have been developed and shown success across various types of animal remains, not only the bone and teeth (Hagelberg et al. 1994; Rohland and Hofreiter 2007) but also hair (Barnes et al. 2002; Gilbert et al. 2004), feces (Poinar et al. 1998), and soft mummified tissues (Pääbo 1985; Ermini et al. 2008; Gamba et al. 2016). In brief, the aDNA extraction procedure generally consists of the following steps: homogenization/powdering, lysis/digestion, purification, and elution. Depending on what type of tissue is being processed, the homogenization step can be conducted in different ways, such as using a dental drill, pestle and mortar, or cryogenic mill. It is not well-known what effect drill speeds and the generation of heat have on the degradation and denaturing of DNA, but a study suggested that drilling at standard drill speeds (c. 1,000 RPM) could generate large heat loadings and thus decreased DNA yields up to 30 times compared to powdering by cryogenic mill (Adler et al. 2011). However, drilling is still often preferred, especially when the specimen's morphological integrity needs to be maintained (O'Rourke et al. 2000). Therefore, using a lower drill speed (e.g., 100 RPM) is recommended (Adler et al. 2011). The lysis step often takes place in a digestion buffer that is used for breaking down cells and releasing the DNA. For example, the digestion buffer for lysing bones often contains EDTA and proteinase K to disintegrate hydroxyapatite and collagen, respectively, whereas that for lysing hair or nail cells often contains sodium dodecyl sulfate (SDS), dithiothreitol (DTT), and proteinase K, which will break down keratin. Modifications to the lysis/digestion step for improving endogenous DNA yield and reducing the exogenous fraction have been widely tested out. For example, a brief predigestion (15–30 min) with EDTA and proteinase K has shown success in increasing proportions of endogenous DNA by several folds (Damgaard et al. 2015). Likewise, increased aDNA yield was observed by using a secondary digestion and decalcification with a lysis buffer containing N-laurylsarcosyl detergent solution (Gamba et al. 2016). Similar detergent solutions have also previously been implemented in aDNA extraction protocols (Richards et al. 1995; Gamba et al. 2014; Pinhasi et al. 2015). Furthermore, it has been found that decontamination pre-treatments of bone powders using phosphate buffer or sodium hypochlorite (bleach) can significantly reduce contaminating microbial DNA content and increase proportions of endogenous DNA [Korlevic et al. 2015; Boessenkool et al. 2017)]. However, this is at the expense of a dramatic reduction in the absolute amount of endogenous DNA (Korlevic et al. 2015, T. Lan unpubl.; Basler et al. 2017). Contrarily, one study on ancient dental calculus observed no significant DNA loss with pre-treatment using bleach or EDTA buffer (Warinner et al. 2014). Another study also indicated there was no significant difference in total DNA yield, endogenous DNA content, and number of unique reads (Nieves-Colón et al. 2017). These inconsistent observations suggest that the influence of pre-treatment decontamination may be sample-dependent, likely because the conditions of ancient samples vary substantially due to distinct preservation environments and tissue types.

Purification is the most critical step that determines DNA recovery rate and contaminant removal efficiency. Based on different purification methods, a series of protocols have been developed. In early aDNA studies in the late 1980s, one of the most commonly adopted protocols was the phenol-chloroform-based method, which isolates DNA molecules from other components by repeated separation into hydrophobic and aqueous phases (Hagelberg et al. 1989; Hagelberg and Clegg 1991). To date, this method has stayed popular due to its ease of application, efficient removal of inhibitors, and low cost (Miller et al. 2008; Keller et al. 2012; Hofmanova et al. 2016). However, it should be noted that both phenol and chloroform are highly toxic, and researchers should apply this method with caution. In 1998, Yang et al. (1998) simplified the purification step by using silica-based spin columns and demonstrated the efficiency for purifying DNA from ancient human bone (hereafter referred to as the column-based protocol). About a decade later, a two-step silica particle-based purification method was developed by Rohland and Hofreiter (2007), where the first step uses silica particles that suspend together with a binding buffer containing a chaotropic salt, guanidine isothiocyanate (GuSCN), and other sodium salts to adsorb the DNA molecules, followed by a second step that removes contaminating chemicals and salts with an ethanol washing step (this method is hereafter referred to as the in-solution silica-based protocol). With greater DNA recovery efficiency and PCR success rate than alternative approaches, this protocol soon became one of the most widely adopted DNA extraction methods in aDNA studies, including ancient genome analysis of Neanderthal (Green et al. 2010; Prufer et al. 2014), Denisovan (Reich et al. 2010; Meyer et al. 2012), ancient human (Fu et al. 2014; Rasmussen et al. 2014; Allentoft et al. 2015), and other mammals (Schubert et al. 2014; Librado et al. 2015). Three years later, Rohland and colleagues simplified the washing and elution steps by using commercially available silica-based spin columns as had been used by Yang et al. (1998), thus largely increasing the sample throughput (Rohland et al. 2010). More recently, Dabney and colleagues further optimized the column-based protocol by using another binding buffer containing elevated concentration of guanidine hydrochloride as the chaotropic salt (Dabney et al. 2013). This optimized column-based protocol has shown not only an increased aDNA yield, particularly in the yield of ultrashort fragments (<50 bp), but also in the recovering of DNA fragments of all sizes (35–150 bp) (Dabney et al. 2013; Gamba et al. 2016). It is probably today one of the most commonly used aDNA extraction methods, and it has been applied in many recent aDNA studies, including genome studies of ancient humans and mammals (Lazaridis et al. 2014; Palkopoulou et al. 2015; Park et al. 2015; Cassidy et al. 2016; Frantz et al. 2016; Jeong et al. 2016). A comparison study of the column-based and in-solution protocols showed the silica column-based methods to perform better than the in-solution method in terms of endogenous DNA yield and their molecular diversity when extracting DNA from ancient bone samples (Gamba et al. 2016). However, the in-solution method may be more efficient when extracting large amounts of DNA as the volume of silica solution can be adjusted more easily without the limitation from spin columns.

While these advances in aDNA extraction protocols have greatly improved the recovery of authentic DNA from highly degraded samples, further studies with additional optimizations and screening of different types of samples, including from lower latitude and warmer regions and less optimal substrates, are needed, especially to address the problem of removal of inhibitors and exogenous human and microbial contamination without loss of endogenous DNA.

3.2 Ancient DNA Extraction from Plants

Depending on sampling location, plant remains, such as seeds, pollen, and wood, are often found charred, waterlogged, desiccated, or mineralized. Particularly, plant seeds (or fruits) and pollen, usually protected by sturdy outer layers, can be extremely resistant to chemical and physical damages, providing an ideal source for long-term DNA preservation (Gugerli et al. 2005; Schlumbaum et al. 2008). A major challenge to isolating DNA from plant remains, however, is that plant tissues are usually rich in secondary by-products, sugars, and other potential PCR inhibitors, which likely interfere with DNA yield and downstream applications (Schlumbaum et al. 2008). Therefore, a lysis buffer containing cetyltrimethylammonium bromide (CTAB)/dodecyltrimethylammonium bromide (DTAB) and polyvinylpyrrolidone (PVP) is commonly used to break down and remove these compounds (Allaby et al. 1997). The use of *N*-phenacylthiazolium bromide (PTB) was suggested to take apart complexes formed by the Maillard reaction (Poinar 2002). Protocols such as CTAB/DTAB methods, PTB-based methods, silica-based methods, or commercially available DNA extraction kits have all been successfully applied to various plant tissue remains, with minor adjustments according to sample conditions (Allaby et al. 1997; Schlumbaum et al. 1998; Dumolin-Lapègue et al. 1999; Blatter et al. 2002; Jaenicke-Despres et al. 2003; Pollmann et al. 2005; Liepelt et al. 2006; Bilgic et al. 2016). Interestingly, a method originally implemented for extracting DNA from ancient hair samples (Gilbert et al. 2004), however, was shown to outperform several other methods in a comprehensive comparison of aDNA extraction methods on non-carbonized archaeobotanical remains (Wales et al. 2014). Furthermore, an extraction protocol combining the use of a PTB/DTT buffer mixture during lysis along with the DNA binding procedure used to recover short molecules from animal bones (Dabney et al. 2013) was recently shown to increase the distribution of ultrashort DNA fragments from herbarium specimens when compared to using conventional CTAB and Qiagen DNeasy Mini Spin Columns (Gutaker et al. 2017). Although genetic studies based on herbarium specimens have been performed for decades, aDNA research on plants from archaeological and paleontological contexts has lagged behind equivalent work on animals. For example, the different preservation methods used for treating herbarium specimens (see Bakker 2018), the often high levels of secondary compounds and polysaccharides in plants as mentioned above, and the difficulties in obtaining DNA from plant remains, such as wood, seeds, fruits, or pollen (see Parducci et al. 2018) all remain a challenge in plant paleogenomics. Continued optimizations and further testing with different types of remains will likely allow for improvement of plant aDNA recovery in the future.

4 Next-Generation Sequencing and Library Preparation

Before the advent of NGS technologies, conventional PCR followed by Sanger sequencing was routinely implemented in aDNA research to obtain DNA sequences. Although it is still widely used for sequencing smaller sets of aDNA fragments or in the initial stages of identifying an unknown specimen or assessing degree of degradation, it is not feasible for large-scale genomic studies due to its limited throughput and high cost. On the contrary, highly parallelized NGS technologies can generate millions to billions of DNA sequence reads, making it possible to obtain whole-genome-scale data within a short period and at much lower cost per base. Also, the fragmented nature of aDNA is well-suited for NGS sequencing because an early step in NGS workflows is often to break the modern DNA molecules into smaller pieces, e.g., through enzyme-based treatment, nebulization,

method, introduced by Illumina, ligates a single, Y-shaped adapter with a T-overhang to both ends of DNA molecules that carry A-overhangs (Fig. 1b) (Bentley et al. 2008). The directionality of 3′ A-overhangs ensures that each molecule is ligated to distinct adapter pairs, therefore avoiding massive loss of DNA due to non-distinct ligation. Importantly, though, blunt-end libraries have nevertheless in some cases produced higher endogenous contents than libraries with AT-overhang (pers. obs., Seguin-Orlando et al. 2013), and it has been shown that library preparation procedures based on AT-overhang adapter ligation can introduce a bias in base composition by limiting the ability to ligate DNA templates starting with thymines and therefore deaminated cytosines (Seguin-Orlando et al. 2013). An enzymatic treatment that uses uracil-DNA glycosylase (UDG) to remove deaminated cytosines (uracils), and endonuclease VIII to cut at the resulting abasic sites, has shown great success in reducing the number of miscoding lesions effectively (Briggs et al. 2010). For quick library screening, partial UDG treatment can be used to repair damaged bases in the interior of DNA molecules while preserving them at the termini for authentication (Rohland et al. 2015). Other adaptations, for example, using heat inactivation of the enzymes rather than spin column-based purification, replacing NaOH with heat to retrieve DNA molecules from streptavidin-coated beads, double indexing, and amplifying subsets of sample in a number of parallel PCR, have also been proposed to reduce DNA loss and increase the conversion efficiency (Maricic and Paabo 2009; Kircher 2012; Fortes and Paijmans 2015a; Carøe et al. 2018).

In contrast to the double-stranded methods, a single-stranded library preparation method that specifically takes account of the nature of aDNA – extremely low quantities of highly fragmented and damaged DNA molecules – was developed for sequencing on the Illumina platform (Fig. 1c) (Gansauge and Meyer 2013) and later adapted to the Ion Torrent sequencing platform (Bennett et al. 2014). This method first ligates single-stranded aDNA molecules to a 3′-biotinylated adapter using a single-stranded ligase, CircLigase, and immobilizes the ligation product to streptavidin beads. Next, a primer complementary to the adapter is used to copy the bound-to-bead template strand, and a second adapter is then joined to the double-stranded molecules by blunt-end ligation. The newly synthesized strand, which is not attached to the beads, is then released by heat denaturation and eluted to allow PCR amplification using barcoded primers (Gansauge and Meyer 2013). A recent upgrade of the single-stranded library preparation, ssDNA2.0 (Fig. 1d), replaces CircLigase with T4 DNA ligase and uses a splinter oligonucleotide with a stretch of random bases that are hybridized to the 3′-biotinylated adapter, making this method less costly and better compatible with automation (Gansauge et al. 2017).

Compared to the double-stranded methods, the single-stranded preparation method offers several advantages. For example, DNA molecules are tightly bound to streptavidin-coated beads during all reaction steps, which avoids the loss of molecules in purification when using silica spin columns or carboxylated beads. Moreover, DNA molecules with single-strand breaks on both strands are entirely lost in double-stranded library preparation, whereas they can be recovered as multiple fragments upon heat denaturation with the single-stranded method (Gansauge and

Meyer 2013). Systematic comparisons of these two library construction strategies have found that the single-stranded method was able to recover more short reads and allowed conversion of a higher ratio of endogenous to exogenous DNA (Bennett et al. 2014). The effectiveness of the single-stranded method has been demonstrated in several aDNA studies (Meyer et al. 2012; Fu et al. 2014; Frantz et al. 2016; Ramos-Madrigal et al. 2016), and it is currently viewed to be the most efficient aDNA library preparation method with conversion efficiencies of about 30–70 (Gansauge and Meyer 2013). However, there are noteworthy limitations to this method as well. For example, it is more costly and time-consuming than double-stranded library preparation, and the conversion efficiency drops for molecules longer than 120 bp (Gansauge and Meyer 2013). Furthermore, studies have found that the benefit of the single-stranded method is inversely proportional to the endogenous content of the sample and/or amount of sequencing undertaken (Barlow et al. 2016; Sandoval-Velasco et al. 2017). Therefore, the double-stranded protocol is generally still the method of choice for many projects, provided that the samples are reasonably well-preserved or sufficient material is available for destructive sampling. For critical or poorly preserved samples, one strategy could be to use the double-stranded protocol first for screening and then switch to the single-stranded protocol for obtaining more data.

4.2 *Targeted Enrichment*

Attaining high sequencing depth and accuracy from ancient samples often requires extensive sequencing when an aDNA library is directly sequenced. This is not only costly but also unfeasible in cases where ancient material is limited. To increase the sequencing focus on the endogenous fraction, targeted enrichment methods based on hybridization (also referred to as hybridization capture or targeted capture) have been implemented to enrich for selected genomic regions, such as organellar genomes, exomes, single chromosome, or even whole nuclear genomes (Briggs et al. 2009; Burbano et al. 2010; Maricic et al. 2010; Carpenter et al. 2013; Fu et al. 2013a, 2016; Castellano et al. 2014; Enk et al. 2014). Enk et al. (2014) showed that in-solution targeted enrichment was able to retrieve over 70% of endogenous DNA from a pre-enrichment library containing less than 5% of endogenous DNA. However, it should be noted that genome enrichment can generate high levels of sequence clonality (fraction of total mapped reads that are clonal duplicates of single original template molecule), which may lead to overestimation of enrichment rates, and thus it is crucial to remove clonal duplicates during the downstream data processing (Avila-Arcos et al. 2011; Ávila-Arcos et al. 2015). The general principle of hybridization-based targeted enrichment is to create a set of bait molecules with high sequence similarity to the target sequence of interest, which then hybridize to the DNA in a library (Fig. 2). The hybridized bait and target are then immobilized and eluted after non-hybridized background fragments are washed away, resulting in a much higher ratio of target versus contaminant DNA retained in

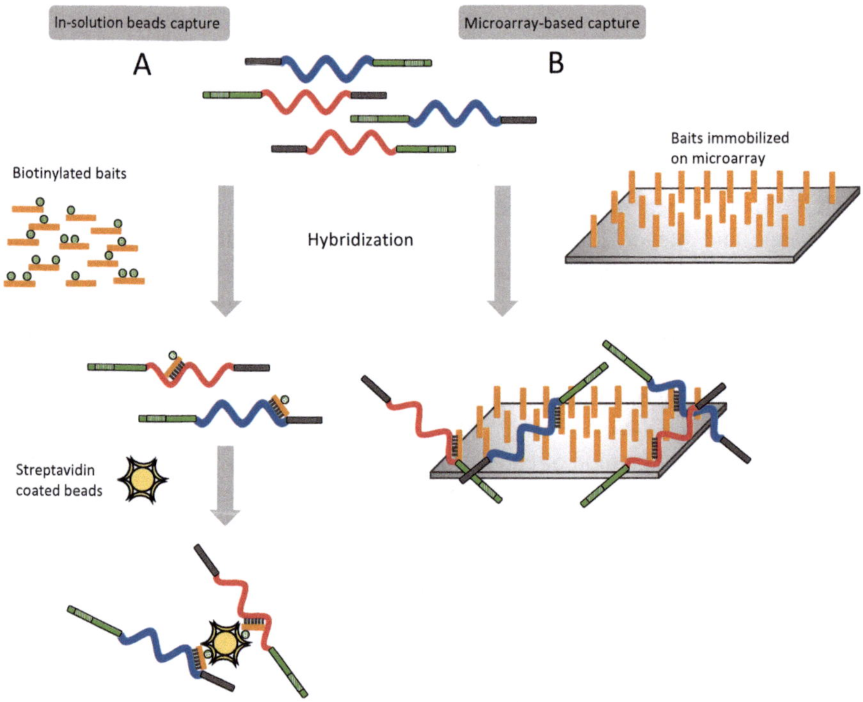

Fig. 2 Schematic of hybridization-based targeted enrichment process. Genomic DNA libraries are either hybridized to biotinylated target-specific baits and captured by streptavidin-coated beads in solution (**a**) or hybridized to baits that immobilized on a microarray surface (**b**). Uncaptured background DNA is then washed away, and the hybridized target DNA is released and eluted for subsequent sequencing

the library, and subsequent sequencing is therefore more cost- and time-effective (Mamanova et al. 2010). Thermodynamically, hybridization capture prefers shorter molecules, effectively introducing a bias against modern contaminating DNA, as these modern DNA fragments are generally expected to be longer than ancient fragments (Hodges et al. 2007). Moreover, hybridization capture tolerates higher number of mismatches than conventional PCR, enabling cross-species capture for non-model organisms or extinct lineages with no prior sequence information available by using baits designed from close (and modern) relatives. For example, mitochondrial enrichment has been performed with up to 25–27% sequence divergence between the probe and target (Peñalba et al. 2014; Paijmans et al. 2016).

Hybridization capture can either be conducted on a solid-state microarray or in solution. The former requires expensive hardware and a vast excess of DNA library over probes, whereas the latter is more cost-effective and uses a smaller quantity of sequencing library (Gnirke et al. 2009). Both methods have already shown great success in various aDNA studies (Burbano et al. 2010; Bos et al. 2014; Enk et al. 2014; Fortes and Paijmans 2015b; Haak et al. 2015; Bos et al. 2016; Duggan Ana et al. 2016; Fu et al. 2016; Paijmans et al. 2016; Spyrou et al. 2016), demonstrating exceptional advantages of the capture approach. Mitogenome enrichment, which has been conducted in many aDNA studies, usually delivers complete mitochondrial genomes with high depth of coverage owing to the abundant mtDNA survived in ancient samples (Krause et al. 2010; Maricic et al. 2010; Fu et al. 2013a, b; Orlando et al. 2013; Vilstrup et al. 2013; Zhang et al. 2013; Sarkissian et al. 2015; Mohandesan et al. 2017; Lan et al. 2017). Whole-genome capture has been successful in enriching Neanderthal (Carpenter et al. 2013), ancient human (Schroeder et al. 2015), and mammoth (Enk et al. 2014) DNA libraries and delivered the full bacterial genome of ancient *Yersinia pestis* strains (Bos et al. 2011; Schuenemann et al. 2011), as well as of historical leprosy strains (Schuenemann et al. 2013). However, obtaining decent depth of coverage through whole-genome capture is currently still very costly, particularly for samples with large genome size (e.g., mammals) and low levels of endogenous DNA content. In addition, it is likely impractical to efficiently custom-design whole-genome baits for extinct species with no genome information or modern DNA available from close relatives due to much higher genomic complexity compared to mitochondrial genomes alone. Alternatively, for these cases, it is more feasible to perform hybridization capture for a subset of the genome, for example, the exome, which is generally more conserved than other, noncoding regions of the genome, thus enabling cross species capture over medium evolutionary distances (Cosart et al. 2011; Bi et al. 2012). This is exemplified by a recent study, where Castellano et al. (2014) assembled complete exomes for three Neanderthals by using human coding exon sequences as baits. Similarly, Fu et al. (2013b) reconstructed all non-repetitive regions of chromosome 21 in an ancient anatomically modern human. Furthermore, McCormack et al. (2016) was able to obtain an average of 4,460 SNPs among 27 bird museum specimens through capture of 5,060 ultraconserved genomic elements (UCEs), a class of highly conserved nuclear markers distributed throughout the genomes of most organisms. Hybridization capture has also been successfully carried out in plant species. For example, da Fonseca et al. (2015) conducted enrichment of the exons of 348 genes from 32 archaeological maize cob samples and obtained an average depth of coverage ~10× on target, and Sánchez Barreiro et al. (2017) custom-designed 20,000 Restriction Enzyme-Associated Loci RNA baits (REALbaits) using genotyping by sequencing (GBS) and discovered 22,813 SNPs from a set of 38 historic ragweed samples.

It is important to note that aDNA hybridization enrichment is very sensitive, and its efficiency is subject to various factors and different capture parameters. Recent studies observed that the enrichment rate (post-enrichment endogenous DNA proportion to the pre-enrichment proportion) varied significantly between individual samples within the same experiment (Carpenter et al. 2013; Enk et al. 2013; Enk et al. 2014). Many factors, including sequence divergence between bait and target, hybridization and washing temperature, target length and concentration, target complexity, bait type and concentration, bait tiling strategy, bait length, all contribute to the enrichment rate (Avila-Arcos et al. 2011; Bodi et al. 2013; Li et al. 2013; Ávila-Arcos et al. 2015). Although studies have been carried out to investigate the

contamination rates, either from mitochondrial genome or nuclear genome. A popular mitogenome-based estimation method utilizes diagnostic sites, which are defined as those where a large dataset of modern human mitochondrial sequences differs from the ancient sample (Green et al. 2006, 2008; Meyer et al. 2012; Olalde et al. 2014; Prufer et al. 2014). A likelihood-based framework has also been applied to co-estimate sequencing error and contamination rates from ancient mitochondrial sequences (Fu et al. 2013a, b). Additionally, bioinformatic tools, such as Schmutzi (Renaud et al. 2015) and contamMix (Fu et al. 2013b), were developed to estimate contamination by quantifying deamination patterns. For autosomal contamination rate estimation, Meyer et al. (2012) developed a maximum likelihood-based method that co-estimate sequence error, contamination, and two population parameters (correlated to divergence and heterozygosity) from the archaic Denisovan genome. This method uses sites that appear fixed and derived in modern human genome as compared to great ape outgroups and assumes that any human contamination will contribute to these derived alleles for which the ancient individual is at least partially ancestral. For samples that are determined as female, a method to estimate contamination from the Y chromosome can be applied to detect male contamination by computing the number of aligned reads in Y-unique regions (Reich et al. 2010; Meyer et al. 2012; Prufer et al. 2014). Similarly, a method that can detect contamination from the X chromosome in male samples was proposed based on the fact that males have only one copy of the X chromosomes; therefore, any heterozygous sites detected on the X chromosome in males would suggest contamination (Korneliussen et al. 2014). For non-hominin samples, detection of modern DNA contamination can be conducted based on mitochondrial genome sequences. Reads that can be mapped to mitochondrial genomes from modern organisms (such as a modern human mitogenome) using certain criteria, e.g., mapping quality over 20 or shorter mapping edit distance, can be determined as a possible source of contamination (Schubert et al. 2012; Greig et al. 2015).

5.2 *Reconstruction of Ancient Genomes*

Reconstructing a genome from sequencing reads can be accomplished either by reference-guided mapping, aligning reads to a known reference genome, or by "de novo" assembly, assembling overlapped reads into longer fragments without the use of a reference genome. In aDNA studies, reference-guided mapping, e.g., using BWA (Li and Durbin 2010) and MIA (Green et al. 2008), is the most commonly used approach, as it performs better on aDNA sequence reads, which are usually short and low in abundance. However, the excessive postmortem DNA damage and high levels of exogenous DNA contamination in aDNA data, together with sequencing errors, can reduce the effectiveness of the mapping algorithm. To improve mapping efficiency and accuracy, several mapping strategies or parameters have been developed, including trimming likely damaged positions at reads termini (Meyer et al. 2012; Schubert et al. 2012; Prufer et al. 2014), merging overlapping short reads into long ones (Meyer et al. 2012; Prufer et al. 2014), disabling the seed option and tolerating up to two "gap opens" when using BWA (Schubert et al. 2012), and performing iterative mapping (Green et al. 2008; Hahn et al. 2013). Also, a combination of de novo assembly followed by mapping to related taxa has successfully reconstructed genomes of ancient pathogenic strain (Rajaraman et al. 2013). Furthermore, a tool designed for mapping divergent short reads, MaxSSmap, allows for mapping ancient reads that are rejected by other mapping tools, such as BWA, and with high accuracy and low error (Turki and Roshan 2014). However, another important limitation of reference-guided mapping is that as the sequence divergence between reference and reads increases, the performance of the algorithm decreases. Therefore, it can only be applied to taxa where reliable reference genomes from closely related representatives are available. For taxa that lack such reference genomic information, it represents an important challenge to reconstruct their genomes, and enhancements on assembly approaches need to be investigated by future studies.

5.3 Other Analytical Tools

After the paleogenomes are reconstructed, various downstream bioinformatic tools are then used to call variants (SNPs) or genotypes and infer evolutionary histories, including evolutionary relationships, demographic fluctuations, and admixture events (Fig. 3). Although this is far from an exhaustive review, here we summarize some of the most commonly used bioinformatic tools for interpreting paleogenomic data, as well as existing problems and concerns with using these tools.

Due to the fact that most downstream genomic analytical tools take SNPs or genotypes as input, the first step of the analytical pipeline is usually to call SNPs or genotypes from genotype likelihoods that are calculated based on the associated mapping and sequencing quality scores of the aligned reads. SNP calling programs such as SAMtools (Li et al. 2009) and the GATK suite (McKenna et al. 2010) are widely adopted in both modern and ancient genomic studies. However, these programs do not specifically account for postmortem damage of aDNA sequences, resulting in excess of C→T and G→A transitions in the variant calling output (Botigue et al. 2017). A recently developed variant caller, aDNA_GenoCaller, which incorporates postmortem damage patterns detected by mapDamage, was able to substantially decrease the overrepresentation of C→T and G→A sites that had been identified by GATK in ancient dog genomic data (Botigue et al. 2017). Another program, ANGSD, in which the algorithm is optimized for low- and medium-coverage data, uses genotype likelihoods or sample allele frequency likelihoods rather than called genotypes to perform downstream analyses, such as admixture inference, population genetic statistics, and principal component analyses (PCA) (Korneliussen et al. 2014).

Inference of phylogenetic relationship can be conducted either based on mitochondrial genomes or autosomal SNP dataset by using conventional phylogenetic tools, such as maximum likelihood-based RAxML (Stamatakis 2014) and PhyML (Guindon et al. 2010) and Bayesian-based BEAST (Bouckaert et al. 2014) and MrBayes (Ronquist and Huelsenbeck 2003). The PCA module in EIGENSTRAT suite (Price et al. 2006) is a commonly used multivariate analysis tool to explore the complex patterns of genetic variation and genetic affinity among individuals or populations. The hidden Markov model-based software PSMC (Li and Durbin 2011) and G-PhoCS (Gronau et al. 2011) have often been implemented to investigate demographic history, such as effective population size fluctuations and population divergence time. It is important to note, however, that PSMC inference may give false signals on recent timescales, which are often of interest in aDNA studies, and for genomes below an average of 20× depth coverage, which is almost always the case for ancient genomes, the false-negative rate (FNR) correction is required for PSMC inference. Population divergence time can also be estimated using BEAST (Vilstrup et al. 2013), although the shear amount of data in genome-scale studies often make Bayesian inference computationally unfeasible. STRUCTURE (Pritchard et al. 2000), ADMIXTURE (Alexander et al. 2009), and NGSAdmix (Skotte et al. 2013) are among the most widely used programs to assess population genetic structure by performing maximum likelihood estimation of individual ancestries from allele frequency differences. Modeling tools such as *f*-statistics [and D-statistics (Green et al. 2010; Durand et al. 2011)] in the AdmixTools package (Patterson et al. 2012) and TreeMix (Pickrell and Pritchard 2012) were developed to infer the genetic affinity and admixture among populations and species. Likewise, another tool, MixMapper (Lipson et al. 2014), is able to infer and visualize population relationship including admixture events that produce the best fit to allele frequency. Similarly, an R package, admixturegraph (Leppälä et al. 2017), was developed recently to evaluate admixture graphs by comparing the fit of the *f*-statistics between different graphs.

It should be noted that the algorithms of most downstream tools listed above were developed for analyses of genome data from modern organisms and they do not take postmortem DNA damage patterns into account. Analyses may be biased and results misleading if a large number of substitutions caused by postmortem DNA damage exist in the input dataset and the number of SNPs is low, as may often be the case with low-coverage ancient genomes. Therefore, one should consider incorporating steps to remove excessive damaged sites in the pipeline and assessing mapping and SNP quality before applying these tools. Furthermore, some bioinformatic tools may be incapable of handling ancient genomes with low depth of coverage. Researchers should review the theoretical background and algorithm of the program to be used and evaluate if its implementation on aDNA data is appropriate. Finally, future efforts should be made on a more standardized "best practice" for processing and analyzing aDNA data. For example, an automated ancient reconstruction pipeline, EAGER (Peltzer et al. 2016), was recently introduced to simplify the analysis of large-scale paleogenomic datasets.

6 Conclusions and Future Perspectives

Recent technical developments have greatly improved the accessibility of aDNA in both quantity and time depth and have expanded the potential for aDNA research from limited individual ancient genome to the scale of population genomics. Although techniques that were developed or adapted to aDNA have shown great successes, efforts should continue to focus on developing, comparing, and optimizing experimental protocols for all steps, in particular, for increasing aDNA retrieval rate, sequencing library conversion efficiency, and targeted enrichment rate. Techniques such as DNA sequencing are likely also to continue to develop in the coming decades; thus, it is essential to tailor new advances to suit the needs of aDNA studies. Meanwhile, many bioinformatic tools have been adapted to interpret genetic information from ancient genomes, and novel tools are being developed to take account for aDNA features. The accumulation of aDNA NGS data will likely provide additional information on postmortem DNA damage patterns and, thus, help to better understand and correct aDNA datasets. Reference-guided mapping, which is currently the most common approach to reconstructing paleogenomes, will likely continue to be implemented in most aDNA studies, but such approaches are not able to detect genomic architectural rearrangements and regions that are no longer present in the modern reference genomes. As sequencing depth increases and more accurate algorithms are applied, we expect that de novo assembly will soon help to reconstruct genomes from extinct populations and species. Finally, the need for a consistent means of processing and analyzing aDNA data among the many bioinformatic tools may drive an effort to standardize the workflow in the near future.

References

Adler CJ, Haak W, Donlon D, Cooper A. Survival and recovery of DNA from ancient teeth and bones. J Archaeol Sci. 2011;38:956–64.

Alexander DH, Novembre J, Lange K. Fast model-based estimation of ancestry in unrelated individuals. Genome Res. 2009;19:1655–64.

Allaby RG, O'Donoghue K, Sallares R, Jones MK, Brown TA. Evidence for the survival of ancient DNA in charred wheat seeds from European archaeological sites. Anc Biomol. 1997;1:119–29.

Allentoft ME, Collins M, Harker D, Haile J, Oskam CL, et al. The half-life of DNA in bone: measuring decay kinetics in 158 dated fossils. Proc Biol Sci. 2012;279:4724–33.

Allentoft ME, Sikora M, Sjogren KG, Rasmussen S, Rasmussen M, et al. Population genomics of Bronze Age Eurasia. Nature. 2015;522:167–72.

Avila-Arcos MC, Cappellini E, Romero-Navarro JA, Wales N, Moreno-Mayar JV, et al. Application and comparison of large-scale solution-based DNA capture-enrichment methods on ancient DNA. Sci Rep. 2011;1:74.

Ávila-Arcos MC, Sandoval-Velasco M, Schroeder H, Carpenter ML, Malaspinas AS, Wales N, Peñaloza F, Bustamante CD, Gilbert MTP. Comparative performance of two whole-genome capture methodologies on ancient DNA Illumina libraries. Methods Ecol Evol. 2015;6:725–34.

Bakker FT. Herbarium genomics: plant archival DNA explored. In: Lindqvist C, Rajora OP, editors. Paleogenomics. Cham: Springer; 2018.

Gamba C, Jones ER, Teasdale MD, McLaughlin RL, Gonzalez-Fortes G, et al. Genome flux and stasis in a five millennium transect of European prehistory. Nat Commun. 2014;5:5257.

Gamba C, Hanghoj K, Gaunitz C, Alfarhan AH, Alquraishi SA, Al-Rasheid KAS, Bradley DG, Orlando L. Comparing the performance of three ancient DNA extraction methods for high-throughput sequencing. Mol Ecol Resour. 2016;16:459–69.

Gansauge M-T, Meyer M. Single-stranded DNA library preparation for the sequencing of ancient or damaged DNA. Nat Protoc. 2013;8:737–48.

Gansauge M-T, Gerber T, Glocke I, Korlević P, Lippik L, Nagel S, Riehl LM, Schmidt A, Meyer M. Single-stranded DNA library preparation from highly degraded DNA using T4 DNA ligase. Nucleic Acids Res. 2017;45:e79.

Gilbert MTP, Wilson AS, Bunce M, Hansen AJ, Willerslev E, et al. Ancient mitochondrial DNA from hair. Curr Biol. 2004;14:R463–4.

Gilbert MTP, Bandelt H-J, Hofreiter M, Barnes I. Assessing ancient DNA studies. Trends Ecol Evol. 2005;20:541–4.

Ginolhac A, Rasmussen M, Gilbert MTP, Willerslev E, Orlando L. mapDamage: testing for damage patterns in ancient DNA sequences. Bioinformatics. 2011;27:2153–5.

Gnirke A, Melnikov A, Maguire J, Rogov P, LeProust EM, Brockman W, Fennell T, Giannoukos G, Fisher S, Russ C. Solution hybrid selection with ultra-long oligonucleotides for massively parallel targeted sequencing. Nat Biotechnol. 2009;27:182–9.

Green RE, Krause J, Ptak SE, Briggs AW, Ronan MT, et al. Analysis of one million base pairs of Neanderthal DNA. Nature. 2006;444:330–6.

Green RE, Malaspinas AS, Krause J, Briggs AW, Johnson PLF, et al. A complete Neandertal mitochondrial genome sequence determined by high-throughput sequencing. Cell. 2008;134:416–26.

Green RE, Krause J, Briggs AW, Maricic T, Stenzel U, et al. A draft sequence of the neandertal genome. Science. 2010;328:710–22.

Greig K, Boocock J, Prost S, Horsburgh KA, Jacomb C, Walter R, Matisoo-Smith E. Complete mitochondrial genomes of New Zealand's first dogs. PLoS One. 2015;10:e0138536.

Gronau I, Hubisz MJ, Gulko B, Danko CG, Siepel A. Bayesian inference of ancient human demography from individual genome sequences. Nat Genet. 2011;43:1031–4.

Gugerli F, Parducci L, Petit RJ. Ancient plant DNA: review and prospects. New Phytol. 2005;166:409–18.

Guindon S, Dufayard J-F, Lefort V, Anisimova M, Hordijk W, Gascuel O. New algorithms and methods to estimate maximum-likelihood phylogenies: assessing the performance of PhyML 3.0. Syst Biol. 2010;59:307–21.

Gutaker RM, Reiter E, Furtwängler A, Schuenemann VJ, Burbano HA. Extraction of ultrashort DNA molecules from herbarium specimens. Biotechniques. 2017;62:76–9.

Haak W, Lazaridis I, Patterson N, Rohland N, Mallick S, et al. Massive migration from the steppe was a source for Indo-European languages in Europe. Nature. 2015;522:207–11.

Hagelberg E, Clegg JB. Isolation and characterization of DNA from archaeological bone. Proc R Soc Lond B Biol Sci. 1991;244:45–50.

Hagelberg E, Sykes B, Hedges R. Ancient bone DNA amplified. Nature. 1989;342:485.

Hagelberg E, Thomas MG, Cook CE Jr, Sher AV, Baryshnikov GF, Lister AM. DNA from ancient mammoth bones. Nature. 1994;370:333–4.

Hahn C, Bachmann L, Chevreux B. Reconstructing mitochondrial genomes directly from genomic next-generation sequencing reads – a baiting and iterative mapping approach. Nucleic Acids Res. 2013;41:e129.

Handt O, Krings M, Ward RH, Pääbo S. The retrieval of ancient human DNA sequences. Am J Hum Genet. 1996;59:368–76.

Hansen HB, Damgaard PB, Margaryan A, Stenderup J, Lynnerup N, Willerslev E, Allentoft ME. Comparing ancient DNA preservation in petrous bone and tooth cementum. PLoS One. 2017;12:e0170940.

Heintzman PD, Soares AE, Chang D, Shapiro B. Paleogenomics. In: Meyers RA, editor. Reviews in cell biology and molecular medicine. Weinheim: Wiley; 2015.

Helgason A, Palsson S, Lalueza-Fox C, Ghosh S, Sigurdardottir S, Baker A, Hrafnkelsson B, Arnadottir L, Thorsteinsdottir U, Stefansson K. A statistical approach to identify ancient template DNA. J Mol Evol. 2007;65:92–102.

Higgins D, Kaidonis J, Townsend G, Hughes T, Austin JJ. Targeted sampling of cementum for recovery of nuclear DNA from human teeth and the impact of common decontamination measures. Investigative Genet. 2013;4:18.

Higuchi R, Bowman B, Freiberger M, Ryder OA, Wilson AC. DNA sequences from the quagga, an extinct member of the horse family. Nature. 1984;312:282–4.

Hodges E, Xuan Z, Balija V, Kramer M, Molla MN, Smith SW, Middle CM, Rodesch MJ, Albert TJ, Hannon GJ. Genome-wide in situ exon capture for selective resequencing. Nat Genet. 2007;39:1522–7.

Hofmanova Z, Kreutzer S, Hellenthal G, Sell C, Diekmann Y, et al. Early farmers from across Europe directly descended from Neolithic Aegeans. Proc Natl Acad Sci U S A. 2016;113:6886–91.

Hofreiter M, Paijmans JLA, Goodchild H, Speller CF, Barlow A, Fortes GG, Thomas JA, Ludwig A, Collins MJ. The future of ancient DNA: technical advances and conceptual shifts. Bioessays. 2015;37:284–93.

Hughey JR, Braga JC, Aguirre J, Woelkerling WJ, Webster JM. Analysis of ancient DNA from fossil corallines (Corallinales, Rhodophyta). J Phycol. 2008;44:374–83.

Jaenicke-Despres V, Buckler ES, Smith BD, Gilbert MTP, Cooper A, Doebley J, Paabo S. Early allelic selection in maize as revealed by ancient DNA. Science. 2003;302:1206–8.

Jeong C, Ozga AT, Witonsky DB, Malmstrom H, Edlund H, et al. Long-term genetic stability and a high-altitude East Asian origin for the peoples of the high valleys of the Himalayan arc. Proc Natl Acad Sci U S A. 2016;113:7485–90.

Jones ER, Gonzalez-Fortes G, Connell S, Siska V, Eriksson A, et al. Upper Palaeolithic genomes reveal deep roots of modern Eurasians. Nat Commun. 2015;6:8912.

Jonsson H, Ginolhac A, Schubert M, Johnson PLF, Orlando L. mapDamage2.0: fast approximate Bayesian estimates of ancient DNA damage parameters. Bioinformatics. 2013;29:1682–4.

Keller A, Graefen A, Ball M, Matzas M, Boisguerin V, et al. New insights into the Tyrolean Iceman's origin and phenotype as inferred by whole-genome sequencing. Nat Commun. 2012;3:698.

Kircher M. Analysis of high-throughput ancient DNA sequencing data. Methods Mol Biol. 2012;840:197–228.

Kistler L, Ware R, Smith O, Collins M, Allaby RG. A new model for ancient DNA decay based on paleogenomic meta-analysis. Nucleic Acids Res. 2017;45:6310–20.

Korlevic P, Gerber T, Gansauge MT, Hajdinjak M, Nagel S, Aximu-Petri A, Meyer M. Reducing microbial and human contamination in DNA extractions from ancient bones and teeth. Biotechniques. 2015;59:87–93.

Korneliussen TS, Albrechtsen A, Nielsen R. ANGSD: analysis of next generation sequencing data. BMC Bioinformatics. 2014;15:356.

Krause J, Briggs AW, Kircher M, Maricic T, Zwyns N, Derevianko A, Paabo S. A complete mtDNA genome of an early modern human from Kostenki, Russia. Curr Biol. 2010;20:231–6.

Lan T, Gill S, Bellemain E, Bischof R, Nawaz MA, Lindqvist C. Evolutionary history of enigmatic bears in the Tibetan Plateau-Himalaya region and the identity of the yeti. Proc R Soc B. 2017;284:20171804.

Lazaridis I, Patterson N, Mittnik A, Renaud G, Mallick S, et al. Ancient human genomes suggest three ancestral populations for present-day Europeans. Nature. 2014;513:409–13.

Leppälä K, Nielsen SV, Mailund T. Admixturegraph: an R package for admixture graph manipulation and fitting. Bioinformatics. 2017;33:1738–40.

Li H, Durbin R. Fast and accurate long-read alignment with Burrows-Wheeler transform. Bioinformatics. 2010;26:589–95.

Li H, Durbin R. Inference of human population history from individual whole-genome sequences. Nature. 2011;475:493–6.

Li H, Handsaker B, Wysoker A, Fennell T, Ruan J, Homer N, Marth G, Abecasis G, Durbin R. The sequence alignment/map format and SAMtools. Bioinformatics. 2009;25:2078–9.

Li C, Hofreiter M, Straube N, Corrigan S, Naylor GJ. Capturing protein-coding genes across highly divergent species. Biotechniques. 2013;54:321–6.

Librado P, Sarkissian CD, Ermini L, Schubert M, Jonsson H, et al. Tracking the origins of Yakutian horses and the genetic basis for their fast adaptation to subarctic environments. Proc Natl Acad Sci U S A. 2015;112:E6889–97.

Liepelt S, Sperisen C, Deguilloux MF, Petit RJ, Kissling R, Spencer M, Beaulieu JL, Taberlet P, Gielly L, Ziegenhagen B. Authenticated DNA from ancient wood remains. Ann Bot. 2006;98:1107–11.

Lindahl T. Instability and decay of the primary structure of DNA. Nature. 1993;362:709–15.

Lipson M, Loh P-R, Patterson N, Moorjani P, Ko Y-C, Stoneking M, Berger B, Reich D. Reconstructing Austronesian population history in island Southeast Asia. Nat Commun. 2014;5:4689.

Mak SST, Gopalakrishnan S, Carøe C, Geng C, Liu S, et al. Comparative performance of the BGISEQ-500 vs Illumina HiSeq2500 sequencing platforms for palaeogenomic sequencing. Gigascience. 2017;6:1–13.

Mamanova L, Coffey AJ, Scott CE, Kozarewa I, Turner EH, Kumar A, Howard E, Shendure J, Turner DJ. Target-enrichment strategies for next-generation sequencing. Nat Methods. 2010;7:111–8.

Margulies M, Egholm M, Altman WE, Attiya S, Bader JS, Bemben LA, Berka J, Braverman MS, Chen Y-J, Chen Z. Genome sequencing in microfabricated high-density picolitre reactors. Nature. 2005;437:376–80.

Maricic T, Paabo S. Optimization of 454 sequencing library preparation from small amounts of DNA permits sequence determination of both DNA strands. Biotechniques. 2009;46:51–2, 54–7.

Maricic T, Whitten M, Paabo S. Multiplexed DNA sequence capture of mitochondrial genomes using PCR products. PLoS One. 2010;5:e14004.

Mascher M, Schuenemann VJ, Davidovich U, Marom N, Himmelbach A, et al. Genomic analysis of 6,000-year-old cultivated grain illuminates the domestication history of barley. Nat Genet. 2016;48:1089–93.

McCormack JE, Tsai WLE, Faircloth BC. Sequence capture of ultraconserved elements from bird museum specimens. Mol Ecol Resour. 2016;16:1189–203.

McKenna A, Hanna M, Banks E, Sivachenko A, Cibulskis K, Kernytsky A, Garimella K, Altshuler D, Gabriel S, Daly M. The Genome Analysis Toolkit: a MapReduce framework for analyzing next-generation DNA sequencing data. Genome Res. 2010;20:1297–303.

Meyer M, Kircher M. Illumina sequencing library preparation for highly multiplexed target capture and sequencing. Cold Spring Harb Protoc. 2010;2010:pdb.prot5448.

Meyer M, Kircher M, Gansauge M-T, Li H, Racimo F, et al. A high-coverage genome sequence from an archaic denisovan individual. Science. 2012;338:222–6.

Miller W, Drautz DI, Ratan A, Pusey B, Qi J, et al. Sequencing the nuclear genome of the extinct woolly mammoth. Nature. 2008;456:387–90.

Miller W, Schuster SC, Welch AJ, Ratan A, Bedoya-Reina OC, et al. Polar and brown bear genomes reveal ancient admixture and demographic footprints of past climate change. Proc Natl Acad Sci. 2012;109:E2382–90.

Mohandesan E, Speller CF, Peters J, Uerpmann HP, Uerpmann M, De Cupere B, Hofreiter M, Burger PA. Combined hybridization capture and shotgun sequencing for ancient DNA analysis of extinct wild and domestic dromedary camel. Mol Ecol Resour. 2017;17:300–19.

Nieves-Colón MA, Ozga AT, Pestle WJ, Cucina A, Tiesler V, Stanton TW, Stone AC. Comparison of two ancient DNA extraction protocols for skeletal remains from tropical environments. bioRxiv. 2017:184119.

Olalde I, Allentoft ME, Sanchez-Quinto F, Santpere G, Chiang CWK, et al. Derived immune and ancestral pigmentation alleles in a 7,000-year-old Mesolithic European. Nature. 2014;507:225–8.

Orlando L, Ginolhac A, Raghavan M, Vilstrup J, Rasmussen M, et al. True single-molecule DNA sequencing of a pleistocene horse bone. Genome Res. 2011;21:1705–19.

Orlando L, Ginolhac A, Zhang GJ, Froese D, Albrechtsen A, et al. Recalibrating Equus evolution using the genome sequence of an early Middle Pleistocene horse. Nature. 2013;499:74–8.

Orlando L, Gilbert MTP, Willerslev E. Reconstructing ancient genomes and epigenomes. Nat Rev Genet. 2015;16:395–408.

O'Rourke DH, Hayes MG, Carlyle SW. Ancient DNA studies in physical anthropology. Ann Rev Anthropol. 2000;29:217–42.

Pääbo S. Molecular cloning of ancient Egyptian mummy DNA. Nature. 1985;314:644–5.

Pääbo S. Ancient DNA: extraction, characterization, molecular cloning, and enzymatic amplification. Proc Natl Acad Sci. 1989;86:1939–43.

Pääbo S, Poinar H, Serre D, Jaenicke-Despres V, Hebler J, Rohland N, Kuch M, Krause J, Vigilant L, Hofreiter M. Genetic analyses from ancient DNA. Annu Rev Genet. 2004;38:645–79.

Paijmans JLA, Fickel J, Courtiol A, Hofreiter M, Forster DW. Impact of enrichment conditions on cross-species capture of fresh and degraded DNA. Mol Ecol Resour. 2016;16:42–55.

Palkopoulou E, Mallick S, Skoglund P, Enk J, Rohland N, et al. Complete genomes reveal signatures of demographic and genetic declines in the woolly mammoth. Curr Biol. 2015;25:1395–400.

Parducci L, Nota K, Wood J. Reconstructing past vegetation communities using ancient DNA from Lake sediments. In: Lindqvist C, Rajora OP, editors. Paleogenomics. Cham: Springer; 2018.

Park SDE, Magee DA, McGettigan PA, Teasdale MD, Edwards CJ, et al. Genome sequencing of the extinct Eurasian wild aurochs, Bos primigenius, illuminates the phylogeography and evolution of cattle. Genome Biol. 2015;16:234.

Patterson N, Moorjani P, Luo Y, Mallick S, Rohland N, Zhan Y, Genschoreck T, Webster T, Reich D. Ancient admixture in human history. Genetics. 2012;192:1065–93.

Peltzer A, Jager G, Herbig A, Seitz A, Kniep C, Krause J, Nieselt K. EAGER: efficient ancient genome reconstruction. Genome Biol. 2016;17

Peñalba JV, Smith LL, Tonione MA, Sass C, Hykin SM, Skipwith PL, McGuire JA, Bowie RC, Moritz C. Sequence capture using PCR-generated probes: a cost-effective method of targeted high-throughput sequencing for nonmodel organisms. Mol Ecol Resour. 2014;14:1000–10.

Pickrell JK, Pritchard JK. Inference of population splits and mixtures from genome-wide allele frequency data. PLoS Genet. 2012;8:e1002967.

Pinhasi R, Fernandes D, Sirak K, Novak M, Connell S, et al. Optimal ancient DNA yields from the inner ear part of the human petrous bone. PLoS One. 2015;10:e0129102.

Poinar HN. The genetic secrets some fossils hold. Acc Chem Res. 2002;35:676–84.

Poinar HN, Hofreiter M, Spaulding WG, Martin PS, Stankiewicz BA, Bland H, Evershed RP, Possnert G, Paabo S. Molecular coproscopy: dung and diet of the extinct ground sloth Nothrotheriops shastensis. Science. 1998;281:402–6.

Poinar HN, Schwarz C, Qi J, Shapiro B, MacPhee RDE, et al. Metagenomics to paleogenomics: large-scale sequencing of mammoth DNA. Science. 2006;311:392–4.

Pollmann B, Jacomet S, Schlumbaum A. Morphological and genetic studies of waterlogged Prunus species from the Roman vicus Tasgetium (Eschenz, Switzerland). J Archaeol Sci. 2005;32:1471–80.

Price AL, Patterson NJ, Plenge RM, Weinblatt ME, Shadick NA, Reich D. Principal components analysis corrects for stratification in genome-wide association studies. Nat Genet. 2006;38:904–9.

Pritchard JK, Stephens M, Donnelly P. Inference of population structure using multilocus genotype data. Genetics. 2000;155:945–59.

Prufer K, Racimo F, Patterson N, Jay F, Sankararaman S, et al. The complete genome sequence of a Neanderthal from the Altai Mountains. Nature. 2014;505:43–9.

Raghavan M, Skoglund P, Graf KE, Metspalu M, Albrechtsen A, et al. Upper Palaeolithic Siberian genome reveals dual ancestry of Native Americans. Nature. 2014;505:87–91.

Rajaraman A, Tannier E, Chauve C. FPSAC: fast phylogenetic scaffolding of ancient contigs. Bioinformatics. 2013;29:2987–94.

Ramos-Madrigal J, Smith BD, Moreno-Mayar JV, Gopalakrishnan S, Ross-Ibarra J, Gilbert MTP, Wales N. Genome sequence of a 5,310-year-old maize cob provides insights into the early stages of maize domestication. Curr Biol. 2016;26:3195–201.

Rasmussen M, Li Y, Lindgreen S, Pedersen JS, Albrechtsen A, et al. Ancient human genome sequence of an extinct Palaeo-Eskimo. Nature. 2010;463:757–62.

Rasmussen M, Anzick SL, Waters MR, Skoglund P, DeGiorgio M, et al. The genome of a Late Pleistocene human from a Clovis burial site in western Montana. Nature. 2014;506:225–9.

Reich D, Green RE, Kircher M, Krause J, Patterson N, et al. Genetic history of an archaic hominin group from Denisova Cave in Siberia. Nature. 2010;468:1053–60.

Renaud G, Slon V, Duggan AT, Kelso J. Schmutzi: estimation of contamination and endogenous mitochondrial consensus calling for ancient DNA. Genome Biol. 2015;16:224.

Richards MB, Sykes BC, Hedges REM. Authenticating DNA extracted from ancient skeletal remains. J Archaeol Sci. 1995;22:291–9.

Rohland N, Hofreiter M. Ancient DNA extraction from bones and teeth. Nat Protoc. 2007;2:1756–62.

Rohland N, Siedel H, Hofreiter M. A rapid column-based ancient DNA extraction method for increased sample throughput. Mol Ecol Resour. 2010;10:677–83.

Rohland N, Harney E, Mallick S, Nordenfelt S, Reich D. Partial uracil – DNA – glycosylase treatment for screening of ancient DNA. Philos T R Soc B. 2015;370:20130624.

Ronquist F, Huelsenbeck JP. MrBayes 3: Bayesian phylogenetic inference under mixed models. Bioinformatics. 2003;19:1572–4.

Sánchez Barreiro F, Vieira FG, Martin MD, Haile J, Gilbert MTP, Wales N. Characterizing restriction enzyme-associated loci in historic ragweed (Ambrosia artemisiifolia) voucher specimens using custom-designed RNA probes. Mol Ecol Resour. 2017;17:209–20.

Sandoval-Velasco M, Lundstrøm IK, Wales N, Ávila-Arcos MC, Schroeder H, Gilbert MTP. Relative performance of two DNA extraction and library preparation methods on archaeological human teeth samples. Sci Technol Archaeol Res. 2017;3:80–8.

Sarkissian CD, Allentoft ME, Avila-Arcos MC, Barnett R, Campos PF, et al. Ancient genomics. Philos T R Soc B. 2015;370:20130387.

Schlumbaum A, Jacomet S, Neuhaus JM. Coexistence of tetraploid and hexaploid naked wheat in a neolithic lake dwelling of Central Europe: evidence from morphology and ancient DNA. J Archaeol Sci. 1998;25:1111–8.

Schlumbaum A, Tensen M, Jaenicke-Despres V. Ancient plant DNA in archaeobotany. Veg Hist Archaeobotany. 2008;17:233–44.

Schroeder H, Avila-Arcos MC, Malaspinas A-S, Poznik GD, Sandoval-Velasco M, et al. Genome-wide ancestry of 17th-century enslaved Africans from the Caribbean. Proc Natl Acad Sci U S A. 2015;112:3669–73.

Schubert M, Ginolhac A, Lindgreen S, Thompson JF, Al-Rasheid KAS, Willerslev E, Krogh A, Orlando L. Improving ancient DNA read mapping against modern reference genomes. BMC Genomics. 2012;13:178.

Schubert M, Ermini L, Sarkissian CD, Jonsson H, Ginolhac A, et al. Characterization of ancient and modern genomes by SNP detection and phylogenomic and metagenomic analysis using PALEOMIX. Nat Protoc. 2014;9:1056–82.

Schuenemann VJ, Bos K, DeWitte S, Schmedes S, Jamieson J, et al. Targeted enrichment of ancient pathogens yielding the pPCP1 plasmid of Yersinia pestis from victims of the Black Death. Proc Natl Acad Sci U S A. 2011;108:E746–52.

Schuenemann VJ, Singh P, Mendum TA, Krause-Kyora B, Jager G, et al. Genome-wide comparison of medieval and modern Mycobacterium leprae. Science. 2013;341:179–83.
Seguin-Orlando A, Schubert M, Clary J, Stagegaard J, Alberdi MT, Prado JL, Prieto A, Willerslev E, Orlando L. Ligation bias in illumina next-generation DNA libraries: implications for sequencing ancient genomes. PLoS One. 2013;8:e78575.
Seguin-Orlando A, Korneliussen TS, Sikora M, Malaspinas AS, Manica A, et al. Genomic structure in Europeans dating back at least 36,200 years. Science. 2014;346:1113–8.
Skoglund P, Northoff BH, Shunkov MV, Derevianko AP, Paabo S, Krause J, Jakobsson M. Separating endogenous ancient DNA from modern day contamination in a Siberian Neandertal. Proc Natl Acad Sci U S A. 2014;111:2229–34.
Skotte L, Korneliussen TS, Albrechtsen A. Estimating individual admixture proportions from next generation sequencing data. Genetics. 2013;195:693–702.
Smith CI, Chamberlain AT, Riley MS, Cooper A, Stringer CB, Collins MJ. Neanderthal DNA: not just old but old and cold? Nature. 2001;410:771–2.
Smith CI, Chamberlain AT, Riley MS, Stringer C, Collins MJ. The thermal history of human fossils and the likelihood of successful DNA amplification. J Hum Evol. 2003;45:203–17.
Spyrou MA, Tukhbatova RI, Feldman M, Drath J, Kacki S, et al. Historical Y. pestis genomes reveal the European Black Death as the source of ancient and modern plague pandemics. Cell Host Microbe. 2016;19:874–81.
Stamatakis A. RAxML version 8: a tool for phylogenetic analysis and post-analysis of large phylogenies. Bioinformatics. 2014;30:1312–3.
Turki T, Roshan U. MaxSSmap: a GPU program for mapping divergent short reads to genomes with the maximum scoring subsequence. BMC Genomics. 2014;15:969.
Vilstrup JT, Seguin-Orlando A, Stiller M, Ginolhac A, Raghavan M, et al. Mitochondrial phylogenomics of modern and ancient equids. PLoS One. 2013;8:e55950.
Wales N, Andersen K, Cappellini E, Ávila-Arcos MC, Gilbert MTP. Optimization of DNA recovery and amplification from non-carbonized archaeobotanical remains. PLoS One. 2014;9:e86827.
Ward R, Stringer C. Standards for research on ancient DNA. Nature. 1997;388:226.
Warinner C, Rodrigues JFM, Vyas R, Trachsel C, Shved N, et al. Pathogens and host immunity in the ancient human oral cavity. Nat Genet. 2014;46:336–44.
Willerslev E, Cooper A. Ancient DNA. Proc R Soc Lond B Biol Sci. 2005;272:3–16.
Yang DY, Eng B, Waye JS, Dudar JC, Saunders SR. Technical note: improved DNA extraction from ancient bones using silica-based spin columns. Am J Phys Anthropol. 1998;105:539–43.
Zhang HC, Paijmans JLA, Chang FQ, Wu XH, Chen GJ, et al. Morphological and genetic evidence for early Holocene cattle management in northeastern China. Nat Commun. 2013;4:2755.

Paleoproteomics: An Introduction to the Analysis of Ancient Proteins by Soft Ionisation Mass Spectrometry

Michael Buckley

Abstract The field of proteomic research, analogous to genomic research, has only recently witnessed a rapid increase in its application to the study of ancient materials. Bone has been the most commonly used archaeological and paleontological resource for recovering biological information. This has most frequently been for ancient genomic analysis, but some of the potential advantages of proteomics lie in its ability to discriminate between sources of the molecules, rather than the particular species or individual. However, proteomes could be considered more dynamic, offering different types of information than otherwise available through DNA analyses. Proteins are also considered to survive for much longer periods of time than substantial lengths of DNA and therefore the development of proteomics allows for the possibility of being able to recover information much further back in time than previously thought possible. In this chapter, the progress of this area called 'paleoproteomics' is reviewed, highlighting some of its greatest achievements but also some of the current limitations in the field across proteins from a range of different materials.

Keywords Ancient proteins · Collagen · Extinct taxa · Paleoproteomics · Phylogenetics · Soft ionisation mass spectrometry

1 Introduction

The field of ancient DNA (aDNA) has seen remarkable revolutions in methodology in recent years, as thoroughly discussed throughout the collection of papers in this book. Yet the most recent development of genomic technologies (Lander 1996; Lockwood et al. 2006), furthered by the advances in next-generation sequencing

M. Buckley (✉)
School of Earth and Environmental Sciences, Manchester Institute of Biotechnology, University of Manchester, Manchester, UK
e-mail: m.buckley@manchester.ac.uk

Charlotte Lindqvist and Om P. Rajora (eds.), *Paleogenomics*,
Population Genomics [Om P. Rajora (Editor-in-Chief)],
https://doi.org/10.1007/13836_2018_50,

(NGS; Mardis 2008), was accompanied by great advances in the analysis of proteins and the study of proteomes (James 1997; Anderson and Anderson 1998) that represent the complete set of proteins in an organism, cell or tissue. The recovery of proteome-wide information is somewhat more methodologically challenging than for genomic information, but substantial progress has been seen in recent years (Breker and Schuldiner 2014). It is these developments and more specifically their application to the study of ancient proteins that will form the focus of this chapter.

It has been shown that the fossilisation potential of organic molecules is extremely variable, as some compounds common in living organisms are extremely rare in the fossil record and only preserved under exceptional conditions (Curry 1988). Given that the constituent nucleic acids of DNA are much less stable than the amino acids which make up proteins, the potential of recovering information from ancient proteins has intrigued scientists for over half a century. Initial work by Abelson (1956) showed that amino acids survive in ancient fossils such as Devonian fish, Jurassic and Cretaceous dinosaurs, and Oligocene to Pliocene horses. It has also long been assumed that the preservation of some biomolecules, such as proteins, is also enhanced by protection within biominerals, e.g. in bones, shells and teeth.

Like DNA, proteins contain fundamental genetic information that is key to molecular phylogenetic reconstruction, but they exist in larger quantities and have much greater potential for preservation (Curry 1988; Logan et al. 1991; Robbins et al. 1993). In addition to their protection within mineral components, bone proteins have been shown to escape chemical or microbial attack and thus complete degradation by hydrolysis when physically encapsulated in hydrophobic networks of humic acids or other refractory organic matter, such as in compost (Zang et al. 2000). Although traces of amino acids have been reported from much earlier time periods, Bada et al. (1999) reported proteins to hydrolyse within 100,000 to 1 million years, whereas others placed this value for proteins in bone at slightly longer for the dominant helical protein collagen (several million years) and substantially longer for particular mineral-binding non-collagenous proteins (NCPs), such as osteocalcin (Nielsen-Marsh 2002). In addition to surviving for potentially much longer periods of time, there is no analogous step in proteomics to the amplification process in typical DNA-based studies, vastly reducing concerns of contamination often associated with ancient biomolecule studies (Pääbo et al. 2004). From protein sequence studies, relationships can be inferred and the numbers of effective amino acid substitutions can be related to base changes in the genetic code of the DNA molecule, as was demonstrated for extinct moas through sequencing by Edman degradation (Huq et al. 1990).

2 Soft Ionisation Mass Spectrometry and Proteomics

Until the early 1990s, one of the most common methods of sequencing peptides was by Edman degradation, which relies on the identification of amino acids chemically cleaved in a stepwise fashion from the amino-terminus of a peptide by reaction with phenylisothiocyanate and cleavage of the resulting phenylthiocarbamyl derivatives (Edman 1956). However, this method often failed when the peptide being analysed possessed an acetylated or otherwise blocked amino-terminus (Wellner et al. 1990). Techniques using mass spectrometry (MS) overcome some of these issues by avoiding a sequential interpretation step. However, this also creates different problems in asserting confidence in the sequences inferred (discussed later).

Mass spectrometry (MS) is an analytical technique in which molecules from within a test sample are converted to gaseous ions that are subsequently separated in a mass spectrometer according to their mass-to-charge ratio (*m/z*) and detected. Depending on the type of inlet and ionisation techniques used, the sample may already exist as ions in solution or it may be ionised in conjunction with its volatilisation or other methods in the ion source. 'Soft ionisation' techniques are where the evaporation and ionisation of the molecular species into the gaseous phase are carried out without extensive fragmentation. The two most common ionisation methods currently used are matrix-assisted laser desorption/ionisation (MALDI), which was first described by Karas and Hillenkamp (1988) and Tanaka et al. (1988), and electrospray ionisation (ESI), first described by Yamashita and Fenn (1984), but are very different. As the name suggests, MALDI relies on the use of a matrix that is co-crystallised with a solution of peptides to form a solid, which is then irradiated with UV light. Although the exact mechanisms are still debated, the matrix is typically thought to absorb the laser energy and then desorb, along with the sample molecules that it passes protons to, into the gaseous phase as neutral as well as ionised molecules. By contrast, ESI is substantially different because it retains the sample in solution until ionisation, which is dispersed by electrospray into a fine charged aerosol by passing through a metal capillary at high voltage. As such this approach more commonly results in multiple charge states that allow larger mass ranges to be observed. Where MALDI is typically more frequently used for high-throughput peptide mass fingerprinting, ESI approaches are more often used for obtaining sequence information, as well as being more useful in the study of more complex sample mixtures.

Even though mass spectrometers can measure the *m/z* of intact proteins, such as in the analysis of ancient osteocalcin (OC) by Ostrom et al. (2000), it is often inferences of protein composition through the analysis of peptides that are analysed in proteomic analyses because they are more soluble and more readily detected (their sequence (tandem) spectra are also typically more complete and therefore appropriate for interpretation). Given that ancient proteins are more likely to suffer diagenetic alterations that cause further difficulties in interpretation, analysis at the peptide level in ancient samples is also likely to be simpler. Although the enzymatic digest of a purified protein would result in a relatively simple analysis, it is often complex mixtures of proteins that are collectively digested and analysed. In order to improve the analysis of large numbers of peptides, several peptide separation techniques can be employed prior to MS analysis, the most common being liquid chromatography. In addition to the ion source, there are two other main components of a mass spectrometer, the mass analyser and the ion detectors. There are four main combinations of mass analysers used in proteomics, particularly with MALDI and ESI: time-of-flight

organs; and it is also a ready source of the key regulatory inorganic ions calcium, magnesium and phosphate. It is a natural composite material, consisting of approximately 70% inorganic material, predominantly the calcium phosphate hydroxyapatite (HAP), and an organic (mostly protein) component that is predominantly the structural protein collagen (Boskey and Posner 1984). The inorganic phase gives the tissue resistance to compression forces and is composed of a carbonate hydroxyapatite with a unit cell formula often generalised as $Ca_{10}(PO_4)_6{\cdot}(OH)_2$.

The organic phase, which gives resistance to tension forces by providing flexibility and forming the matrix upon and within which mineral crystals are grown, accounts for 25–30% (by weight), of which collagen predominates accounting for approximately 90% (by weight) of the constituents in the organic matrix (Millard 2001). Collagens are the most abundant structural protein in the animal kingdom, and of the more than 27 types of collagen, the fibrous collagen type I (hereafter written as 'collagen (I)') is prevalent, particularly in bone. Fibrous proteins are notably insoluble in water due to the large number of hydrophobic residues in their primary structure, and so collagen is a very good candidate for preservation in the burial environment. The remaining proteins are considered 'non-collagenous proteins' and have a range of functions and biochemical properties. Bone proteins have been extracted by various methods over the last few decades but typically involve a decalcification step to remove the mineral phase (e.g. by hydrochloric acid or the 'softer' chelating approach of ethyldiaminetetraacetic acid), producing an acid-soluble fraction that can be studied and an acid-insoluble fraction that is typically extracted with a denaturing buffer and/or further gelatinised (heated) to convert the predominantly collagen fraction into gelatine, also releasing the many NCPs associated with collagen (e.g., Wadsworth and Buckley 2014). These fractions would then be enzymatically digested, often following reduction and alkylation of the disulphide bonds present in many NCPs, and purified for proteomic analysis in a relatively routine manner (e.g., using solid phase extraction cartridges).

3.1.1 Bone Collagen

The basic collagen (I) molecule is approximately 300 nm in length and 1.5 nm in diameter (Kadler et al. 1996). It consists of three α-chains, two identical α1 (I) chains and one genetically different α2 (I) chain (Vuorio and de Crombrugghe 1998). Each helically wound alpha chain contains 338 triplets of amino acids constructed from repeating Gly-Xaa-Yaa triplets, where Xaa and Yaa can be any amino acid but are frequently the amino acids proline and hydroxyproline (Hulmes 1992). The glycine at every third position of each chain is a prerequisite for the folding of the three chains into such a tight triple helix (Fig. 2a). The pyrrolidine ring of proline introduces a left-handed twist in the peptide backbone of each α-chain, placing glycine residues into the centre of a triple helix as the tight packing of the protein strands can accommodate no other residue. The relatively uninterrupted triple helix of collagen (I) molecules is flanked by short non-helical telopeptides; the telopeptides, which do not have a repeating Gly-Xaa-Yaa structure and do not

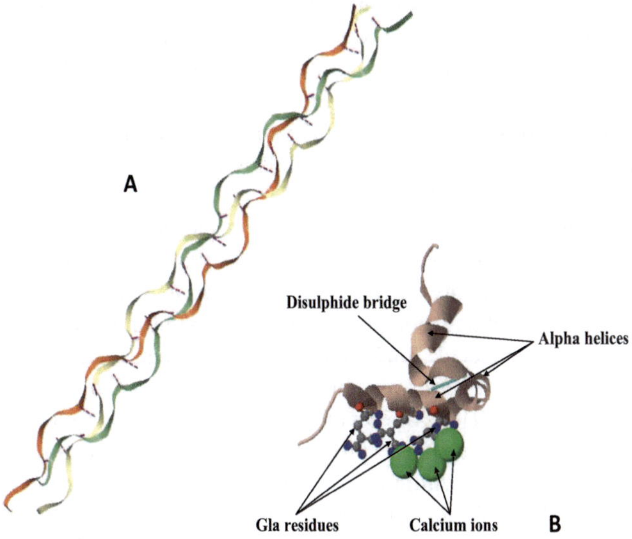

Fig. 2 Illustration of (**a**) the collagen triple helix with internal hydrogen bonds and (**b**) the docking of porcine OC onto the HAP, showing the affinity of the three Gla residues at positions 17, 21 and 24 to bind to the calcium ions (in green) of the HAP (created in jmol using 1k6f and 1q3 pdb files)

adopt a triple helical conformation, are essential for fibril formation (Kadler et al. 1996).

The collagen molecule also has some unusual post-translational modifications (PTMs), which are the hydroxylation of some proline and lysine residues. The assembly of types I, II and III collagen into fibrils is accompanied by the formation of inter- and intramolecular covalent cross-links between chains which confer high tensile and mechanical strength on the fibrils (Kadler et al. 1996). These covalent cross-links inherently prevent the collagen molecules from sliding past each other under stress; in doing so they provide tensile strength and stability to the collagen fibril (Bailey 2001). Collagen is also known for possessing additional types of cross-links involving saccharides that make the molecules of collagen much more stable at higher temperatures. These early glycation products are reversible and do not accumulate in most proteins. In long-lived proteins like collagen, these glycation products are able to undergo a series of reactions that result in more persistent advanced glycation end products (AGEs or Maillard products).

3.1.2 Non-collagenous Proteins (NCPs) in Bone

Several families of proteins associated with the collagen matrix are involved in regulation of the mineralised process, although recent research into the characterisation of bone matrix has emphasised its complexity, with Schreiweis et al. (2007)

identifying 133 proteins and Jiang et al. (2007) showing as many as 2,479 unique proteins associated with bone. However, some of these proteins have multiple functions beyond their role in mineralisation (Zhu et al. 2007); they include phosphorylated proteins, proteoglycans, glycoproteins, and gamma-carboxy-glutamic acid-containing (Gla) proteins with one, the small mineral-binding protein OC, the most abundant of all NCPs found in bone making up approximately 20% by weight of the total NCPs (Hauschka 1986).

Jiang et al. (2007) used proteomics to confirm the presence of various types of proteins in bone. Not only were many bone-specific proteins identified, but this study also showed that many of the NCPs were present in minute amounts. For example, the bone morphogenetic protein (BMP) content in bone tissue was ~2 ng/kg wet bone tissue (Jiang et al. 2007). The suite of proteins, i.e. the proteome, can be categorised in several different ways, but usually this is done by typical subcellular location and/or function. There are several software packages, such as STRING and GeneMapper, that match up these properties and identify functional correlations between proteins observed within a set of given proteins. Among the bone-specific proteins, Jiang et al. (2007) identified many embedded in the matrix, including OC, osteonectin (SPARC), bone sialoprotein, fibronectin, MGP, BMPs, growth factors, cytokines and proteoglycans (like perlecan and biglycan). Other NCPs that are not specific to bone, such as osteopontin, were also identified. The identified proteins exhibited a broad spectrum of functions, including the control of cell proliferation, cell-matrix interactions and mediation of hydroxyapatite deposition. Several serum-derived proteins, including serum albumin, haemoglobin, myoglobin and alpha-2-Heremans-Schmid glycoprotein (A2HS), are reported to bind to the mineral component (Triffitt et al. 1976; Weiner et al. 1976). As an example, two identified proteins, biglycan and creatine kinase, are involved in bone growth and differentiations (Nogami et al. 1987; Wallace et al. 2006). Furthermore, many proteins associated with bone matrix degradation, such as cathepsin, matrix metalloproteinases (MMPs) and plasminogen, were also observed (Jiang et al. 2007). By comparison, the lipid component of bone makes up only about 0.1% (wt) of the tissue, three-quarters of this is triglyceride (triacylglycerol), with the rest being predominantly cholesterol (Williams and Elliot 1989).

The vitamin K-dependent Ca^{2+}-binding protein OC is the most abundant NCP of adult bone, which is secreted by osteoblasts and present in bone matrix and serum (Colombo et al. 1993). Isolated osteocalcins are small acidic proteins (pI≈4.0) containing 46–50 amino acid residues; three of these amino acids are Gla residues at positions 17, 21 and 24 (Hauschka et al. 1989) that appear to be responsible for the Ca^{2+}-binding properties of OC to phospholipid vesicles (Gendreau et al. 1989) and hydroxyapatite (Hauschka and Carr 1982; Delmas et al. 1984; Hauschka and Wians 1989). The affinity of OC for bone mineral is supported by the clustering of these calcium-binding sites on the molecules' surface; this derives from the presence of the doubly charged Gla residues and charged Asp residue all being concentrated in a common surface region on the same side of the molecule (Frazao et al. 2005).

The Gla helix domain is likely to be of ancient evolutionary origin as it occurs in a subset of proteins and peptides which interact in a particular fashion with Ca^{2+} ions and Ca^{2+} mineral surfaces. The staggering of Gla residues in primary sequence by three or four positions allows them to project from one face of the alpha helix (Fig. 2b), seen in both OC and MGP in bone (Price and Williamson 1985), and elsewhere such as the blood proteins prothrombin and Factor X (Nelsestuen et al. 1974; Schwalbe et al. 1989). Due to this ability to bind to the HAP, various studies have suggested a physiological role for OC as a matrix signal for the recruitment and differentiation of osteoclasts (Malone 1982; Glowacki et al. 1991; Liggett et al. 1994).

3.2 Information Content of Ancient Bone Proteins

Although proteins from fossil bone specimens have been analysed for decades, first by amino acid composition analyses (Abelson 1959; Jones and Vallentyne 1960), then by immunological approaches (Prager et al. 1980; Armstrong et al. 1983; Lowenstein and Ryder 1985) and finally by sequencing approaches (Huq et al. 1989), the first applications of proteomic methods were not until the start of this century (Ostrom et al. 2000). However, rather than looking at collagen, in part due to the complexity in sequencing such a large molecule with so many variable modifications and relatively limited variable sequence information, Ostrom et al. (2000) analysed osteocalcin (OC); not only is OC the most abundant of the NCPs, but it can be readily isolated using a simple and cheap solid-phase extraction (SPE) protocol (e.g., Colombo et al. 1993). In brief, the solution from the demineralisation of bone could be applied to this SPE cartridge, and once allowed to pass through, subsequent volumes of increasingly organic solvent would be applied, removing many other proteins and peptides but isolating the OC (Buckley et al. 2008a). This could then be dried out in order to remove organic solvent and once resuspended can then be spotted onto a MALDI target plate for analysis (further methods of concentration could also be applied in the case of fossils with low protein abundance such as by HPLC). Although initial studies were mainly limited to taxa with well-characterised protein sequences available, such as horse (*Equus caballus*; Ostrom et al. 2006) and bison (*Bison priscus*; Nielsen-Marsh et al. 2002), they demonstrated the potential to recover protein sequence information in a way that could potentially yield phylogenetic information in addition to investigating the survival of protein secondary structure (Ostrom et al. 2006).

Where the OC isolation procedure could be considered a type of 'top-down' proteomics (although the definition of this has changed with particular use of the ion trap), which could use but do not depend upon probability matching algorithms against a database of known sequences, 'bottom-up' or 'shotgun proteomics' methods that do rely on these have become increasingly applied due to the amount of information that they can generate. However, this is at the cost of confidence in the results, with increasing sensitivity and more frequent detection of laboratory contaminants. Such methods were applied to the reports of peptides recovered from a 68-million-year-old fossil of *Tyrannosaurus rex* (Asara et al. 2007), but there continues to be a debate regarding their authenticity (Buckley et al. 2008b; Kaye et al. 2008). Most importantly, it was collagen peptides from a bottom-up approach that were reported, rather than the outcome of a top-down approach such as OC

(Buckley et al. 2016). It is also clear that when the sequences of both collagen chains are concatenated, phylogenetic topologies are relatively consistent with those recovered from more in-depth DNA analyses, at least in mammals (Fig. 3). Although the species information in many of the other proteins described in this chapter have typically not been as widely studied to date it is clear that many of them are likely to provide as much if not more information (Table 1). For example, of the 214 amino acids making up the alpha-1S casein of cattle (*Bos taurus*), only 181 (note an insertion of 8 amino acids) are the same in sheep (*Ovis aries*; i.e. ~88%) whereas for collagen these are close to 99%. Interestingly, although assessing the percentage variability of different protein types could be considered meaningful for the potential phylogenetic information available via paleoproteomics, due to the variable sequence coverage between proteins it is not as much as would be assumed. Despite being able to recognise as much as 80–90% sequence coverage for collagen in sub-fossil bone samples, this is typically much lower for many of the other proteins present (see Buckley and Wadsworth 2014) with many of the protein identifications reported for the other tissues typically being based on only a few proteins each. However, the greater amenability to species identification in the absence of the most relevant sequences (i.e. cross-species proteomics) lies in the ability to detect those peptides containing the amino acid substitutions between particular species. Although this has been reported as reasons for observed absences in peptide matches of extinct taxa (e.g. Asara et al. 2007), many of the proteins observed in bone are highly conserved enough for this to be mitigated by 'error-tolerant'-based approaches (which allow for a single amino acid change to occur per peptide within the database searches). However, it is particularly encouraging that even though there are only two amino acids in myoglobin that separate *Bos* from *Ovis* (Table 1), taxa that diverged between 18.5 and 25 Ma (Hassanin et al. 2012; Nomura et al. 2013), Solazzo et al. (2008) were able to distinguish between more closely related seal taxa (Fulton and Strobeck 2010). This clearly has implications on the importance of protein function as a major influence in the rate of sequence change across different groups.

Table 1 Similarity of protein sequences between cattle (*Bos taurus*) and sheep (*Ovis aries*)

Protein	No. amino acids conserved	Total amino acids	Percentage identity (%)
Casein alpha S1	189	214	88
Beta-casein	206	224	91
Keratin 4	529	559	94
Beta-lactoglobulin	169	180	93
Albumin	560	607	92
Fetuin	334	364	91
Myoglobin	152	154	98
Collagen alpha 1(I)	1,050	1,056	99
Collagen alpha 2(I)	1,022	1,038	98

3.4 Age-Related Information from Proteins and their Post-translational Modifications (PTMs)

In addition to the species information, protein sequencing via proteomics can offer types of information that are not as easily obtained via DNA-based methods, particularly in the form of tissue source, which is important when studying the composition of artefacts or food residues (Tokarski et al. 2006). However, some protein fragments are only present at particular stages of development that can be identified in the archaeological record [e.g. in animal skins; (Brandt et al. 2014)], potentially improving our understanding of husbanding practices. The relative abundances of particular proteins, particularly some serum NCPs present in bone, appear to be potentially informative of relative biological age (e.g. Procopio et al. 2017; Sawafuji et al. 2017), but the extent to which this would be applicable in archaeological material thousands of years old remains unclear. Beyond sequence and relative abundance interpretation, a key advantage of proteomics over earlier sequencing methods is the ease at which structural changes could be observed, particularly through PTM analysis. Early mass spectrometric investigations for information beyond sequences from ancient bone were initially detected through focussed experimentation and direct observation from top-down approaches (e.g. MALDI; Ostrom et al. 2000). But more recent work in this area has concentrated on the study of some of the more common naturally occurring PTMs, such as asparagine and glutamine deamidation, as a means of relative age estimation of archaeological material. Although this field of research is in its infancy and currently appears to have poor chronologic resolution (Doorn et al. 2012; Wilson et al. 2012) due to the highly variable factors that affect such decay, the potential use of deamidation measurements to clearly discriminate between endogenous and exogenous material appears to be relatively robust (Buckley et al. 2017). Yet there is a range of other PTMs in ancient bone that are frequently detected using bottom-up approaches through the use of error-tolerant-type searches (e.g. against a database of known PTMs such as UniMod). These PTMs include but are not limited to oxidation, carbonylation, glycation and cross-linking as the most likely to affect ancient remains, but to date they have largely only been described in a qualitative manner.

3.5 Contamination and Sequence Mismatches: A Dinosaur Protein Case Study

In addition to the limitations in taxonomic resolution, which pales in comparison to what can be achieved using the latest aDNA methods (particularly with NGS), proteomic methods greatly suffer from the fact that it is not a true form of sequencing, unlike its predecessor Edman degradation. Current proteomic methods are not determining the primary structure (amino acid sequence) of a protein or peptide in a sequential manner that can be considered as robust, but it does so by probability matching against expected in silico results. Although this probability-based matching vastly improves the speed at which spectra can be matched against public databases, it means that it cannot be relied upon as readily when residues of indistinguishable mass (i.e., that are isobaric) are present. This is particularly problematic with collagen due to the presence of hydroxylated proline residues that are similar in mass to leucine and isoleucine. This issue is further complicated when trying to determine amino acid substitutions of alanine to serine with a nearby variable oxidation (of proline, lysine or methionine, but where proline residues form approximately one fifth of collagen). Although these issues can be managed to some extent, but it should be remembered that the more advanced proteomic methods, particularly the bottom-up approaches, have been designed to identify the protein composition of proteomes, rather than being optimised to achieve near complete protein sequence coverages. However, where the interest lies more in phylogenetic reconstruction or species determination rather than proteome composition, it is the cruder 'top-down' approaches that are more robust, where tissues dominated by low numbers of high-abundance proteins, such as bone, can be characterised with relative ease.

What continues as the most hotly debated issue in paleoproteomics revolves around the authenticity of peptides claimed as originating from Mesozoic dinosaur fossils. Over a decade ago, a team of biochemists and palaeontologists reported on peptides observed from extractions of fossilised material of a 68-million-year-old *Tyrannosaurus rex* (Asara et al. 2007). Concerns about this initial report were quickly raised, with suggestions of potential contamination from laboratory practice and/or specimen handling (Buckley et al. 2008a) or from bacterial biofilm colonisation (Kaye et al. 2008). In subsequent years, much effort was put against the biofilm hypothesis (Schweitzer et al. 2013, 2016), but little attention has been paid to discounting contamination. This is largely because within only a few years, the same team reported on the finds of peptides from an even older dinosaur fossil, this time an ~80-million-year-old *Brachylophosaurus canadensis* specimen (Schweitzer et al. 2009). The aspect that appeared to rule out contamination was the report of a peptide sequence unique to both dinosaurs. However, recently it has been pointed out that every single peptide in both of these studies could be matched to modern ostrich, reference material used in their laboratory, with much greater confidence than could be placed on their own, unique identifications (Buckley et al. 2017). Furthermore, the single peptide that was considered unique to both dinosaurs in the 2009 study was earlier matched to a different sequence [by a match to the homologous peptide in chicken (*Gallus gallus*)], clearly highlighting issues in the application of proteomics to ancient taxa (e.g. Fig. 4). It is widely accepted that such applications need to be standardised, perhaps in a similar manner as is carried out for proteomic analyses of modern tissues, i.e. if multiple peptides need to be considered essential for the identification of a protein in a given sample, then the same should be expected for sequences in extinct taxa. Intriguingly, members of the same team recently published additional peptides reportedly from the *Brachylophosaur* fossil with more rigorous cleaning of the mass spectrometry instrumentation (Schroeter et al. 2017). However, this time they did not recover the unique dinosaur peptide but

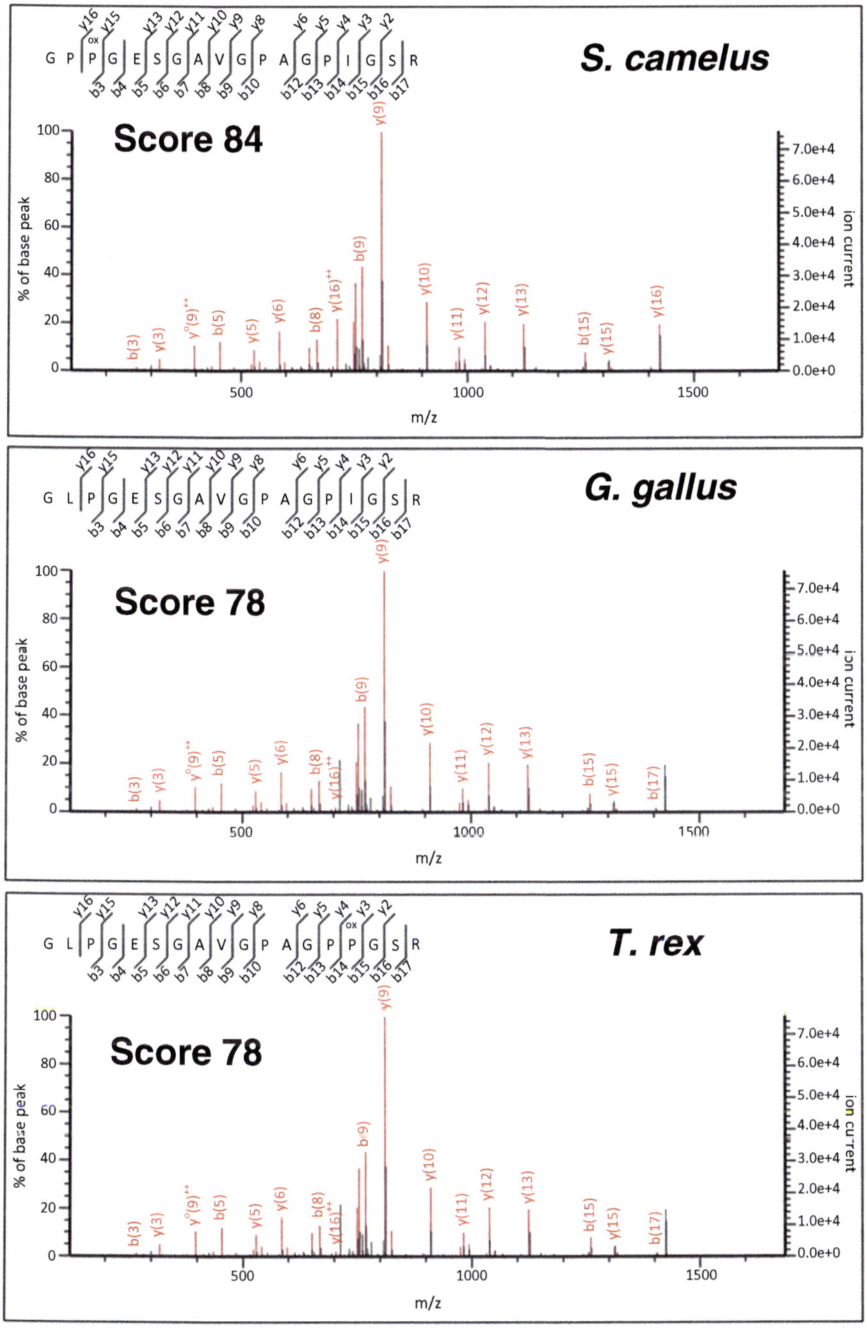

Fig. 4 Example tandem mass spectra showing the results from a digest of ostrich bone collagen and three similar scoring peptides with slight differences in sequence position and PTM, including the ostrich sequence (top), chicken sequence (middle) and proposed unique dinosaur sequence (bottom). Data taken from Buckley et al. (2017) and re-searched against SwissProt and a local database containing the ostrich (*Struthio camelus*) sequence

References

Abelson PH. Paleobiochemistry. Sci Am. 1956;195(1):83–92.
Abelson PH. Geochemistry of organic substances. In: Abelson PH, editor. Researches in geochemistry, vol. 1. Chichester: Wiley; 1959. p. 79–103.
Agnolin FL, Chimento NR. Afrotherian affinities for endemic South American “ungulates”. Mamm Biol. 2011;**76**(2):101–8.
Anderson NL, Anderson NG. Proteome and proteomics: new technologies, new concepts, and new words. Electrophoresis. 1998;19(11):1853–61.
Armstrong WG, Halstead LB, Reed FB, Wood L. Fossil proteins in vertebrate calcified tissues. Philos Trans R Soc Lond B Biol Sci. 1983;B301(1106):301–43.
Asara JM, Schweitzer MH, Freimark LM, Phillips M, Cantley LC. Protein sequences from mastodon and *Tyrannosaurus rex* revealed by mass spectrometry. Science. 2007;316 (5822):280–5.
Bada J, Wang X, Hamilton H. Preservation of key biomolecules in the fossil record: current knowledge and future challenges. Philos Trans R Soc Lond B Biol Sci. 1999;354(1379):77–86.
Bailey AJ. Molecular mechanisms of ageing in connective tissues. Mech Ageing Dev. 2001;122 (7):735–55.
Boskey AL, Posner AS. Bone structure, composition, and mineralization. Orthop Clin North Am. 1984;15(4):597–612.
Brandt LØ, Schmidt AL, Mannering U, Sarret M, Kelstrup CD, Olsen JV, et al. Species identification of archaeological skin objects from Danish bogs: comparison between mass spectrometry-based peptide sequencing and microscopy-based methods. PLoS One. 2014;9(9):e106875.
Breker M, Schuldiner M. The emergence of proteome-wide technologies: systematic analysis of proteins comes of age. Nat Rev Mol Cell Biol. 2014;15(7):453–64.
Brown TA. How ancient DNA may help in understanding the origin and spread of agriculture. Philos Trans R Soc Lond B Biol Sci. 1999;354(1379):89–98.
Buckley M. A molecular phylogeny of Plesiorycteropus reassigns the extinct mammalian order ‘Bibymalagasia’. PLoS One. 2013;8(3):e59614.
Buckley M. Ancient collagen reveals evolutionary history of the endemic South American ‘ungulates’. Proc Biol Sci. 2015;282(1806):20142671.
Buckley M, Wadsworth C. Proteome degradation in ancient bone: diagenesis and phylogenetic potential. Palaeogeogr Palaeoclimatol Palaeoecol. 2014;416:69–79.
Buckley M, Anderung C, Penkman K, Raney BJ, Gotherstrom A, Thomas-Oates J, et al. Comparing the survival of osteocalcin and mtDNA in archaeological bone from four European sites. J Archaeol Sci. 2008a;35(6):1756–64.
Buckley M, Walker A, Ho SY, Yang Y, Smith C, Ashton P, et al. Comment on “Protein sequences from mastodon and Tyrannosaurus rex revealed by mass spectrometry”. Science. 2008b;4 (319):33c.
Buckley M, Collins M, Thomas-Oates J, Wilson JC. Species identification by analysis of bone collagen using matrix-assisted laser desorption/ionisation time-of-flight mass spectrometry. Rapid Commun Mass Spectrom. 2009;23(23):3843–54.
Buckley M, Larkin N, Collins M. Mammoth and Mastodon collagen sequences; survival and utility. Geochim Cosmochim Acta. 2011;75(7):2007–16.
Buckley M, Melton ND, Montgomery J. Proteomics analysis of ancient food vessel stitching reveals >4000-year-old milk protein. Rapid Commun Mass Spectrom. 2013;27(4):531–8.
Buckley M, Fraser S, Herman J, Melton N, Mulville J, Pálsdóttir A. Species identification of archaeological marine mammals using collagen fingerprinting. J Archaeol Sci. 2014;41:631–41.
Buckley M, Gu M, Shameer S, Patel S, Chamberlain A. High-throughput collagen fingerprinting of intact microfaunal remains; a low-cost method for distinguishing between murine rodent bones. Rapid Commun Mass Spectrom. 2016;30:1–8.

Buckley M, Harvey V, Chamberlain A. Species identification and decay assessment of Late Pleistocene fragmentary vertebrate remains from Pin Hole Cave (Creswell Crags, UK) using collagen fingerprinting. Boreas. 2017; https://doi.org/10.1111/bor.12225.

Cappellini E, Jensen LJ, Szklarczyk D, Ginolhac A, da Fonseca RA, Stafford TW Jr, et al. Proteomic analysis of a pleistocene mammoth femur reveals more than one hundred ancient bone proteins. J Proteome Res. 2011;11(2):917–26.

Colombo G, Fanti P, Yao CH, Malluche HH. Isolation and complete amino acid sequence of osteocalcin from canine bone. J Bone Miner Res. 1993;8(6):733–43.

Curry GB. Amino acids and proteins from fossils. In: Eglinton G, Curry GB, editors. Molecular evolution and the fossil record. Knoxville, TN: Paleontological Society; 1988. p. 20–33.

Delmas PD, Tracy RP, Riggs BL, Mann K. Identification of the non collagenous proteins of bovine bone by two-dimensional gel electrophoresis. Calcif Tissue Int. 1984;36:308–16.

Demarchi B, Hall S, Roncal-Herrero T, Freeman CL, Woolley J, Crisp MK, et al. Protein sequences bound to mineral surfaces persist into deep time. Elife. 2016;5:e17092.

Di Donato A, Filippone E, Ercolano MR, Frusciante L. Genome sequencing of ancient plant remains: findings, uses and potential applications for the study and improvement of modern crops. Front Plant Sci. 2018;9:441.

Doorn NL, Wilson J, Hollund H, Soressi M, Collins MJ. Site-specific deamidation of glutamine: a new marker of bone collagen deterioration. Rapid Commun Mass Spectrom. 2012;26 (19):2319–27.

Edman P. Mechanism of the phenyl isothiocyanate degradation of peptides. Nature. 1956;177 (4510):667–8.

Frazao C, Simes DC, Coelho R, Alves D, Williamson MK, Price PA, et al. Structural evidence of a fourth Gla residue in fish osteocalcin: biological implications. Biochemistry. 2005;44 (4):1234–42.

Fulton TL, Strobeck C. Multiple markers and multiple individuals refine true seal phylogeny and bring molecules and morphology back in line. Proc R Soc Lond B Biol Sci. 2010;277 (1684):1065–70.

Gendreau MA, Krishnaswamy S, Mann KG. The interaction of bone Gla protein (osteocalcin) with phospholipid vesicles. J Biol Chem. 1989;264(12):6972–8.

Glowacki J, Rey C, Glimcher MJ, Cox KA, Lian J. A role for osteocalcin in osteoclast differentiation. J Cell Biochem. 1991;45(3):292–302.

Hassanin A, Delsuc F, Ropiquet A, Hammer C, Jansen van Vuuren B, Matthee C, et al. Pattern and timing of diversification of Cetartiodactyla (Mammalia, Laurasiatheria), as revealed by a comprehensive analysis of mitochondrial genomes. C R Biol. 2012;335(1):32–50.

Hauschka PV. Osteocalcin: the vitamin-K dependent Ca-binding protein of bone matrix. Haemostasis. 1986;16:258–72.

Hauschka PV, Carr SA. Calcium-dependant a-helical structure in osteocalcin. Biochemistry. 1982;21:2538–47.

Hauschka PV, Wians FH. Osteocalcin-hydroxyapatite interaction in the extracellular organic matrix of bone. Anat Rec. 1989;224:180–8.

Hauschka PV, Lian JB, Cole DE, Gundberg CM. Osteocalcin and matrix Gla protein: vitamin K-dependent proteins in bone. Physiol Rev. 1989;69(3):990–1047.

Hedges SB, Schweitzer MH. Detecting dinosaur DNA. Science. 1995;268(5214):1191–2.

Hollemeyer K, Altmeyer W, Heinzle E, Pitra C. Matrix-assisted laser desorption/ionization time-of-flight mass spectrometry combined with multidimensional scaling, binary hierarchical cluster tree and selected diagnostic masses improves species identification of Neolithic keratin sequences from furs of the Tyrolean Iceman Oetzi. Rapid Commun Mass Spectrom. 2012;26 (16):1735–45.

Hulmes GM. The collagen superfamily – diverse structures and assemblies. Essays Biochem. 1992;27:49–67.

Huq N, Tseng A, Chapman G. Partial amino acid sequence of osteocalcin from an extinct species of ratite bird. Biochem Int. 1989;21(3):491–6.

Huq NL, Tseng A, Chapman GE. Partial amino acid sequence of osteocalcin from an extinct species of ratite bird. Biochem Int. 1990;21:491–6.

James P. Protein identification in the post-genome era: the rapid rise of proteomics. Q Rev Biophys. 1997;30(04):279–331.

Jiang X, Ye M, Jiang X, Liu G, Feng S, Cui L, et al. Method development of efficient protein extraction in bone tissue for proteome analysis. J Proteome Res. 2007;6(6):2287–94.

Jones JD, Vallentyne JR. Biogeochemistry of organic matter. Geochim Cosmochim Acta. 1960;21:1–34.

Kadler KE, Holmes DF, Trotter JA, Chapman J. Collagen fibril formation. Biochem J. 1996;316:1–11.

Karas M, Hillenkamp F. Laser desorption ionization of proteins with molecular masses exceeding 10,000 daltons. Anal Chem. 1988;60(20):2299–301.

Kaye TG, Gaugler G, Sawlowicz Z. Dinosaurian soft tissues interpreted as bacterial biofilms. PLoS One. 2008;3(7):e2808.

Lander ES. The new genomics: global views of biology. Science. 1996;274(5287):536.

Lendaro E, Ippoliti R, Bellelli A, Brunori M, Zito R, Citro G, et al. On the problem of immunological detection of antigens in skeletal remains. Am J Phys Anthropol. 1991;86(3):429–32.

Liggett WH Jr, Lian JB, Greenberger JS, Glowacki J. Osteocalcin promotes differentiation of osteoclast progenitors from murine long-term bone marrow cultures. J Cell Biochem. 1994;55 (2):190–9.

Lockwood WW, Chari R, Chi B, Lam WL. Recent advances in array comparative genomic hybridization technologies and their applications in human genetics. Eur J Hum Genet. 2006;14(2):139–48.

Logan G, Collins M, Eglinton G. Preservation of organic biomolecules. In: Allison PA, Briggs DEG, editors. Taphonomy releasing the data locked in the fossil record, vol. 9. New York: Plenum; 1991. p. 1–24.

Lowenstein JM, Ryder OA. Immunological systematics of the extinct quagga (Equidae). Experientia. 1985;41(9):1192–3.

MacPhee RD. Morphology, adaptations, and relationships of Plesiorycteropus: and a diagnosis of a new order of eutherian mammals. Bulletin of the AMNH; no. 220. 1994.

Malone JD. Recruitment of osteoclast precursors by purified bone matrix constituents. J Cell Biol. 1982;92(1):227–30.

Mardis ER. Next-generation DNA sequencing methods. Annu Rev Genomics Hum Genet. 2008;9:387–402.

Matrix Science. 2016. http://www.matrixscience.com/. Accessed 28 Apr 2018.

Millard A. Deterioration of bone. In: Pollard AM, Brothwell D, editors. Handbook of archaeological sciences. New York: Wiley; 2001.

Nelsestuen GL, Zytkovicz TH, Howard JB. The mode of action of vitamin K identification of caboxyglutamic acid as a component of prothrombin. J Biol Chem. 1974;249(19):6347–50.

Nielsen-Marsh C. Biomolecules in fossil remains-multidisciplinary approach to endurance. Biochemist. 2002;24(3):12–4.

Nielsen-Marsh CM, Ostrom PH, Gandhi H, Shapiro B, Cooper A, Hauschka PV, et al. Exceptional preservation of bison bones >55 ka as demonstrated by protein and DNA sequences. Geology. 2002;30(12):1099–102.

Nogami HMD, Oohira A, Ogasawara NMD. Levels of creatine kinase activity in cartilage of tubular and nontubular bone in relation to pathogenesis of achondroplasia. Clin Orthop Relat Res. 1987;(219):308–12.

Nomura K, Yonezawa T, Mano S, Kawakami S, Shedlock AM, Hasegawa M, et al. Domestication process of the goat revealed by an analysis of the nearly complete mitochondrial protein-encoding genes. PLoS One. 2013;8(8):e67775.

Ostrom PH, Schall M, Gandhi H, Shen TL, Hauschka PV, Strahler JR, et al. New strategies for characterizing ancient proteins using matrix-assisted laser desorption ionization mass spectrometry. Geochim Cosmochim Acta. 2000;64(6):1043–50.

Ostrom PH, Gandhi H, Strahler JR, Walker AK, Andrews PC, Leykam J, et al. Unraveling the sequence and structure of the protein osteocalcin from a 42 ka fossil horse. Geochim Cosmochim Acta. 2006;70(8):2034–44.

Pääbo S, Poinar H, Serre D, Jaenicke-Després V, Hebler J, Rohland N, Kuch M, Krause J, Vigilant L, Hofreiter M. Genetic analyses from ancient DNA. Annu Rev Genet. 2004;38:645–79.

Prager EM, Wilson AC, Lowenstein JM, Sarich VM. Mammoth albumin. Science. 1980;209:287–9.

Price PA, Williamson MK. Primary structure of bovine matrix Gla protein, a new vitamin K-dependent bone protein. J Biol Chem. 1985;260(28):14971–5.

Procopio N, Chamberlain AT, Buckley M. Intra- and interskeletal proteome variations in fresh and buried bones. J Proteome Res. 2017;16(5):2016–29.

Robbins LL, Muyzer G, Brew K. Macromolecules form living and fossil biominerals; Implications for the establishment of molecular phyolgenies. In: Engle MH, Macko SA, editors. Organic geochemistry. New York: Plenum; 1993. p. 799–816.

Roepstorff P, Fohlman J. Letter to the editors. Biol Mass Spectrom. 1984;11(11):601.

Rybczynski N, Gosse JC, Harington CR, Wogelius RA, Hidy AJ, Buckley M. Mid-Pliocene warm-period deposits in the High Arctic yield insight into camel evolution. Nat Commun. 2013;4:1550.

Sawafuji R, Cappellini E, Nagaoka T, Fotakis AK, Jersie-Christensen RR, Olsen JV, et al. Proteomic profiling of archaeological human bone. R Soc Open Sci. 2017;4(6):161004.

Schreiweis MA, Butler JP, Kulkarni NH, Knierman MD, Higgs RE, Halladay DL, et al. A proteomic analysis of adult rat bone reveals the presence of cartilage/chondrocyte markers. J Cell Biochem. 2007;101:466–76.

Schroeter ER, DeHart CJ, Cleland TP, Zheng W, Thomas PM, Kelleher NL, Bern M, Schweitzer MH. Expansion for the *Brachylophosaurus canadensis* collagen I sequence and additional evidence of the preservation of cretaceous protein. J Proteome Res. 2017;16(2):920–32.

Schwalbe RA, Ryan J, Stern DM, Kisiel W, Dahlback B, Nelsestuen GL. Protein structural requirements and properties of membrane binding by gamma-carboxyglutamic acid-containing plasma proteins and peptides. J Biol Chem. 1989;264(34):20288–96.

Schweitzer MH, Zheng W, Organ CL, Avci R, Suo Z, Freimark LM, Lebleu VS, Duncan MB, Vander Heiden MG, Neveu JM, Lane WS. Biomolecular characterization and protein sequences of the Campanian hadrosaur *B. canadensis*. Science. 2009;324(5927):626–31.

Schweitzer MH, Zheng W, Cleland TP, Bern M. Molecular analyses of dinosaur osteocytes support the presence of endogenous molecules. Bone. 2013;52(1):414–23.

Schweitzer MH, Moyer AE, Zheng W. Testing the hypothesis of biofilm as a source for soft tissue and cell-like structures preserved in dinosaur bone. PLoS One. 2016;11(2):e0150238.

Solazzo C, Fitzhugh WW, Rolando C, Tokarski C. Identification of protein remains in archaeological potsherds by proteomics. Anal Chem. 2008;80(12):4590–7.

Solazzo C, Courel B, Connan J, Van Dongen BE, Barden H, Penkman K, Taylor S, Demarchi B, Adam P, Schaeffer P, Nissenbaum A. Identification of the earliest collagen-and plant-based coatings from Neolithic artefacts (Nahal Hemar cave, Israel). Sci Rep. 2016;6:31053.

Stewart NA, Gerlach RF, Gowland RL, Gron KJ, Montgomery J. Sex determination of human remains from peptides in tooth enamel. Proc Natl Acad Sci. 2017;114(52):13649–54.

Tanaka K, Waki H, Ido Y, Akita S, Yoshida Y, Yoshida T. Protein and polymer analyses up to m/z 100,000 by laser ionization time-of-flight mass spectrometry. Rapid Commun Mass Spectrom. 1988;2(8):151–3.

Tokarski C, Martin E, Rolando C, Cren-Olivé C. Identification of proteins in renaissance paintings by proteomics. Anal Chem. 2006;78(5):1494–502.

Triffitt JT, Gebauer U, Ashton BA, Owen ME, Reynolds JJ. Origin of plasma alpha2-HS-glycoprotein and its accumulation in bone. Nature. 1976;262(5565):226–7.

Vuorio E, de Crombrugghe B. The family of collagen genes. Annu Rev Biochem. 1998;59:837–72.

Wadsworth C, Buckley M. Proteome degradation in fossils: investigating the longevity of protein survival in ancient bone. Rapid Commun Mass Spectrom. 2014;28(6):605–15.

aRNA (Rollo 1985; Fordyce et al. 2013a; Rollo et al. 1991; Smith et al. 2014a, b; Venanzi and Rollo 1990; Ng et al. 2014) have appeared in the peer-reviewed literature over the same 30-year period (Table 1) that saw aDNA work grow in leaps and bounds to a current total of almost 6,000 research publications. Perhaps ironically, aRNA was initially at the forefront of archaeogenetics in the field's early years; but, due to an early lack of "interesting" data and an assumption of poorer general preservation, its study went on hiatus from the early 1990s until this decade. Possibly due to the additional precautions known to be essential to working with fresh RNA, such as strictly RNase-free conditions and much colder storage temperatures, aRNA was almost subconsciously dismissed as a recalcitrant and often fruitless molecule. This, as we will explore further, perhaps rests on two preconceptions about aRNA: its limited usefulness and unavailability, both of which may perhaps be incorrect.

From the outset, a major goal of aDNA research teams was to explore the evolutionary process at the molecular level; aDNA was a gift in the form of an evolutionary photo album, reducing the need to infer ancestral or extinct sequences but instead providing a relatively accurate snapshot of past genomes. As a result, several recent publications have rewritten early hominid evolution (Vernot and Akey 2015; Seguin-Orlando et al. 2014; Fu et al. 2014; Meyer et al. 2012), and similarly methodological work has clarified the evolutionary trajectories of numerous plants

Table 1 Summary of research publications into ancient RNA (not including reviews or critiques)

Year	Reference	Organism	Notes
1985	Rollo (1985)	Cress	Accidental; total nucleic acids from old tissue
1990	Venanzi and Rollo (1990)	Humans, maize, cress	Contesting previous criticisms but with unwitting reference to DNA damage patterns
1991	Rollo et al. (1991)	Maize	Generic nucleic acid extraction, similar to Rollo (1985)
1997	Fraile et al. (1997)	Tobamovirus	Evidence of tobamovirus infection in museum tobacco leaves
1999	Castello et al. (1999)	Tobamovirus	Ice core RT-PCR detection; contested
2006	Zhang et al. (2006)	Influenza	Lake ice-frozen influenza; RT-PCR detection; contested as contamination
2013	Fordyce et al. (2013a)	Maize	Complete transcriptome of maize kernels; proof of principle
2013	Guy (2013)	Peach latent mosaic viroid	Viriod RNA from a 50-year-old peach leaf tissue
2014	Smith et al. (2014a)	Barley stripe mosaic virus	First complete RNA genome
2014	Smith et al. (2014b)	Barley	Identification of siRNA acting in the plant RNA-directed DNA methylation pathway
2014	Ng et al. (2014)	Caribou, bacterial	Partial RNA viral genome in fecal matter
2016	Keller et al. (2017)	Human	Regulatory miRNA discovered in Tyrolean Iceman; soft tissues
2017	Smith et al. (2017)	Barley	Regulatory miRNA in desiccated barley seeds

(Allaby et al. 2015; Palmer et al. 2012; Paris 2016) and animals (Evin et al. 2015; Skoglund et al. 2015; Orlando et al. 2013) through the lens of both natural and human-mediated selection (i.e., domestication syndrome). When reconstructing phylogenies or genomic mutations at this level, RNA is not necessarily useful, since none of its eukaryotic incarnations give any more information than the genomic DNA sequences from which they originated. In fact, when one considers the transcriptome model and the absence of introns, RNA contains substantially less useful information than DNA for this purpose (viruses and viroids with RNA-based genomes are the exception to this rule, as will be discussed later in the chapter).

Phylogenetic analysis however is not the only use of ancient sequence data. Increasingly, paleogenomics is attempting to unravel cellular processes, mediated by the genome, as they originally occur(ed) in real time. The stresses that cause responses, the factors that lead to mutations, and a host of intricacies of interaction are all potential areas of investigation, and ones in which aRNA may provide more accurate and powerful insights than DNA. The different classes (such as messenger, regulatory, ribosomal, etc.), and importantly amounts, of RNA molecules can tell us a great deal about what was going on in the genome at the last moment before death or dormancy and thus what kind of environmental conditions were being experienced. The challenge here is unraveling those different classes, either through sequence or other chemical markers, to gain a truer picture.

The utility of aRNA is further compounded by its availability. The general robustness of DNA in comparison to RNA is well documented and almost an extension of the biochemical dogma surrounding ancient biomolecules. As early researchers noted, possibly underestimating of the gravity of their observation, biochemical preservation of nucleic acids in seeds is not a simple model and that the "trend may be general" for ancient biomolecules to break down into ultra-small fragments (Rollo 1985). In terms of chemical stability, different pattern-specific degradation rates should be expected between RNA and DNA (Willerslev et al. 2004); see Sect. 3. However it is the release of RNases during autolytic decomposition in many tissues, which promotes the general degradation of RNA over DNA (Huynen et al. 2012). This naturally implies that near-ideal preservation conditions would be essential for anything more than trace levels of aRNA survival. Since these early studies, aRNA extractions have not generally been attempted from tissues under conditions that are now known to be conducive to aDNA persistence. During this time, however, a number of successful attempts at germinating archaeological seeds (Yashina et al. 2012; Sallon et al. 2008) suggested core RNA components involved in germination must be capable of sufficient survival under the correct conditions.

In this chapter, we will discuss older and post-hiatus research into aRNA, promising recent research following some paradigm-shifting discoveries in transcriptomics, and suggest steps toward a potential new synthesis on its utility.

2 A Brief History of aRNA Study

The number of studies that have focused primarily on aRNA is sufficiently small that a comprehensive review can be given in this section (see Table 1). This was not destined to be the case however; following the groundbreaking recovery of the first aDNA sequences from a preserved Quagga in 1984 (Higuchi et al. 1984), attention almost immediately turned to other ancient biomolecules. Only the next year, aRNA extraction from ancient cress seeds was a fortuitous accident in what was an attempt to indiscriminately extract nucleic acid from plant material (Rollo 1985), the RNA itself being identified by molecular hybridization. Similar results derived from preferential extraction of RNA in a general nucleic acid extraction method were subsequently observed using maize kernels (Rollo et al. 1991), causing disagreement over the relative proportions of depolymerized, modified DNA (Pääbo 1986; Rogan and Salvo 1990) versus unmodified RNA. The controversy stemmed from the ubiquitous presence of uracil in ancient nucleic acids, originally part of the robust argument (Venanzi and Rollo 1990) for the increased survivability of RNA over DNA.

However, the presence of uracil in archaeogenomic samples is now largely attributed to a breakdown process of DNA; in the presence of water molecules, hydrolytic deamination of cytosines results in uracil. This process occurs readily at overhanging ends of fragmented, double-stranded DNA molecules (of which there are many in a typical aDNA sample) and so cannot be attributed solely to the presence of RNA. Although this was empirically demonstrated relatively early on (Pääbo 1989) and several years later in detail (Hofreiter et al. 2001), and despite the interesting sidenote that DNase-free RNase removed the majority of nucleic acids of all types from mummified maize kernels, the study of aRNA became very much a secondary concern to aDNA. Again, this was presumably due to its perceived (lack of) information value and an unfortunate propensity in academic science to neglecting the publication of negative results. There were no further explicit aRNA research papers published for several years, the only interim mention being a revisiting of existing archaeobotanical extraction methods (Rollo et al. 1994).

The next piece of research to explicitly investigate aRNA appeared in 1999, following the detection of tomato mosaic tobamovirus genomic RNA in a glacial ice cores ranging from 5,000 to 140,000 years old (Castello et al. 1999). While the clean-lab procedures and controls common in today's paleogenomic labs were not followed, and similarities to modern strains persisted at all strata, the potential for aRNA from pathogen genomes, as opposed to relatively uninformative transcriptomes, was rapidly becoming apparent as a viable source of ecological, genomic, and pathogenic information in a previously untapped biomolecular resource. An elegant hypothesis of atmospheric recycling of viruses from melted glacial water was put forward to explain the somewhat concerning similarity between modern and ancient sequences, touted as much the same process as is commonly observed with bacterial and fungal spores. However, this hypothesis apparently failed to gain traction in the academic community, and aRNA was again not studied widely for several years, until a similar study also claiming to have evidence for ancient viral RNA genomes was published, detailing influenza A in Siberian lake ice (Zhang et al. 2006). The authors here proposed a similar hypothesis of host-/carrier-mediated recycling resulting from freeze/thaw cycles of the lakes, but this was met with skepticism, and the research was widely discredited as being laboratory contamination (Worobey 2008) on the basis of suspicious levels of similarity to modern laboratory strains in a supposedly rapidly evolving genome. Simultaneous questions were also raised about the earlier tobamovirus work, on the same basis, and by virtue that few of the (by now quite famous) criteria for ancient DNA authenticity (Cooper and Poinar 2000) were applied to either study.

In their 2004 review of nucleic acid potential in permafrost conditions (Willerslev et al. 2004), the authors noted a general expectation of decreased RNA survival in the archaeological record, in particular with reference to fragmentation (see Sect. 3.1). Pertinently, molecular fragment size does not automatically define a molecule as "ancient." In fact, no particular criterion does; "ancient," after all, is a subjective term, as is "historical." Historical DNA and RNA are terms often used to describe samples not necessarily of an archaeological context but found in some other biological repository such as museum or herbarium. Since the majority of degradation patterns and techniques for extracting and analyzing these materials are identical to those of unequivocally "ancient" materials (and a distinct lack of alternative ancient RNA!), it is not inappropriate to discuss examples of aRNA work from herbaria here. Recent research into general biomolecular breakdown processes advocate that the majority of fragmentation events occur within the first few years, the remaining degradation plateauing and thereafter defined by long-term environment (Kistler et al. 2017). While older RNA has been examined from seeds, which themselves are adapted specifically for long-term stability of cellular machinery including nucleic acid components, it is encouraging that RNA from softer tissues can survive for at least decades after death. A 2013 study identified amplifiable viroid RNA of peach latent mosaic viroid in 50-year-old leaf tissue (Guy 2013), showing amplicon lengths within reasonable expectations based on observations of similarly aged DNA molecules. A 1997 indirect attempt to identify aRNA from similar (although older) tissue samples suggested that complete virions can remain intact, enough to still be infectious, for a century or more (Fraile et al. 1997). While an inoculation method from herbaria lesions such as this would not meet the typical criteria for the study of ancient biomolecules (Cooper and Poinar 2000), the results and negative controls provide encouraging evidence of aRNA persistence.

Further evidence of aRNA persistence came about the following year, when a partial aRNA viral genome was sequenced from a permafrost environment, albeit a surprising one (Ng et al. 2014). The presence of a plant virus in caribou feces allowed insight not only into the paleoecology of northwest Canada but also into the survivability of a dogmatically "fragile" molecule in the presence of a substance replete with microbial activity and all its associated enzymatic activity. The presence of enzymes from the caribou digestive tract and its microflora suggests that permafrost environments have the potential to negate certain decompositional processes and allow RNA to survive for long periods.

microbial decomposition processes (Keller et al. 2017). While most of these examples are relatively recent and have only been possible by utilizing next-generation sequencing technologies, the latter perhaps represents a reevaluation of the "RNA survival dogma."

3.1 Migration and Loss

The most noticeable characteristic of ancient DNA and RNA is its availability – or, more precisely, lack thereof. The recoverable quantity of endogenous nucleic acids per unit mass of tissue from ancient material is usually significantly depleted from its original in vivo levels, and only in exceptional cases will it approach a level that is comparable to modern material. There are several possible mechanisms underlying this, and the reality is probably at least some degree a reflection of them all.

First, as previously mentioned, much of this loss in actual recovery can be attributed to the ultrashort fragments in the theoretical fragment size distribution due to the limitations of isolation chemistry. The majority of current methods employed for DNA recovery rely on precipitation of DNA from solution in a combination of chaotropic salt and alcohol, followed by binding to a silicon dioxide matrix, and this isolation of RNA is chemically identical (Poeckh et al. 2008). Methods have been refined over the years to allow recovery of smaller and smaller fragments (Dabney et al. 2013), down to less than 15 bases, although the very smallest fragments of the theoretical distribution are never recovered using current methods.

Second, it is to be expected that at least some of the original nucleic acids will have been degraded into derivatives that are no longer recognizable as the original nucleic acids. This type of time-dependent degradation is a complex issue but likely to be a function of various factors such as temperature, humidity, tissue type, surrounding pH, microbial activity, and even background radiation. Paleogenomicists generally find congruent evidence to support this (for at least DNA); for example, burnt grains often have much less recoverable material than desiccated equivalents, and material from permafrost environments generally gives better results than those from tropical conditions. Specific decay processes can also influence the levels of DNA and RNA independently, particularly the often more ubiquitous presence of RNases in many tissues and microorganisms (Guy 2013). All these processes are likely, in some fashion, to be an extension of the issues discussed in this section – and also intrinsic to the problem of recovering ultrashort fragments.

Finally, the diffusion of molecules through the matrix of their deposition environment(s), away from their source, greatly reduces the amount of available genetic information. Often in ancient DNA literature, particularly where a next-generation sequencing approach has been taken, there is at least a passing mention of the "endogenous content," usually represented as a percentage of total reads. "Endogenous" in this case refers to DNA sequences that likely belong to the organism being studied, unless dealing with a metagenomic assembly, and "exogenous" refers to everything else (or sequences of such low complexity they could belong to a wide range of organisms, including the one being studied). In a situation where significant molecular movement occurs, the endogenous DNA diffuses away from its tissue of origin and is in turn replaced by other DNA from the surrounding environment and thus may result in low endogenous content. The extent of this diffusion is again determined by several environmental factors including temperature, humidity, and tissue type. For example, in a small, reasonably closed system such as a seed with intact pericarp, one might expect greater levels of endogenous DNA and RNA than, for example, porous bone. Quite often this is the case, although again dependent on other factors such as water percolation and temperature. Several studies have shown increased endogenous content where liquid water is largely absent, for example, from desiccated (Palmer et al. 2012) and permafrozen (Mouttham et al. 2015) environments, which contrasts with the results from humid environments (Pinhasi et al. 2015). When dealing with the nucleic acid content of a metagenomic assemblage (such as a soil sediment core), the genomic data generated might have only a superficial bearing on the physical mass of identifiable macrofossil species contained within it (Smith et al. 2015); however the anaerobic conditions of truly waterlogged material, while potentially lacking in endogenous ancient DNA, often slow other diagenetic processes detailed later (Brown et al. 2015).

It is important to note that as with the other factors mentioned, we are forming hypotheses of RNA availability based primarily on empirical evidence of DNA as a proxy. Other than differential types of chemical breakdown, the limited evidence of aRNA behavior has so far been as expected; however, we may have to alter our expectations of aRNA as more evidence becomes available.

3.2 *Fragmentation*

Like DNA, the primary structural support element of RNA lies in its phosphate backbone. Breakage of both strands is a requirement for full DNA molecule fragmentation, and so one might be forgiven for expecting a greater rate of fragmentation in a single-stranded molecule such as RNA. Primarily, RNA's 2′ hydroxyl (OH) group, which DNA does not have, has the potential to induce strand cleavage by hydrolyzing its own adjacent phosphodiester bond (Fordyce et al. 2013b). This action can be further compounded when catalyzed by certain cations (Lindahl 1967) such as calcium, which may be ubiquitous in, for example, a skeletal assemblage. Indeed, a review of the potential for DNA and RNA survivability in permafrost conditions (Willerslev et al. 2004) outlined succinctly the expectation of a generally elevated degradation rate of RNA compared to DNA.

However, empirical data from truly ancient plant material suggests that in some circumstances, the opposite may be the case (Fordyce et al. 2013a). The reason is possibly down to the fact that, in practicality, RNA by itself is quite often not entirely single stranded. Since the principle of complementary base pairing still very much applies to it, RNA has a propensity to form secondary structure, spontaneously folding back in on itself and creating de facto strings of base pairs from sequence regions with enough complementarity to each other (Zuker et al. 1999). Secondary structure formation is thought to effect the rate of phosphodiester bond hydrolysis (Fordyce et al. 2013b), seen by the greater persistence in highly secondary structure-forming RNA types such as ribosomal RNA compared to messenger RNA transcripts (Laing and Draper 1994). In fact, it is this ability to form secondary structures which has the emergence of the microRNA regulatory pathway (see later).

While it seems that exact relative rates of degradation cannot be estimated due to a dearth of data, Fordyce et al. provide a detailed review of biochemical interactions contributing to RNA breakdown (Fordyce et al. 2013b), and Willerslev et al. also noted that the presence of the 2′ hydroxyl group in RNA should, in theory, increase its inherent lability. Expected fragmentation patterns are especially pertinent, as recent research (Kistler et al. 2017) has noted that the theoretical distribution of DNA degradation, which should follow an exponential curve increasing toward small fragments, is not seen in recovered material. Instead such fragmentation often follows a lognormal distribution (Renaud et al. 2017), the "missing" upper end of which can be explained by inefficiencies in isolation protocols (that is to say, ultrashort fragments having insufficient mass for salt-assisted precipitation and binding to a silica medium, the basic principles around which the vast majority of nucleic acid extractions are carried out).

A further compounding factor for modeling aRNA decay is the variable abundance, length, and types found in a cell. A typical eukaryotic nuclear genome has a standardized size and copy number (depending on ploidy, although this is two for the majority of species), and mitochondria again have a standard size but variable copy number. RNA species however are much, much smaller and vary wildly in size in vivo compared to their DNA counterparts. A small regulatory RNA at its smallest is around 18 nucleotides (nt), while a large transcript can be as long as 100,000 nt in length. The oft-abundant ribosomal RNAs are usually in between at a few thousand nt. To compound things further, the copy number of RNA varies according to tissue type, age, and even the immediacy of the organism's environment; under stress, a cell may be producing more regulatory RNAs or making more transcript for a certain gene. Since the smallest physically recoverable size of RNA is around the small RNA size, disentangling "real" small RNA from the breakdown product of a larger transcript is fraught with problems.

Congruent with expectations for DNA (Kistler et al. 2017), excessive strand breakage alone renders nucleic acids unusable in terms of both recovery and analysis; while the physical mass of aRNA in such a breakdown process may not change, at least within a closed system, resultant ultrashort ($<$10 nt) molecules are not recoverable by most extraction protocols. If they were, the reduced sequence complexity renders such fragments highly prone to false reference alignments and thus misidentification. On the other hand, base cleavage from the backbone (depurination) events are predicted to occur at a slower rate in RNA than DNA, potentially allowing a greater proportion of usefully sized RNA molecules to be competent for sequencing.

3.3 Deamination

Cytosine deamination, the loss of the amine group on a cytosine to produce uracil, is arguably one of the most characterized and discussed lesions of ancient DNA, even to the point of its presence being seen as a proxy for authenticity (Briggs et al. 2007). This particular lesion, at least in ancient DNA, is (probably) limited to exposed, single-stranded ends ("overhangs") where the complementary strand is broken at the phosphodiester backbone and the terminal end no longer has enough entropy to sustain the hydrogen bonds between complementary bases. When reading these newly converted uracils, many of the polymerases involved in sequencing library preparation treat uracils as thymine and so incorporate adenine as the complementary base. Following several rounds of PCR, cloned molecules of those containing uracil now contain thymine and are read by sequencers as such. During subsequent data analysis when mapping reads to a reference genome, the patterns of deamination signals can be characterized by virtue of the large numbers of sequencing reads using the program mapDamage (Ginolhac et al. 2011). They typically manifest as cytosine > uracil (read as thymine) misincorporations at the 5′ end of the molecule. At the 3′ end, we see a "mirror image" misincorporation; overhanging uracils are paired with adenine during the "strand repair" step of library construction (whereas an intact cytosine would normally pair with guanine). During PCR, the "misincorporated" adenines show as a guanine > adenine mismatch when mapped to the reference sequence.

How does this play out when the target molecule is single stranded, as are the majority of RNA classes? We know that cytosines in ancient RNA can become deaminated in the same way as DNA, but interestingly, cytosine deamination damage patterns are not randomized or constant across the strand as one would expect from a molecule which is entirely single stranded. In the small RNA fraction of archaeological barley, the authors discovered distinct misincorporation patterns at both ends of the sequenced molecule, with significantly fewer in the middle region (Smith et al. 2014a). Secondary structure formation, as detailed above, may "protect" mid-sequence cytosines while leaving terminal nucleotides exposed.

An emerging phenomenon in ancient DNA is the observation of a different type of deamination, this time from 5-methyl-cytosine (5mC) to thymine. The deamination reaction has much the same chemistry as C > U transitions, tends to occur at overhangs, and in sequencing shows up as the same C > T modification. However, it cannot be distinguished from the deamination product of an unmethylated cytosine, unless steps are taking in library building, such as removing uracils (UDG treatment), or using a polymerase that stalls at uracils (i.e., only reporting sequences that do not contain deaminated cytosines). Where this takes place, a similar algorithm to mapDamage, EpiPALEOMIX (Hanghøj et al. 2016) can be used to identify sites that have previously been methylated cytosine.

Exactly how this phenomenon could be applied to ancient RNA is not entirely clear – yet – but could have applications. Cytosine methylation akin to that of DNA (5mC) is known to occur in certain types of noncoding RNA, such as ribosomal,

targeting of either mature short RNAs or the terminal ends of other transcripts. In both cases, the original type of RNA can be deduced from the sequence data itself if the organism's transcriptome is well characterized. With this method, not only can non-hyper-redundant RNA sequences be retrieved, but their expression levels give a direct snapshot of in vivo processes that were taking place perimortem and, by extension, what sort of pressures were present.

5 Regulatory RNA

Eukaryotic genomes are regulated by networks of interactions between transcripts, proteins, and the genome itself. One such actor in a (relatively) new interaction pathway is that of the small RNA, 18–24 nt RNA molecules which themselves appear in two distinct flavors: short interfering RNA (siRNA) and microRNA (miRNA). Both have been discovered to persist in archaeological material, under very different preservation conditions.

MicroRNAs are involved in specific, often highly conserved regulatory pathways and as such are derived from specific genes within a genome. miRNA gene primary transcripts form a stem-loop (or "hairpin"), double-stranded structure from which the mature miRNA is processed by a series of protein complexes. DICER-like proteins cleave the dsRNA into its mature length, and the complementary strand is degraded, leaving only a single-stranded mature miRNA. This is then incorporated into a second set of Argonaute (AGO) proteins to form the RISC (RNA-induced silencing complex). Depending on the specifics of the RISC, the mature miRNA is directed to its genomic or transcriptomic target to enact its function, which is usually downregulation of messenger RNA by interception through complementary base pairing followed by cleavage of the transcript by the AGO protein. As is the case with regulatory networks, the miRNA target may itself be a downregulator such as a transcription factor, and so increased miRNA expression may not simply equate to reduced protein-coding gene expression. This however is a simplified summary; for a detailed overview of the miRNA biogenesis pathway, see Winter et al. (2009). Relevant to this chapter, however, is that these miRNAs are now known to be recoverable not only from ancient material (Keller et al. 2017; Smith et al. 2017) but from mammalian soft tissues that would have a priori not been suitable candidates due to autolytic releases of RNases. The first study of this kind (taken from various tissues of Ötzi, the permafrost-preserved Tyrolean "iceman") showed that tissue-specific miRNA profiles are recoverable from ancient material, validating the principle of aRNA recovery for purposes beyond the genomic. The second, taken from desiccated barley grain from ancient Egypt, demonstrated that differential, environmentally induced profiles can be similarly reconstructed and as such give a real-time in vivo snapshot of adaptive processes. Even with these two conditions being at almost opposite ends of the spectrum of archaeological record (although both known to be conducive to aDNA survival), they represent access to information that cannot be achieved using ancient DNA alone.

The second type of small RNA, siRNA, is of similar size although differs in function and biogenesis. Typically, siRNA directs methyltransferases to genomic targets, resulting in suppression of gene expression since transcriptases often falter at methylated sites. siRNA can also be incorporated into the RISC complex to neutralize transcripts and other RNA molecules. Theoretically, siRNA can be produced from any RNA molecule; any transcript can be made double stranded using endogenous polymerases, and the resulting dsRNA can be processed into a targeting complex, similar to miRNA. Ancient siRNA sequences, isolated from the same archaeological barley as mentioned above, have been shown to correlate with genomic sequences showing elevated methylation (Smith et al. 2014b) in what is likely a stress-induced response in plants known as RNA-directed DNA methylation (RdDM).

Pertinently, siRNAs do not have to be endogenous to the genome facilitating their biogenesis (Snead and Rossi 2010). As previously mentioned, any RNA molecule can become a template for siRNA, regardless of its origin. A known immune response in plants is to produce siRNA deriving directly from the genome of an invading pathogen, allowing the RISC complex to be directed to that genome and prevent the pathogen's functions, such as protein production or replication. In some cases, as detailed below, this could be key to discovering exogenous RNA of interest.

6 RNA Genomes

To date, only one aRNA genome has been characterized (Smith et al. 2014a). Sequence fragments of ancient barley stripe mosaic virus (BSMV) were detected when performing routine metagenomics on sRNA sequence data. Interestingly, many of the small sequences used to reconstruct the genome were in the size ranges expected for siRNA, suggesting that at least some of these fragments were not necessarily fragmented genomic components but the result of siRNA biogenesis as a defense mechanism by the infected host. This was, essentially, a serendipitous finding; equally fortunate is the fact that the typical RNA genome, limited only to viruses (or at least as far as we know), is only between 3 and 30 Kb in length, making reconstruction from a relatively small NGS dataset a fairly straightforward process.

The concept of a genomic "arms race" between pathogen and host, particularly where humans and staple crops are concerned, has long been a concern of the medical and agricultural communities (Stahl and Bishop 2000). One tool we may have to address these concerns is an understanding of the evolution of pathogens that we may be able to predict their evolutionary trajectories given a certain set of circumstances. Being in possession of more ancient or archaic strains of pathogens can give a wider basis in which to incorporate into existing models. Several human pathogenic genomes have been characterized from ancient DNA (Bos et al. 2015; Schuenemann et al. 2011; Müller et al. 2014), but the general lack of aRNA data means that viral genomes have not. Indeed, previously discovered but possibly inauthentic ancient viruses (Castello et al. 1999; Zhang et al. 2006) would have benefitted from being sequenced using NGS technology; like aDNA, telltale patterns of cytosine deamination are known to occur toward the ends of aRNA molecules, which is an obvious control for authenticity. Perhaps if RNA work becomes routine for samples of an appropriate preservation condition, more pathogen genomes may be recovered.

7 Endogenous Transcriptomics

While small RNA molecules can tell us about how the genome is being regulated, ribosomal and transfer RNA, due to their inherent ubiquity, can tell us relatively little. The final aspect of the transcriptome, however, messenger RNA, can potentially provide verification for aspects of the regulatory processes identified from small RNA. As with complete aRNA genomes, only one empirical study assessing the complete transcriptome using high-throughput sequencing has been published (Fordyce et al. 2013a). This represents an untapped resource which, when combined with other RNA classes, can give truly new insights into evolution, domestication, and drivers of these such as human interaction and paleoclimate.

8 Technical Considerations for aRNA

To conclude, we recap some of the technical considerations that should be taken into account when performing this kind of work. This should not be taken as anything akin to "criteria of authenticity" since those considerations should be a given when working with ancient biomolecules, however, but the following minor points can make big differences.

8.1 Isolation of aRNA

At the risk of stating the obvious, all extraction and pre-PCR laboratory work should take place in a strictly controlled setting, in the same fashion as with ancient DNA. Reagents, tubes, etc. should be certified RNase-free, and it is especially important to bear this in mind when making one's own reagents. Unlike modern samples, however, the release of endogenous RNases from sample tissue during predigestion (i.e., crushing/grinding) is probably not such an issue due to the likely degradation of these enzymes over time. This however should not be a given, so samples that possibly contain active RNases (e.g., historical/herbaria tissues) should be flash frozen in liquid nitrogen first, to minimize potential degradation.

Analyses based on extracted RNA often require prior removal of any co-extracted DNA. There are several ways to do this, but no single method is ideal. DNase treatment is routinely used for modern tissues, but the fragmentary nature of ancient DNA limits the number of cleavage sites available to DNase (Sutton and Brown 1997) and so reduces its efficiency. Compensating for this using extended incubation may prove detrimental as DNase will preferentially degrade RNA where a suitable DNA substrate is unavailable (Smith 2012). The second method is to extract total nucleic acids in acidic (pH 4.8) phenol. Since DNA is slightly less acidic than RNA, lowering the pH of the organic solvent encourages DNA molecules to move out of the aqueous phase toward the interphase, leaving RNA alone in the aqueous phase (Chomczynski and Sacchi 2006). While this works well to eliminate DNA, it is time-consuming, reduces overall yield of RNA, and involves working with a dangerous substance. The removal of other coprecipitates often seen from ancient tissues can be achieved using repeat organic extractions.

Since many protocols for size-based selection of RNA are designed to capture small (>18 nt) RNA, no particular modified protocols geared specifically to ancient, fragmented RNA are required. Generally, they work along similar principles to ancient DNA, being bound to a silica matrix in the presence of chaotropic salts such as guanidinium thiocyanate.

8.2 NGS Library Building

As we described earlier, the 3′ hydroxyl group of endogenously processed small RNA can be taken advantage of for enrichment by using a pre-adenylated adapter for NGS library construction. This however has limitations where RNA fragments do not include a 3′ OH group where nonenzymatic shearing has taken place, and the high copy number of redundant rRNA and tRNA makes extracting meaningful information difficult. There are no protocols available at present to directly address this, although certain types of molecular enrichment might be a possibility with refinement to capture RNA instead of DNA. For the time being, we can take advantage of the ever-increasing output of existing NGS platforms and simply discard redundant data, albeit at a significantly higher cost per base than DNA.

8.3 Cytosine Deamination

As with aDNA, cytosines in aRNA are susceptible to hydrolytic deamination that converts them to uracil. Again, like aDNA, this tends to occur toward the termini of the molecule where cytosines are likely to be exposed on a single strand. Unlike aDNA, however, where UDG (uracil-DNA-glycolase) treatment can be employed to remove uracils and repair the abasic sites (Briggs et al. 2010), RNA cannot be subject to uracil removal since uracil is a canonical base. This poses a problem in

de los Rios A, Ramirez R, Estévez P. RNase in Lasallia hispanica and Cornicularia normoerica: multiplicity of electromorphs and activity changes during a hydration-dehydration cycle. J Exp Bot. 1996;47(12):1927–33.

Elbashir SM, Lendeckel W, Tuschl T. RNA interference is mediated by 21- and 22-nucleotide RNAs. Genes Dev. 2001;15(2):188–200.

Evin A, et al. Unravelling the complexity of domestication: a case study using morphometrics and ancient DNA analyses of archaeological pigs from Romania. Philos Trans R Soc Lond B Biol Sci. 2015;370(1660):20130616.

Fabre A-L, et al. An efficient method for long-term room temperature storage of RNA. Eur J Hum Genet. 2014;22(3):379–85.

Fordyce SL, et al. Deep Sequencing of RNA from ancient maize kernels. PLoS One. 2013a;8(1): e50961.

Fordyce SL, et al. Long-term RNA persistence in postmortem contexts. Investig Genet. 2013b;4(1):7.

Fraile A, et al. A century of tobamovirus evolution in an Australian population of Nicotiana glauca. J Virol. 1997;71(11):8316–20.

Fu Q, et al. Genome sequence of a 45,000-year-old modern human from western Siberia. Nature. 2014;514(7523):445–9.

Ginolhac A, et al. mapDamage: testing for damage patterns in ancient DNA sequences. Bioinformatics. 2011;27(15):2153–5.

Guy PL. Ancient RNA? RT-PCR of 50-year-old RNA identifies peach latent mosaic viroid. Arch Virol. 2013;158(3):691–4.

Hanghøj K, et al. Fast, accurate and automatic ancient nucleosome and methylation maps with epiPALEOMIX. Mol Biol Evol. 2016;33:3284–98.

Hansen AJ, et al. Crosslinks rather than strand breaks determine access to ancient DNA sequences from frozen sediments. Genetics. 2006;173(2):1175–9.

Harris ME, Christian EL. RNA crosslinking methods. Methods Enzymol. 2009;468:127–46.

Higuchi R, et al. DNA sequences from the quagga, an extinct member of the horse family. Nature. 1984;312(5991):282–4.

Hofreiter M, et al. DNA sequences from multiple amplifications reveal artifacts induced by cytosine deamination in ancient DNA. Nucleic Acids Res. 2001;29(23):4793–9.

Huang Y, Li L. DNA crosslinking damage and cancer – a tale of friend and foe. Transl Cancer Res. 2013;2(3):144–54.

Hussain S, et al. Characterizing 5-methylcytosine in the mammalian epitranscriptome. Genome Biol. 2013;14(11):215.

Huynen L, Millar CD, Lambert DM. Resurrecting ancient animal genomes: The extinct moa and more. BioEssays. 2012;34(8):661–9.

Keller A, et al. miRNAs in ancient tissue specimens of the Tyrolean Iceman. Mol Biol Evol. 2017;34:793–801.

Kistler L, Ware R, Smith O, Collins MJ, Allaby RG. A new general model for ancient DNA decay based on paleogenomic meta-analysis. Nucleic Acids Res. 2017;45(11):6310–20.

Laing LG, Draper DE. Thermodynamics of RNA folding in a conserved ribosomal RNA domain. J Mol Biol. 1994;237(5):560–76.

Lindahl T. Irreversible heat inactivation of transfer ribonucleic acids. J Biol Chem. 1967;242(8):1970–3.

Meyer M, et al. A high-coverage genome sequence from an archaic Denisovan individual. Science. 2012;338(6104):222–6.

Mouttham N, et al. Surveying the repair of ancient DNA from bones via high-throughput sequencing. BioTechniques. 2015;59(1):19.

Müller R, Roberts CA, Brown TA. Biomolecular identification of ancient Mycobacterium tuberculosis complex DNA in human remains from Britain and continental Europe. Am J Phys Anthropol. 2014;153(2):178–89.

Ng TFF, et al. Preservation of viral genomes in 700-y-old caribou feces from a subarctic ice patch. Proc Natl Acad Sci. 2014;111(47):16842–7.
Orlando L, et al. Recalibrating Equus evolution using the genome sequence of an early Middle Pleistocene horse. Nature. 2013;499(7456):74–8.
Pääbo S. Molecular genetic investigations of ancient human remains. Cold Spring Harb Symp Quant Biol. 1986;51:441–6.
Pääbo S. Ancient DNA: extraction, characterization, molecular cloning, and enzymatic amplification. Proc Natl Acad Sci U S A. 1989;86(6):1939–43.
Palmer SA, et al. Archaeogenomic evidence of punctuated genome evolution in Gossypium. Mol Biol Evol. 2012;29(8):2031–8.
Paris HS. Overview of the origins and history of the five major cucurbit crops: issues for ancient DNA analysis of archaeological specimens. Veg Hist Archaeobotany. 2016;25:405–14.
Pinhasi R, et al. Optimal ancient DNA yields from the inner ear part of the human petrous bone. PLoS One. 2015;10(6):e0129102.
Poeckh T, et al. Adsorption and elution characteristics of nucleic acids on silica surfaces and their use in designing a miniaturized purification unit. Anal Biochem. 2008;373(2):253–62.
Renaud G, et al. gargammel: a sequence simulator for ancient DNA. Bioinformatics. 2017;33:577–9.
Rogan PK, Salvo JJ. Study of nucleic acids isolated from ancient remains. Am J Phys Anthropol. 1990;33(S11):195–214.
Rogers SO, Bendich AJ. Ribosomal RNA genes in plants: variability in copy number and in the intergenic spacer. Plant Mol Biol. 1987;9(5):509–20.
Rollo F. Characterisation by molecular hybridization of RNA fragments isolated from ancient (1400 B.C.) seeds. Theor Appl Genet. 1985;71(2):330–3.
Rollo F, Venanzi FM, Amici A. Nucleic acids in mummified plant seeds: biochemistry and molecular genetics of pre-Columbian maize. Genet Res. 1991;58(3):193–201.
Rollo F, Venanzi FM, Amici A. DNA and RNA from ancient plant seeds. In: Herrmann B, Hummel S, editors. Ancient DNA: recovery and analysis of genetic material from paleontological, archaeological, museum, medical, and forensic specimens. New York: Springer; 1994. p. 218–36.
Sallon S, et al. Germination, genetics, and growth of an ancient date seed. Science. 2008;320(5882):1464.
Schaefer M, et al. RNA cytosine methylation analysis by bisulfite sequencing. Nucleic Acids Res. 2009;37(2):e12.
Schuenemann VJ, et al. Targeted enrichment of ancient pathogens yielding the pPCP1 plasmid of Yersinia pestis from victims of the Black Death. Proc Natl Acad Sci. 2011;108(38):E746–52.
Seguin Orlando A, et al. Paleogenomics. Genomic structure in Europeans dating back at least 36,200 years. Science. 2014;346(6213):1113–8.
Skoglund P, et al. Ancient wolf genome reveals an early divergence of domestic dog ancestors and admixture into high-latitude breeds. Curr Biol. 2015;25(11):1515–9.
Smith O. Small RNA-mediated regulation, adaptation and stress response in barley archaeogenome. PhD thesis, School of Life Sciences, University of Warwick; 2012.
Smith O, et al. A complete ancient RNA genome: identification, reconstruction and evolutionary history of archaeological Barley Stripe Mosaic Virus. Sci Rep. 2014a;4:4003.
Smith O, et al. Genomic methylation patterns in archaeological barley show de-methylation as a time-dependent diagenetic process. Sci Rep. 2014b;4:5559.
Smith O, et al. Sedimentary DNA from a submerged site reveals wheat in the British Isles 8000 years ago. Science. 2015;347(6225):998–1001.
Smith O, et al. Small RNA activity in archaeological barley shows novel germination inhibition in response to environment. Mol Biol Evol. 2017;34(10):2555–62.
Snead NM, Rossi JJ. Biogenesis and function of endogenous and exogenous siRNAs. Wiley Interdiscip Rev RNA. 2010;1(1):117–31.

Spanò C, Buselli R, Grilli I. Dormancy and germination in wheat embryos: ribonucleases and hormonal control. Biol Plant. 2008;52(4):660.
Squires JE, et al. Widespread occurrence of 5-methylcytosine in human coding and non-coding RNA. Nucleic Acids Res. 2012;40(11):5023–33.
Stahl EA, Bishop JG. Plant-pathogen arms races at the molecular level. Curr Opin Plant Biol. 2000;3(4):299–304.
Sutton DH, Brown T. The dependence of DNase I activity on the conformation of oligodeoxynucleotides. Biochem J. 1997;321(2):481–6.
Tuschl T, et al. Modified RNA ligase for efficient 3′ modification of RNA. Google Patents; 2014.
Venanzi FM, Rollo F. Mummy RNA lasts longer. Nature. 1990;343(6253):25–6.
Vernot B, Akey JM. Complex history of admixture between modern humans and Neandertals. Am J Hum Genet. 2015;96(3):448–53.
Willerslev E, Cooper A. Review paper. Ancient DNA. Proc R Soc B Biol Sci. 2005;272(1558):3–16.
Willerslev E, Hansen AJ, Poinar HN. Isolation of nucleic acids and cultures from fossil ice and permafrost. Trends Ecol Evol. 2004;19(3):141–7.
Winter J, et al. Many roads to maturity: microRNA biogenesis pathways and their regulation. Nat Cell Biol. 2009;11(3):228–34.
Worobey M. Phylogenetic evidence against evolutionary stasis and natural abiotic reservoirs of influenza A virus. J Virol. 2008;82(7):3769–74.
Yashina S, et al. Regeneration of whole fertile plants from 30,000-y-old fruit tissue buried in Siberian permafrost. Proc Natl Acad Sci U S A. 2012;109(10):4008–13.
Zhang G, et al. Evidence of influenza A virus RNA in Siberian lake ice. J Virol. 2006;80(24):12229–35.
Zuker M, Mathews DH, Turner DH. Algorithms and thermodynamics for RNA secondary structure prediction: a practical guide. In: Barciszewski J, Clark BFC, editors. RNA biochemistry and biotechnology. Dordrecht: Springer; 1999. p. 11–43.

Ancient Epigenomics

Kristian Hanghøj and Ludovic Orlando

Abstract Recent molecular and computational advances in ancient DNA research have revealed that genome-scale epigenetic information can be retrieved from subfossil material. In fact, it appears that particular features of the chromatin, and its regulatory epigenetic marks at the time of death, are preserved within ancient DNA extracts. The characterization of this additional layer of information, which represents an interface between the genome and the environment and is not coded within modifications of the DNA sequence itself, opens new horizons for ancient DNA research. At the individual level, ancient epigenetic marks can provide novel molecular phenotypes of the age at death, diet restriction, and other stress conditions, including sociocultural changes. At the population level, they can complement classical inference based on genetic information to reveal the regulatory changes underlying divergence, speciation, and extinction. Exploiting such information will nonetheless be challenging, due to the nature of epigenetic data, which can vary across cell types, tissues, sex, and age, and be significantly influenced by genetic variation and environmental exposure.

Keywords Ancient DNA · Cytosine deamination · DNA methylation · Epigenome · Nucleosome protection

K. Hanghøj · L. Orlando (✉)
Centre for GeoGenetics, Natural History Museum of Denmark, Copenhagen, Denmark

Laboratoire AMIS, CNRS UMR 5288, Université de Toulouse, Université Paul Sabatier (UPS), Toulouse, France
e-mail: ludovic.orlando@univ-tlse3.fr

Charlotte Lindqvist and Om P. Rajora (eds.), *Paleogenomics*, Population Genomics [Om P. Rajora (Editor-in-Chief)], https://doi.org/10.1007/13836_2018_18,

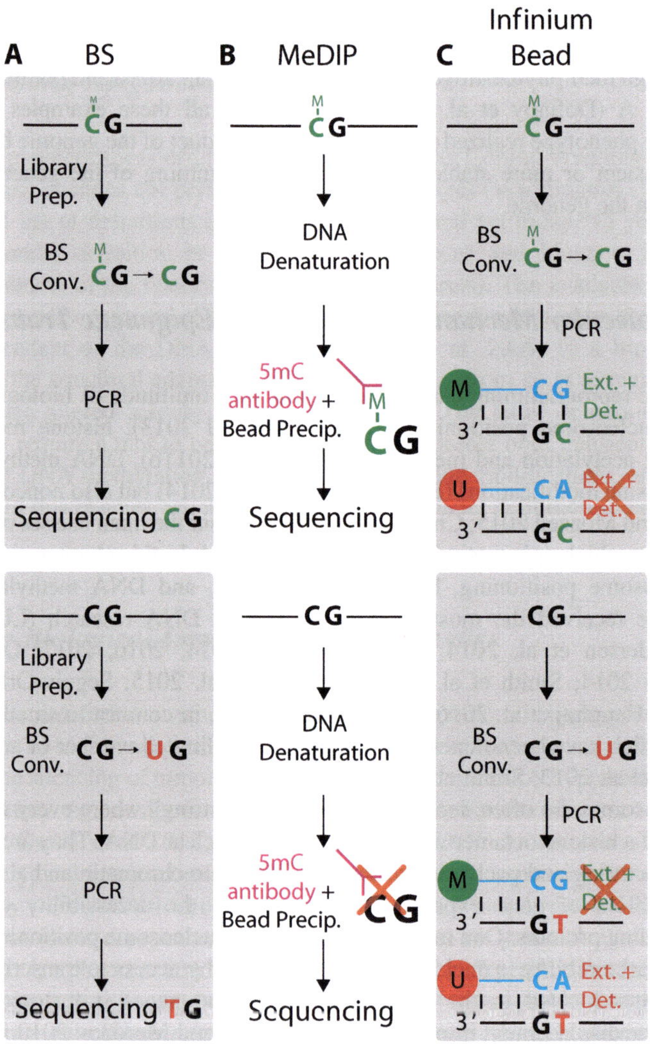

Fig. 1 Molecular methods for characterizing methylation profiles in fresh cells and tissues. The top (bottom) row illustrates the procedure on a single methylated (un-methylated) cytosine. (**a**) Bisulfite sequencing (BS) leverages on the deamination of un-methylated cytosines into uracils then sequenced as thymines. Methylated cytosines are not affected by the treatment. Three popular protocols, namely, reduced representation bisulfite sequencing (RRBS), whole-genome bisulfite sequencing (WGBS), and Infinium BeadChip (see Fig. **c**), use bisulfite treatment to detect methylation marks. In the economical RRBS protocol, the DNA is cleaved with a restriction enzyme prior to library building in contrast to a simple sonication step in the WGBS protocol. (**b**) Methylation DNA immunoprecipitation (MeDIP) enriches methylated DNA molecules using an antibody specific for 5-methylcytosine. First, sonicated DNA is denatured. This improves the binding affinity of the antibody. DNA molecules rich in methylated cytosines (top row) are more likely to bind to the antibody than DNA molecules with zero or few methylated CpGs (bottom row). Bound DNA molecules are recovered with magnetic beads and sequenced. (**c**) Infinium bead methylation arrays

predominantly found in a CpG dinucleotide context, and high (low) methylation levels in promoter regions are often, but not always, associated with low (high) gene expression (Ball et al. 2009).

1.4 Detecting Epigenetic Marks

Recent technological advances, uncovering the complexity of histone modifications and their relation to chromatin compaction (Bell et al. 2011b) or the differences present in DNA methylation profiles between tissues and cell types (Petropoulos et al. 2016), have facilitated the characterization of epigenetic marks at the genome scale. As a result, several initiatives, such as the ENCODE (Encyclopedia Of DNA Elements) consortium and Roadmap Epigenomics Mapping Consortium, have undertaken to map functional elements, including epigenetic marks, in the genome, providing nucleosome occupancy maps and DNA methylation profiles at the single nucleotide level across a large number of human cell types (Bernstein et al. 2010; ENCODE Project Consortium 2012). Similar initiatives exist in mice and other animals, especially domesticated (the FAANG consortium, standing for Functional Annotation of Animal Genomes; Andersson et al. 2015), as well as in yeast (Pokholok et al. 2005) and plants, in particular model species (Chodavarapu et al. 2010).

DNA methylation is perhaps the epigenetic mark that has received the most attention due to the early development of high-throughput assays. Once treated with sodium bisulfite, un-methylated cytosines can be converted into uracils and sequenced as thymine analogs. As methylated epialleles are resistant to this conversion, un-methylated alleles can be revealed post-sequencing by tracking C $\rightarrow$ T changes within bona fide sequence contexts. In the human genome, methylation of cytosines primarily occurs in the ~28 million CpG dinucleotides, of which ~70–80% are methylated (Ehrlich et al. 1982); however, methylated epialleles outside CpG dinucleotides have also been identified (He and Ecker 2015). In other species, methylated epialleles are found in different sequence contexts, e.g., CHG and CHH in plants (Gehring 2016). Combined with high-throughput DNA sequencing (HTS), bisulfite treatment can provide methylation maps at the single nucleotide level (Fig. 1a).

Fig. 1 (continued) share the bisulfite treatment step with BS. However, library building and sequencing the BS-treated DNA molecules are replaced with a BeadChip technology. This method uses two different types of beads per CpG site, M (green) for detection of methylated cytosines and U (red) for detection of un-methylated cytosines. Extension and detection (Ext. + Det.) occur only when no mismatches are present in the dinucleotide. This leads to a ratio of fluorescence per CpG site that reflects the methylation level of the site. Legend: *BS Conv* bisulfite conversion, *Library Prep* library preparation, *Bead Precip* bead precipitation, *Ext. + Det.* extension and detection

Other more economical approaches are also available. In reduced representation bisulfite sequencing (RRBS), the DNA is first enzymatically cleaved to over-represent CpG dinucleotides toward the template ends, before DNA libraries are built and bisulfite converted (Gu et al. 2011). Following HTS, methylation maps at a single nucleotide resolution are achieved, although not for the whole genome but only within restriction footprints (Fig. 1a). MRE-Seq (methylation-sensitive restriction enzyme sequencing) is also based on enzymatic restriction, except that the enzymes used act in a methylation-sensitive manner (Maunakea et al. 2010). Instead of bisulfite treatment, methylation DNA immunoprecipitation (MeDIP) makes use of antibodies to target DNA regions containing methylated CpGs (Weber et al. 2005) prior to applying high-throughput DNA detection methods, including DNA microarrays or HTS (Fig. 1b).

Finally, sequencing-free methods such as the Infinium HumanMethylation450 BeadChip distributed by Illumina have provided a fast and cost-effective alternative to generate DNA methylation profiles at ~480,000 CpG in the human genome, including ~99% of RefSeq annotated genes. This assay is, however, mostly biased to CpG island-rich promoters (Allum et al. 2015) and thereby lacks information on distal regulatory elements, such as enhancers, insulators, and transposable elements (Bauer et al. 2016) (the recent implementation of this technology improves the probe ascertainment bias present in the ~480 k loci by covering ~900 k loci). Here, for each target locus, two specific DNA probes are linked to beads: the first terminates with a CpG dinucleotide, whereas the second terminates with a CpA dinucleotide (Fig. 1c). Following DNA bisulfite treatment, methylated CpGs remain intact, allowing the perfect annealing of target DNA molecules and the bead associated to the CpG probe and their detection through a single-base extension reaction. The same goes for the other bead type, however this time with one mismatch preventing extension and, thus, detection. The situation is reversed at un-methylated loci as CpGs are bisulfite converted into TpG, matching CpA probes.

1.5 Factors Influencing Epigenetic Changes

With such technologies, the epigenomic variation present in human individuals, especially DNA methylation, could start to be characterized (Lister et al. 2009; Fraser et al. 2012; Heyn et al. 2013; Fagny et al. 2015; Zhang et al. 2015), in particular in relation to the etiology of most common, complex diseases. However, part of the variation observed between whole-blood methylomes could be tracked back to the respective contribution of the different cell lineages (Houseman et al. 2012; Galanter et al. 2017) and, thus, to the health status of the individuals. The raw epigenomic variation observed between individuals must thus be corrected for cell composition effects, or ideally be measured on flow-sorted pure cell types, before being interpreted biologically. In addition to showing cell-specific epigenetic profiles, a number of loci also show age-dependent methylation profiles (so-called clock CpGs) and even offer molecular methods for estimating the age of individuals

(Horvath 2013; Horvath et al. 2016). The discrepancy between this age estimated on the basis of methylation profiles and the real biological age seems to be a powerful predictor of the age of death, with individuals with older-than-expected methylation clocks showing increased risks of death (Marioni et al. 2015; Chen et al. 2016).

Importantly, epigenomic variation is not decoupled from genetic variation since particular alleles appear preferentially associated with specific methylation and/or chromatin states. So-called methylation quantitative trait loci (meQTLs) precisely refer to those loci showing genetic variants influencing methylation state and are commonplace in those epigenome-wide association studies (EWAS; Rakyan et al. 2011), where genome-wide genotyping information is also collected (e.g., Galanter et al. 2017). The detected association between the epigenetic state of a locus and a complex phenotype and/or disease is thus not necessarily causal but could be due to their indirect association with specific genetic variants.

Given the importance of the genetic influence on epigenomic variation, it is perhaps not surprising that hundreds to thousands of population-specific CpGs (so-called pop-CpGs), showing fixed methylation profiles in lymphoblastoid cell lines from different human population backgrounds, could be identified (Fraser et al. 2012; Heyn et al. 2013). At such loci, it is possible to separate individuals according to their geographic origins. A substantial fraction of the DNA methylation is, however, not associated with genetic variation, but mediated through complex genotype-environment effects (Fraser et al. 2012), environmental influences (Heyn et al. 2013), and epigenetic drift (Taudt et al. 2016).

1.6 Environment-Driven Epigenetic Changes

In addition to cell lines, important epigenomic differences have been found in the whole-blood methylomes of individuals living in the Central African belt, in particular between rainforest hunter-gathering pygmies and various farmer groups (Fagny et al. 2015). While demonstrating the importance of genetic variation in shaping epigenetic profiles, the study also emphasized the importance of the environment, such as current habitat and lifestyle. DNA methylation differences between hunter-gatherer and farmer groups living in forest environments were indeed strongly associated with nearby genetic variation, which segregated in those groups since they split some ~60,000 years ago (Patin et al. 2009). However, whether they lived in urban/rural habitats or in the rainforest, the genetically homogeneous group of farmers showed significant methylation differences, principally at immunity-related genes. Similar findings have been reported for the GALA II cohort including 573 Latino children, where approximately one third of the methylation variation associated with ethnicity at about ~1,000 loci could not be explained by genetic ancestry alone and was likely resulting from nongenetic factors such as economic, sociocultural, and environmental exposure (Galanter et al. 2017).

The influence of the environment in shaping epigenetic profiles has been documented in a large number of studies (e.g., Galanter et al. 2017), mostly relying

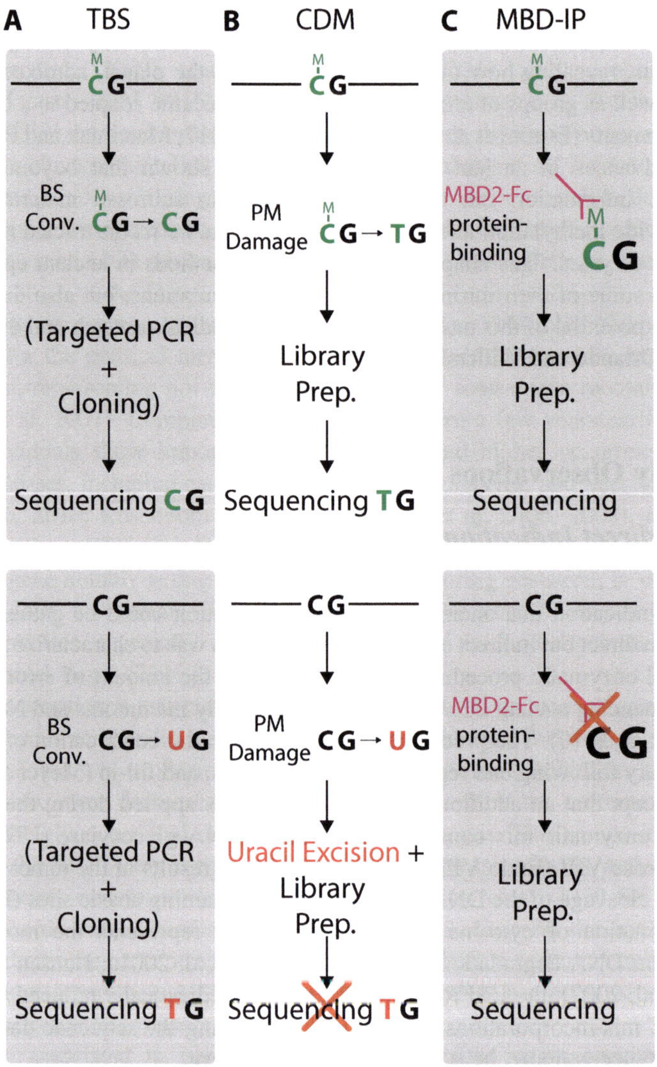

Fig. 2 Current molecular methods for characterizing methylation profiles in ancient specimens. The top (bottom) row illustrates the procedure on a single methylated (un-methylated) cytosine. (**a**) Cytosine deamination mapping (CDM) exploits the distribution of postmortem deamination damage introduced at CpG sites. Methylated cytosines are deaminated to thymines, instead of uracils for un-methylated cytosines. Uracils are excised prior to sequencing, leaving CpG → TpG conversion rates in the sequencing data as a proxy for those cytosines that were methylated premortem. See Fig. 3 for a schematic illustration of the MS proxy. (**b**) Targeted bisulfite (TBS) is similar to bisulfite sequencing (Fig. 1a), except that BS-treated DNA templates are amplified and cloned prior to sequencing. (**c**) Methyl-binding domain immunoprecipitation (MDB-IP) enriches for DNA molecules with methylated CpGs using MBD2-Fc proteins. DNA molecules with multiple methylated CpGs are more likely to be sequenced (top row) than DNA molecules with zero or few methylated CpGs (bottom row). DNA molecules bound to MBD2-Fc can be recovered using magnetic beads prior to library preparation and sequencing. Legend: *PM* postmortem, *BS Conv* bisulfite conversion, *Library Prep* library preparation

un-methylated CpGs indeed produces UpGs within ancient DNA templates, most of which get eliminated by the UNG + Endo VIII treatment. However, the same degradation reaction converts methylated CpGs (mCpGs) into TpGs, which remain intact following UNG + Endo VIII treatment and, thus, leave CpG → TpG mis-incorporations within the sequence data (Fig. 2a).

Consistent with this model, at CpG islands, where methylation levels are low, the CpG → TpG mis-incorporation rate posttreatment was more than halved compared to the rest of the genome. Of note, the remaining CpG → TpG mis-incorporations were mostly found toward read starts (and their complementary CpG → CpA toward read ends) as postmortem cytosine deamination takes place ~100 times faster within overhangs (Lindahl 1993). At that stage, it appeared that the analysis of CpG → TpG mis-incorporation patterns could provide information about DNA methylation landscapes in ancient genomes; however, the analysis was not performed as the amount of sequencing data generated was not sufficient.

2.2 *Direct Observations*

Bisulfite sequencing experiments on Late Pleistocene steppe bison bones provided the first evidence that ancient methylation epialleles can be identified (Llamas et al. 2012). In contrast to whole-genome bisulfite sequencing (WGBS where HTS is applied following bisulfite treatment of DNA extracts (Fig. 1a)), a targeted approach was followed through the PCR amplification, cloning, and sequencing of regions selected for their known methylation patterns (Fig. 2b). In particular, four multicopy hypermethylated transposons and two single-copy imprinted genes show differential methylation of parental alleles. The latter could only be successfully amplified in one ~30,500-year-old specimen, showing sufficient levels of DNA preservation. Importantly, the sequencing data obtained from 8 to –20 clones per PCR product showed consistent CpG → TpG mis-incorporation patterns. Not only did multicopy DNA fragments show the expected high CpG → TpG levels for all samples investigated, but also at imprinted single-copy genes, the ~30,500-year-old specimen showed a ~50:50 mixture of non-converted haplotypes carrying only CpG alleles, indicative of methylation, and converted haplotypes carrying only TpG alleles. Ancient DNA templates, thus, still carry a fraction of intact mCpGs, and bisulfite sequencing of ancient DNA could, thus, reveal methylation epialleles in ancient specimens.

The methodology is not, however, without technical limitations. Following postmortem deamination, mCpGs are naturally converted into TpGs, in a process similar to the conversion of un-methylated CpGs by sodium bisulfite. In situations where the postmortem deamination of cytosines is extreme and DNA is treated with sodium bisulfite, both methylated and un-methylated epialleles can give rise to CpG → TpG conversions in the sequence data. Additionally, sodium bisulfite conversion is generally performed at high concentrations (typically 5 M). Although this insures that high conversion rates are achieved, this also results in an extensive degradation of the DNA material, reducing the number of viable molecules by an order of magnitude or more (Grunau et al. 2001). The assay, thus, generally requires high quantities of high-quality DNA, which is rarely available for ancient samples. Perhaps not surprisingly, PCR attempts failed for all but one of the six Late Pleistocene bone extracts following bisulfite treatment (Llamas et al. 2012). This sample likely showed exceptional DNA preservation as >500 bp mitochondrial DNA fragments could be successfully amplified prior to bisulfite treatment (successful mitochondrial amplification is generally limited to the ~100 bp range with similar material; see Hofreiter et al. 2001b for a review).

Further work by Smith and colleagues using similar technology on 30 human samples from North America spanning the last 5 millennia confirmed that DNA preservation levels strongly impacted the accuracy of the methodology, with low amounts of starting DNA material resulting in high variability of methylation percentage site-wise (Smith et al. 2015). These authors suggested a DNA concentration of 15 pg/uL as a minimum threshold, although this number is derived from quantitative PCR estimates of a monomorphic 64 bp ABO7 fragment and is thus likely only indicative. It is indeed dependent on the relative amount of present-day human contamination sources present in the extracts examined, as well as the shape of the size distribution of endogenous templates. Finally, the authors observed that the ancient human bones analyzed showed lower methylation levels than forensic bones at the LINE element investigated (Smith et al. 2015). This holds true even when restricting the analyses to the best-preserved material, suggesting that postmortem cytosine deamination might significantly have converted mCpGs into TpGs, thereby inflating the amounts of signatures of un-methylated alleles. Of note, the contribution of postmortem deamination of un-methylated cytosines into UpGs is likely to be relatively more limited, as the deamination decay rate at un-methylated cytosines appears much slower than at methylated cytosines (Smith et al. 2014b). Nonetheless, differences in DNA preservation levels remain a potential important confounding factor to consider in analyses aiming at identifying changes in methylation levels, or differentially methylated regions, along temporal and/or geographical time series.

3 The First Ancient Methylomes

3.1 The Paleo-Eskimo Saqqaq

Four years after the potential of nucleotide mis-incorporation patterns for the recovery of ancient methylation information was unveiled (Briggs et al. 2010), sufficient sequencing data from a few ancient individuals were available to reconstruct the first ancient methylation maps at the genome scale. The first release of such ancient methylomes (Pedersen et al. 2014) was exploiting the ~3.5 billion of reads that were generated in 2010 while sequencing the first ancient human genome (Rasmussen et al. 2010). The specimen consisted of a hair tuft from an individual from Southwest Greenland, radiocarbon dated to ~3,600–4,170 years ago, and belonging to the Saqqaq Paleo-Eskimo culture.

In contrast to the methodology presented by Briggs and colleagues (Briggs et al. 2010), the data were not generated following UNG + Endo VIII treatment. However, the type of DNA library constructed together with the DNA polymerase used for its amplification also allowed the recovery of methylation signature through the analysis of nucleotide mis-incorporation patterns (Pedersen et al. 2014). Here, standard Illumina DNA libraries were prepared, where the end-repair reaction first introduces 3′-A overhangs that are used for ligating DNA adapters showing 3′-T overhangs. Following ligation, a simple PCR is performed to restore sufficient amounts of DNA libraries for Illumina sequencing. While a handful of DNA libraries were amplified using *Hifi* DNA polymerase, most of the sequencing data came from DNA libraries amplified by the *Phusion* DNA polymerase, then part of the standard Illumina library building process. One important feature of the *Phusion* DNA polymerase is that it cannot bypass uracil residues, leaving behind the pool of DNA strands affected by postmortem cytosine deamination (Fig. 2a). DNA strands copied from such strands are, however, available for sequencing but do not carry the typical C → T (C → U) mis-incorporation but its complementary G → A. Of course, the complementary C → T is restored at such site following the copy of G → A mis-incorporations during further PCR cycles. Due to the 5′ → 3′ directionality of DNA strands, only these strands carrying C → T are produced during the bridge-PCR amplification generating Illumina sequencing clusters, and sequencing reactions can only generate complementary sequences, thus including G → A and not C → T mis-incorporations (Seguin-Orlando et al. 2015b). Therefore, the combination of the library type and its amplification by the *Phusion* results in the almost complete absence of C → T mis-incorporation near read starts (as opposed to what was seen with other libraries constructed following blunt-end ligation, be *Phusion* amplified or not). This was empirically verified in the original publication reporting that the Saqqaq genome as ancient 16S rDNA templates PCR amplified with the *Phusion* showed very limited C → T mis-incorporation (Rasmussen et al. 2010). What was however overlooked in this publication is that the limited fraction C → T mis-incorporations observed could be exploited to reconstruct the first genome-wide methylation map.

Again, the trick consisted of realizing what happens to mCpGs during postmortem deamination: they get converted into TpGs and not UpGs as un-methylated CpGs do. Therefore, the CpG → TpG mis-incorporations observed in the Saqqaq sequence data deriving from *Phusion*-amplified libraries built following 3′-A/T ligation offer a single signature of past methylated epialleles. Of course, those mCpGs that were not deaminated after death are still sequenced at CpGs (except for rare sequencing errors); therefore, the signal-to-noise ratio obtained for tracking methylation signatures is purely driven by underlying postmortem deamination rates. Luckily, these are faster in methylated cytosines than un-methylated ones (Smith et al. 2014b) and all the more so within overhangs (Lindahl 1993). Therefore, a simple statistic providing the fraction of CpG → TpG conversions observed at read starts could offer a proxy of past DNA methylation levels at each CpG site (the statistics could also be modified to account for complementary CpG → CpA conversions at read ends). Given that not all sites are expected to be deaminated and that the per site coverage is relatively limited and highly variable across genomic regions (e.g., the *Phusion* DNA polymerase favors GC-rich and short DNA templates (Dabney and Meyer 2012)), calculating the so-called methylation scores (MS) of CpG → TpG conversion counts within regions (100 bp, 1 kb, and 2 kb) could offer higher sensitivity and more robust measurements of regional methylation levels (Pedersen et al. 2014).

A number of observations validated the proposed estimates of regional methylation levels. First, these were minimal in the mitochondrial genome and other regions known to be hypomethylated in the nuclear genome, such as CpG islands (CGI), compared to exons and introns. Second, such variations were only seen in CpG dinucleotide contexts, in line with methylation preferentially affecting such sites. Third, estimated methylation scores recapitulated the known inverse relationship between the methylation levels and promoter %GC content and CpG density. Fourth, methylation scores within the 2 kb region flanking CTCF-binding sites showed the expected ~200 bp periodicity. Finally, when compared to present-day methylomes from different tissues, the Saqqaq methylation clustered together with hair, as expected from the source material used in DNA extraction (Rasmussen et al. 2010). It thus demonstrated for the first time that genuine genome-wide regional methylation maps could be derived from nucleotide mis-incorporation patterns. These contrast with WGBS data providing single nucleotide resolution (Lister et al. 2009; Krueger et al. 2012).

The possibility to calculate MS at each and every region of the genome showing sufficient sequence coverage opened for potentially important applications (Fig. 3a). First, the methylation levels measured at clock CpGs are known to offer a rather precise estimate of the age of a given individual (Horvath 2013). Yet, determining the age at death of ancient individuals is often difficult. Available skeletal morphological characters, such as bone degeneration indexes, generally loose in precision as age increases (Brooks and Suchey 1990), and dentition age atlas show relatively good accuracy, although only in subadult individuals (AlQahtani et al. 2014). Therefore, calculating MS at clock CpGs and leveraging on a known age model of changes in methylation levels at such sites holds the potential to provide estimates of the age at death from ancient individuals. Applied to the Saqqaq data at 2 CpG sites where such an age model was calibrated for hair tissues, this approach returned consistent estimates ranging from 44 to 69 years (Pedersen et al. 2014). Of note, in this analysis, MS were first calculated within 2 kb regions centered on both CpGs (as methylation levels are significantly correlated at such distances (Eckhardt et al. 2006; Down et al. 2008)) and further converted into absolute measures of methylation levels based on a linear model relating the Saqqaq MS within ~182,000 2 kb regions and absolute methylation levels experimentally measured in the hair of five living individuals. This was consistent with the clustering of the Saqqaq methylome and hair tissues.

In a second application, the Saqqaq MS scores were calculated within gene bodies and promoters in an attempt to predict gene expression levels in the ancient hair. While genes showing constitutive expression exhibit high levels of methylation within gene bodies, their promoters are generally hypomethylated (Ball et al. 2009). In contrast, genes showing low expression levels generally show promoter

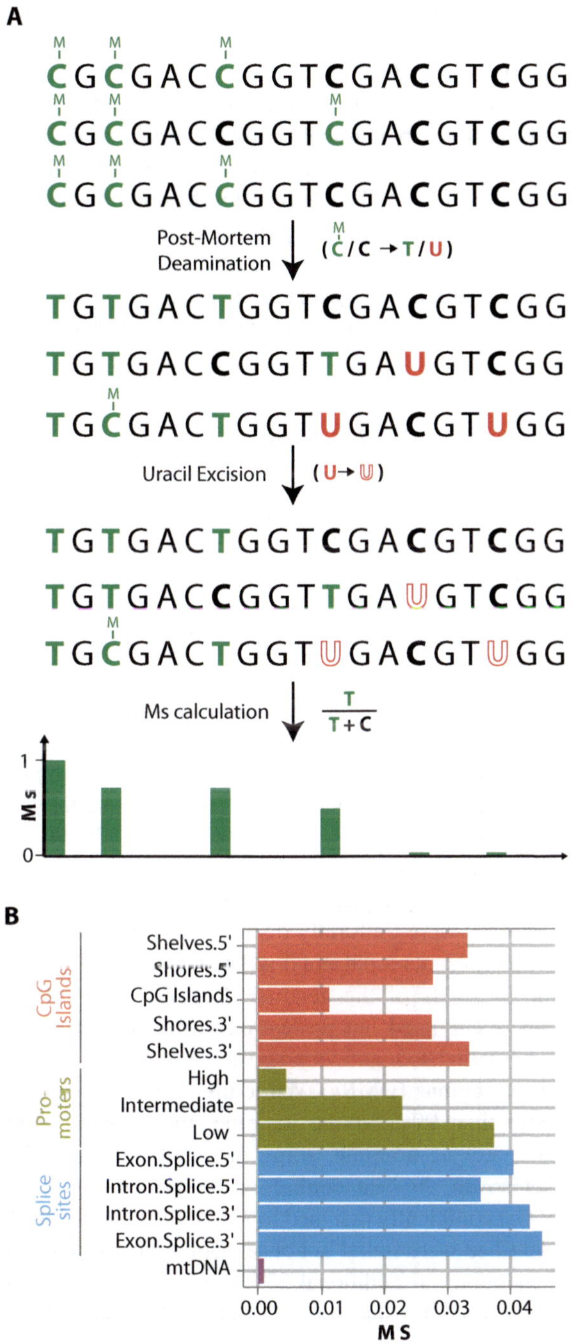

Fig. 3 Current computational methods for calculating methylation scores (MS). (**a**) Postmortem deamination converts methylated cytosines (green) into thymines (green), whereas un-methylated cytosines (black bold) are turned into uracils (red). Not all cytosines are converted as postmortem deamination is a stochastic process. MS is calculated as the frequency of deaminated cytosines per site, estimated here through the fraction of CpG → TpG conversions observed in the sequence data. (**b**) MS recapitulates expected methylation patterns across a series of genomic regions showing low cross-tissue variance. The plot is generated from the sequencing data underlying the 7,000-year-old Stuttgart sample (Lazaridis et al. 2014) using epiPALEOMIX (Hanghøj et al. 2016)

hypermethylation. Therefore, the ratio of methylation levels within gene bodies and promoters provide a simple proxy for gene expression levels. Calculated on all annotated gene models showing sufficient sequence coverage, these ratios showed significant correlation with the expression levels measured from RNA-Seq data in living hair follicle cells (Pedersen et al. 2014). Hair-specific keratins were found within the genes predicted with high-expression values, together with other proteins known to be essential for the hair structure and mechanical properties. This approach, thus, holds the potential to predict the potential functional consequences of epigenetic changes measured in ancient individuals. It, however, still requires further work as the correlation measured was only moderate and the contribution of different cell types in tissues other than hair has not been evaluated.

3.2 Archaic Hominins

Soon after the Saqqaq hair methylome was published, the bone methylomes of two archaic hominins, a Denisovan and a Neanderthal, were characterized (Gokhman et al. 2014), building on the high-coverage data generated while sequencing their genomes (Meyer et al. 2012; Prüfer et al. 2014). The methodology was based on the UNG + Endo VIII treatment described by Briggs and colleagues (Briggs et al. 2010), and methylation scores were calculated using a proxy similar in principle to MS, except that the value is returned for each CpG (and not within windows), accounting for the methylation levels detected in 43 and 59 neighboring CpGs for the Neanderthal and Denisovan, respectively. Here again, the proxy was validated by its ability to recapitulate methylation levels experimentally measured in living osteoblast cells and a diversity of other tissues at 3,804 promoters showing constitutive hypomethylation.

The Saqqaq sequence data were further analyzed to show that cytosine deamination rates were uniformly distributed along the genome, confirming that no deamination hotspots (coldspots) inflate (reduce) the local methylation scores inferred for specific parts of the genome. Regions where the archaic bone methylomes are hypermethylated but present-day osteoblasts (taken as a surrogate for bones) are hypomethylated, and vice-versa, could readily be interpreted as differentially methylated regions (DMRs). While the three methylomes show consistency throughout more than 99% of the genome, the authors identified ~1,100 such DMRs in each archaic hominin, some of which reflected changes that occurred along the Denisovan, the Neanderthal, or the present-day human branch and some that could not be confidently assigned to any of those branches. Importantly, none of the DMRs identified were found in housekeeping genes, which further validated the approach. Generating an additional 37 osteoblast and bone methylome profiles at ~27,000 CpGs providing data for ~5% of the DMRs identified, the authors could confirm that the differences observed within DMRs were mostly fixed within present-day humans. More than half of the remaining ~95% DMRs showed stable methylation levels across a range of tissues (even though these were collected from individuals representing a range of sex, ethnicities, and ages) and different in the archaic hominins. This final set of regions provided, thus, reliable DMR candidates between archaic hominins and present-day humans. Interestingly, genes located within DMR candidates are twice as likely to be disease related in present-day humans than those located outside DMR candidates, and a third of these are associated with neurological and psychiatric disorders.

The final set of DMR candidates included a number of interesting loci. In particular, three DMRs were found within the HOXD cluster, and two additional DMRs were located within 25 kb and might act as putative enhancers. The HOXD cluster contains essential gene determinants of limb morphogenesis, and mice loss of functions of HOXD genes include abnormal digit formation (Monge et al. 2003), shortening of the zeugopod (Davis and Capecchi 1994), and fusion of proximal row of the carpus (Favier et al. 1995). Interestingly, Neanderthal and present-day humans harbor a range of postcranial skeletal differences, some of which might have originated from fixed epigenetic changes during the divergence between these lineages. Looking at whether the targets of transcription factors present in the identified DMRs are also overrepresented in DMR candidates, the authors found MEIS1, a key regulator of limb development, interacting with some of the HOXD genes. The number of differentially methylated transcription factors whose targets were overrepresented in DMR candidates was higher than expected by chance alone, suggesting that coordinated epigenetic changes at such loci could have significantly affected expression profiles within a single evolutionary event.

Recent work has improved the list of DMR candidates above, especially for those changes underpinning the origins of our own species (Gokhman et al. 2017). This was made possible through the sequencing of additional high-quality genomes from ancient individuals, including anatomically modern humans and a Neanderthal specimen, as well as the characterization of genome-wide bone methylomes from present-day individuals and chimpanzee outgroups. In this work, a first set of DMR candidates were listed from the comparison of the sequence data underlying the genome of a ~50,000-year-old anatomically modern individual (Ust'Ishim) and those from archaic hominins. The list was then refined in a second step considering as valid only those candidates showing methylation profiles in present-day individuals similar to Ust'Ishim. This two-step procedure not only helped eliminate possible artifactual signatures deriving from age, ethnicity, and sex differences (as the set of present-day individuals spanned a whole range of such factors) but also from methodological differences in collecting ancient and present-day data. Depending on the combination of filtering criteria used by the authors, approximately ~880–1,600 DMRs were identified where epigenetic reprogramming took place along the evolutionary branch of anatomically modern humans. Only a fifth overlapped with the DMR candidates previously reported (Gokhman et al. 2014). Moreover, it is surprising that Ust'Ishim, being an anatomically modern human, shares methylation profiles with the archaic hominins in the HOXD cluster (Hanghøj et al. 2016; Gokhman et al. 2017), especially given that previous analyses suggested a major epigenetic reprogramming of the HOXD cluster in anatomically modern humans, after their divergence from archaic hominins (Gokhman et al. 2014). Further work is thus needed to assess the exact impact of methylation changes at the HOXD cluster on the anatomy of human limbs.

The new list of candidates appears to be significantly enriched in genes affecting the voice and associated with the development of the vocal tract. Additionally, the genes affecting the vocal tract and the flattening of the face, such as COL2A1 and ACAN, but also their key regulator SOX9, show a higher fraction of CpGs that are differentially methylated in anatomically modern humans. Similar enrichment for the vocal tract was found when considering the so-called human accelerated regions (HARs) instead of the DMR candidates. HARs represent regions that accumulated higher-than-expected mutational changes in humans vs other vertebrates, suggesting that some of the epigenetic changes detected might be derived from their strong statistical association with some preferential alleles. Regardless of the underlying mechanism, the findings suggest that genes involved in the vocal tract have concentrated both genetic and epigenetic changes along the evolutionary branch leading to anatomically modern humans, which echoes our unique capacity for speech. Importantly, these results assume that DMRs identified in bones are also propagated to the larynx.

Finally, another study identified that one of the top-selected regions in the genome of present-day Greenlandic Inuits was likely acquired by introgression from Denisovan-like ancestors (Racimo et al. 2016). The two genes present in the region, TBX15 and WARS2, are associated with the differentiation and distribution of fat tissues through the body. More specifically, present-day Inuits show subtle changes in their methylation profile depending on whether they carry at least one copy of the introgressed block or no copies. In particular, one single CpG was found to show significantly reduced methylation level in carriers of the introgressed block. Outside the introgressed block, the authors also identified a DMR around the TSS of TBX15, consisting of 16 CpGs, where carriers showed lower methylation levels compared to noncarriers. Finally, they found that gene expression of TBX15 was 22% higher in individuals carrying the introgressed block (Racimo et al. 2016). These findings open for the possibility that the introgression of the Denisovan genomic region overlapping TBX15 and WARS2 was associated with a reprogramming of the surrounding regulatory layer, through changes in DNA methylation patterns.

3.3 General Trends

At the time the first ancient methylomes were characterized, only a handful of high-quality ancient genomes had been sequenced (Der Sarkissian et al. 2015a). The situation has drastically changed now that over 100 ancient genomes over 1X average depth of coverage are available (see Orlando et al. 2015; Llamas et al. 2017; Marciniak and Perry 2017 and references therein). Most of these belong to the archaic hominin and present-day human lineages, but an increasing number consist of nonhuman animals, including horses (Orlando et al. 2013; Jónsson et al. 2014;

Schubert et al. 2014; Der Sarkissian et al. 2015b; Librado et al. 2015, 2017), aurochs (Park et al. 2015; Braud et al. 2017), dogs and wolves (Skoglund et al. 2015; Frantz et al. 2015), woolly mammoths (Miller et al. 2008; Lynch et al. 2015; Palkopoulou et al. 2015), and polar bear (Miller et al. 2012), and plants (Ramos-Madrigal et al. 2016). Taken together, they make it possible to assess whether DNA methylation signatures of similar quality can be obtained from a broader range of sample types, environment, and experimental procedures.

The development of the computational software epiPALEOMIX greatly facilitated this task (Fig. 3a). Unlike ROAM (Gokhman et al. 2014), which is restricted to the inference of ancient methylation patterns in archaic hominins, epiPALEOMIX can handle any read alignment file against any reference genome (Hanghøj et al. 2016). It automates MS calculations within user-defined regions and can accommodate sequencing data generated from a range of experimental procedures, including those described above as well as others such as single-stranded DNA libraries (Gansauge and Meyer 2013). Applied to the sequence data underlying 35 ancient genomes, epiPALEOMIX revealed a series of interesting features.

First, ancient teeth, and not just hair and bones, show expected DNA methylation profiles at regions used as controls such as CGIs and their shores and shelves (as exemplified in Fig. 3b). Additionally, MS scores within 1,500 bp windows centered on CpG sites show strong correlation with the methylation levels experimentally measured in a range of cell types and tissues, with hair, osteoblast, and tooth gingival generally showing the highest correlation coefficients. Second, methylation signatures can be recovered in the absence of UNG + Endo VIII treatment (Briggs et al. 2010; Gokhman et al. 2014) or 3′-A/T ligation and *Phusion* amplification (Rasmussen et al. 2010; Pedersen et al. 2014). This is due to the fact that DNA library constructs based on 3′-A/T overhang ligation exhibit the same molecular properties as those amplified with the *Phusion* polymerase when amplified with other DNA polymerases that cannot bypass uracils, such as the AccuPrime Pfx (Seguin-Orlando et al. 2015b). Additionally, the difference in cytosine deamination rates between methylated and un-methylated alleles is sufficient to generate higher CpG → TpG conversions, thus, mis-incorporations at methylated CpGs. In this case, the background MS score found along the genome is higher, and the difference observed between hypo- and hypermethylated regions is accordingly reduced. It follows that the correlation coefficients observed between MS and methylation levels experimentally measured in the closest matching cell type and/or tissue are inferior to those observed on with sequencing data generated following UNG + Endo VIII treatment and 3′-A/T ligation and *Phusion* amplification. Lastly, attempts to estimate the age at death from methylation scores around clock CpGs were deceptive as most ancient individuals were estimated to be older than 50 years when they died, which appears highly unlikely in past human groups. Likewise, the ~12,800-year-old Antzick specimen was estimated to have died when he was 34 years old, despite the bones clearly belonging to a 1–2-year-old infant (Rasmussen et al. 2014). The methodology thus requires further improvement before strong inference of past demographic parameters can be confidently done. Among possible current limitations are whether the age model used, which was built from present-day populations, remains valid for past populations, all exposed to very different environmental conditions, diet, and pathogens. Additionally, how robust the inference is to different contributions of various cell types is presently unknown. Yet, the osseous material extracted from different individuals likely shows different cell-type balances. Preliminary work on the teeth of 21 present-day individuals showed that the age estimated from the methylation levels measured at 5–13 CpGs located in three genes (ELOVL2, FHL2, and PENK) shows far less precision for the dentin, an odontoblast-rich material, than those estimated from both cementum and pulp, both of which enclose various other cell types, such as cementocytes, cementoblasts, blood vessels, and even nerves and other soft tissues (Giuliani et al. 2015).

In contrast to modern methylation maps with single-base resolution, computational proxies on sequencing data recovered from an ancient specimen, such as MS, calculate regional methylation levels. Two different regional-based approaches have been applied, one with fixed window size and a dynamically changing window size. The fixed window sizes used in Pedersen et al. (2014) and Hanghøj et al. (2016) were chosen to reflect a high level of positive correlation of methylation levels within 1–2 kb (Eckhardt et al. 2006; Bell et al. 2011a). The dynamic window size, used in Gokhman et al. (2014), depends on the local CpG density of the reference genome and, obviously, the user-defined number of CpGs required per window. To find the optimal number of CpGs per window, correlation tests between the two archaic hominins and present-day RRBS methylation maps from bones were conducted. This translated in a median block size of 3,755 bp and 5,257 bp, respectively (Gokhman et al. 2014).

Despite using fixed or dynamic window sizes, future computational proxies will move toward single-base resolution with more sequencing data. However, a universal, definitive depth-of-coverage threshold necessary to reconstruct high-quality DNA methylation maps cannot be provided. This is so because both the underlying amount of DNA damage and the number of independent DNA molecules available for sequencing impact the quality of the reconstructed methylation map. From already published methylomes, we know that expected methylation profiles at control regions like CGIs, their shores and shelves can be recovered from genome sequence at ~2X average depth of coverage and above (Hanghøj et al. 2016). However, the identification of DMRs was so far only possible from deeper sequence data, consisting of ~7X to ~52X depth of coverage (Gokhman et al. 2017).

4 The First Ancient Nucleosome Maps

DNA methylation is the epigenetic mark that has received most attention thus far, and some even argued that ancient epigenomics "would likely be limited to analysis of cytosine methylation" (Llamas et al. 2012). Even though short peptides survive in subfossil material over a hundreds of thousand years (Orlando et al. 2013) to a few million years (Schweitzer et al. 2014; Demarchi et al. 2016) (older traces have been claimed (Schweitzer et al. 2007, 2009), but remain debated (Buckley et al. 2008)),

periodicity within nucleosomal dyads (Struhl and Segal 2013) (Fig. 4c). Finally, CTCF-binding regions also show strong nucleosome phasing within the 2 kb flanking the CTCF-binding site, when the latter is bound to the CTCF protein (so-called occupied CTCF-binding regions) (Fig. 4d) (Kelly et al. 2012). It is noteworthy that strong phasing of occupied CTCF binding regions is expected not only for nucleosomes but also for DNA methylation, with peaks of DNA methylation corresponding to linker DNA and valleys of DNA methylation corresponding to nucleosome dyads (Fig. 4d) (Fu et al. 2008).

All these patterns were observed on the Saqqaq Paleo-Eskimo sequence data (Pedersen et al. 2014), which demonstrated for the first time that nucleosomes, and probably transcription factors and other proteins interacting with DNA, leave their footprint on ancient DNA fragments. Therefore, even in the absence of nucleosomes, genome-wide nucleosome maps can be reconstructed in the past. All the sequence patterns presented above are automated within epiPALEOMIX (Hanghøj et al. 2016) to facilitate and standardize their analysis in ancient genomes. These provide for each position along the genome a nucleosome score, which can be viewed as a probability that a nucleosome dyad is centered at that given position and/or a quality measure of the prediction. The software also embeds procedures for correcting depth-of-coverage profiles for %GC variation introduced by the DNA polymerases used during DNA library amplification (Dabney and Meyer 2012) and the sequencing technology itself (Quail et al. 2008; Ross et al. 2013). Applied to the sequence data underlying 35 ancient genomes, these analyses revealed that most of the expected patterns listed above can be found in other tissues than the hair, bones, and teeth in particular, and sometimes up to ~40,000–50,000 years (Hanghøj et al. 2016). However, not all these patterns could be found in all specimens, as some were virtually lost in some individuals. This suggests that DNA fragmentation footprints can be, but are not necessarily, present in subfossil material, potentially due to a combination of factors such as differences in DNA preservation, balance of cell types (mixing different nucleosome landscapes), cell death processes (in particular necrosis vs apoptosis), and more. Despite the variability observed across samples, the authors identified that expected nucleosome and methylation patterns can be found within occupied CTCF-binding regions. Although the required sequencing effort to recover this signal is heavily dependent on the amount of cytosine deamination observed, two million reads mapped within the CTCF-binding regions ±2 kb showed sufficient to recover the signal for all human samples analyzed in Hanghøj et al. (2016). Given that the expected patterns are not found in modern DNA, they can be used as additional, novel authentication criteria.

5 Implications

5.1 Sequencing Ancient DNA

In addition to providing an additional layer of information and novel approaches toward sequence authentication, the survival of epigenetic marks and/or footprints on ancient DNA has several practical consequences while gathering ancient DNA information. First, nucleosome protection is an important factor driving patterns of DNA fragmentation after death, leading to an overrepresentation of nucleosome-protected regions in the sequence data, and by extension in regions involved in any type of DNA-protein interaction. Leveraging on the higher molecular complexity of such regions could be beneficial to target-enrichment approaches, the efficacy of which is limited by a number of factors, including the template size distribution and the library redundancy (Carpenter et al. 2013; Cruz-Dávalos et al. 2016). Now that experimental procedures have been developed to target enrich up to a few millions of loci in a test tube (Haak et al. 2015; Mathieson et al. 2015), we suggest that the probes be placed in regions known to be associated with nucleosomes in osseous material once (1) quality is controlled for other parameters in conditioning probe/template annealing (e.g., sequence entropy, self-annealing, %GC content, etc.) and (2) ascertained according to the research question addressed.

Similarly, leveraging on the presence of DNA methylation marks through methods similar to MeDIP (Weber et al. 2005) (Fig. 1b) also holds the potential to fractionize ancient DNA extracts according to their methylation levels. In addition to revealing ancient methylation landscapes, this could also help enrich ancient DNA extracts for endogenous DNA. In particular, except for hair, and specific osseous material such as petrosal bones (Pinhasi et al. 2015) and tooth cementum (Damgaard et al. 2015), ancient DNA extracts correspond to a metagenomic mixture of endogenous templates derived from the host itself and exogenous templates (Green et al. 2006). The latter generally represent the majority of the DNA fragments extracted and mostly derive, in situations where contamination by present-day individuals is limited, from environmental microbes, most of which of bacterial origins (Der Sarkissian et al. 2014; Louvel et al. 2016). Shotgun sequencing is, thus, not a cost-effective alternative for sequencing the host genome, in contrast to target enrichment, which involves DNA annealing to probes covering preselected regions of interest (Mathieson et al. 2015). However, since DNA methylation in bacterial genomes takes place in different sequence contexts than in vertebrates (Murray et al. 2012), using antibodies specific to mCpGs or protein constructs grafted with methyl-binding domains (MBD) from vertebrate methyl transferases could help in immunoprecipitation and thereby enrich for hypermethylated vertebrate DNA. While the approach would bias sequencing data against hypomethylated regions such as mitochondrial DNA and CGIs, it might still offer a cost-effective alternative to shotgun sequencing for recovering genome-scale data from ancient individuals.

Seguin-Orlando and colleagues tested this idea on a range of samples, spanning a whole range of tissues (soft vs hard), preservation contexts, and time (Seguin-Orlando et al. 2015a). Sequencing both the immunoprecipitated (MBD+) and the supernatant fraction (MBD−), they showed that the positions covered in the former achieve higher depth of coverage than the latter, which overall improves the quality of the sequencing data (Fig. 2c). However, as expected since methylation mostly takes place within CpG contexts in vertebrates, the covered regions following immunoprecipitation exhibit strong %GC and CpG bias and are depleted in the hypomethylated mitochondrial genome. Importantly, the size distribution of endogenous MBD+ templates is significantly shifted upward, suggesting that the majority of ultrashort DNA templates cannot develop strong enough MBD interactions to be immunoprecipitated. This thus limits the method applicability to only those specimens where DNA fragmentation is limited. The method should also not be recommended in case of extensive deamination since MBDs show affinity to mCpGs but not to their deaminated TpG forms. The method might be more adequate when using other enzymatic domains and/or antibodies targeting non-vertebrate methylated bases, such as 6-methyl-adenines in various sequence contexts (Murray et al. 2012), which show more limited deamination rates than methylated cytosines.

Nonetheless, Smith and colleagues applied MBD-based enrichment with success to the DNA extracts from barley seeds from Qasr Ibrim spanning the last 2,800 years (Smith et al. 2014b). As all samples originated from the same excavation site, they were in an ideal situation to quantify the degradation kinetics of mCpG sites. For all samples but one, the decay in the total amount of DNA methylation genome-wide was found to follow a single-order kinetics, indicating an overall half-life for mCpG $\rightarrow$ TpG conversions of approximately 1,500 years. That, combined with the loss resulting from fragmentation at mCpG sites, implies that less than 1% of the mCpGs present in seeds would remain intact following ~5,000 years of preservation in a Qasr Ibrim-like environment. In these conditions, the indirect methods exploiting patterns of CpG $\rightarrow$ TpG mis-incorporation would offer a better alternative than MBD-based approaches to characterize genome-wide DNA methylation landscapes. The same goes for enrichment methods selecting uracil-containing DNA strands as part of an early experimental step during library building (Gansauge and Meyer 2014). By focusing on damaged templates, this method can help eliminate a substantial fraction of non-damaged modern DNA contamination, however, at the likely cost to underrepresenting those genomic regions showing high methylation levels (and where mostly TpGs are formed instead of UpGs following cytosine deamination). Interestingly, the one barley sample that did not follow the expected kinetics showed a total amount of DNA methylation almost three times higher than expected based on its archeological age alone (Smith et al. 2014b). This sample could further be demonstrated as infected by the barley stripe mosaic virus (Smith et al. 2014a), and the unusual genome methylation levels observed are likely part of the viral stress response, which in plants involves global genome methylation (Boyko et al. 2007).

5.2 DNA Damage Models

Despite DNA methylation marks and nucleosome protection (or any protection mediated through DNA-protein interactions) affecting postmortem DNA decay, no statistical model of DNA damage currently available takes such factors into account. There is yet ample evidence that C $\rightarrow$ T nucleotide mis-incorporation rates are inflated within CpG contexts (Seguin-Orlando et al. 2015a, b; Smith et al. 2014b; Skoglund et al. 2015; Hanghøj et al. 2016). Despite this, standard models of postmortem DNA damage assume different decay rates for those cytosines located within overhangs or in double-stranded regions, regardless of their methylation state (Briggs et al. 2007; Jónsson et al. 2013). Likewise, current statistical models of postmortem DNA fragmentation consider the process to be random (Allentoft et al. 2012), despite growing evidence for nucleosome protection. Obtaining accurate estimates of postmortem DNA decay is, however, instrumental for data authentication and exploring the temporal limits of DNA survival. It is also essential for limiting the amounts of errors present in ancient DNA data (e.g., see Jónsson et al. 2013; Kousathanas et al. 2017), in particular for determining which among the TpG dinucleotides observed at one CpG site correspond to true genetic variants or degradation by-products of methylated alleles. This will require the development of new models incorporating the knowledge gained from the study of cytosine methylation and nucleosome protection. Such models should ideally co-estimate not only the probability of each genotype at a given CpG site (their so-called genotype likelihood) but also their probability to be methylated or not.

5.3 *Tissue Specificity and Scarcity*

Successful genome-scale analyses of ancient specimens are generally limited to three types of subfossil material, namely, bone, hair, and tooth. The remaining tissues, including all soft tissues, are usually too degraded, although a few exceptions exist (Keller et al. 2012; Jónsson et al. 2014; Seguin-Orlando et al. 2015a, b). While the DNA sequence of every somatic cell from a given individual is identical (with the exception of a few somatic mutations), the epigenetic regulatory layers of two differentiated cells can be quite distinct. In fact, the epigenetic regulatory machinery functions as a codriver for the very differentiation from stem cell to specialized cell. Therefore, comparing the methylomes of ancient individuals, as recovered from osseous and hair material, to those reconstructed from other fresh tissues and/or individual cell types is cumbersome. From the perspective of ancient epigenetics, the limited panel of tissues available to ancient DNA analysis, combined with some intertissue epigenetic variation, remains the key challenge. However, recent research using present-day methylation data has shown that approximately 75% of methylated cytosines display cross-tissue stability (Liu et al. 2016). Additionally, functionally and structurally similar tissues show more similar methylation profiles than two distant ones (Ziller et al. 2013). How much this similarity will suffice to deduce the real biological and potential medical impact of epigenetic differences identified in ancient individuals remains to be addressed.

As an attempt to address this issue, Gokhman and colleagues have proposed that parsimonious reasoning might be helpful (Gokhman et al. 2016). Their concept is simple but relies on the availability of well-characterized methylation maps for the tissue of potential biological interest (e.g., the brain cortex) and the tissue of the ancient specimen (e.g., bones), both in a present-day sister group (e.g., modern humans) to the ancient group analyzed (e.g., Neanderthals) and a close evolutionary outgroup (e.g., chimpanzees). Given the evolutionary tree retracing the phylogenetic relationships of the groups analyzed, the parsimony principle helps predict the methylation state of the ancient individual for tissues other than that originally sequenced. For instance, at a DMR where modern humans are hypomethylated in all somatic tissues and Neanderthals were hypermethylated in bones, finding that chimpanzee somatic tissues are also hypomethylated would suggest that the hypermethylation seen in Neanderthals is specific to their bones and that other somatic tissues, such as the brain cortex, were hypomethylated. Conversely, should the chimpanzee somatic tissues be hypermethylated, the epigenetic change in the bone methylome likely consisted in an elementary reprogramming of the region in the modern human lineage.

6 Conclusion and Future Perspectives

The field of ancient epigenomics is only commencing and yet genome-wide methylation and nucleotide maps of over 35 ancient humans and animal specimens have been obtained (Gokhman et al. 2014; Pedersen et al. 2014; Hanghøj et al. 2016). We can confidently predict that this number will increase exponentially in the forthcoming years, due to ever-increasing sequencing capacities and a relative standardization of the molecular methods used for sequencing whole genomes. In particular, some forms of UNG + Endo VIII treatment can still leave enough damage signatures for data authentication while drastically reducing the impact of damage-driven nucleotide mis-incorporations on downstream analyses (Rohland et al. 2015). Such approaches are increasingly popular and generate sequence data enabling both genome sequencing and methylation mapping (Hanghøj et al. 2016). This illustrates perhaps best what is the most central difference in the methods currently used for characterization ancient and present-day epigenomes: the former mainly exploit the signatures left by the natural process of postmortem DNA decay, while the latter generally requires additional chemical treatment (e.g., with sodium bisulfite) or sophisticated immunoprecipitation techniques (e.g., Chip-Seq).

The characterization of ancient epigenomes is not devoid of specific pending problems. For example, the factors contributing to the preservation of nucleosome footprints are still unknown. To which extent variable cell-type balances participate in the epigenetic differences observed between individuals? If important, can this be statistically modeled using methods similar to what the software SourceTracker does to identify the relative contributions of different mixtures of metagenomic sources (Knights et al. 2011)? Or could this simply be mitigated by developing experimental standards aimed at sampling the same bone region across individuals? Given the extent of epigenetic variation between individuals, and the importance of parameters, such as age, sex, genetic background, etc., which archeological contexts will provide the necessary population-wide material to capture the true extent past epigenetic variation? And which fraction of important biological changes affecting major past transitions, may these be cultural, epidemiological, or environmental, will be reflected in the epigenome of the bones and teeth, the most abundant type of subfossil material present in the archeological record?

We are confident that at least a number, if not most, of these limitations can be satisfactorily addressed. This will then open new perspectives on past evolutionary crises and their potential impact on the epigenome. In our opinion, the following questions will likely receive great attention in the near future: what has been the epigenetic impact of major climate changes such as the Last Glacial Maximum on megafauna epigenomes? Were such changes the same in survivors and lineages that went extinct? What has been the epigenetic impact of the Neolithic and Industrial revolutions and how such possible changes still impact our health today? Which epigenetic changes accompanied animal domestication and the artificial selection of breeds? This chapter mainly focused on ancient epigenetic signatures in ancient mammals, mostly ancient human individuals, given that most of the available data have been generated from such organisms. Fascinating work has been performed in plant seeds however (Smith et al. 2014b), and together with the survival of small RNA molecules (Fordyce et al. 2013), another essential type of epigenetic marks (Collins et al. 2011), this material might also yield instrumental advances in the field of ancient epigenomics.

Box 1: Transgenerational Epigenetic Inheritance

Transgenerational epigenetic inheritance has recently been reported in a number of organisms, including plants, nematodes, and eutherian mammals (see Richards 2006; Daxinger and Whitelaw 2012; Heard and Martienssen 2014 for reviews). In the latter, the mother and the offspring share similar environments during pregnancy. Environmentally induced epigenetic changes at that stage could shape similar epigenetic landscapes in mothers and offspring (F_1 generation) and even in their future offspring (F_2 generation) as the primordial germ cells are determined early in the development. As a consequence, transgenerational epigenetic inheritance can only be claimed if traits nongenetically determined are maintained throughout a minimum of three consecutive generations (Anway and Skinner 2006). This has been shown, for example, for DNA methylation changes apparently persistent for up to the F_4 generation in rats prenatally exposed to high doses of vinclozolin, a common pesticide for the agricultural industry, in the absence of further exposure in subsequent generations (Anway et al. 2005). The mechanism underlying this phenomenon is unknown but involves the transfer of molecules other than DNA during fertilization, such as noncoding RNAs (Chen et al. 2016), which can contribute to reestablish modified methylation landscapes at specific loci despite the two waves of global genome demethylation in the germline and also early postfertilization (Hackett and Surani 2013; Heard and Martienssen 2014). Regardless of the underlying biological mechanisms, transgenerational epigenetic inheritance holds the potential for Lamarckian models of evolution, where environment-driven traits are transmitted across generations (Danchin et al. 2011).

References

Allentoft ME, Collins M, Harker D, et al. The half-life of DNA in bone: measuring decay kinetics in 158 dated fossils. Proc Biol Sci. 2012;279:4724–33. https://doi.org/10.1098/rspb.2012.1745.

Allis CD, Jenuwein T. The molecular hallmarks of epigenetic control. Nat Rev Genet. 2016;17:487–500. https://doi.org/10.1038/nrg.2016.59.

Allum F, Shao X, Guénard F, et al. Characterization of functional methylomes by next-generation capture sequencing identifies novel disease-associated variants. Nat Commun. 2015;6:7211. https://doi.org/10.1038/ncomms8211.

AlQahtani SJ, Hector MP, Liversidge HM. Accuracy of dental age estimation charts: Schour and Massler, Ubelaker and the London Atlas. Am J Phys Anthropol. 2014;154:70–8. https://doi.org/10.1002/ajpa.22473.

Andersson L, Archibald AL, Bottema CD, et al. Coordinated international action to accelerate genome-to-phenome with FAANG, the functional annotation of animal genomes project. Genome Biol. 2015;16:57. https://doi.org/10.1186/s13059-015-0622-4.

Anway MD, Skinner MK. Epigenetic transgenerational actions of endocrine disruptors. Endocrinology. 2006;147:S43–9. https://doi.org/10.1210/en.2005-1058.

Anway MD, Cupp AS, Uzumcu M, Skinner MK. Epigenetic transgenerational actions of endocrine disruptors and male fertility. Science. 2005;308:1466–9. https://doi.org/10.1126/science.1108190.

Bai L, Morozov AV. Gene regulation by nucleosome positioning. Trends Genet. 2010;26:476–83. https://doi.org/10.1016/j.tig.2010.08.003.

Ball MP, Li JB, Gao Y, et al. Targeted and genome-scale strategies reveal gene-body methylation signatures in human cells. Nat Biotechnol. 2009;27:361–8. https://doi.org/10.1038/nbt.1533.

Bauer T, Trump S, Ishaque N, et al. Environment-induced epigenetic reprogramming in genomic regulatory elements in smoking mothers and their children. Mol Syst Biol. 2016;12:861.

Bell JT, Pai AA, Pickrell JK, et al. DNA methylation patterns associate with genetic and gene expression variation in HapMap cell lines. Genome Biol. 2011a;12:R10. https://doi.org/10.1186/gb-2011-12-1-r10.

Bell O, Tiwari VK, Thoma NH, Schübeler D. Determinants and dynamics of genome accessibility. Nat Rev Genet. 2011b;12:554–64. https://doi.org/10.1038/nrg3017.

Berger SL, Kouzarides T, Shiekhattar R, Shilatifard A. An operational definition of epigenetics. Genes Dev. 2009;23:781–3. https://doi.org/10.1101/gad.1787609.

Bernstein BE, Stamatoyannopoulos JA, Costello JF, et al. The NIH roadmap epigenomics mapping consortium. Nat Biotechnol. 2010;28:1045–8. https://doi.org/10.1038/nbt1010-1045.

Bird A. Perceptions of epigenetics. Nat Lond. 2007;447:396–8. https://doi.org/10.1038/nature05913.

Boyko A, Kovalchuk I. Genetic and epigenetic effects of plant-pathogen interactions: an evolutionary perspective. Mol Plant. 2011;4:1014–23. https://doi.org/10.1093/mp/ssr022.

Boyko A, Kathiria P, Zemp FJ, et al. Transgenerational changes in the genome stability and methylation in pathogen-infected plants: (virus-induced plant genome instability). Nucleic Acids Res. 2007;35:1714–25. https://doi.org/10.1093/nar/gkm029.

Braud M, Magee DA, Park SDE, et al. Genome-wide microRNA binding site variation between extinct wild aurochs and modern cattle identifies candidate microRNA-regulated domestication genes. Front Genet. 2017;8:3. https://doi.org/10.3389/fgene.2017.00003.

Briggs AW, Stenzel U, Johnson PLF, et al. Patterns of damage in genomic DNA sequences from a Neandertal. Proc Natl Acad Sci U S A. 2007;104:14616–21. https://doi.org/10.1073/pnas.0704665104.

Briggs AW, Stenzel U, Meyer M, et al. Removal of deaminated cytosines and detection of in vivo methylation in ancient DNA. Nucleic Acids Res. 2010;38:e87. https://doi.org/10.1093/nar/gkp1163.

Brogaard K, Xi L, Wang J-P, Widom J. A map of nucleosome positions in yeast at base-pair resolution. Nature. 2012;486:496–501. https://doi.org/10.1038/nature11142.

Brooks S, Suchey JM. Skeletal age determination based on the os pubis: a comparison of the Acsádi-Nemeskéri and Suchey-Brooks methods. Hum Evol. 1990;5:227–38. https://doi.org/10.1007/BF02437238.

Buckley M, Walker A, Ho SYW, et al. Comment on "Protein sequences from mastodon and Tyrannosaurus rex revealed by mass spectrometry". Science. 2008;319:33. https://doi.org/10.1126/science.1147046. Author reply 33.

Carpenter ML, Buenrostro JD, Valdiosera C, et al. Pulling out the 1%: whole-genome capture for the targeted enrichment of ancient DNA sequencing libraries. Am J Hum Genet. 2013;93:852–64. https://doi.org/10.1016/j.ajhg.2013.10.002.

Castillo-Fernandez JE, Spector TD, Bell JT. Epigenetics of discordant monozygotic twins: implications for disease. Genome Med. 2014;6:60. https://doi.org/10.1186/s13073-014-0060-z.

Chen Q, Yan W, Duan E. Epigenetic inheritance of acquired traits through sperm RNAs and sperm RNA modifications. Nat Rev Genet. 2016;17:733–43. https://doi.org/10.1038/nrg.2016.106.

Chodavarapu RK, Feng S, Bernatavichute YV, et al. Relationship between nucleosome positioning and DNA methylation. Nature. 2010;466:388–92. https://doi.org/10.1038/nature09147.

Chua EYD, Vasudevan D, Davey GE, et al. The mechanics behind DNA sequence-dependent properties of the nucleosome. Nucleic Acids Res. 2012;40:6338–52. https://doi.org/10.1093/nar/gks261.

Collins LJ, Schönfeld B, Chen XS. The epigenetics of non-coding RNA. In: Tollefsbol T, editor. Handbook of epigenetics: the new molecular and medical genetics. London: Academic; 2011. p. 49–61.

Cruz-Dávalos DI, Llamas B, Gaunitz C, et al. Experimental conditions improving in solution target enrichment for ancient DNA. Mol Ecol Resour. 2016;17(3):508–22. https://doi.org/10.1111/1755-0998.12595.

Dabney J, Meyer M. Length and GC-biases during sequencing library amplification: a comparison of various polymerase-buffer systems with ancient and modern DNA sequencing libraries. BioTechniques. 2012;52:87–94. https://doi.org/10.2144/000113809.

Dabney J, Knapp M, Glocke I, et al. Complete mitochondrial genome sequence of a middle Pleistocene cave bear reconstructed from ultrashort DNA fragments. Proc Natl Acad Sci U S A. 2013;110:15758–63. https://doi.org/10.1073/pnas.1314445110.

Damgaard PB, Margaryan A, Schroeder H, et al. Improving access to endogenous DNA in ancient bones and teeth. Sci Rep. 2015;5:11184. https://doi.org/10.1038/srep11184.

Danchin É, Charmantier A, Champagne FA, et al. Beyond DNA: integrating inclusive inheritance into an extended theory of evolution. Nat Rev Genet. 2011;12:475–86. https://doi.org/10.1038/nrg3028.

Daniel FI, Cherubini K, Yurgel LS, et al. The role of epigenetic transcription repression and DNA methyltransferases in cancer. Cancer. 2011;117:677–87. https://doi.org/10.1002/cncr.25482.

Davis AP, Capecchi MR. Axial homeosis and appendicular skeleton defects in mice with a targeted disruption of hoxd-11. Development. 1994;120:2187–98.

Daxinger L, Whitelaw E. Understanding transgenerational epigenetic inheritance via the gametes in mammals. Nat Rev Genet. 2012;13:153–62. https://doi.org/10.1038/nrg3188.

de Rooij SR, Painter RC, Phillips DIW, et al. Impaired insulin secretion after prenatal exposure to the Dutch famine. Diabetes Care. 2006;29:1897–901. https://doi.org/10.2337/dc06-0460.

de Rooij SR, Wouters H, Yonker JE, et al. Prenatal undernutrition and cognitive function in late adulthood. Proc Natl Acad Sci U S A. 2010;107:16881–6. https://doi.org/10.1073/pnas.1009459107.

Deans C, Maggert KA. What do you mean, "epigenetic"? Genetics. 2015;199:887–96. https://doi.org/10.1534/genetics.114.173492.

Demarchi B, Hall S, Roncal-Herrero T, et al. Protein sequences bound to mineral surfaces persist into deep time. elife. 2016;5:e17092. https://doi.org/10.7554/eLife.17092.

Der Sarkissian C, Ermini L, Jónsson H, et al. Shotgun microbial profiling of fossil remains. Mol Ecol. 2014;23:1780–98. https://doi.org/10.1111/mec.12690.

Der Sarkissian C, Allentoft ME, Ávila-Arcos MC, et al. Ancient genomics. Philos Trans R Soc Lond Ser B Biol Sci. 2015a;370:20130387. https://doi.org/10.1098/rstb.2013.0387.

Der Sarkissian C, Ermini L, Schubert M, et al. Evolutionary genomics and conservation of the endangered Przewalski's horse. Curr Biol. 2015b;25:2577–83. https://doi.org/10.1016/j.cub.2015.08.032.

Dolinoy DC. The agouti mouse model: an epigenetic biosensor for nutritional and environmental alterations on the fetal epigenome. Nutr Rev. 2008;66:S7–S11. https://doi.org/10.1111/j.1753-4887.2008.00056.x.

Dolinoy DC, Weidman JR, Waterland RA, Jirtle RL. Maternal genistein alters coat color and protects Avy mouse offspring from obesity by modifying the fetal epigenome. Environ Health Perspect. 2006;114:567–72.

Dolinoy DC, Huang D, Jirtle RL. Maternal nutrient supplementation counteracts bisphenol A-induced DNA hypomethylation in early development. Proc Natl Acad Sci U S A. 2007;104:13056–61. https://doi.org/10.1073/pnas.0703739104.

Down TA, Rakyan VK, Turner DJ, et al. A Bayesian deconvolution strategy for immunoprecipitation-based DNA methylome analysis. Nat Biotechnol. 2008;26:779–85. https://doi.org/10.1038/nbt1414.

Eckhardt F, Lewin J, Cortese R, et al. DNA methylation profiling of human chromosomes 6, 20 and 22. Nat Genet. 2006;38:1378–85. https://doi.org/10.1038/ng1909.

Ehrlich M, Gama-Sosa MA, Huang LH, et al. Amount and distribution of 5-methylcytosine in human DNA from different types of tissues of cells. Nucleic Acids Res. 1982;10:2709–21.

ENCODE Project Consortium. An integrated encyclopedia of DNA elements in the human genome. Nature. 2012;489:57–74. https://doi.org/10.1038/nature11247.

Ermini L, Der Sarkissian C, Willerslev E, Orlando L. Major transitions in human evolution revisited: a tribute to ancient DNA. J Hum Evol. 2015;79:4–20. https://doi.org/10.1016/j.jhevol.2014.06.015.

Fagny M, Patin E, MacIsaac JL, et al. The epigenomic landscape of African rainforest hunter-gatherers and farmers. Nat Commun. 2015;6:10047. https://doi.org/10.1038/ncomms10047.

Favier B, Meur ML, Chambon P, Dollé P. Axial skeleton homeosis and forelimb malformations in Hoxd-11 mutant mice. Proc Natl Acad Sci. 1995;92:310–4.

Feil R, Fraga MF. Epigenetics and the environment: emerging patterns and implications. Nat Rev Genet. 2012;13:97–109. https://doi.org/10.1038/nrg3142.

Fordyce SL, Kampmann M-L, van Doorn NL, Gilbert MTP. Long-term RNA persistence in postmortem contexts. Investig Genet. 2013;4:7. https://doi.org/10.1186/2041-2223-4-7.

Fraga MF, Ballestar E, Paz MF, et al. Epigenetic differences arise during the lifetime of monozygotic twins. Proc Natl Acad Sci U S A. 2005;102:10604–9. https://doi.org/10.1073/pnas.0500398102.

Frantz LAF, Schraiber JG, Madsen O, et al. Evidence of long-term gene flow and selection during domestication from analyses of Eurasian wild and domestic pig genomes. Nat Genet. 2015;47:1141–8. https://doi.org/10.1038/ng.3394.

Fraser HB, Lam LL, Neumann SM, Kobor MS. Population-specificity of human DNA methylation. Genome Biol. 2012;13:R8. https://doi.org/10.1186/gb-2012-13-2-r8.

Fu Y, Sinha M, Peterson CL, Weng Z. The insulator binding protein CTCF positions 20 nucleosomes around its binding sites across the human genome. PLoS Genet. 2008;4:e1000138. https://doi.org/10.1371/journal.pgen.1000138.

Galanter JM, Gignoux CR, Oh SS, et al. Differential methylation between ethnic sub-groups reflects the effect of genetic ancestry and environmental exposures. elife. 2017;6:e20532. https://doi.org/10.7554/eLife.20532.

Gamba C, Hanghøj K, Gaunitz C, et al. Comparing the performance of three ancient DNA extraction methods for high-throughput sequencing. Mol Ecol Resour. 2016;16:459–69. https://doi.org/10.1111/1755-0998.12470.

Gansauge M-T, Meyer M. Single-stranded DNA library preparation for the sequencing of ancient or damaged DNA. Nat Protoc. 2013;8:737–48. https://doi.org/10.1038/nprot.2013.038.

Gansauge M-T, Meyer M. Selective enrichment of damaged DNA molecules for ancient genome sequencing. Genome Res. 2014;24:1543–9. https://doi.org/10.1101/gr.174201.114.

Gehring M. Prodigious plant methylomes. Genome Biol. 2016;17:197. https://doi.org/10.1186/s13059-016-1065-2.

Giuliani C, Cilli E, Bacalini MG, et al. Inferring chronological age from DNA methylation patterns of human teeth. Am J Phys Anthropol. 2015;159(4):585–95. https://doi.org/10.1002/ajpa.22921.

Gokhman D, Lavi E, Prüfer K, et al. Reconstructing the DNA methylation maps of the Neandertal and the Denisovan. Science. 2014;344:523–7. https://doi.org/10.1126/science.1250368.

Gokhman D, Meshorer E, Carmel L. Epigenetics: it's getting old. Past meets future in Paleoepigenetics. Trends Ecol Evol. 2016;31(4):290–300. https://doi.org/10.1016/j.tree.2016.01.010.

Gokhman D, Agranat-Tamir L, Housman G, et al. Recent regulatory changes shaped human facial and vocal anatomy. bioRxiv. 2017;106955. https://doi.org/10.1101/106955.

Green RE, Krause J, Ptak SE, et al. Analysis of one million base pairs of Neanderthal DNA. Nature. 2006;444:330–6. https://doi.org/10.1038/nature05336.

Grunau C, Clark SJ, Rosenthal A. Bisulfite genomic sequencing: systematic investigation of critical experimental parameters. Nucleic Acids Res. 2001;29:E65.

Gu H, Smith ZD, Bock C, et al. Preparation of reduced representation bisulfite sequencing libraries for genome-scale DNA methylation profiling. Nat Protoc. 2011;6:468–81. https://doi.org/10.1038/nprot.2010.190.

Haak W, Lazaridis I, Patterson N, et al. Massive migration from the steppe was a source for Indo-European languages in Europe. Nature. 2015;522:207–11. https://doi.org/10.1038/nature14317.

Hackett JA, Surani MA. Beyond DNA: programming and inheritance of parental methylomes. Cell. 2013;153:737–9. https://doi.org/10.1016/j.cell.2013.04.044.

Hackman DA, Farah MJ, Meaney MJ. Socioeconomic status and the brain: mechanistic insights from human and animal research. Nat Rev Neurosci. 2010;11:651–9. https://doi.org/10.1038/nrn2897.

Hanghøj K, Seguin-Orlando A, Schubert M, et al. Fast, accurate and automatic ancient nucleosome and methylation maps with epiPALEOMIX. Mol Biol Evol. 2016;33(12):3284–98. https://doi.org/10.1093/molbev/msw184.

Hansen A, Willerslev E, Wiuf C, et al. Statistical evidence for miscoding lesions in ancient DNA templates. Mol Biol Evol. 2001;18:262–5.

He Y, Ecker JR. Non-CG methylation in the human genome. Annu Rev Genomics Hum Genet. 2015;16:55–77. https://doi.org/10.1146/annurev-genom-090413-025437.

Heard E, Martienssen RA. Transgenerational epigenetic inheritance: myths and mechanisms. Cell. 2014;157:95–109. https://doi.org/10.1016/j.cell.2014.02.045.

Heijmans BT, Tobi EW, Stein AD, et al. Persistent epigenetic differences associated with prenatal exposure to famine in humans. Proc Natl Acad Sci U S A. 2008;105:17046–9. https://doi.org/10.1073/pnas.0806560105.

Heyn H, Moran S, Hernando-Herraez I, et al. DNA methylation contributes to natural human variation. Genome Res. 2013;23:1363–72. https://doi.org/10.1101/gr.154187.112.

Ho S-M, Johnson A, Tarapore P, et al. Environmental epigenetics and its implication on disease risk and health outcomes. ILAR J. 2012;53:289–305. https://doi.org/10.1093/ilar.53.3-4.289.

Hofreiter M, Jaenicke V, Serre D, et al. DNA sequences from multiple amplifications reveal artifacts induced by cytosine deamination in ancient DNA. Nucleic Acids Res. 2001a;29:4793–9.

Hofreiter M, Serre D, Poinar HN, et al. Ancient DNA. Nat Rev Genet. 2001b;2:353–9. https://doi.org/10.1038/35072071.

Holoch D, Moazed D. RNA-mediated epigenetic regulation of gene expression. Nat Rev Genet. 2015;16:71–84. https://doi.org/10.1038/nrg3863.

Horvath S. DNA methylation age of human tissues and cell types. Genome Biol. 2013;14:R115. https://doi.org/10.1186/gb-2013-14-10-r115.

Horvath S, Gurven M, Levine ME, et al. An epigenetic clock analysis of race/ethnicity, sex, and coronary heart disease. Genome Biol. 2016;17:171. https://doi.org/10.1186/s13059-016-1030-0.

Houseman EA, Accomando WP, Koestler DC, et al. DNA methylation arrays as surrogate measures of cell mixture distribution. BMC Bioinformatics. 2012;13:86. https://doi.org/10.1186/1471-2105-13-86.

Jin H, Rube HT, Song JS. Categorical spectral analysis of periodicity in nucleosomal DNA. Nucleic Acids Res. 2016;44(5):2047–57. https://doi.org/10.1093/nar/gkw101.

Jónsson H, Ginolhac A, Schubert M, et al. mapDamage2.0: fast approximate Bayesian estimates of ancient DNA damage parameters. Bioinformatics. 2013;29:1682–4. https://doi.org/10.1093/bioinformatics/btt193.

Jónsson H, Schubert M, Seguin-Orlando A, et al. Speciation with gene flow in equids despite extensive chromosomal plasticity. Proc Natl Acad Sci U S A. 2014;111:18655–60. https://doi.org/10.1073/pnas.1412627111.

Keller A, Graefen A, Ball M, et al. New insights into the Tyrolean Iceman's origin and phenotype as inferred by whole genome sequencing. Nat Commun. 2012;3:698.

Kelly TK, Liu Y, Lay FD, et al. Genome-wide mapping of nucleosome positioning and DNA methylation within individual DNA molecules. Genome Res. 2012;22:2497–506. https://doi.org/10.1101/gr.143008.112.

Kelman Z, Moran L. Degradation of ancient DNA. Curr Biol. 1996;6:223.

Knights D, Kuczynski J, Charlson ES, et al. Bayesian community-wide culture-independent microbial source tracking. Nat Methods. 2011;8:761–3. https://doi.org/10.1038/nmeth.1650.

Kogan SB, Kato M, Kiyama R, Trifonov EN. Sequence structure of human nucleosome DNA. J Biomol Struct Dyn. 2006;24:43–8. https://doi.org/10.1080/07391102.2006.10507097.

Kousathanas A, Leuenberger C, Link V, et al. Inferring heterozygosity from ancient and low coverage genomes. Genetics. 2017;205:317–32. https://doi.org/10.1534/genetics.116.189985.

Krueger F, Kreck B, Franke A, Andrews SR. DNA methylome analysis using short bisulfite sequencing data. Nat Methods. 2012;9:145–51. https://doi.org/10.1038/nmeth.1828.

Lam LL, Emberly E, Fraser HB, et al. Factors underlying variable DNA methylation in a human community cohort. Proc Natl Acad Sci U S A. 2012;109(Suppl 2):17253–60. https://doi.org/10.1073/pnas.1121249109.

Lazaridis I, Patterson N, Mittnik A, et al. Ancient human genomes suggest three ancestral populations for present-day Europeans. Nature. 2014;513:409–13. https://doi.org/10.1038/nature13673.

Li B, Carey M, Workman JL. The role of chromatin during transcription. Cell. 2007;128:707–19. https://doi.org/10.1016/j.cell.2007.01.015.

Librado P, Der Sarkissian C, Ermini L, et al. Tracking the origins of Yakutian horses and the genetic basis for their fast adaptation to subarctic environments. Proc Natl Acad Sci U S A. 2015;112:E6889–97. https://doi.org/10.1073/pnas.1513696112.

Librado P, Gamba C, Gaunitz C, et al. Ancient genomic changes associated with domestication of the horse. Science. 2017;356:442–5. https://doi.org/10.1126/science.aam5298.

Lindahl T. Instability and decay of the primary structure of DNA. Nature. 1993;362:709–15. https://doi.org/10.1038/362709a0.

Lister R, Ecker JR. Finding the fifth base: genome-wide sequencing of cytosine methylation. Genome Res. 2009;19:959–66. https://doi.org/10.1101/gr.083451.108.

Lister R, Pelizzola M, Dowen RH, et al. Human DNA methylomes at base resolution show widespread epigenomic differences. Nature. 2009;462:315–22. https://doi.org/10.1038/nature08514.

Liu H, Liu X, Zhang S, et al. Systematic identification and annotation of human methylation marks based on bisulfite sequencing methylomes reveals distinct roles of cell type-specific hypomethylation in the regulation of cell identity genes. Nucleic Acids Res. 2016;44:75–94. https://doi.org/10.1093/nar/gkv1332.

Llamas B, Holland ML, Chen K, et al. High-resolution analysis of cytosine methylation in ancient DNA. PLoS One. 2012;7:e30226. https://doi.org/10.1371/journal.pone.0030226.
Llamas B, Willerslev E, Orlando L. Human evolution: a tale from ancient genomes. Philos Trans R Soc Lond Ser B Biol Sci. 2017;372:1713. https://doi.org/10.1098/rstb.2015.0484.
Louvel G, Der Sarkissian C, Hanghøj K, Orlando L. metaBIT, an integrative and automated metagenomic pipeline for analysing microbial profiles from high-throughput sequencing shotgun data. Mol Ecol Resour. 2016;16:1415–27. https://doi.org/10.1111/1755-0998.12546.
Lumey LH, Stein AD, Kahn HS, et al. Cohort profile: the Dutch Hunger winter families study. Int J Epidemiol. 2007;36:1196–204. https://doi.org/10.1093/ije/dym126.
Lynch VJ, Bedoya-Reina OC, Ratan A, et al. Elephantid genomes reveal the molecular bases of Woolly Mammoth adaptations to the arctic. Cell Rep. 2015;12:217–28. https://doi.org/10.1016/j.celrep.2015.06.027.
Marciniak S, Perry GH. Harnessing ancient genomes to study the history of human adaptation. Nat Rev Genet. 2017;18(11):659–74. https://doi.org/10.1038/nrg.2017.65.
Marioni RE, Shah S, McRae AF, et al. DNA methylation age of blood predicts all-cause mortality in later life. Genome Biol. 2015;16:25. https://doi.org/10.1186/s13059-015-0584-6.
Mathieson I, Lazaridis I, Rohland N, et al. Eight thousand years of natural selection in Europe. 2015. https://doi.org/10.1101/016477.
Maunakea AK, Nagarajan RP, Bilenky M, et al. Conserved role of intragenic DNA methylation in regulating alternative promoters. Nature. 2010;466:253–7. https://doi.org/10.1038/nature09165.
McGowan PO, Sasaki A, D'Alessio AC, et al. Epigenetic regulation of the glucocorticoid receptor in human brain associates with childhood abuse. Nat Neurosci. 2009;12:342–8. https://doi.org/10.1038/nn.2270.
Meyer M, Kircher M. Illumina sequencing library preparation for highly multiplexed target capture and sequencing. Cold Spring Harb Protoc. 2010;2010:db.prot5448. https://doi.org/10.1101/pdb.prot5448.
Meyer M, Kircher M, Gansauge M-T, et al. A high-coverage genome sequence from an archaic Denisovan individual. Science. 2012;338:222–6. https://doi.org/10.1126/science.1224344.
Miller W, Drautz DI, Ratan A, et al. Sequencing the nuclear genome of the extinct woolly mammoth. Nature. 2008;456:387–90. https://doi.org/10.1038/nature07446.
Miller W, Schuster SC, Welch AJ, et al. Polar and brown bear genomes reveal ancient admixture and demographic footprints of past climate change. PNAS. 2012;109(36):E2382–90. https://doi.org/10.1073/pnas.1210506109.
Monge I, Kondo T, Duboule D. An enhancer-titration effect induces digit-specific regulatory alleles of the HoxD cluster. Dev Biol. 2003;256:212–20.
Morey C, Avner P. Genetics and epigenetics of the X chromosome. Ann N Y Acad Sci. 2010;1214:E18–33. https://doi.org/10.1111/j.1749-6632.2010.05943.x.
Murgatroyd C, Patchev AV, Wu Y, et al. Dynamic DNA methylation programs persistent adverse effects of early-life stress. Nat Neurosci. 2009;12:1559–66. https://doi.org/10.1038/nn.2436.
Murray IA, Clark TA, Morgan RD, et al. The methylomes of six bacteria. Nucleic Acids Res. 2012;40:11450–62. https://doi.org/10.1093/nar/gks891.
Nielsen R, Akey JM, Jakobsson M, et al. Tracing the peopling of the world through genomics. Nature. 2017;541:302–10. https://doi.org/10.1038/nature21347.
Orlando L, Willerslev E. Evolution. An epigenetic window into the past? Science. 2014;345:511–2. https://doi.org/10.1126/science.1256515.
Orlando L, Ginolhac A, Zhang G, et al. Recalibrating Equus evolution using the genome sequence of an early middle Pleistocene horse. Nature. 2013;499:74–8. https://doi.org/10.1038/nature12323.
Orlando L, Gilbert MTP, Willerslev E. Reconstructing ancient genomes and epigenomes. Nat Rev Genet. 2015;16:395–408. https://doi.org/10.1038/nrg3935.
Painter RC, de Rooij SR, Bossuyt PM, et al. Early onset of coronary artery disease after prenatal exposure to the Dutch famine. Am J Clin Nutr. 2006;84:322–7.

Painter RC, Osmond C, Gluckman P, et al. Transgenerational effects of prenatal exposure to the Dutch famine on neonatal adiposity and health in later life. BJOG Int J Obstet Gynaecol. 2008;115:1243–9. https://doi.org/10.1111/j.1471-0528.2008.01822.x.

Palkopoulou E, Mallick S, Skoglund P, et al. Complete genomes reveal signatures of demographic and genetic declines in the woolly mammoth. Curr Biol. 2015;25:1395–400. https://doi.org/10.1016/j.cub.2015.04.007.

Park SDE, Magee DA, McGettigan PA, et al. Genome sequencing of the extinct Eurasian wild aurochs, Bos primigenius, illuminates the phylogeography and evolution of cattle. Genome Biol. 2015;16:234. https://doi.org/10.1186/s13059-015-0790-2.

Patin E, Laval G, Barreiro LB, et al. Inferring the demographic history of African farmers and pygmy hunter-gatherers using a multilocus resequencing data set. PLoS Genet. 2009;5: e1000448. https://doi.org/10.1371/journal.pgen.1000448.

Pedersen JS, Valen E, Velazquez AMV, et al. Genome-wide nucleosome map and cytosine methylation levels of an ancient human genome. Genome Res. 2014;24:454–66. https://doi.org/10.1101/gr.163592.113.

Pembrey M, Saffery R, Bygren LO, et al. Human transgenerational responses to early-life experience: potential impact on development, health and biomedical research. J Med Genet. 2014;51:563–72. https://doi.org/10.1136/jmedgenet-2014-102577.

Petropoulos S, Panula SP, Schell JP, Lanner F. Single-cell RNA sequencing: revealing human pre-implantation development, pluripotency and germline development. J Intern Med. 2016;280:252–64. https://doi.org/10.1111/joim.12493.

Pinhasi R, Fernandes D, Sirak K, et al. Optimal ancient DNA yields from the inner ear part of the human petrous bone. PLoS One. 2015;10:e0129102. https://doi.org/10.1371/journal.pone.0129102.

Plongthongkum N, Diep DH, Zhang K. Advances in the profiling of DNA modifications: cytosine methylation and beyond. Nat Rev Genet. 2014;15:647–61. https://doi.org/10.1038/nrg3772.

Pokholok DK, Harbison CT, Levine S, et al. Genome-wide map of nucleosome acetylation and methylation in yeast. Cell. 2005;122:517–27. https://doi.org/10.1016/j.cell.2005.06.026.

Portales-Casamar E, Lussier AA, Jones MJ, et al. DNA methylation signature of human fetal alcohol spectrum disorder. Epigenetics Chromatin. 2016;9:25. https://doi.org/10.1186/s13072-016-0074-4.

Prüfer K, Racimo F, Patterson N, et al. The complete genome sequence of a Neanderthal from the Altai Mountains. Nature. 2014;505:43–9. https://doi.org/10.1038/nature12886.

Quail MA, Kozarewa I, Smith F, et al. A large genome center's improvements to the Illumina sequencing system. Nat Methods. 2008;5:1005–10. https://doi.org/10.1038/nmeth.1270.

Racimo F, Gokhman D, Fumagalli M, et al. Archaic adaptive introgression in TBX15/WARS2. Mol Biol Evol. 2016;34(3):509–24. https://doi.org/10.1093/molbev/msw283.

Radford EJ, Ito M, Shi H, et al. In utero effects. In utero undernourishment perturbs the adult sperm methylome and intergenerational metabolism. Science. 2014;345:1255903. https://doi.org/10.1126/science.1255903.

Rakyan VK, Down TA, Balding DJ, Beck S. Epigenome-wide association studies for common human diseases. Nat Rev Genet. 2011;12:529–41. https://doi.org/10.1038/nrg3000.

Ramakrishnan V. Histone structure and the organization of the nucleosome. Annu Rev Biophys Biomol Struct. 1997;26:83–112. https://doi.org/10.1146/annurev.biophys.26.1.83.

Ramos-Madrigal J, Smith BD, Moreno-Mayar JV, et al. Genome sequence of a 5,310-year-old maize cob provides insights into the early stages of maize domestication. Curr Biol. 2016;26:3195–201. https://doi.org/10.1016/j.cub.2016.09.036.

Rasmussen M, Li Y, Lindgreen S, et al. Ancient human genome sequence of an extinct Palaeo-Eskimo. Nature. 2010;463:757–62. https://doi.org/10.1038/nature08835.

Rasmussen M, Anzick SL, Waters MR, et al. The genome of a late Pleistocene human from a Clovis burial site in western Montana. Nature. 2014;506:225–9. https://doi.org/10.1038/nature13025.

Rebelo AP, Williams SL, Moraes CT. In vivo methylation of mtDNA reveals the dynamics of protein-mtDNA interactions. Nucleic Acids Res. 2009;37:6701–15. https://doi.org/10.1093/nar/gkp727.

Rhee I, Jair KW, Yen RW, et al. CpG methylation is maintained in human cancer cells lacking DNMT1. Nature. 2000;404:1003–7. https://doi.org/10.1038/35010000.

Richards EJ. Inherited epigenetic variation – revisiting soft inheritance. Nat Rev Genet. 2006;7:395–401. https://doi.org/10.1038/nrg1834.

Rohland N, Harney E, Mallick S, et al. Partial uracil-DNA-glycosylase treatment for screening of ancient DNA. Philos Trans R Soc Lond Ser B Biol Sci. 2015;370:20130624. https://doi.org/10.1098/rstb.2013.0624.

Roseboom T, de Rooij S, Painter R. The Dutch famine and its long-term consequences for adult health. Early Hum Dev. 2006;82:485–91. https://doi.org/10.1016/j.earlhumdev.2006.07.001.

Ross MG, Russ C, Costello M, et al. Characterizing and measuring bias in sequence data. Genome Biol. 2013;14:R51. https://doi.org/10.1186/gb-2013-14-5-r51.

Schubert M, Jónsson H, Chang D, et al. Prehistoric genomes reveal the genetic foundation and cost of horse domestication. Proc Natl Acad Sci U S A. 2014;111:E5661–9. https://doi.org/10.1073/pnas.1416991111.

Schweitzer MH, Suo Z, Avci R, et al. Analyses of soft tissue from Tyrannosaurus rex suggest the presence of protein. Science. 2007;316:277–80. https://doi.org/10.1126/science.1138709.

Schweitzer MH, Zheng W, Organ CL, et al. Biomolecular characterization and protein sequences of the Campanian hadrosaur B. canadensis. Science. 2009;324:626–31. https://doi.org/10.1126/science.1165069.

Schweitzer MH, Schroeter ER, Goshe MB. Protein molecular data from ancient (>1 million years old) fossil material: pitfalls, possibilities and grand challenges. Anal Chem. 2014;86:6731–40. https://doi.org/10.1021/ac500803w.

Seguin-Orlando A, Gamba C, Der Sarkissian C, et al. Pros and cons of methylation-based enrichment methods for ancient DNA. Sci Rep. 2015a;5:11826. https://doi.org/10.1038/srep11826.

Seguin-Orlando A, Hoover CA, Vasiliev SK, et al. Amplification of TruSeq ancient DNA libraries with AccuPrime Pfx: consequences on nucleotide misincorporation and methylation patterns. Sci Technol Archaeol Res. 2015b;1(1):1–9.

Skoglund P, Ersmark E, Palkopoulou E, Dalén L. Ancient wolf genome reveals an early divergence of domestic dog ancestors and admixture into high-latitude breeds. Curr Biol. 2015;25:1515–9. https://doi.org/10.1016/j.cub.2015.04.019.

Slotkin RK, Martienssen R. Transposable elements and the epigenetic regulation of the genome. Nat Rev Genet. 2007;8:272–85. https://doi.org/10.1038/nrg2072.

Smith O, Clapham A, Rose P, et al. A complete ancient RNA genome: identification, reconstruction and evolutionary history of archaeological Barley Stripe Mosaic Virus. Sci Rep. 2014a;4:4003. https://doi.org/10.1038/srep04003.

Smith O, Clapham AJ, Rose P, et al. Genomic methylation patterns in archaeological barley show de-methylation as a time-dependent diagenetic process. Sci Rep. 2014b;4:5559. https://doi.org/10.1038/srep05559.

Smith RWA, Monroe C, Bolnick DA. Detection of cytosine methylation in ancient DNA from five native american populations using bisulfite sequencing. PLoS One. 2015;10:e0125344. https://doi.org/10.1371/journal.pone.0125344.

Strahl BD, Allis CD. The language of covalent histone modifications. Nature. 2000;403:41–5. https://doi.org/10.1038/47412.

Struhl K, Segal E. Determinants of nucleosome positioning. Nat Struct Mol Biol. 2013;20:267–73. https://doi.org/10.1038/nsmb.2506.

Susser E, Kirkbride JB, Heijmans BT, et al. Maternal prenatal nutrition and health in grandchildren and subsequent generations. Annu Rev Anthropol. 2012;41:577–610. http://www.annualreviews.org/doi/10.1146/annurev-anthro-081309-145645. Accessed 28 Feb 2017.

Taudt A, Colomé-Tatché M, Johannes F. Genetic sources of population epigenomic variation. Nat Rev Genet. 2016;17:319–32. https://doi.org/10.1038/nrg.2016.45.

Tobi EW, Lumey LH, Talens RP, et al. DNA methylation differences after exposure to prenatal famine are common and timing- and sex-specific. Hum Mol Genet. 2009;18:4046–53. https://doi.org/10.1093/hmg/ddp353.

Triantaphyllopoulos KA, Ikonomopoulos I, Bannister AJ. Epigenetics and inheritance of phenotype variation in livestock. Epigenetics Chromatin. 2016;9:31. https://doi.org/10.1186/s13072-016-0081-5.

Valouev A, Johnson SM, Boyd SD, et al. Determinants of nucleosome organization in primary human cells. Nature. 2011;474:516–20. https://doi.org/10.1038/nature10002.

Waddington CH. The epigenotype. Endeavour. 1942;1:18–20.

Weber M, Davies JJ, Wittig D, et al. Chromosome-wide and promoter-specific analyses identify sites of differential DNA methylation in normal and transformed human cells. Nat Genet. 2005;37:853–62. https://doi.org/10.1038/ng1598.

Zhang W, Spector TD, Deloukas P, et al. Predicting genome-wide DNA methylation using methylation marks, genomic position, and DNA regulatory elements. Genome Biol. 2015;16:14. https://doi.org/10.1186/s13059-015-0581-9.

Ziller MJ, Gu H, Müller F, et al. Charting a dynamic DNA methylation landscape of the human genome. Nature. 2013;500:477–81. https://doi.org/10.1038/nature12433.

Part II
Paleogenomics Case Studies: From Ancient Pathogens to Primates

Ancient Pathogens Through Human History: A Paleogenomic Perspective

Stephanie Marciniak and Hendrik N. Poinar

Abstract Ancient bacterial and viral genomes provide a window into the tempo, chronology, and rate of evolutionary processes of pathogens that have accompanied humans throughout history, from catastrophic pandemics (e.g., *Yersinia pestis* and the Black Death) to diseases associated with "everyday" morbidity and mortality (e.g., tuberculosis, leprosy, hepatitis B virus). Excitingly, the scope of pathogens that can be explored using ancient DNA methods is expanding, largely due to advances in the recovery of these typically minute molecular fractions. Increasingly, ancient DNA is applied to the study of rapidly or slowly evolving pathogens across significant time transects (hundreds to thousands of years ago) enabling us to investigate ancient genomic diversity through a comparative lens that can potentially inform our understanding of how a pathogen has changed over time. In this chapter, we highlight the impact of changing molecular strategies in recovering and analyzing ancient pathogen genomes alongside the wealth of information within the historical record that both informs and challenges the framework used to explore pathogens and human disease in the past. The power of ancient DNA to detect the signatures of ancient pathogens is also tempered by recognized limitations in characterizing the relative "outcome" of complex human-pathogen interactions in diverse archaeological contexts.

S. Marciniak
McMaster Ancient DNA Centre, Department of Anthropology, McMaster University, Hamilton, ON, Canada

Department of Anthropology, Pennsylvania State University, University Park, PA, USA
e-mail: szm316@psu.edu

H. N. Poinar (✉)
McMaster Ancient DNA Centre, Department of Anthropology, McMaster University, Hamilton, ON, Canada

Michael G. DeGroote Institute for Infectious Disease Research and the Department of Biochemistry, McMaster University, Hamilton, ON, Canada

Humans and the Microbiome Program, Canadian Institute for Advanced Research, Toronto, ON, Canada
e-mail: poinarh@mcmaster.ca

Charlotte Lindqvist and Om P. Rajora (eds.), *Paleogenomics*,
Population Genomics [Om P. Rajora (Editor-in-Chief)],
https://doi.org/10.1007/13836_2018_52,

Keywords Ancient DNA · Evolutionary history · Human-pathogen interactions

1 Introduction

The evolutionary and demographic history of human pathogens is underscored by interactions with biological, ecological, and cultural factors that shape the epidemiology of disease in a specific historical space whether on the scale of the fourteenth-century CE Black Death pandemic across Europe (Benedictow 2004) or widespread disease burdens, such as leprosy from fourth-century BCE to fourteenth-century CE Europe (Schuenemann et al. 2018a). A multidisciplinary (or consilient) approach that combines the expertise of historians, archaeologists, geneticists, and bioarchaeologists (to name a few) in collaborative research provides a humanistic and scientific framework to study these dimensions of past human-disease interactions (Newfield and Labuhn 2017). An increasingly integrative component of exploring human disease in the past, in a holistic fashion, is the added use of ancient DNA (aDNA) to recover these microbes from ancient human skeletal remains, by providing a time stamp for implicating specific pathogens in their respective spatio-temporal and historical contexts (Fig. 1).

The unique opportunity of harnessing ancient DNA to explore the human history of disease shows the complementarity of molecular strategies to reconstruct pathogen evolutionary biology (Bos et al. 2011; Wagner et al. 2014), illuminate aspects of epidemiology (Duggan et al. 2016), or provide insight into disease-associated microbes, rather than solely infectious agents (Devault et al. 2017; Maixner et al. 2014). In this chapter, we briefly discuss investigating disease-associated pathogens

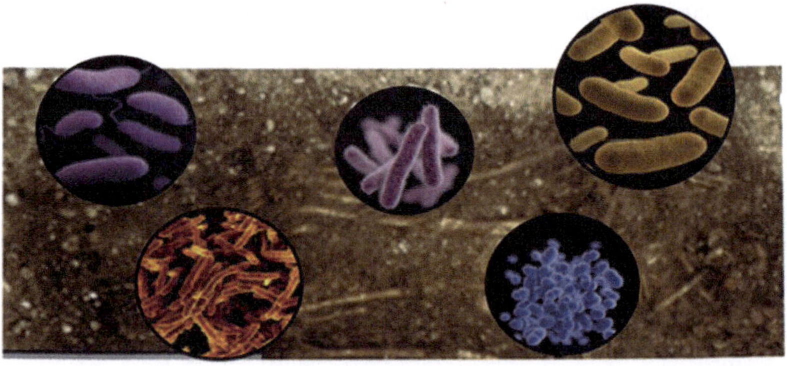

Fig. 1 Diverse microbes are found within human archaeological assemblages, and ancient DNA provides a means to recover these molecular signatures. Image: a shared burial of individuals from the Aschheim-Bajuwarenring cemetery, Bavaria, Germany (Wagner et al. 2014). (burial) – Content providers: CDC/James Archer and NIAID (National Institute of Allergy and Infectious Diseases) Permission to use the burial/skeleton given by Hans Volpert, after (Wagner et al. 2014)

in the archaeological record; the trajectory of ancient pathogen genomics from the earliest polymerase chain reaction (PCR) studies to the current high-throughput sequencing strategies, alongside the theoretical and methodological considerations in integrating ancient DNA strategies to explore human disease in the past; as well as highlighting exciting future prospects in ancient pathogen genomics.

2 Disease-Associated Pathogens and the Archaeological Record

The archaeological record is dominated by chronic and long-term infections with latent or episodic activity (e.g., tuberculosis, leprosy) due to higher probabilities of developing skeletal lesions as the duration of pathogenic activity ranges from months to years (Roberts and Manchester 2005). On the other hand, acute infections of rapid mortality and/or spontaneous recovery (e.g., bubonic plague, malaria, smallpox) are rarely identified as no skeletal traces of disease remain (Ortner 2003; Roberts and Manchester 2005; Steinbock 1976). The identification of specific disease-associated processes in human skeletal remains is challenging, however, because a disease will not manifest the same way in all individuals, while the finite responses of bone to pathogenic stimuli (e.g., proliferation, destruction, or a combination) may result in similar "disease" expression (Roberts and Manchester 2005; Steinbock 1976). This is further complicated by the fact that potentially only 5–20% of individuals will show pathological changes from infectious diseases, with responses ranging from non-specific or even absent due to heterogeneous susceptibility (Ortner 2003; Roberts and Manchester 2005; Steinbock 1976; Wood et al. 1992).

There is potential to find ancient pathogen DNA in affected or unaffected skeletal samples and their associated contexts as the response to disease (as manifested skeletally) is often complex and individualistic (Baron et al. 1996), resulting in a dichotomous nature of disease interpretation, where the presence of pathological lesions may represent healthy individuals surviving the disease, rather than less healthy individuals (termed the Osteological Paradox) (DeWitte and Stojanowski 2015; Wood et al. 1992). However, an absence of lesions may also indicate elimination of the pathogen prior to skeletal involvement or, contrarily, those dying from acute disease or other causes, such as trauma and accidents (Milner and Boldsen 2017; Wood et al. 1992). These interconnected issues were highlighted in a study of seven human skeletal remains from Wharram Percy (England, tenth to sixteenth century CE) to investigate whether rib lesions (which occur in 1–8% of clinical scenarios) could be used in archaeological assemblages to not only identify tuberculosis ("the disease") but also the causative mycobacterial agent (Mays et al. 2002). However, there was no consistent association in recovering *Mycobacterium tuberculosis* DNA from ribs displaying clear tuberculosis lesions (Mays et al. 2002).

Although the methods used in this study are not applied today, the results suggested a complex interaction between host immunity, the route of tuberculosis infection (e.g., inhalation, lymphatic), and other respiratory diseases in a particular community (Mays et al. 2002).

Further, the individual response to infection is also framed by the susceptibility to disease (heterogeneity and frailty; Wood et al. 1992), as well as interaction with the causative microbe itself. This was demonstrated in the absence of *Yersinia pestis* molecular signatures in a subset of 61 putative plague victims from documented plague pits (thirteenth- to seventeenth-century CE France, England, and Denmark), suggesting differential pathogen dissemination and recoverability (e.g., *Y. pestis* caused the infection but did not colonize the dental pulp that was subsampled) (Gilbert et al. 2004). This finding contrasted with the previously successful identification of *Y. pestis* signatures in historical plague assemblages in southern France (Drancourt and Raoult 2002). As such, for blood-borne pathogens, varying levels of bacteremia or septicemia may confound the detection of a causative agent postmortem. Although a known pathogen may be associated with an assemblage, the level of infection around the time of death can impact the relative "abundance" of ancient pathogen DNA within the total host DNA, and thus detection may be very selective (i.e., down to the "luck" of specific sampling).

The presence and interaction of circulating pathogens with human hosts, in a given environment, create a dynamic pattern of disease. The complexity of the human response to these insults broadly impacts the successful retrieval of ancient DNA and in turn the identification of the "disease." This work can be informed by the availability of distinct evidentiary sources (i.e., osteological, archaeological, or historical) for a given locale, enabling the design of a molecular strategy for detecting a particular pathogen(s). For example, although rare, evidence correlating a specific pathogen to a given context may include diagnostic skeletal or tissue changes (e.g., kyphosis and vertebral collapse indicative of tuberculosis; Mays et al. 2001), catastrophic burials connected with the historical record where a large number of individuals die over a short time period due to epidemics or plagues (e.g., *Yersinia pestis* and the Black Death; Spyrou et al. 2016; Bos et al. 2011; Drancourt et al. 1998), or written historical sources documenting symptomology associated with a presumed mass mortality event, such as the Antonine Plague that swept through the Roman Empire (165–180 CE) (Fears 2004; Littman and Littman 1973). However, it is more commonly encountered that the presence and distribution of pathogens are not known for a particular (singular) locale in the past or do not provide a consensus about diseases present at the time, thereby requiring the prioritization of contextually relevant pathogens informed by a range of evidentiary sources (i.e., literary, archaeological, or paleoenvironmental). For example, the causative agents of the Antonine Plague are variably interpreted from the surviving literary evidence recorded by the Roman physician Galen (second century CE) as smallpox, measles, bubonic plague, yellow fever, typhus, typhoid fever, or meningitis (Cunha and Cunha 2008; Littman and Littman 1973), which on its own challenges the prioritization of a specific pathogen.

In cases without prior evidence of disease or a particular pathogen, investigating multiple putative pathogens all at once to identify the most likely candidate may more accurately represent the complexity of disease in archaeological assemblages. Because the disease process is often also multicausal (i.e., not a relationship of a singular pathogen to a singular disease) (Dutour 2013), emphasizing the contribution of coinfecting as well as co-circulating (within the environment) microbes to morbidity and mortality may provide a more complete understanding of disease history. For example, the frequency and distribution of each disease in a given environment depends on pathogen-specific factors (infectivity, virulence, mode of transmission), ecological factors (climate, urbanization), and pathogen-pathogen interactions (frequency and distribution of all other diseases) (Faure 2014; Grmek 1969, p. 1476). However, regardless of a single versus a multiple targeted approach used to identify ancient pathogens, they remain limited by current completeness of representative databases for the identification and confirmation of putative pathogenic agents.

3 Using Ancient DNA to Reveal the Imprints of Pathogens

Ancient DNA (aDNA), that is, the recovery of DNA from samples that are "old" (which is a relative term that is often used indiscriminately in the field), is typically characterized by short fragments (medians between 30 and 60 base pairs or bp), damaged toward their 5′ and 3′ ends due to the complex nature of hydrolysis (termed deamination and listed as C to T and G to A changes in sequence read data), and embedded in a dynamic molecular pool (e.g., environmental, microbial, or modern sources) (Fig. 2). Postmortem processes cause DNA damage through molecular and chemical degradation, while exogenous microbial constituents leaching from the burial environment broadly contribute to highly variable preservation (even within the same individual) that is not correlated with time (Allentoft et al. 2012; Kistler et al. 2017, Molak and Ho 2011). Ultimately, the endogenous DNA fraction typically ranges from less than 1–5%, with the pathogen fraction far less than 1% of the overall DNA constituents (Burbano et al. 2010; Carpenter et al. 2013; Devault et al. 2014b).

Despite these obstacles in working with pathogen aDNA, early studies demonstrated the capability of recovering pathogen DNA in the absence of skeletal lesions, such as tuberculosis (Baron et al. 1996; Zink et al. 2001), greatly leveraging the power of molecular strategies to identify pathogens in the past. Although current techniques have dramatically shifted the capability to access the minute pathogen fraction sequestered in ancient human skeletal remains, the pre-genomic area was a boon for providing the foundation of pathogen genomic research.

While detecting ancient pathogens using PCR-based approaches (i.e., >80 bp) was limited, mostly by the median fragment sizes of ancient DNA (~30–40 bp), there were a few key advantages. PCR is highly sensitive, and assays could be focused on repetitive or multi-copy genes/regions (i.e., *IS6110* for *Mycobacterium*

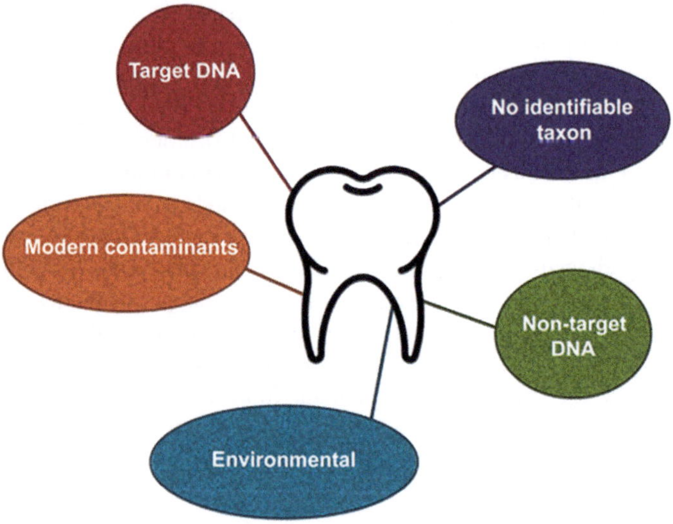

Fig. 2 Ancient DNA is often a complex mixture of DNA from different sources, ranging from the endogenous DNA from the specimen itself to modern and ancient DNA contaminants. There are also constituents that are not taxonomically identifiable, as current genomic databases are not representative of the totality of microbes. (tooth DNA components) – Tooth by Creaticca Creative Agency from the Noun Project

tuberculosis) enabling identification of pathogen signatures from *Mycobacterium tuberculosis* (Salo et al. 1994), *Mycobacterium leprae* (Haas et al. 2000), *Treponema pallidum pallidum* (Kolman et al. 1999), and *Yersinia pestis* (Drancourt et al. 1998). Unfortunately, given the ubiquity of bacteria, confirming the "uniqueness" of a specific pathogen signature requires obtaining sufficient spans of authentic genomic DNA, and this is problematic as PCR amplification is biased toward longer DNA fragments, which are more commonly associated with modern environmental and/or laboratory microbes which have similar genomic content (Knapp and Hofreiter 2010).

One of the key criteria in working with ancient pathogen DNA is to clearly differentiate aDNA from modern contaminants (e.g., miscoding lesions due to postmortem damage or sequence error) and false positives (e.g., a priori targeting genetically diverse regions) (Harkins et al. 2015), which is challenging with PCR-based strategies. For example, the primers used to capture 18S rRNA for human *Plasmodium* spp. in fifth-century CE infants from Lugnano (Italy) can also amplify non-specific DNA from a variety of eukaryotic organisms (mostly fungi), which requires sequencing to differentiate positive signals from microbial "noise" (Sallares and Gomzi 2001). Although they may be specific to a given pathogen and informed by modern genomic studies, the use of repetitive sequences (e.g., *IS6110* in Thierry et al. (1990), *IS1081* in van Soolingen et al. (1992)) present in several copies per genome remains a challenge. First, the DNA of interest is embedded in a complex molecular pool of fragmented DNA from exogenous contaminants and endogenous non-pathogenic/nontarget microorganisms that likely overwhelm the endogenous DNA in this mixture. For example, mobile insertion elements associated with *Mycobacterium tuberculosis* [e.g., *IS6110* 1,316 bp and *IS1081* 1,324 bp (Thierry et al. 1990; van Soolingen et al. 1992)] and dispersed repetitive elements associated with *Mycobacterium leprae* [*RLEP* 689 bp (Woods and Cole 1990)] are present in 6–28 copies per genome (Thierry et al. 1990; van Soolingen et al. 1992; Woods and Cole 1990). Although such repetitive elements have been applied with success in the identification of these two pathogens as part of multidisciplinary molecular and paleopathological approaches (Mays et al. 2001; Taylor et al. 2013), the specificity of the molecular identification is complicated due to the similarity of these mobile elements in other mycobacterial agents (e.g., *M. bovis*, *M. microti*, *M. africanum*) or extraneous soil microorganisms. Probe-based quantitative PCR methods targeting multiple loci in the *Mycobacterium tuberculosis* complex have been used to prioritize archaeological skeletal remains putatively containing mycobacterial DNA (i.e., insertion elements and the *rpoB* gene) (Harkins et al. 2015). This work demonstrates a rigorous detection strategy that emphasizes assessing sample inhibition, employing specific and sensitive primer/probe sets, and using multiple confirmatory replicates to ameliorate some of the obstacles associated with detecting pathogen DNA via quantitative PCR-based techniques (Harkins et al. 2015).

The shift to high-throughput sequencing (HTS) facilitated greater access to the endogenous fraction of aDNA by the massive parallelization of sequencing (millions of molecules at once) and by allowing retrieval of short DNA targets, representing the bulk of the endogenous DNA fraction in fossil remains (Knapp and Hofreiter 2010; Millar et al. 2008). The capability of HTS to sequence short reads associated with aDNA is complemented by laboratory-based innovations in the extraction and recovery of extremely short DNA (>15 bp) molecules alongside maximizing the conversion of those fragments into DNA libraries (i.e., repairing nicks or damaged ends of molecules) (Dabney et al. 2013; Gansauge and Meyer 2013; Meyer and Kircher 2010), as well as authenticating them based on their characteristic damage patterns (Briggs et al. 2007; Brotherton et al. 2007; Jonsson et al. 2013; Kistler et al. 2015) (Fig. 3). However, in order to make use of such innovations for ancient pathogen genomics (i.e., maximizing the yield of endogenous pathogen DNA), selecting the most "representative" sample from an individual specimen remains an important consideration, with a focus on substrates such as teeth, which are vascularized and relatively resistant to postmortem degradation (e.g., subsampling the dental pulp, calculus, or roots) (Adler et al. 2011; Drancourt et al. 1998; Warinner et al. 2014). DNA extraction strategies have relied heavily on phenol-chloroform (Hagelberg and Clegg 1991) in the past but are now based mostly on chaotropic salts and silica (Dabney et al. 2013; Höss and Pääbo 1993) to solubilize and digest bound DNA molecules. These new methods, like the old ones, remain

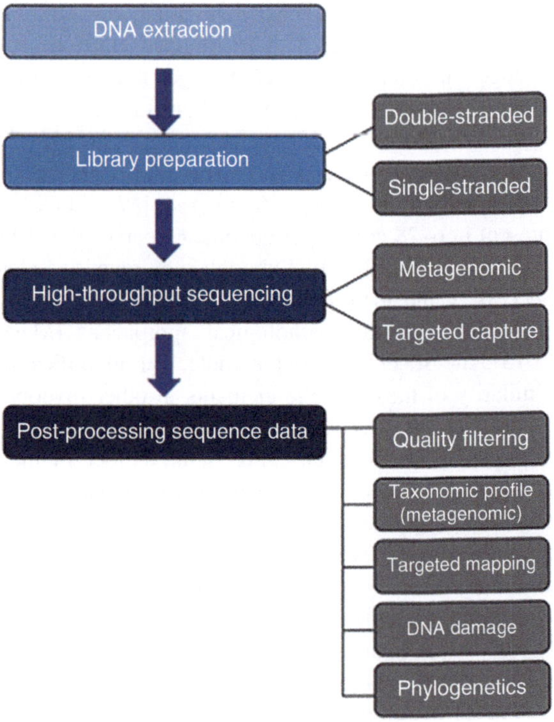

Fig. 3 Representation of a typical experimental workflow to extract and analyze DNA from archaeological human specimens

inherently lossful, and the postmortem processes that degrade DNA may facilitate its sequestration in physiological spaces that complicate its extractability (Campos et al. 2012; Geigl 2002). To render DNA extracts sequenceable further requires conversion into platform-specific "libraries" (e.g., repairing the ends of the DNA, attaching universal adaptors and unique barcodes) that are prepared from either double- or single-stranded DNA molecules (Gansauge et al. 2017; Meyer and Kircher 2010). While heavily damaged DNA fragments may not be incorporated into the double-stranded DNA libraries (e.g., nicks, abasic sites, miscoding lesions), the "selection" of "undamaged" molecules may ease their bioinformatic identification (Barlow et al. 2016), while single-stranded DNA libraries may incorporate more damaged molecules typical of aDNA samples (e.g., less than 30 bp) (Gansauge et al. 2017; Wales et al. 2015; Ávila-Arcos et al. 2015). Despite the uncertainty of aDNA recovery from sample selection to constructing sequenceable aDNA libraries, the scalability of sequencing strategies to specific research questions is one of the most exciting applications in the high-throughput sequencing era.

The advantages of pathogen research in the high-throughput sequencing era is particularly notable with metagenomic (shotgun) sequencing as a strategy to profile a more representative portion of the entirety of the molecular constituents of an extract, without biasing the experiment by selecting a target (e.g., specific genes, genomes, or loci), which is beneficial in contexts where there is equivocal evidence for the historical presence (or absence) of particular diseases. However, endogenous pathogen molecules are generally taxonomic constituents in low abundance and are often undetectable (e.g., less than 0.005% of human *Plasmodium* species reads in Marciniak et al. 2016), so it is hardly cost-effective to sequence such metagenomic samples to great depths (beyond 30× coverage) in order to access the pathogen fraction in a diagnostic or identifiable way (Devault et al. 2014b; Whatmore 2014). Yet, exceptional specimens, such as calcified nodules, have shown the success of this approach, including 6.5-fold coverage (0.48% of shotgun reads) of a *Brucella melitensis* genome from an adult male (fourteenth-century CE, Sardinia, Italy) showing the antiquity of this zoonotic infection in a historical context associated with animal husbandry, such as sheep and goats (Kay et al. 2014). Similarly, *Staphylococcus saprophyticus* recovered from an adult female from thirteenth-century CE Troy (33–66% sequences identified via shotgun) shows genetic similarity to strains from livestock, suggesting a different reservoir for this bacterial infection than the present-day (Devault et al. 2017).

As a workaround for accessing the pathogen constituents in these substantial metagenomic datasets, bioinformatics tools provide an additional strategy to screen for specific pathogens in a manner that is the least biased of the PCR and capture-based approaches. Various metagenomic pipelines or classifiers are key parts of authenticating ancient microbial DNA. These include alignment-based methods, such as MALT (Herbig et al. 2016) and BLAST (Altschul et al. 1990), as well as alignment-free methods such as MEGAN (Huson et al. 2007), KRAKEN (Wood and Salzberg 2014), and MetaPHlan (Segata et al. 2012) (for an overview of their applicability to ancient DNA, see Warinner et al. 2017). For example, the MEGAN ALignment Tool (MALT) creates metagenomic profiles by specifically assigning reads to bacterial taxa in a customizable database (Herbig et al. 2016) and was recently applied to identify *Salmonella enterica* (365–659 reads taxonomically assigned in three samples) from a sixteenth-century CE epidemic burial in Oaxaca (Mexico), where subsequent downstream capture enabled 3–96X average coverage of *S. enterica* genomes (Vågene et al. 2018).

A successful yet biased alternative approach to recovering pathogen aDNA draws on the capture or enrichment of core (or target) genomic regions, with the objective of increasing the proportion of the target relative to the nontarget background (Gnirke et al. 2009). The approach uses probes (sometimes referred to as "baits") that are designed to hybridize to specific genomic targets that are selected a priori, using in-solution probes or solid platforms (microarrays) (Burbano et al. 2010; Carpenter et al. 2013; Enk et al. 2014). Commercially synthesized custom probe sets are scalable (hundreds of kilobases to megabases), flexible, and cost-effective; however, custom microarrays can obtain higher probe density (hundreds of thousands to millions of probes) to enable the capture of larger target regions (Orlando et al. 2015). However, there is currently little data testing differences in sensitivity and selectivity between the two methods. Within ancient pathogen genomics, microarrays are increasingly applicable in identifying metagenomic constituents of archaeological samples (Devault et al. 2014b), while parallel pathogen detection provides a strategy to explore co-circulating and coinfecting agents in a single individual (Bos et al. 2014). Capturing as many unique target molecules (sensitivity) at the cost of reduced complexity (specificity) is one strategy to maximize the informational content of aDNA, and technical parameters (e.g., hybridization temperature/time, bait design) are further scalable to achieve this (Ávila-Arcos et al. 2015; Cruz-Dávalos et al. 2016). To reduce the loss of target, hybridization temperature and time are suggested as 50–60°C for 16–48 h with bait tiling density as 2x to facilitate the capture of degraded DNA that is diverse ("unique"), but this is at a cost of increasing the non-specificity of the reads captured (Cruz-Dávalos et al. 2016; Enk 2015). Typically target aDNA molecules are much shorter than the baits (e.g., 80 bp) and occur at a lower frequency, which requires more moderate hybridization temperatures to ensure the bound aDNA is not melted away from the bait alongside longer incubation times, which try to ensure the aDNA molecules encounter in the baits (Enk 2015). A drawback of capture strategies is that only the specific target is recovered (although there is a tolerance ~15–20% for divergence from the bait sequence identity), novel or divergent regions will not be identified, and no capture strategy recovers all of the target genomic regions due to the kinetics of the process itself (e.g., targets encountering the baits, removing non-specifically bound molecules). In some cases, ancestrally designed genome baits can be useful for more distantly related taxa or ones that are perceived to have diverged over longer time periods or evolve quickly (viral pathogens) (Delsuc et al. 2016).

The improvements in not just identification of pathogens but also the extent of genomic information accessible enabled a shift in the historical contextualization of pathogens from simply the presence or absence (i.e., partial or full genomes/genes) to increasingly hypothesis-driven research situated within evolutionary frameworks, due to the expanding range of diverse pathogens that have been recovered thus far (e.g., DNA viruses including human parvovirus in Mühlemann et al. (2018a), hepatitis B virus in Krause-Kyora et al. (2018a), and low-pathogen load bacterial infections, such as *Treponema pallidum pallidum* in Schuenemann et al. (2018b)) (Fig. 4). Reconstructing *Yersinia pestis* strains from the Black Death in southern Europe revealed genetic similarity to the initial wave from 1348 in London, suggesting long term persistence of plague in the Mediterranean littoral and a possible historical demographic scenario suggesting diffusion from the Mediterranean into Northern Europe moving eastward into Russia and China (Bos et al. 2016; Spyrou et al. 2016). Similarly, the resolution of ancient pathogen genomics has revealed that the seventeenth-century CE Variola virus genome (18X coverage) recovered from a Lithuanian child mummy pushed back the timing of pre-vaccination Variola genomes as 200 years prior to eradication, which provides an opportunity for further study on the antiquity of this virus in earlier time periods (e.g., fourth-century China) and the accumulation of genetic changes over the course of its interactions with human demographic history (Duggan et al. 2016).

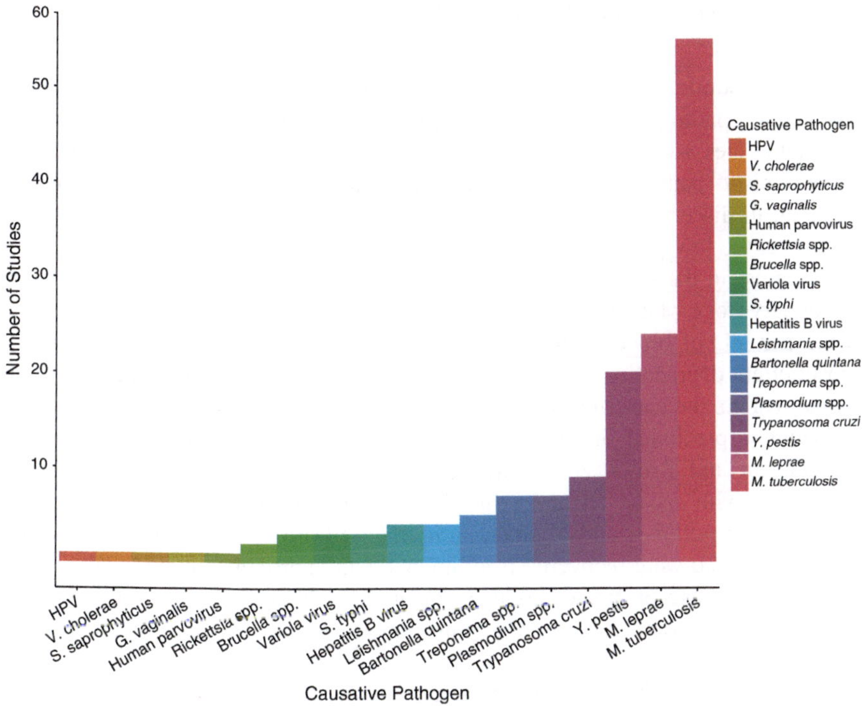

Fig. 4 A representation of the diverse pathogens investigated from ancient DNA studies (e.g., PCR, metagenomic, hybridization capture). The *y*-axis represents only the number of studies that investigated a pathogen, not the number of genomes/genomic data recovered. The plot was generated with the R ggplot2 package (Wickham 2016)

4 Theoretical and Methodological Considerations in Applying Paleogenomic Approaches to Study Pathogens in the Past

Although dramatic improvements are evident in the field of ancient pathogen genomics, such as an expanding diversity of causative microbes recovered from archaeological contexts (e.g., bacteria, parasites, and DNA and RNA viruses) and enhanced temporal resolution of pathogen evolutionary histories (e.g., hepatitis B virus in Patterson Ross et al. 2018; Mühlemann et al. 2018b; Krause-Kyora et al. 2018a), crucial considerations remain regarding differential pathogen postmortem preservation and heterogeneity in the human response to microbes when integrating ancient DNA methods to investigate the presence and/or evolutionary history of a particular pathogen(s).

4.1 Differential Pathogen Recoverability

Beyond postmortem degradation, issues of pathogenesis may confound detection of the minute pathogen fraction, and not all pathogens are equally retrievable from ancient tissues, such as infections confined to soft tissues (e.g., enteric pathogens) or opportunistic pathogens (e.g., commensals that can become pathogenic and are difficult to discern as they stem from an individual's microbial composition). For example, the low-pathogen load of *Treponema pallidum pallidum* (causative agent of syphilis) in the tertiary stage of the disease (bone involvement) contrasts to the secondary stage when bacteremia is at its peak (e.g., up to ~1,900 copies of the *polA* gene per mL of blood; Cruz et al. 2010) as well as its biological sensitivity to postmortem degradation (e.g., the absence of a lipopolysaccharide outer cell membrane renders the bacterium vulnerable to destruction from exogenous sources) has previously precluded its replicable identification in ancient human remains (Bouwman and Brown 2005; Kolman et al. 1999; von Hunnius et al. 2007). However, with recent advances in HTS technology, *Treponema* spp. genomes (*T. pallidum pallidum* and *T. pallidum pertenue*) from seventeenth- to nineteenth-century Mexico were recovered from infants with approximately 57–93% of reads with 3X coverage or greater (Schuenemann et al. 2018b). With further work, the evolutionary history and origin of this important human pathogen are open to exploration.

Most enteric pathogens (e.g., bacteria, viruses, parasites) impact the intestinal system (i.e., not blood-borne) causing varied clinical manifestations (e.g., inflammation, vomiting, fever, diarrhea) and are restricted to soft tissues without hematogenous spread (Kolling et al. 2012) complicating recovery from skeletal samples and thus may benefit from using preserved soft tissues to maximize detection. However, finding the most likely tissue with highest-pathogen load can greatly facilitate the contextualization of genomic information. For example, a reconstructed nineteenth-century *Vibrio cholerae* genome (15X coverage) from a preserved intestinal subsample from an archived medical collection was identified as causing the second cholera pandemic and, although similar to the modern biotype, exhibited potentially greater pathogenicity due to the presence of CTX virions (Devault et al. 2014a). Similarly, the independent recovery of another enteric pathogen, *Salmonella enterica* serovar Paratyphi C strain (which is the causative agent of bacterial enteric fever), using teeth from archaeological samples from sixteenth-century CE Mexico (Vågene et al. 2018) and 1200 ± 50 CE Norway (Zhou et al. 2017) suggests the increasing potential to not only integrate ethnohistorical context in the investigation of diseases during times of demographic, cultural, and social change but also the applicability of these ancient genomes to contribute to timing the evolution of the human-specific pathogenic strains.

Opportunistic "pathogens" or bacteria that normally inhabit a variety of niches in the human body (e.g., the gastrointestinal tract, oral cavity, or skin) without necessarily causing visible disease unless there are physiological perturbations (stress) are also important facets of exploring human health in the past. One such example is dental calculus (calcified dental plaque) from archaeological humans, where recent work has shown the shifting composition of the oral microbiota during the transition from hunting and gathering to farming (~10,000 years ago) (Adler et al. 2013), as well as the antiquity of putative antibiotic resistance genes associated with the oral microbiota in archaeological human remains (Germany, 950–1200 CE) (Warinner et al. 2014). Similarly, abscesses or stones (kidney, urinary, placental, etc.) also provide exceptional records of bacterial evolution. Recently, the recovery of *Staphylococcus saprophyticus* and *Gardnerella vaginalis* (~300X and ~60X genomic coverage) from calcified abscesses (2–3 cm in diameter), found among the skeletal remains of a female from Byzantine Troy (790–860 years BP), provides the first evidence for maternal sepsis in the archaeological record. Insight gleaned from this pregnancy-related infection suggests pathological outcomes are dynamically shaped by host interactions with not only their own microbiota but also environmental factors (e.g., ecological, zoonotic, social) that interact in the pathway of disease causation and susceptibility (Devault et al. 2017).

As the archaeological record is variably constituted by mummified tissue and skeletal remains, the characterization of pathogens draws on optimizing the choice of individual sample to maximize pathogen detection and recovery informed by a microbe's pathogenesis. Hematogenous infections, such as *Yersinia pestis* or *Plasmodium falciparum*, or those that are relatively protected from destruction in host cells, such as bacilli associated with *M. tuberculosis* and *M. leprae*, may be detectable in teeth, which facilitate (but do not guarantee) the sequestration and preservation of ancient molecules (Adler et al. 2011; Drancourt et al. 1998). Infections confined to soft tissue without hematogenous spread, such as the Variola virus or *Vibrio cholerae*, have greater chances of recovery in mummified or preserved tissue, rather than skeletal remains as shown in recent work (Biagini et al. 2012; Devault et al. 2014a; Duggan et al. 2016). Although there is no consistent correlation between a pathological skeletal lesion and the successful recovery of ancient DNA, it may also be relevant to prioritize additional subsampling regions depending on the disease-associated pathogen under investigation and the relative "completeness" of the skeletal remains, such as vertebral areas, abscesses or ribs for tuberculosis (Faerman et al. 1997; Mays et al. 2001), rhinomaxillary area and bones of the hands or feet for leprosy (Montiel et al. 2003), and ribs or vertebrae for brucellosis (Mutolo et al. 2012).

In sum, the recovery of surviving ancient pathogen DNA depends not only on the implementation of a rigorous protocol, various suggested criteria for authentication, and specific genomic targets but also understanding disease pathogenesis (i.e., localization or dissemination of microbes throughout infection, immunity responses) to help select the most representative sample (e.g., soft tissue or teeth) for optimal pathogen detection. Thus, it is crucial to tailor the recovery of these pathogenic remnants from a complex microbial background, whether these are the molecular components of the pathogen itself (e.g., whole genome, plasmids, proteins) as previously mentioned or the by-products of a pathogen infection (e.g., mycolic acids, antigens). The immunological detection of antigens and/or the identification of organic biomolecules (i.e., lipids) may confirm exposure to infection, but not necessarily pathogen presence itself (Tran et al. 2011). *Mycobacteria*-specific mycolic and mycocerosic acid lipid biomarkers differentiate infected from non-infected archaeological samples with or without skeletal lesions (Redman et al. 2009), as well as distinguishing among leprosy and tuberculosis (Minnikin et al. 2011). Similarly, *P. falciparum* antigens, such as histidine-rich protein-2-antigen (PfHRP-2), have been positively detected in mummies (Bianucci et al. 2008; Miller et al. 1994; Rabino Massa et al. 2000). However, whether detecting antigens or lipid biomarkers, there are challenges due to potential cross-reactivity between molecules, the effects of postmortem degradation, and the sensitivity of detection (Tran et al. 2011), which require consideration in demonstrating the authenticity of results.

4.2 Heterogeneity Underlying Pathogen Detection

The heterogeneity of the human response to microbes (susceptibility/resistance) in conjunction with behavioral, social, ecological, and other biological factors contribute to the local "microbial reservoir" (i.e., the particular pathogens and non-pathogens that co-circulate at a given time in a particular location) alongside varied patterns of morbidity and mortality (Fig. 5). This aspect of heterogeneity is unknown in the archaeological record, as assemblages remain a select sample of individuals (i.e., age-related survival, incomplete recovery of remains, differential preservation) (Milner and Boldsen 2017; Saunders et al. 1995). As such, there are two significant themes that underscore working with ancient pathogens, first that absence of evidence is not evidence of absence, and second, pathogen signatures are not equivalent to cause of death as "live" microbes cannot be differentiated from

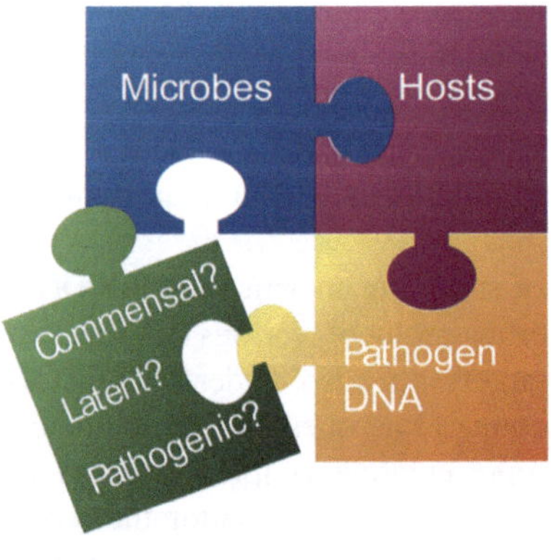

Fig. 5 Ancient DNA provides molecular information about the pathogen itself, which can be framed in relation to an individual from a given locale, such as highlighting causal pathways that may have impacted disease (e.g., ecological, social, biological). But the outcome of these interactions (e.g., latent, infectious, commensal) is beyond the scope of inferences about relative "health" in the past. (puzzle) – CC0 1.0 Universal (CC0 1.0) Public Domain Dedication

"dead" ones (i.e., whether an infection has been cleared or not). For example, a pandemic such as Black Death that is associated with mass burials/plague pits putatively suggests an agent (*Y. pestis*) contributed to widespread mortality; however, non-catastrophic contexts, which are more frequently encountered, may encapsulate a range of "day-to-day" microbes/infections from bacterial (e.g., *M. tuberculosis*, *M. leprae*) to parasitic (e.g., malaria or hookworm), which preclude correlations of a single pathogen signature to a singular cause of death.

Accordingly, this heterogeneity in susceptibility and the complexity of the disease experience in past human communities are not as direct as one disease being caused by one pathogen (Dutour 2013). There are interactive causalities that embed these pathogen-pathogen interactions and impact the frequency and distribution of other diseases in a given context, as one disease may hinder the effects of another (antagonistic), facilitate the invasion and "contagiousness" of another (synergistic), or do not influence one another (Gonzalez et al. 2010; Singer 2010). For example, gastrointestinal diseases caused by *Vibrio cholerae*, *Shigella* spp., and *Salmonella* spp. may be worsened by a malaria infection (*P. vivax* or *P. falciparum*) (Cunnington 2012), potentially due to the similar ecological conditions that can sustain these enteric pathogens and malaria simultaneously, such as malaria-endemic areas in France that had heightened morbidity and mortality during cholera outbreaks (Dubreuil and Rech 1836; Faure 2014). Similarly, individuals with existing tuberculosis infections may face exacerbated consequences if also infected with malaria, as perturbations to the innate immune system response from the *Plasmodium* parasite leads to excessive pro-inflammatory responses in the lungs and over-activation of immune cells (e.g., macrophages), increasing the mycobacterial load (Mueller et al. 2012). For example, PCR-based analyses putatively identified *M. tuberculosis* (*IS6110*) and *P. falciparum* (apical membrane antigen, merozoite surface protein) coinfections in Egyptian mummies ($n = 4$, 806 BC to 124 AD) that was attributed to crop cultivation on marshy land alongside increased population density that potentially proliferated both of these diseases (Lalremruata et al. 2013). These complex interactions create dynamic "pathogen pools" that are entangled from an investigatory perspective where the co-circulation and coexistence of chronic infections, acute diseases, and opportunistic infections are intertwined not only with one another but within dynamic biosocial contexts (e.g., disease ecology, human-environment interactions).

There is a further related consideration that by using the term "pathogen," it denotes a causative agent of disease with an inferred consistency of symptomology and infective properties (e.g., pathogenicity, virulence), but pathogens are microbes with the capacity to "harm" a host, and it is the outcome of this interaction that is expressed in a susceptible host as disease, latency, or commensalism (Casadevall and Pirofski 2002). Drawing into this interaction between the host and microbe has led to further methodological avenues to explore disease in the past by harnessing proteomic or protein-based analyses as well as the landscape of human immunity.

5 Beyond Ancient DNA: Leveraging Other (Non-DNA) Biomolecules to Explore Disease in the Past

Ancient proteomics is gaining ground as a strategy to explore disease by identifying pathogen-expressed proteins using high-resolution mass spectrometry (Warinner et al. 2014; Corthals et al. 2012). The protein fraction is expected to preserve "better" than DNA, which degrades more rapidly while also providing a potentially much greater time depth (Cappellini et al. 2014). For example, metagenomic and metaproteomic analyses of dental calculus from a medieval specimen (Germany, 950–1200 CE) indicated antigenic gingipain proteins associated with *Porphyromonas gingivalis* (Warinner et al. 2014). Similarly, mouth swabs from a 500-year-old Inca mummified individual revealed proteins associated with severe airway inflammation in response to a bacterial infection (e.g., apolipoprotein) (Corthals et al. 2012). Recently, peptides identified by mass spectrometry, belonging to *Y. pestis*, were isolated from Medieval death registries (sixteenth century CE, Milano) (D'Amato et al. 2018). However, mass spectrometry detection of peptide fragments is less sensitive than high-throughput sequencing techniques because the expression level of the proteins depends on the stage of disease, and it is also crucial to modify proteomics methods for use on ancient samples that are vastly different from modern samples (e.g., in terms of degradation and damage). Using a proteomic approach on mummified lung tissue from Hungarian individuals (eighteenth to nineteenth century) positive and negative for *M. tuberculosis* aDNA did not yield proteins unique to the pathogen, emphasizing the importance of considering modern contamination and degradation in such analyses (Hendy et al. 2016). Although obstacles remain to be addressed, a proteomics strategy may be applied to RNA viruses or other non-easily attainable pathogen remains, such as *Plasmodium*, where the stage-specific protein expression is well-characterized (Florens et al. 2002).

An area of further interest is to understand past pathogens to inform the present, since our pathogens also represent the history of our exposure and interactions with them. Aspects of the innate immune system (the first line of defense against microbes) and adaptive immunity (immediate response to pathogens that are not heritable) are important in driving the evolution of human resistance or susceptibility to pathogens (Barreiro and Quintana-Murci 2010). The evolutionary history of our adaptations to pathogens has left its mark in our DNA, the most remarkable and currently well-defined being the immune responses to malaria infection (e.g., thalassemia, glucose-6-phosphate dehydrogenase deficiency) (Kwiatkowski 2005). Recent paleogenomic work suggests malaria may not have been as strong a selective force due to its emergence as a recent severe human pathogen, as the allele frequencies of malaria resistance variants in a cross-section of 224 individuals from the Upper Paleolithic to post-Roman period were similar to present-day populations (Gelabert et al. 2017). However, a greater representation of ancient human genomes from Italy and across a wider temporal transect would be beneficial to further explore the spectrum of changes in the frequency of these variants. Although the dynamics of the host-pathogen interaction are also responsive to the local context (e.g., climate, physical environment, demography, social factors), it is increasingly possible to explore the timeframe of human interaction with pathogens over the long term through the innate immune system or short-term adaptive immunity (Barnes et al. 2011). For example, susceptibility to lepromatous leprosy is mediated by human leukocyte antigen allele DRB1*15:01 in modern populations, and medieval skeletons from a leprosarium in Denmark (1270–1550 AD) that tested positive for the *M. leprae* bacterium also showed a significant association of this HLA allele, which may have impaired the immune response to *M. leprae*, leading to increased disease susceptibility (Krause-Kyora et al. 2018b).

6 Conclusions and Future Perspectives

Inferring how a specific "pathogen" impacted a particular individual(s) or the epidemiological landscape remains a challenging goal for ancient pathogen research. By integrating a multidisciplinary approach that is truly collaborative and exemplifies the depth of knowledge that can be interleaved by other experts (archaeologists, classicists, historians, epidemiologists, bioinformaticians) can we arrive at a truly novel syntheses of the dynamic "disease" past. We are steadily moving beyond solely using ancient DNA as the endpoint of showing a particular pathogen was present at a given point in time, toward emphasizing the contextualization of such molecular signatures and expanding the range of pathogens that can be detected, as recent work has shown that it is increasingly possible to gain insight into not only pathogens that are "invisible" in the archaeological record (i.e., DNA viruses) but also a pathogen's evolutionary history (e.g., virulence, genetic diversity), as well as the adaptive landscape of human responses to disease.

Although the early years of aDNA pathogen work focused on particular "culprits," such as leprosy, tuberculosis, and plague, we are seeing increased representation of less canonical pathogens, not easily identified in the archaeological and skeletal record (e.g., smallpox, malaria, salmonella, and *Staphylococcus* spp.), which has significant implications in framing the potential co-circulating pathogens that impacted everyday life in the past as well as the unique biosocial context of disease experienced by those individuals' suffering. Methodological and technological innovations in the recovery, identification, authentication, and phylogenetic assessment of ancient pathogens have propelled the field forward, such that pathogens that were previously "unrecoverable" using traditional PCR-based strategies can now illuminate our understanding of pathogen diversity and evolution across diverse spatiotemporal contexts. However, underlying the exciting shifts in ancient pathogen DNA work, there remain challenges, largely the varied heterogeneity and susceptibility of individuals to disease that complicates inferences of the biological consequences of a pathogen or suite of pathogens may have had on an individual's relative "health." Also, the absence of recovering a pathogen signature does not equate to it being absent historically, which necessarily tempers interpretations on the geographical distribution or prevalence of a disease in the past. Despite the challenges, it remains crucial to have collaborations among historians, epidemiologists, archaeologists, bioarchaeologists, and geneticists (to name a few), such that inferences of disease and causative agents are comprehensively framed within the unique context that humans and their pathogens coexist and survive.

References

Adler CJ, Haak W, Donlon D, Cooper A. Survival and recovery of DNA from ancient teeth and bones. J Archaeol Sci. 2011;38(5):956–64. https://doi.org/10.1016/j.jas.2010.11.010.

Adler CJ, Dobney K, Weyrich LS, Kaidonis J, Walker AW, Haak W, et al. Sequencing ancient calcified dental plaque shows changes in oral microbiota with dietary shifts of the Neolithic and Industrial revolutions. Nat Genet. 2013;45(4):450–5. https://doi.org/10.1038/ng.2536.

Allentoft ME, Collins M, Harker D, Haile J, Oskam CL, Hale ML, et al. The half-life of DNA in bone: measuring decay kinetics in 158 dated fossils. Proc R Soc Lond B Biol Sci. 2012;279(1748):4724–33. http://rspb.royalsocietypublishing.org/content/early/2012/10/05/rspb.2012.1745?sid%3Dabb89d94-00f1-431b-8863-c62996e35478=.

Altschul SF, Gish W, Miller W, Myers EW, Lipman DJ. Basic local alignment search tool. J Mol Biol. 1990;215(3):403–10. https://doi.org/10.1016/S0022-2836(05)80360-2.

Ávila-Arcos MC, Sandoval-Velasco M, Schroeder H, Carpenter ML, Malaspinas A-S, Wales N, et al. Comparative performance of two whole-genome capture methodologies on ancient DNA Illumina libraries. Methods Ecol Evol. 2015;6(6):725–34. https://doi.org/10.1111/2041-210X.12353.

Barlow A, Gonzalez Fortes GM, Dalen L, Pinhasi R, Gasparyan B, Rabeder G, et al. Massive influence of DNA isolation and library preparation approaches on palaeogenomic sequencing data. bioRxiv. 2016; https://doi.org/10.1101/075911. http://biorxiv.org/content/early/2016/09/19/075911.abstract.

Barnes I, Duda A, Pybus OG, Thomas MG. Ancient urbanization predicts genetic resistance to tuberculosis. Evolution. 2011;65(3):842–8. https://doi.org/10.1111/j.1558-5646.2010.01132.x.

Baron H, Hummel S, Herrmann B. *Mycobacterium tuberculosis* complex DNA in ancient human bones. J Archaeol Sci. 1996;23(5):667–71. https://doi.org/10.1006/jasc.1996.0063.

Barreiro LB, Quintana-Murci L. From evolutionary genetics to human immunology: how selection shapes host defence genes. Nat Rev Genet. 2010;11(1):17–30. https://doi.org/10.1038/nrg2698.

Benedictow OJ. The Black Death, 1346-1353: the complete history. Woodbridge: Boydell Press; 2004.

Biagini P, Thèves C, Balaresque P, Géraut A, Cannet C, Keyser C, et al. Variola virus in a 300-year-old Siberian mummy. N Engl J Med. 2012;367(21):2057–9. https://doi.org/10.1056/NEJMc1208124.

Bianucci R, Mattutino G, Lallo R, Charlier P, Jouin-Spriet H, Peluso A, et al. Immunological evidence of *Plasmodium falciparum* infection in an Egyptian child mummy from the Early Dynastic Period. J Archaeol Sci. 2008;35(7):1880–5. https://doi.org/10.1016/j.jas.2007.11.019.

Bos KI, Schuenemann VJ, Golding GB, Burbano HA, Waglechner N, Coombes BK, et al. A draft genome of *Yersinia pestis* from victims of the Black Death. Nature. 2011;478(7370):506–10. https://doi.org/10.1038/nature10549.

Bos KI, Jäger G, Schuenemann VJ, Vågene ÅJ, Spyrou MA, Herbig A, et al. Parallel detection of ancient pathogens via array-based DNA capture. Philos Trans R Soc Lond B Biol Sci. 2014;370(1660). http://rstb.royalsocietypublishing.org/content/370/1660/20130375.full.

Bos KI, Herbig A, Sahl J, Waglechner N, Fourment M, Forrest SA, et al. Eighteenth century *Yersinia pestis* genomes reveal the long-term persistence of an historical plague focus. elife. 2016;5:e12994. https://doi.org/10.7554/eLife.12994.

Bouwman AS, Brown TA. The limits of biomolecular palaeopathology: ancient DNA cannot be used to study venereal syphilis. J Archaeol Sci. 2005;32(5):703–13. https://doi.org/10.1016/j.jas.2004.11.014.

Briggs AW, Stenzel U, Johnson PLF, Green RE, Kelso J, Prüfer K, et al. Patterns of damage in genomic DNA sequences from a Neandertal. Proc Natl Acad Sci U S A. 2007;104(37):14616–21. https://doi.org/10.1073/pnas.0704665104.

Brotherton P, Endicott P, Sanchez JJ, Beaumont M, Barnett R, Austin J, Cooper A. Novel high-resolution characterization of ancient DNA reveals C > U-type base modification events as the sole cause of post mortem miscoding lesions. Nucleic Acids Res. 2007;35(17):5717–28. https://doi.org/10.1093/nar/gkm588.

Burbano HA, Hodges E, Green RE, Briggs AW, Krause J, Meyer M, et al. Targeted investigation of the Neandertal genome by array-based sequence capture. Science. 2010;328(5979):723–5. http://science.sciencemag.org/content/328/5979/723.

Campos PF, Craig OE, Turner-Walker G, Peacock E, Willerslev E, Gilbert MTP. DNA in ancient bone – where is it located and how should we extract it? Ann Anat. 2012;194(1):7–16. https://doi.org/10.1016/j.aanat.2011.07.003.

Cappellini E, Collins MJ, Gilbert MTP. Unlocking ancient protein palimpsests. Science. 2014;343(6177):1320–2. http://science.sciencemag.org/content/343/6177/1320.

Carpenter ML, Buenrostro JD, Valdiosera C, Schroeder H, Allentoft ME, Sikora M, et al. Pulling out the 1%: whole-genome capture for the targeted enrichment of ancient DNA sequencing libraries. Am J Hum Genet. 2013;93(5):852–64. https://doi.org/10.1016/j.ajhg.2013.10.002.

Casadevall A, Pirofski L-A. What is a pathogen? Ann Med. 2002;34(1):2–4. https://doi.org/10.1080/078538902317338580.

Corthals A, Koller A, Martin DW, Rieger R, Chen EI, Bernaski M, et al. Detecting the immune system response of a 500 year-old Inca mummy. PLoS One. 2012;7(7):e41244. https://doi.org/10.1371/journal.pone.0041244.

Cruz AR, Pillay A, Zuluaga AV, Ramirez LG, Duque JE, Aristizabal GE, et al. Secondary syphilis in Cali, Colombia: new concepts in disease pathogenesis. PLoS Negl Trop Dis. 2010;4(5):e690. https://doi.org/10.1371/journal.pntd.0000690.

Cruz-Dávalos DI, Llamas B, Gaunitz C, Fages A, Gamba C, Soubrier J, et al. Experimental conditions improving in-solution target enrichment for ancient DNA. Mol Ecol Resour. 2016;17(3):508–22. https://doi.org/10.1111/1755-0998.12595.

Cunha CB, Cunha BA. Great plagues of the past and remaining questions. In: Raoult D, Drancourt M, editors. Paleomicrobiology: past human infections. Berlin: Springer; 2008. p. 1–20.

Cunnington AJ. Malaria and susceptibility to other infections. London School of Hygiene & Tropical Medicine; 2012. https://doi.org/10.17037/PUBS.00901045.

D'Amato A, Zilberstein G, Zilberstein S, Compagnoni BL, Righetti PG. Of mice and men: traces of life in the death registries of the 1630 plague in Milano. J Proteome. 2018; https://doi.org/10.1016/j.jprot.2017.11.028.

Dabney J, Knapp M, Glocke I, Gansauge M-T, Weihmann A, Nickel B, et al. Complete mitochondrial genome sequence of a Middle Pleistocene cave bear reconstructed from ultrashort DNA fragments. Proc Natl Acad Sci U S A. 2013;110(39):15758–63. https://doi.org/10.1073/pnas.1314445110.

Delsuc F, Gibb GC, Kuch M, Billet G, Hautier L, Southon J, et al. The phylogenetic affinities of the extinct glyptodonts. Curr Biol. 2016;26(4):R155–6. https://doi.org/10.1016/j.cub.2016.01.039.

Devault AM, Golding GB, Waglechner N, Enk JM, Kuch M, Tien JH, et al. Second-pandemic strain of *Vibrio cholerae* from the Philadelphia cholera outbreak of 1849. N Engl J Med. 2014a;370(4):334–40. https://doi.org/10.1056/NEJMoa1308663.

Devault AM, McLoughlin K, Jaing C, Gardner S, Porter TM, Enk JM, et al. Ancient pathogen DNA in archaeological samples detected with a Microbial Detection Array. Sci Rep. 2014b;4:4245. https://doi.org/10.1038/srep04245.

Devault AM, Mortimer TD, Kitchen A, Kiesewetter H, Enk JM, Golding GB, et al. A molecular portrait of maternal sepsis from Byzantine Troy. elife. 2017;6:e20983. https://doi.org/10.7554/eLife.20983.

DeWitte SN, Stojanowski CM. The Osteological Paradox 20 years later: past perspectives, future directions. J Archaeol Res. 2015;23(4):397–450. https://doi.org/10.1007/s10814-015-9084-1.

Drancourt M, Raoult D. Molecular insights into the history of plague. Microbes Infect. 2002;4(1):105–9. https://doi.org/10.1016/S1286-4579(01)01515-5.

Drancourt M, Aboudharam G, Signoli M, Dutour O, Raoult D. Detection of 400-year-old *Yersinia pestis* DNA in human dental pulp: an approach to the diagnosis of ancient septicemia. Proc Natl Acad Sci. 1998;95(21):12637–40. https://doi.org/10.1073/pnas.95.21.12637.

Dubreuil J, Rech A. Rapport sur le Choléra-Morbus Asiatique qui a Régné dans le midi de la France en 1835. Montpellier: Martel; 1836.

Duggan AT, Perdomo MF, Piombino-Mascali D, Marciniak S, Poinar D, Emery MV, et al. 17th century Variola virus reveals the recent history of smallpox. Curr Biol. 2016;26(24):3407–12. https://doi.org/10.1016/j.cub.2016.10.061.

Dutour O. Paleoparasitology and paleopathology. Synergies for reconstructing the past of human infectious diseases and their pathocenosis. Int J Paleopathol. 2013;3(3):145–9. https://doi.org/10.1016/j.ijpp.2013.09.008.

Enk JM. Time, temperature, and tiling density when capturing ancient DNA. In: Plant & Animal Genome Conference XXIII. 2015.

Enk JM, Devault AM, Kuch M, Murgha YE, Rouillard J-M, Poinar HN. Ancient whole genome enrichment using baits built from modern DNA. Mol Biol Evol. 2014;31(5):1292–4. https://doi.org/10.1093/molbev/msu074.

Faerman M, Jankauskas R, Gorski A, Bercovier H, Greenblatt CL. Prevalence of human tuberculosis in a medieval population of Lithuania studied by ancient DNA analysis. Anc Biomol. 1997;1:205–14.

Faure E. Malarial pathocoenosis: beneficial and deleterious interactions between malaria and other human diseases. Front Physiol. 2014;5:441. https://doi.org/10.3389/fphys.2014.00441.

Fears JR. The plague under Marcus Aurelius and the decline and fall of the Roman Empire. Infect Dis Clin N Am. 2004;18(1):65–77.

Florens L, Washburn MP, Raine JD, Anthony RM, Grainger M, Haynes JD, et al. A proteomic view of the *Plasmodium falciparum* life cycle. Nature. 2002;419(6906):520–6. https://doi.org/10.1038/nature01107.

Gansauge M-T, Meyer M. Single-stranded DNA library preparation for the sequencing of ancient or damaged DNA. Nat Protoc. 2013;8(4):737–48. https://doi.org/10.1038/nprot.2013.038.

Gansauge M-T, Gerber T, Glocke I, Korlević P, Lippik L, Nagel S, et al. Single-stranded DNA library preparation from highly degraded DNA using T4 DNA ligase. Nucleic Acids Res. 2017;45(10):e79. https://doi.org/10.1093/nar/gkx033.

Geigl E-M. On the circumstances surrounding the preservation and analysis of very old DNA. Archaeometry. 2002;44(3):337–42. https://doi.org/10.1111/1475-4754.t01-1-00066.

Gelabert P, Olalde I, de-Dios T, Civit S, Lalueza-Fox C. Malaria was a weak selective force in ancient Europeans. Sci Rep. 2017;7(1):1377. https://doi.org/10.1038/s41598-017-01534-5.

Gilbert MTP, Cuccui J, White W, Lynnerup N, Titball RW, Cooper A, Prentice MB. Absence of *Yersinia pestis*-specific DNA in human teeth from five European excavations of putative plague victims. Microbiology. 2004;150(2):341–54. https://doi.org/10.1099/mic.0.26594-0.

Gnirke A, Melnikov A, Maguire J, Rogov P, LeProust EM, Brockman W, et al. Solution hybrid selection with ultra-long oligonucleotides for massively parallel targeted sequencing. Nat Biotechnol. 2009;27(2):182–9. https://doi.org/10.1038/nbt.1523.

Gonzalez J-P, Guiserix M, Sauvage F, Guitton J-S, Vidal P, Bahi-Jaber N, et al. Pathocenosis: a holistic approach to disease ecology. EcoHealth. 2010;7(2):237–41. https://doi.org/10.1007/s10393-010-0326-x.

Grmek MD. Préliminaires d'une étude historique des maladies. Annales. Histoire, Sciences Sociales. 1969;24:1437–83.

Haas CJ, Zink A, Pálfi G, Szeimies U, Nerlich AG. Detection of leprosy in ancient human skeletal remains by molecular identification of Mycobacterium leprae. Am J Clin Pathol. 2000;114(3):428–36. https://doi.org/10.1093/ajcp/114.3.428.

Hagelberg E, Clegg JB. Isolation and characterization of DNA from archaeological bone. Proc R Soc Lond Ser B Biol Sci. 1991;244(1309) https://doi.org/10.1098/rspb.1991.0049. http://rspb.royalsocietypublishing.org/content/244/1309/45.abstract.

Harkins KM, Buikstra JE, Campbell T, Bos KI, Johnson ED, Krause J, Stone AC. Screening ancient tuberculosis with qPCR: challenges and opportunities. Philos Trans R Soc Lond B Biol Sci. 2015;370(1660) https://doi.org/10.1098/rstb.2013.0622. http://rstb.royalsocietypublishing.org/content/370/1660/20130622.

Hendy J, Collins M, Teoh KY, Ashford DA, Thomas-Oates J, Donoghue HD, et al. The challenge of identifying tuberculosis proteins in archaeological tissues. J Archaeol Sci. 2016;66:146–53. https://doi.org/10.1016/j.jas.2016.01.003.

Herbig A, Maixner F, Bos KI, Zink A, Krause J, Huson DH. MALT: fast alignment and analysis of metagenomic DNA sequence data applied to the Tyrolean Iceman. bioRxiv. 2016; https://doi.org/10.1101/050559. http://biorxiv.org/content/early/2016/04/27/050559.abstract.

Höss M, Pääbo S. DNA extraction from Pleistocene bones by a silica-based purification method. Nucleic Acids Res. 1993;21(16):3913–4. http://www.ncbi.nlm.nih.gov/pubmed/8396242.

Huson DH, Auch AF, Qi J, Schuster SC. MEGAN analysis of metagenomic data. Genome Res. 2007;17(3):377–86. https://doi.org/10.1101/gr.5969107.

Jonsson H, Ginolhac A, Schubert M, Johnson PLF, Orlando L. mapDamage2.0: fast approximate Bayesian estimates of ancient DNA damage parameters. Bioinformatics. 2013;29(13):1682–4. https://doi.org/10.1093/bioinformatics/btt193.

Kay GL, Sergeant MJ, Giuffra V, Bandiera P, Milanese M, Bramanti B, et al. Recovery of a medieval *Brucella melitensis* genome using shotgun metagenomics. MBio. 2014;5(4): e01337–14. https://doi.org/10.1128/mBio.01337-14.

Kistler L, Smith O, Ware R, Momber G, Bates R, Garwood P, et al. Thermal age, cytosine deamination and the veracity of 8,000 year old wheat DNA from sediments. bioRxiv. 2015; https://doi.org/10.1101/032060. http://biorxiv.org/content/early/2015/11/18/032060.abstract.

Kistler L, Ware R, Smith O, Collins M, Allaby RG. A new model for ancient DNA decay based on paleogenomic meta-analysis. Nucleic Acids Res. 2017; https://doi.org/10.1093/nar/gkx361.

Knapp M, Hofreiter M. Next generation sequencing of ancient DNA: requirements, strategies and perspectives. Genes. 2010; https://doi.org/10.3390/genes1020227.

Kolling G, Wu M, Guerrant RL. Enteric pathogens through life stages. Front Cell Infect Microbiol. 2012;2:114. https://doi.org/10.3389/fcimb.2012.00114.

Kolman CJ, Centurion-Lara A, Lukehart SA, Owsley DW, Tuross N. Identification of *Treponema pallidum* subspecies *pallidum* in a 200-year-old skeletal specimen. J Infect Dis. 1999;180(6):2060–3. https://doi.org/10.1086/315151.

Krause-Kyora B, Susat J, Key FM, Kuhnert D, Bosse E, Immel A, et al. Neolithic and medieval virus genomes reveal complex evolution of hepatitis B. elife. 2018a;7:e36666.

Krause-Kyora B, Nutsua M, Boehme L, Pierini F, Pedersen DD, Kornell S-C, et al. Ancient DNA study reveals HLA susceptibility locus for leprosy in medieval Europeans. Nat Commun. 2018b;9:1569.

Kwiatkowski DP. How malaria has affected the human genome and what human genetics can teach us about malaria. Am J Hum Genet. 2005;77(2):171–92. https://doi.org/10.1086/432519.

Lalremruata A, Ball M, Bianucci R, Welte B, Nerlich AG, Kun JFJ, Pusch CM. Molecular identification of falciparum malaria and human tuberculosis co-infections in mummies from the Fayum Depression (Lower Egypt). PLoS One. 2013;8(4):e60307. https://doi.org/10.1371/journal.pone.0060307.

Littman RJ, Littman ML. Galen and the Antonine Plague. Am J Philol. 1973;94(3):243–55. https://doi.org/10.2307/293979.

Maixner F, Thomma A, Cipollini G, Widder S, Rattei T, Zink A. Metagenomic analysis reveals presence of *Treponema denticola* in a tissue biopsy of the Iceman. PLoS One. 2014;9(6): e99994. https://doi.org/10.1371/journal.pone.0099994.

Marciniak S, Prowse TL, Herring DA, Klunk J, Kuch M, Duggan AT, et al. *Plasmodium falciparum* malaria in 1st–2nd century CE southern Italy. Curr Biol. 2016; https://doi.org/10.1016/j.cub.2016.10.016.

Mays S, Taylor GM, Legge AJ, Young DB, Turner-Walker G. Paleopathological and biomolecular study of tuberculosis in a medieval skeletal collection from England. Am J Phys Anthropol. 2001;114(4):298–311. https://doi.org/10.1002/ajpa.1042.

Mays S, Fysh E, Taylor GM. Investigation of the link between visceral surface rib lesions and tuberculosis in a Medieval skeletal series from England using ancient DNA. Am J Phys Anthropol. 2002;119(1):27–36. https://doi.org/10.1002/ajpa.10099.

Meyer M, Kircher M. Illumina sequencing library preparation for highly multiplexed target capture and sequencing. Cold Spring Harb Protoc. 2010;2010(6):pdb.prot5448. https://doi.org/10.1101/pdb.prot5448.

Millar CD, Huynen L, Subramanian S, Mohandesan E, Lambert DM. New developments in ancient genomics. Trends Ecol Evol. 2008;23(7):386–93. https://doi.org/10.1016/j.tree.2008.04.002.

Miller RL, Ikram S, Armelagos GJ, Walker R, Harer WB, Shiff CJ, et al. Diagnosis of *Plasmodium falciparum* infections in mummies using the rapid manual ParaSight-F test. Trans R Soc Trop Med Hyg. 1994;88(1):31–2. https://doi.org/10.1016/0035-9203(94)90484-7.

Milner GR, Boldsen JL. Life not death: epidemiology from skeletons. Int J Paleopathol. 2017;17:26–39. https://doi.org/10.1016/j.ijpp.2017.03.007.

Minnikin DE, Besra GS, Lee OY-C, Spigelman M, Donoghue HD. The interplay of DNA and lipid biomarkers in the detection of tuberculosis and leprosy in mummies and other skeletal remains. In: Gill-Frerking H, Rosendahl W, Zink A, editors. Yearbook of Mummy Studies, vol. 1. Munich: Verlag Dr. Friedrich Pfeil; 2011. p. 109–14.

Molak M, Ho SYW. Evaluating the impact of post-mortem damage in ancient DNA: a theoretical approach. J Mol Evol. 2011;73(3–4):244–55. https://doi.org/10.1007/s00239-011-9474-z.

Montiel R, Garcia C, Canadas MP, Isidro A, Guijo JM, Malgosa A. DNA sequences of *Mycobacterium leprae* recovered from ancient bones. FEMS Microbiol Lett. 2003;226(2):413–4. https://doi.org/10.1016/S0378-1097(03)00617-7.

Mueller A-K, Behrends J, Hagens K, Mahlo J, Schaible UE, Schneider BE. Natural transmission of Plasmodium berghei exacerbates chronic tuberculosis in an experimental co-infection model. PLoS One. 2012;7(10):e48110. https://doi.org/10.1371/journal.pone.0048110.

Mühlemann B, Margaryan A, Damgaard P, Allentoft ME, Vinner L, Hansen AJ, et al. Ancient human parvovirus B19 in Eurasia reveals its long-term association with humans. Proc Natl Acad Sci U S A. 2018a; https://doi.org/10.1073/pnas.1804921115.

Mühlemann B, Jones TC, Damgaard P, Allentoft ME, Shevnina I, Logvin A, et al. Ancient hepatitis B viruses from the Bronze Age to the Medieval period. Nature. 2018b;577:418–23.

Mutolo MJ, Jenny LL, Buszek AR, Fenton TW, Foran DR. Osteological and molecular identification of brucellosis in ancient Butrint, Albania. Am J Phys Anthropol. 2012;147(2):254–63. https://doi.org/10.1002/ajpa.21643.

Newfield TP, Labuhn I. Realizing consilience in studies of pre-instrumental climate and pre-laboratory disease. J Interdiscip Hist. 2017;48(2):211–40.

Orlando L, Gilbert MTP, Willerslev E. Reconstructing ancient genomes and epigenomes. Nat Rev Genet. 2015;16(7):395–408. https://doi.org/10.1038/nrg3935.

Ortner DJ. Identification of pathological conditions in human skeletal remains. 2nd ed. San Diego: Academic; 2003.

Patterson Ross Z, Klunk J, Fornaciari G, Giuffra V, Duchêne S, Duggan AT, et al. The paradox of HBV evolution as revealed from a 16th century mummy. PLoS Pathog. 2018;14(1):e1006750. https://doi.org/10.1371/journal.ppat.1006750.

Rabino Massa E, Cerutti N, Marin A, Savoia D. Malaria in Ancient Egypt: paleoimmunological investigation on predynastic mummified remains. Chungará (Arica). 2000;32(1):7–9. https://doi.org/10.4067/S0717-73562000000100003.

Redman JE, Shaw MJ, Mallet AI, Santos AL, Roberts CA, Gernaey AM, Minnikin DE. Mycocerosic acid biomarkers for the diagnosis of tuberculosis in the Coimbra Skeletal Collection. Tuberculosis. 2009;89(4):267–77. https://doi.org/10.1016/j.tube.2009.04.001.

Roberts CA, Manchester K. Archaeology of disease. Stroud: Sutton Publishing; 2005.

Sallares R, Gomzi S. Biomolecular archaeology of malaria. Anc Biomol. 2001;3:195–213.

Salo WL, Aufderheide AC, Buikstra J, Holcomb TA. Identification of *Mycobacterium tuberculosis* DNA in a pre-Columbian Peruvian mummy. Proc Natl Acad Sci. 1994;91(6):2091–4. http://www.pnas.org/content/91/6/2091.abstract.

Saunders SR, Herring DA, Boyce G. Can skeletal samples accurately represent the living populations they come from? The St. Thomas' cemetery site, Belleville, Ontario. In: Grauer AL, editor. Bodies of evidence: reconstructing history through skeletal analysis. New York: Wiley-Liss; 1995. p. 68–89.

Schuenemann V, Avanzi C, Krause-Kyora B, Seitz A, Herbig A, Inskip S, et al. Ancient genomes reveal a high diversity of *Mycobacterium leprae* in medieval Europe. PLoS Pathog. 2018a;14(5):e1006997.

Schuenemann V, Lankapalli AK, Barquera R, Nelson EA, Hernandez DI, Alonzo VA, et al. Historic *Treponema pallidum* genomes from Colonial Mexico retrieved from archaeological remains. PLoS Negl Trop Dis. 2018b;12(6):e0006447.

Segata N, Waldron L, Ballarini A, Narasimhan V, Jousson O, Huttenhower C. Metagenomic microbial community profiling using unique clade-specific marker genes. Nat Methods. 2012;9(8):811–4. https://doi.org/10.1038/nmeth.2066.

Singer M. Pathogen-pathogen interaction. Virulence. 2010;1(1):10–8. https://doi.org/10.4161/viru.1.1.9933.

Spyrou MA, Tukhbatova RI, Feldman M, Drath J, Kacki S, Beltrán de Heredia J, et al. Historical *Y. pestis* genomes reveal the European Black Death as the source of ancient and modern plague pandemics. Cell Host Microbe. 2016;19(6):874–81. https://doi.org/10.1016/j.chom.2016.05.012.

Steinbock RT. Paleopathological diagnosis and interpretation: bone diseases in ancient human populations. Springfield: Charles C Thomas; 1976.

Taylor GM, Tucker K, Butler R, Pike AWG, Lewis J, Roffey S, et al. Detection and strain typing of ancient *Mycobacterium leprae* from a medieval leprosy hospital. PLoS One. 2013;8(4):e62406. https://doi.org/10.1371/journal.pone.0062406.

Thierry D, Brisson-Noël A, Vincent-Lévy-Frébault V, Nguyen S, Guesdon JL, Gicquel B. Characterization of a *Mycobacterium tuberculosis* insertion sequence, IS6110, and its application in diagnosis. J Clin Microbiologyrobiol. 1990;28(12):2668–73. http://www.ncbi.nlm.nih.gov/pubmed/2177747.

Tran T-N-N, Aboudharam G, Raoult D, Drancourt M. Beyond ancient microbial DNA: nonnucleotidic biomolecules for paleomicrobiology. BioTechniques. 2011;50(6):370–80. https://doi.org/10.2144/000113689.

Vågene ÅJ, Herbig A, Campana MG, Robles García NM, Warinner C, Sabin S, et al. *Salmonella enterica* genomes from victims of a major sixteenth-century epidemic in Mexico. Nat Ecol Evol. 2018; https://doi.org/10.1038/s41559-017-0446-6.

van Soolingen D, Hermans PW, de Haas PE, van Embden JD. Insertion element IS1081-associated restriction fragment length polymorphisms in *Mycobacterium tuberculosis* complex species: a reliable tool for recognizing *Mycobacterium bovis* BCG. J Clin Microbiol. 1992;30(7):1772–7. http://www.ncbi.nlm.nih.gov/pubmed/1352785.

von Hunnius TE, Yang D, Eng B, Waye JS, Saunders SR. Digging deeper into the limits of ancient DNA research on syphilis. J Archaeol Sci. 2007;34(12):2091–100. https://doi.org/10.1016/j.jas.2007.02.007.

Wagner DM, Klunk J, Harbeck M, Devault A, Waglechner N, Sahl JW, et al. *Yersinia pestis* and the Plague of Justinian 541–543 AD: a genomic analysis. Lancet Infect Dis. 2014;14(4):319–26. https://doi.org/10.1016/S1473-3099(13)70323-2.

Wales N, Carøe C, Sandoval-Velasco M, Gamba C, Barnett R, Samaniego JA, et al. New insights on single-stranded versus double-stranded DNA library preparation for ancient DNA. BioTechniques. 2015;59(6):368–71. https://doi.org/10.2144/000114364.

Warinner C, Rodrigues JFM, Vyas R, Trachsel C, Shved N, Grossmann J, et al. Pathogens and host immunity in the ancient human oral cavity. Nat Genet. 2014;46(4):336–44. https://doi.org/10.1038/ng.2906.

Warinner C, Herbig A, Mann A, Fellows Yates JA, Weiß CL, Burbano HA, et al. A robust framework for microbial archaeology. Annu Rev Genomics Hum Genet. 2017;18(1):321–56. https://doi.org/10.1146/annurev-genom-091416-035526.

Whatmore AM. Ancient-pathogen genomics: coming of age? MBio. 2014;5(5):e01676–14. https://doi.org/10.1128/mBio.01676-14.

Wickham H. Ggplot 2: elegant graphics for data analysis. New York: Springer; 2016.

Wood DE, Salzberg SL. Kraken: ultrafast metagenomic sequence classification using exact alignments. Genome Biol. 2014;15(3):R46. https://doi.org/10.1186/gb-2014-15-3-r46.

Wood JW, Milner GR, Harpending HC, Weiss KM. The Osteological Paradox: problems of inferring prehistoric health from skeletal samples. Curr Anthropol. 1992;33(4):343–70. https://doi.org/10.1086/204084.

Woods SA, Cole ST. A family of dispersed repeats in *Mycobacterium leprae*. Mol Microbiol. 1990;4(10):1745–51. https://doi.org/10.1111/j.1365-2958.1990.tb00552.x.

Zhou Z, Lundstrøm I, Tran-Dien A, Duchêne S, Alikhan N-F, Sergeant MJ, et al. Millennia of genomic stability within the invasive Para C Lineage of *Salmonella enterica*. bioRxiv. 2017; https://doi.org/10.1101/105759. http://www.biorxiv.org/content/early/2017/02/14/105759.

Zink A, Haas CJ, Reischl U, Szeimies U, Nerlich AG. Molecular analysis of skeletal tuberculosis in an ancient Egyptian population. J Med Microbiol. 2001;50(4):355–66. https://doi.org/10.1099/0022-1317-50-4-355.

Paleovirology: Viral Sequences from Historical and Ancient DNA

Kyriakos Tsangaras and Alex D. Greenwood

Abstract Paleovirology, the study of viruses from historical or ancient samples, is a relatively unexplored but promising subfield of ancient DNA (aDNA) and paleogenomic research. Analysis of viruses, even over short historical timescales, can provide information on virus evolution and biology that may be difficult or impossible to obtain from examining current viral diversity. This is in part because the rapid evolution and proneness to extinction of strains of many viruses can quickly obscure their origins. Though exceptionally difficult to characterize from ancient DNA or RNA extracts, reports on the successful analysis of historical and ancient viruses have been steadily accumulating. In this chapter, we summarize the successes and failures in this new emerging field.

Keywords 1918 influenza virus · Ancient DNA · Ancient viromes · Biosafety · Giant viruses · Hepatitis · Paleovirology · Poxviruses · Retroviruses

1 Introduction

Paleogenomics has had a demonstrable impact on the genomic study of multicellular organisms and, to a lesser degree, on viruses, bacteria, and parasites (Harkins and Stone 2015; Heintzman et al. 2006). However, paleogenomics could have a particularly profound impact on virology. Viruses exhibit molecular evolutionary rates far higher than any other organism (Gojobori et al. 1990; Drummond et al. 2003;

K. Tsangaras
Department of Translational Genetics, The Cyprus Institute of Neurology and Genetics, Nicosia, Cyprus

A. D. Greenwood (✉)
Department of Wildlife Diseases, Leibniz Institute for Zoo and Wildlife Research, Berlin, Germany

Department of Veterinary Medicine, Freie Universität Berlin, Berlin, Germany
e-mail: greenwood@izw-berlin.de

Charlotte Lindqvist and Om P. Rajora (eds.), *Paleogenomics*, Population Genomics [Om P. Rajora (Editor-in-Chief)],
https://doi.org/10.1007/13836_2018_19,

Duchene et al. 2014). Even the most slowly evolving large DNA viruses exhibit effective population sizes far larger than other microbes, meaning that larger percentages of viral particles will successfully produce new infectious virus and therefore often exhibit macroevolutionary changes over short timescales. The consequence of such rapid evolution is that phylogenetic inference and molecular dating, when at all possible, is subject to extremely high error rates (Drummond et al. 2003). Another consequence is that it is unclear if historical viruses or strains involved in disease in the past are closely related to modern isolates or represent rare or extinct variants. Therefore, obtaining viral sequences that are old or ancient have the potential to upend current tenets of virology in comparable or perhaps to a greater degree than observed in the field of human paleogenomics. This chapter will summarize the work to date that has employed paleovirology to identify and characterize viruses involved in historic epidemics and to answer biological questions that cannot be addressed in full by examining current viral biodiversity.

To date, paleovirology has been coined as a term for the analysis of exogenous viruses that have integrated and invaded host genomes in the past (Aswad and Katzourakis 2012). However, this is a somewhat inaccurate term as the integrated viruses do not stop evolving but rather adopt the mutation rate of the host (Slater et al. 2016). Sequencing of viruses from historical or ancient samples represents viruses that existed at a given time with no subsequent accumulation of mutations. Therefore, in the context of this chapter, paleovirology refers exclusively to viral sequences obtained from historical or ancient samples using paleogenomic techniques.

Paleovirology is underdeveloped compared to other paleogenomic fields. This is not because of lack of interest or chance. Many important viruses have RNA genomes and are generally unstable outside of a living host which limits the possibility of studying them (Katzourakis 2013). Another challenge is that only high-titer viruses are likely to be detected, a pertinent limitation for viral discovery from freshly collected material. For example, results from whole genome studies from infected and highly viremic patients suggest that approximately 10 viral reads per 25 million sequenced reads will be observed (Wylie et al. 2015). The problems are further exacerbated in paleogenomic studies by the low concentrations, extreme degradation, and modification of nucleic acids extracted from historical or ancient samples. Unlike mitochondrial DNA or nuclear DNA which are present in most host cells, even in viremic individuals, there will be less viral nucleic acid as a percentage of the total than host nucleic acids. Another limitation, particularly with respect to early Holocene- and Pleistocene-age samples, is that viruses do not commonly infect nor can they be isolated from bone, the most common source of nucleic acids from older samples. Even in the case of historical samples, museums rarely maintain tissues in museum collections but rather keep skins and skeletons while disposing of the internal organs. Therefore, samples appropriate for viral analysis are rare in most existing collections, and with the exception of permafrost and some cold caves, it is

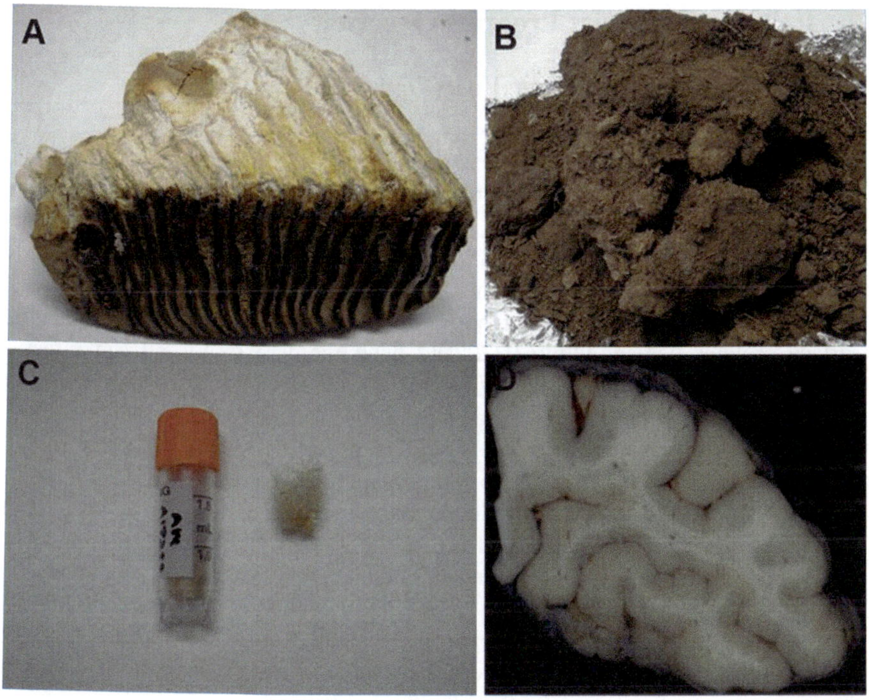

Fig. 1 Examples of the types of samples employed in paleovirological analysis. Viral sequences, particularly retroviral, have been examined from Pleistocene-age fossils such as panel (**a**) woolly mammoth (*Mammuthus primigenius*) bones (Greenwood et al. 2001; Zhou et al. 2009, sample provided by Dick Mol) and panel (**b**) extinct giant ground sloth (*Mylodon darwinii*) coprolites (sample from the Natural History Museum London, described in Slater et al. 2016). Younger historical samples represented in museum collections, often as bone or skin, have been investigated for various viruses including mid-nineteenth-century koala (*Phascolarctos cinereus*) skins shown in panel (**c**) (Avila-Arcos et al. 2013; Tsangaras et al. 2014, sample provided by the Australian Museum). More recent historical but equally difficult samples for analysis are represented by formalin-fixed or paraffin-embedded samples (a brain section of Knut the polar bear, *Ursus maritimus*, of the Berlin Zoological Garden is shown in panel (**d**), photo provided by Claudia Szentiks). Examples of such embedded samples in paleovirological work include the 1918 influenza genome sequencing (Taubenberger et al. 1997)

unlikely that tissue samples or preserved feces appropriate for attempting virological analysis will be common in targeted field expeditions. Examples of sample types used in paleovirological work are shown in Fig. 1. Because of these challenges, there has been less emphasis on paleovirology than other subdisciplines of paleogenomics. Here we review the literature that represents the successful, and in some cases controversial, investigations that have taken place thus far and the future potential of this line of investigation.

2 Paleovirogenomics Progress Overview

2.1 1918 H1N1 Influenza Virus

Influenza viruses are negative-sense negative-strand RNA viruses with segmented genomes that contain eight to nine gene segments; they belong to the *Orthomyxoviridae* family and have the ability to infect a variety of vertebrate animals including humans. The 1918 influenza virus caused the greatest historically recorded viral pandemic and is estimated to have infected approximately one third of the world's population. Influenza type A virus strains are distinguished by the viral surface proteins hemagglutinin (HA) and neuraminidase (NA) and can infect a variety of warm-blooded mammals with aquatic birds serving as the natural reservoir of most known subtypes (Bouvier and Palese 2008; Taubenberger and Morens 2006, 2008). The 1918 pandemic virus was an H1N1 strain. The pandemic spread in two waves; the first wave occurred in the spring of 1918 with relatively mild symptoms displayed in infected individuals. The second wave, which occurred in August of 1918, was more virulent and spread across the globe in less than 6 months. The pandemic claimed the life of more than 40 million people worldwide with unusually high fatality rates among young adults (Taubenberger and Morens 2006; Taubenberger et al. 2000). For 80 years it remained unclear what the pathogen was and why it was so uniquely virulent (Taubenberger and Morens 2006). In fact at the time of the pandemic, the cause of the flu was still unknown with the bacterium *Haemophilus influenzae* suggested as the most likely candidate (Noymer and Garenne 2000).

The characterization of the pandemic-causing virus was one of the first paleovirology success stories, and this was achieved despite the fact that the influenza genome is composed of RNA which is generally less stable than single- or double-stranded DNA (Katzourakis 2013; Tsangaras and Greenwood 2012). In 1995, a research team obtained 14 formalin-fixed paraffin-embedded samples from the Armed Forces Institute of Pathology from soldiers who died in 1918, all exhibiting symptoms associated with the influenza pandemic. RNA was amplified using reverse transcription polymerase chain reaction (RT-PCR) for 11 of 14 samples. The 11 samples were screened for RNA remnants of the hemaglutinin, neuramidase, nucleoprotein, matrix 1 protein, and matrix protein 2 genes of the influenza virus. Of the 11 samples, one was positive for fragments tested. Sequencing of the amplified fragments revealed a novel influenza sequence that was clearly distinguishable from the control strains used and from all characterized influenza strains at the time (Taubenberger and Morens 2006; Taubenberger et al. 1997).

Following the initial successful identification of 1918 influenza partial genome sequences, Reid et al. (1999) were able to sequence the entire HA gene from three independent samples: the formalin-fixed paraffin-embedded sample from which the virus was originally identified and lung tissue samples from two other pandemic victims. One lung sample was formalin-fixed paraffin-embedded, while the other was extracted from tissues retrieved from a corpse exhumed from a permafrost grave in a 1918 influenza pandemic area in Brevig Mission, Alaska. Sequencing of the three complete genes of HA revealed little variation among the three distinct individuals, an indication that the same virus was responsible for the three fatalities (Taubenberger et al. 2000). Phylogenetic analysis of the HA gene placed the 1918 influenza virus within the mammalian influenza H1N1 clade and indicated that the closest relative of the virus was the oldest known classical swine flu strain. However, despite the mammalian grouping, the pandemic sequence had many avian influenza-like features. Unfortunately, sequence of the HA gene could not explain the extreme observed lethality of the 1918 influenza virus (Reid et al. 1999).

The permafrost-derived tissue sample from Brevig Mission, Alaska, pandemic victim was used subsequently to sequence and study the entire 1918 H1N1 genome. Initial sequencing of the coding region was performed using RT-PCR of 22 overlapping fragments. Phylogenetic analyses using all the influenza genes indicated that the pandemic virus was likely of avian origin with accumulation of the necessary mutations to adapt to mammals (Taubenberger et al. 2005; Tumpey et al. 2005). Using the complete sequence of the virus and reverse genetics, Tumpey et al. (2005) were able to reconstruct infectious virus in mouse models to study the biology of the 1918 strain. Reconstructed virus was extremely virulent using different combinations of the eight influenza genes with less virulent strains in mouse models. Several experiments were performed using combinations of 1918 influenza genes along with modern strains in an effort to identify the genes or even the amino acids that caused the high pathogenicity and virulence of the 1918 strain. Results based on the experiments with the recombinant viruses indicated that specific changes in the HA, NA, and polymerase subunit 1 (PB1) genes were key elements of virus adaptation to mammalian hosts and for the 1918 influenza viruses pathogenicity (Tumpey et al. 2005; Pappas et al. 2008). None of this information would be available to science without a paleovirological approach. The 1918 influenza studies were the first to demonstrate the potential of paleovirology to obtain information not available when employing the study of currently circulating viruses.

2.2 *Retroviruses*

Retroviruses are enveloped positive-sense single-stranded RNA viruses that can be found in all vertebrates. Retroviruses are unique among the viral groups due to their replication cycle that involves conversion of the viral genetic material from RNA to DNA and the integration of the resulting genome into the host genome as a provirus (Jern and Coffin 2008; Weiss 2006). The proviral genome includes four genes (*gag*, *pro*, *pol*, and *env*) flanked by long terminal repeats (LTRs) that are generated during the reverse transcription process (Jern and Coffin 2008). Cross-species transmission of simian T-cell leukemia virus 1 (STLV-1) and simian immunodeficiency virus (SIV) retroviruses to humans has given rise to the human pathogens HTLV-1, HIV-1, and HIV-2 which are global epidemics in the case of the HIVs. STLV-1 and SIV are known to infect a wide range of nonhuman hominoids, and cross-species transmission is likely frequent in Africa (Kalish et al. 2005; VandeWoude and Apetrei 2006; Wolfe et al. 2005). Calvignac et al. (2008) examined African green monkey (*Chlorocebus* sp.) museum specimens from the early twentieth century in an effort to identify ancient STLV-1 and SIV proviral sequences. Potential discovery of archival STLV-1 or SIV sequences would allow for a better understanding of the genetic diversity and evolution of these viruses which can evolve at a very rapid rate. Six bone museum specimens were screened using PCR for the presence of retroviral LTRs and the STLV-1 px gene in dedicated aDNA laboratories. Positive results obtained from one out of the six specimens tested identified an STLV-1e strain-related virus and represented the first clear positive results of STLV proviral identification from bone samples dated to the early twentieth century (Calvignac et al. 2008). STLV-1e is known to circulate in the region of Africa from which the museum specimen derived, and the finding of historical STLV-1e in this region is consistent with the current biogeographical distribution.

HIV-1 is a lentivirus responsible for the global acquired immunodeficiency syndrome (AIDS) epidemic. However, the detailed origin of the virus and how it became an epidemic remains obscure. Archival materials obtained from Kinshasa, Democratic Republic of Congo (DRC), were used to study the time of origin and the timescale of HIV-1 viral evolution. Tissue blocks from 1958 to 1960 were screened for viral nucleic acids using RT-PCR. One sample out of 27 screened was positive. Phylogenetic analysis was performed on the Kinshasa sample obtained sequences (DRC 1960), on the sequences obtained from paraffin samples collected from AIDS-positive patients between 1981 and 1997, and on reference sequences from the Los Alamos National Laboratory HIV sequence database. The Bayesian phylogenetic analysis placed the DRC 1960 obtained sequences close to the ancestral node of the subtype lineage forming a monophyletic clade with the modern DRC sequences. Statistical modeling using the DRC 1960 sequences and the oldest obtained HIV 1 sequence dated to 1959 (ZR 1959) provided an estimate for the most recent common ancestor of the HIV-1 group M entering the human population between 1884 and 1924, much earlier than previously believed (Worobey et al. 2008). However, the dates matched well with the founding of Kinshasa, a major capital city that, at the time estimated for the emergence of HIV, was experiencing exponential human population growth representing ideal conditions for the establishment of a pathogen primarily spread by sexual transmission.

Gilbert et al. (2007) also used modern and archival material to examine the origin and emergence of HIV-1 subtype B virus, the predominant HIV strain in most countries. A popular hypothesis was that the pandemic HIV strain emerged in the United States and spread to Haiti via sex tourism in the 1970s. For the study, the group used 117 previously published HIV-1 sequences obtained from around the globe and four Haitian sequences from HIV-1 positive archival material dated between 1982 and 1983. Bayesian phylogenetic analysis of the retroviral *env* gene placed the archival sequences in a basal position within the subtype B phylogeny. The result suggested that HIV-1 subtype B virus first arrived in and spread from Haiti sometime in 1966, before expanding globally in three subsequent transmission events to Trinidad and Tobago, North America, and finally globally. Phylogenetic estimates indicate that HIV-1 subtype B was in circulation in the United States for at least a decade before AIDS was first recognized (Gilbert et al. 2007). The initial emergence in Haiti is better explained by commerce between Haiti and the DRC in the 1960s than sex tourism from the United States.

A recent study on the emergence of HIV-1 subtype B in North America screened approximately 2,000 archival samples dated from 1978 to 1979 for the presence of HIV-1 virus (Worobey et al. 2016). It was previously thought that the HIV epidemic began in the United States in California and specifically the San Francisco (SF) male homosexual population beginning with "patient zero," an AIDS patient who was identified in the popular press. The samples studied, including a sample from "patient zero," were initially collected from homosexual cohort patients from New York City (NYC) and San Francisco (SF) who had been screened for hepatitis B virus. Western blot analysis of the archival samples indicated that 3.7% were positive for HIV-1. RT-PCR efforts to obtain sequence data using HIV-1-specific primers failed to provide positive results. Failure to obtain amplifiable data is not unexpected especially when using archival specimens, and it is most likely due to long-term storage or/and the limited amount of viral RNA present. To overcome the above limitation, Worobey et al. (2016) developed an approach called RNA jackhammering that uses multiplex priming to reverse transcribe and amplify the viral regions of interest and then sequence at high throughput. The researchers were able to fully sequence and assemble HIV-1 genomes from three SF and five NYC patient samples, the oldest viral sequences determined from samples outside of Africa. Bayesian phylogenetic analysis of the eight viral genomes demonstrated that the sequences did not cluster with the older African viral data as was expected. SF and NYC viral genomes instead clustered within the diverse Caribbean clade of HIV-1. Comparison of the archival genomes demonstrated limited genetic diversity among the SF obtained sequences. The results suggest the SF strains originated from a single introduction from NYC around 1976. Molecular analysis of the archival HIV-1 genomes revealed a series of founder effects with data suggesting the HIV-1 epidemic spread from Africa to the Caribbean around 1967, from the Caribbean to NYC sometime during 1971, and from NYC to SF around 1976 (Worobey et al. 2016). Patient zero was determined to have had no involvement in the origin or original transmission of HIV in the United States. Determining the date of transmission of HIV to humans, the dispersal of HIV from the Caribbean and the origin and dispersal within the United States would not have been possible without paleovirological approaches.

Retroviruses that integrate into the host germ line are called endogenous retroviruses (ERVs) and can become permanent genomic elements inherited as Mendelian traits (Tsangaras et al. 2014). ERVs offer the opportunity to examine viral evolutionary history that spans millions of years (Katzourakis 2013). Greenwood et al. (2001) used fossil material from extinct woolly mammoth (*Mammuthus primigenius*), living elephants (*Loxodonta* sp., *Elephas maximus*), and extant members of the Uranotheria (hyraxes; Procaviidae, manatees, and dugongs; Sirenia) to examine the foamy virus like ERV-L group of ERVs across time and taxa. The initial hypothesis based on data from ungulates and carnivores indicated that low copy number of ERV-L sequences should be present in the uranotheres (elephants, sirenians, and hyraxes). Amplification of ERV-L sequences from samples was achieved using PCR with specific primers targeting two regions of the polymerase gene. Long PCR products were amplified from modern samples, while only the shorter region was amplifiable from the woolly mammoth samples. Ungulates and carnivores exhibit relatively low diversity for retroelements of this class. Cloning and sequencing of the amplifiable products demonstrate that contrary to the initial expectations of low ERV-L diversity, ERV-L sequence was highly diverse among both extant and extinct members of the Uranotheria. Species- and mammoth-specific ERV-L sequences were identified. Novel ERV-L sequences were obtained from each taxon indicating that increasing the sampling would result in higher number of unique sequences (Greenwood et al. 2001). The sequencing of the woolly mammoth genome and bioinformatics screening of repetitive elements revealed that mammoth had a higher overall retroviral and retroelement diversity than extant elephants and that many of the lineages of ERVs went extinct with the mammoth further supporting the findings of Greenwood et al. (2001) of high ERV-L diversity in the woolly mammoth genome (Zhao et al. 2009).

Using extant sloth (Folivora) samples, a recent study concluded that sloth endogenized foamy virus (SloEFV) invaded the sloth common ancestor genome approximately 39 million years ago before the estimated divergence of three- and two-toed sloths but after the separation of the Folivora clade (Katzourakis et al. 2009). In an effort to determine the retroviral SloEFV macroevolutionary patterns within folivorans, DNA was extracted from modern and museum specimens. Museum specimens included coprolite samples from the extinct Northrotheridae and Mylodontidae clades, while modern samples for all extant sloth species were analyzed. The SloEFV polymerase gene was enriched from the sample DNA extracts using a targeted enrichment hybridization capture technique and high-throughput sequencing (HTS). A total of 198 SloEFV polymerase gene sequences were obtained from which 26 belong to the extinct Mylodontidae clade, while the Northrotheridae clade samples failed to produce retroviral sequences. Phylogenetic analysis of the retroviral data indicated that some sequences predated the sloth lineages examined reflecting events that occurred sometime in the last 100 Myr. However, some sequences appeared to be lineage specific with the Choloepus clade exhibiting the majority of these private SloEFV lineages (Slater et al. 2016). The results demonstrated complex macroevolutionary patterns of lineage proliferation and loss that are not detectable examining extant sloths in isolation and which do not follow the evolutionary patterns of other loci in the genomes of sloths.

The majority of ERVs are remnants of retroviral infection that occurred in the distant past. The koala retrovirus (KoRV) is the only known retrovirus that is currently in the process of invading the host germ line (Tarlinton et al. 2006). KoRV was initially identified as an ERV based on common integration sites among koalas (*Phascolarctos cinereus*) from northern Australia. Unlike most endogenous viruses, KoRV appeared to be able to produce infectious particles. KoRV prevalence forms a gradient from northern Australia to southern Australia with 100% prevalence in northern Australia and decreasing prevalence and decreasing genomic copy number moving south with island populations such as Kangaroo Island with predominantly KoRV-negative koalas. The gradient of prevalence suggests the endogenization process is at a very early stage and therefore represents a unique opportunity to examine how exogenous retroviruses colonize mammalian genomes. Avila-Arcos et al. (2013) examined northern koala museum specimens dated from the mid-nineteenth century to the 1980s to examine the evolution and endogenization of KoRV over time. Using a multiplex PCR strategy coupled with HTS, the researchers were able to determine the full-length sequence of the retroviral *env* gene for five samples with collection dates spanning 120 years. All KoRV *env* sequences and functional motifs that affect viral infectivity were conserved among the koala museum specimens tested. KoRV prevalence was 100% in 28 northern Australian museum samples from the late nineteenth century with the very few polymorphisms identified and only two that spanned more than one sample date. The data suggests that KoRV evolves slowly and was already widespread in the early 1800s. The results also indicated that koalas have been exposed to infection for more than a century and likely for much longer, recently estimated to be as long as 50,000 years (Avila-Arcos et al. 2013; Ishida et al. 2015). This also suggests endogenization is a relatively slow process, that substantial time is required to fix ERVs in a population, and that evolutionary pressure on the invading retrovirus is surprisingly weak.

A subsequent study examined full KoRV genomes from northern Australia modern and museum specimens including the ones examined in Avila-Arcos et al. (2013). Samples were enriched for KoRV using targeted enrichment hybridization capture and HTS. One hundred and thirty-eight polymorphisms were identified with the majority of them being present in the modern sequenced sample. Regardless of viral gene, none of the identified polymorphisms were in the previously reported functional motifs that affect viral infectivity, which supported the results of Avila-Arcos et al. (2013). Protein model analysis of the coding regions indicated that non-synonymous radical mutations corresponding to large physiochemical difference were significantly elevated in the *env* region when compared with the other viral genes. This could reflect antiviral immune pressure in the ENV protein of KoRV. Viral integration sites were also captured and demonstrated that only 7% of the identified integration sites were present in more than one koala. This lack of common integration sites among koalas coupled with the prevalence data in Avila-Arcos et al. (2013) is strong evidence for the early stage of endogenization of KoRV as few KoRVs are fixed or at high frequency among even northern koalas where prevalence of KoRV is comprehensive (Tsangaras et al. 2014).

2.3 *Anelloviruses*

Anelloviruses are small, circular, non-enveloped, negative-sense, single-stranded DNA viruses that cause chronic viral infections with no disease association. In an effort to identify viral sequences from archeological remains that were older than a century, Bedarida et al. (2011) screened dental pulp samples for the presence of anelloviruses. The *Anelloviridae* family is a good candidate for paleovirological analysis from archaeological remains due to their environmental stability, high prevalence in humans, ease of detection, and presence in a variety of biological locations. The absence of known host genome integration events for this viral group would suggest that identified sequences would derive from viral and not host DNA. DNA extraction and beta globin screening using polymerase chain reaction was performed on 21 dental pulp samples that were obtained from a mass grave in Kaliningrad dated to 1812. Five out of 21 samples were positive for beta globin, an indication that host DNA was still present. A second PCR was performed on the five samples using *Anelloviridae*-specific primers. One of the five beta globin-positive samples yielded a product. Sequencing and phylogenetic analysis of the positive sample identified an ancient anellovirus sequence that was basal to the *Gammatorquevirus* group of anelloviruses with a genetic distance of 23% from the closest modern sequence. The calculated mutation rate of the anellovirus genome is 7×10^{-4} substitutions per site per year indicating that modern contamination is unlikely to explain the obtained sequences (Duffy et al. 2008; Umemura et al. 2002). Anellovirus identification from archaeological samples was the first report of viral sequences from a dental pulp extract and the first identification of non-genomically integrated viral sequences in archaeological material (Bedarida et al. 2011).

2.4 Hepatitis Viruses

Among the most human health-relevant viruses are hepatitis B and C viruses with infection often becoming chronic and 500 million people affected worldwide. Hepatitis viruses damage the liver and can lead to cirrhosis in infected individuals. Although they share common routes of transmission and symptoms, hepatitis viruses belong to different viral groups, have different nucleic acid genome types, and differ in geographical prevalence (Alter 2006). Hepatitis B virus is a highly infectious double-stranded DNA virus that belongs to the hepadnavirus family of viruses; while hepatitis C virus is a positive-sense single-stranded RNA virus that belongs to the *Flaviviridae* family (Pybus et al. 2007; Kahila Bar-Gal et al. 2012).

Hepatitis B virus (HBV) is endemic in many countries worldwide. It can be transmitted by sexual intercourse as well as vertically from mother to child (Franco et al. 2012). The majority of the hepatitis B-infected individuals worldwide are almost exclusively subtype C (>95%). Liver tissue available from a Korean mummy dating to the sixteenth century was extracted and the DNA amplified by PCR with overlapping primers targeting the entire HBV genome. The researchers were able to reconstruct the full HBV genome with this approach. Phylogenetic analysis placed the recovered HBV genome in the subtype C clade clustering with viral sequences obtained from Japan and China. The conclusion drawn was that HBV most likely originated either in China or Japan and spread to Korea. Further sequence analysis of the ancient strain revealed that it evolved at least 3,000–100,000 years ago and represents a viral strain that was in circulation 400–500 years ago (Kahila Bar-Gal et al. 2012).

Another recent study from a mid-sixteenth-century child mummy from Italy employed whole genome shotgun Illumina sequencing in an effort to identify potential pathogens present in the DNA from the extracted tissues (Patterson Ross et al. 2018). Preliminary results indicated the presence of HBV reads in tissues tested. Following the initial results, the team developed an in-solution hybridization enrichment approach targeting the whole genome of HBV which yielded substantial HBV sequence with the expected cytosine deamination that is characteristic of ancient DNA fragments. The data from all tissues were pooled, and reconstruction of the entire genome of the HBV from the ancient mummy was achieved. Phylogenetic analysis of the mummy HBV genome revealed a close relationship with modern HBV genotype D. Further analysis indicated that HBV evolution from both the Korean and Italian mummies is characterized by lack of temporal structure resulting in grouping of the ancient strains with modern. The resulting phylogenetic pattern indicated that HBV diversified before the sixteenth century (Patterson Ross et al. 2018).

Hepatitis C virus (HCV) was first identified in 1989 with little epidemiological data initially available. Historical sequences were initially uncommon due to the recent discovery of the virus with the earliest dated sub-genomic sequence characterized from a sample dated to 1976. To obtain viral dynamic and evolutionary information regarding the course of the HCV epidemic over time, PCR and sequencing was applied by Gray et al. (2013). HCV-positive archived samples from 1953 were identified by antibody screening of bio-banked serum samples. PCR screening of the archival positive samples resulted in two subgenomic HCV subtype 1b sequences that represented the oldest evidence of HCV infection to date. Phylogenetic analysis of the two sequences placed them in different clades with both sequences basal to modern taxa. Pairwise nucleotide diversity of the two 1953 obtained sequences indicated they were very closely related to modern reference strains suggesting HCV subtype 1b was in circulation for some time before 1953. The origin date of the most recent common ancestor of HCV subtype 1b in the United States based on the 1953 obtained sequence was estimated between 1874 and 1926 (Gray et al. 2013).

2.5 *Poxviruses*

The *Poxviridae* is a group of enveloped double-stranded DNA viruses that infect and cause disease in many vertebrates and arthropods (Bolte et al. 1999). Human smallpox is caused by the variola virus of the *Orthopoxvirus* genus. Variola virus has caused several devastating epidemics around the globe with mortality rates of up to 30%. During the last century, smallpox epidemics claimed the life of 300–500 million people and were responsible for 10% of global deaths from 1900 to 1980 (Thèves et al. 2014). Following vaccination programs that lasted almost 200 years, the virus was finally eradicated in the 1980s with the last reported case being in Somalia in 1977 (Biagini et al. 2012; Thèves et al. 2014). The high mortality associated with smallpox infections and with the few historical biological samples available limit the knowledge about diversity and evolution of this pathogen. A mass grave discovered in Yakutia, Eastern Siberia, Russia, dated between the seventeenth and eighteenth centuries has provided information on the evolutionary history of smallpox. Biological samples obtained from two out of five frozen mummies from the grave revealed the presence of iron inclusions in the pulmonary tissues, an indication of preserved blood resulting from a potential hemorrhagic episode. PCR screening using poxvirus-specific primers revealed the presence of sequences from different genes of the variola genome in one of the mummified bodies. Bayesian analysis of the Siberian pox sequences was able to extend the origin of smallpox viral strains back to at least 120 AD. Phylogenetic analysis of the obtained Siberia variola sequences clustered them with the 18 available human sequences but not within the two most prevalent contemporary variola clades. The results suggest that the Siberian poxvirus could be a progenitor or close relative of modern smallpox viral strains (Biagini et al. 2012).

In a recent study, Duggan et al. (2016) examined the evolutionary history of variola virus by screening the mummified remains of a young child of undetermined sex found within a crypt of the Dominican Church of the Holy Spirit of Vilnius, Lithuania, dated between 1643 and 1665 AD. DNA extracted from the remains was initially screened for polyomaviruses, but BLAST analysis of the sequencing results indicated that the majority of viral hits belonged to the variola virus. Variola sequences were then enriched by hybridization capture using modern variola virus sequences as baits. Hybridization capture and HTS resulted in full genome enrichment of the ancient variola virus with an average coverage of 18-fold per base pair. Viral assembly of the ancient variola genome (VD21) and comparison with the modern viral strains indicated a strong conservation of gene content and arrangement with no major rearrangements present. Phylogenetic analysis of the VD21 genome using camelpox and taterapox as outgroups placed the archival sequence basal to all previously sequenced orthopoxvirus strains with the closest being the ancient Siberia variola sequence previously reported by Biagini et al. (2012). The results suggest that all the variola genomes compared shared a common ancestor between 1588 and 1645 AD. The common ancestor predated the vaccination efforts that started in 1796 and also predated the variola minor and major clade divergence that is estimated to have occurred between 1734 and 1793. The discovery and characterization of the VD21 genome provided a new epidemiological calibration point for smallpox and that the variola virus lineages responsible for historical smallpox outbreaks are relatively young and have only been in circulation for approximately 200 years (Duggan et al. 2016).

Monkeypox virus (MPXV) is a zoonotic virus that infects multiple hosts including humans and is considered the most medically important orthopoxvirus since the eradication of smallpox virus in 1977. MPXV in humans causes similar clinical symptoms as smallpox with the first reported case in the Democratic Republic of Congo in the 1970s. Since then there has been a dramatic increase of reported cases of MPXV in humans with no known clear source of infection. The majority of human-related cases are attributed to contact with wildlife, with rodents being the most likely candidate reservoir species. Several rodent families including *Funisciurus*, *Heliosciurus*, and *Cricetomys* have been implicated in the transmission of MPXV, but the exact species is still unknown. To examine historical MPXV infection, Tiee et al. (2018) screened 1,038 museum skin samples of five *Funisciurus* species dating from 1899 to 1993 for viral DNA. Overall 9% (93/1,038) of the specimens were found positive for MPXV with sequencing indicating that all obtained sequences belonged to the Congo Basin strain. Viral DNA was amplifiable from five species with the oldest sample from 1899 indicating that the virus was circulating in *Funisciurus* over 115 years ago. MPXV prevalence analysis among the five species tested positive indicated that *F. anerythrus* and *F.congicus* were more often infected suggesting that these species may have played a major role in the interspecific transmission of the virus (Tiee et al. 2018).

Avipoxviruses are pathogens that belong to the *Chordopoxvirinae* subfamily of the *Poxviridae*. Avipoxviruses infect birds, with their common symptoms being lesions on the feet, beak, and the tissue surrounding the eyes. Virions from avipoxviral infection are known to be persistent in the environment and have the ability to be vectored by insects (Bolte et al. 1999). Parker et al. (2011) used museum specimens from the Galapagos Islands dated between 1891 and 1906 to examine the potential correlation between human inhabitation of the islands and the distribution of the virus. Histopathological examination of 226 specimens identified 59 candidate infected samples from six islands. PCR screening followed by direct sequencing using two avipoxvirus gene primer sets on the 59 samples identified 21 avipoxvirus-infected specimens. Avipoxvirus-positive specimen distributions correlated with islands that were human inhabited at the time of collection, with only one sample being collected from an island that was not inhabited by humans at the time. The results suggest that avipoxvirus dispersion involved humans and estimate the potential period of viral introduction to the islands occurred in the late 1890s (Parker et al. 2011).

2.6 Papillomaviruses

The *Papillomaviridae* are a large family of non-enveloped double-stranded DNA viruses that can cause benign or malignant proliferation of the skin and mucosa (Rector et al. 2007). Skin papillomas associated with these viruses were described in texts dating back to the first century AD. During the examination of 38 mummies found in the church of Saint Domenico Maggiore in Naples dated between the fifteenth and sixteenth century, a female mummy was discovered showing evidence of syphilitic "gumma," a noncancerous skin growth, an indication of a syphilis infection. Further examination for other sexually transmitted diseases revealed papillary skin lesions suggesting the presence of anogenital warts. DNA extracted from the skin of the mummy was screened by PCR for the presence of human papillomavirus (HPV) viral infection. An amplified DNA fragment obtained from the mummy was cloned and sequenced confirming infection of the mummy with two HPV strains. The HPV 18 strain, one of the most prevalent viral strains that is associated with cervical cancer occurrence, and the JC9813 strain, a putative novel HPV with low oncogenic potential, were discovered (Fornaciari et al. 2003; Holman et al. 2014). The two HPV sequences represent the first characterized papillomavirus sequences identified from 449-year-old mummies.

2.7 *Plant Viruses*

Paleovirology research is not limited to mammalian pathogens. Barley stripe mosaic virus (BSMV) is a very well-characterized positive-sense single-stranded RNA virus that can infect a variety of plants. The virus has no vector and is transmitted on pollen and seeds. BSMV was first identified and characterized in the 1950s with the earliest record of its existence within the last 100 years. To study the molecular evolution of BSMV, RNA was extracted from archaeological barley grains collected from southern Egypt. RT-PCR was performed on extracted grain RNA targeting BSMV genes. One of the samples yielded amplicons indicating the presence of BSMV virus. Whole genome HTS of PCR-positive RNA extract and genome assembly identified a 600–900 year-old genome of BSMV. Phylogenetic analysis placed the reconstructed ancient virus basal to the modern BSMV clade. Viral lineage comparison with modern strains indicate that the BSMV virus most likely originated in the Middle East or North Africa and spread along historical trade routes. The genome revealed a slower evolutionary rate than what was previously estimated from modern viral sequences, which explains why the virus appears to be much younger than it is when the ancient sequence is excluded. Smith et al. (2014) were able to reconstruct the first archaeological complete viral genome to date and better estimate the mutation rate of an economically important viral pathogen (Smith et al. 2014).

2.8 *Giant Viruses*

Viral screenings of Pleistocene-age samples lead to the identification, revival, and characterization of a new giant virus family member. The newly discovered virus was named *Pithovirus* due to its amphorae-like shape. Pithoviruses are double-stranded DNA viruses that infect *Acanthamoeba*, and they were isolated from a 30,000-year-old permafrost layer from Siberia. Legendre et al. (2014) in an effort to identify an infectious virus from Siberian permafrost samples inoculated *Acanthamoeba castellanii* cultures with permafrost extract. Light microscopy screening of the inoculated cultures identified replicating viral ovoid particles.

Transmission electron microscopy of the ovoid particles demonstrated the amphorae-like shape typical of pithoviruses. The virus exhibits a replication cycle and genomic features of icosahedral nucleocytoplasmic DNA viruses. Sequencing of the pithovirus revealed a 600-kb-long AT-rich, extremely high repeat containing genome with gene content similar to *Iridovirus* and *Marseillevirus*. Pithoviruses represent the oldest eukaryotic viruses revived to date and suggest extreme survival is possible in permafrost for some viral groups. The results of the study demonstrate the potential to revive and study extinct pathogens. However, the study also suggests that harmful pathogens could be released due to climate warming and other anthropogenic processes such as mining.

3 Ancient Viromes

Viromes represent the full viral component of a species' or environment's microbial content (Lecuit and Eloit 2013). Viral metagenomics that are based on sequencing analysis of all viral genomes available in a sample have promoted the characterization of the viral diversity in modern samples. Using virome sequencing approaches that involved the isolation of viral particles from the sample of interest followed by a shotgun sequencing strategy, Appelt et al. (2014) generated a viral metagenome from a fourteenth-century coprolite sample that was found in a closed barrel in a Middle Age excavation site in Namur, Belgium. Viral-like particles were isolated using centrifugation followed by filtration to remove contaminants. Resulting materials were visualized using electron microscopy to verify the presence of viral-like particles in the samples. DNA extraction of the isolated material was performed followed by HTS on a 454 sequencer. The majority of generated sequences were of unknown origin. The sequences that could be characterized belonged to DNA viruses that infect archaea, eukaryotes, and bacteria. A large proportion of the characterized viral sequences belonged to bacteriophage of the *Siphoviridae* family of double-stranded viruses. The *Siphoviridae* family identification is consistent with modern stool samples. Analyses of the data from the first ancient human virome also revealed the conservation of metabolic function of viral communities identified when compared to modern human fecal viromes (Appelt et al. 2014).

More recently, caribou (*Rangifer tarandus*) fecal material frozen for the last 5000 years and obtained in the Selwyn Mountains, Canada, was examined for viral diversity. DNA was extracted from frozen fecal pellets dated to 700, 2920, 3070, and 3230 years before present, respectively. Nucleic acids obtained were reverse transcribed and HTS on a 454 sequencer. Two novel viral sequences were identified from the 700 year-old fecal material. A DNA virus that was named ancient caribou feces associated virus (aCFV) and a single RNA virus that was named ancient Northwestern Territories cripavirus (aNCV) were characterized. Phylogenetic analysis indicated that aCFV is a distant relative of plant-infecting geminiviruses and fungi-infecting *Sclerotinia* sclerotiorum hypovirulence associated DNA virus 1, while aNCV grouped within the insect-infecting *Cripavirus* viral group. Ng et al. (2014) hypothesized that the viral sequences originated from plant material that the caribou ingested. An aCFV infectivity study was also performed indicating that the virus could still infect the model plant *Nicotiana benthamiana*. Virome data sequenced from the fecal material confirms that intact viral genomes can be obtained from naturally cryopreserved samples under certain conditions, and like the permafrost giant viruses, the viruses can remain infectious (Ng et al. 2014).

4 Controversy

The study of aDNA, a field that is more than 30 years old, has had a profound impact on molecular biology, microbiology, molecular evolution, and genomics (Hofreiter et al. 2015; Rizzi et al. 2012; Tsangaras and Greenwood 2012). aDNA research led to the development of the field of paleovirology, but the discipline progression has not been without its pitfalls. Paleovirology studies have suffered from contamination of laboratory reagents, from environmental microorganisms, or from specimen mishandling (Orlando et al. 2015). Authentication criteria were developed that all aDNA studies should adhere to in order to minimize potential contamination. The gold standards for authentication criteria are physically isolated work areas, inclusion of negative controls, reproduction of results obtained from a sample using the same and different DNA extracts, appropriate molecular behavior (larger amplicons should yield weaker products), and results should be independently reproduced in a separate laboratory (Cooper and Poinar 2000; Tsangaras and Greenwood 2012). The development of stringent authentication criteria can minimize artifacts and has unleashed the potential of aDNA in fields such as paleovirology (Cooper and Poinar 2000; Rizzi et al. 2012; Willerslev and Cooper 2005). Even though the authentication criteria were developed before the availability of high-throughput sequencing, many of the criteria are still vital to ensure the avoidance of contamination. For example, isolated and dedicated ancient DNA work areas remain crucial. Sequencing of negative controls is particularly critical in virus research due to potential reagent contamination (Moustafa et al. 2017). Reproducibility and absence of viral sequences in associated remains or samples are also important criteria. PCR and multiplex PCR approaches that were used before the development of HTS were limited to the number of regions they targeted and the sequence output they produced, making the above PCR-related authenticity criteria a necessity for the reliability of the results. Development of HTS, however, circumvents many of the criteria associated with PCR issues, such as expected molecular behavior of PCR and the need to clone PCR products or quantitate sample-specific DNA. The ability of HTS to sequence the entire genome or large number of fragments that came from the same targeted region provides the ability in the majority of cases to distinguish real ancient fragments from contaminants. Ancient DNA fragments have unique features (e.g., shorter length, postmortem fragmentation observed in depurinated sites, and C-T misincorporation patterns at the ends of DNA fragments) that make them distinguishable from modern DNA fragments enabling their correct classification

(Llamas et al. 2017). Despite more widely spread adherence to most or all of the authentication criteria, there have been controversial results published especially in the early years of the field, for example, the claim of identification of HTLV-1 sequence from an Andean mummy (Tsangaras and Greenwood 2012).

In spite of the great success of the 1918 influenza pandemic strain characterization from archival human material, the lack of adherence to the aDNA authentication criteria resulted in controversial historical avian influenza virus results. Wild waterfowl museum samples collected between 1915 and 1919 were screened for influenza A using a similar approach to that employed for human samples. An HA gene sequence was reported for a sample dating to 1917. The sequence of the HA-positive sample was closely related to that of modern avian influenza viruses rather than to the 1918 pandemic virus. The results were interpreted as representing relative genetic diversity stasis of avian influenza over the past 85 years in waterfowl. Furthermore, the results of the study suggested that the pandemic H1N1 virus did not acquire its HA gene from wild waterfowl (Fanning et al. 2002). However, reanalysis of the results demonstrated that the 1917 archival avian sequence was nearly identical to four modern avian sequences obtained from birds in Ohio in 1999. Phylogenetic analysis of the avian archival sequence and modern sequences revealed a molecular clock pattern of increased genetic distance based on collection date (Worobey 2008). The result suggests that bird influenza does not remain in genetic stasis but has high mutational and replication rates and evolves in a clocklike fashion (Chen and Holmes 2006). Thus even if a very strong selection pressure had been experienced by the archival sequence, it is highly unlikely that it would be identical to modern sequences. Worobey (2008) was able to demonstrate that the 1917 wild waterfowl avian sequence was most likely derived from a modern laboratory contaminant (Worobey 2008).

Human T-cell leukemia virus 1 (HTLV-I) isolation from a 1,500-year-old Andean mummy is also a controversial find. HTLV-I is a retrovirus that can cause leukemia or lymphoma in infected individuals. In an effort to investigate the distribution of HTLV-I virus in South America, Li et al. (1999) isolated DNA from a 1,500-year-old Andean mummy. Using PCR amplification and Sanger sequencing, the researchers reported the isolation of a partial px gene and LTR sequence from the ancient extracts (Li et al. 1999). The results of the study though were disputed due to the lack of rigorous controls and the failure to adhere to the aDNA authentication criteria. Phylogenetic and molecular clock analysis by Gessain et al. (2000) and Vandamme et al. (2000) using the HTLV-I archival data illustrated that the retrieved sequences were inconsistent with an ancient origin and most likely represent contamination with modern HTLV-1 DNA (Gessain et al. 2000; Vandamme et al. 2000).

Plant paleovirology has produced controversial studies such as the report of 140,000-year-old tomato mosaic tobamovirus sequences. Tomato mosaic tobamovirus (ToMV) is a very stable positive-sense single-stranded RNA virus that can be found in clouds, water, and soil and on plants. Castello et al. (1999) hypothesized that the virus could therefore be isolated from ancient glacial ice. Samples dated back 140000 years were obtained from Greenland and screened using RT-PCR for ToMV genes. The ToMV-specific primers were able to produce product for 17 out of 30 meltwaters that were screened. Sequencing of the obtained products revealed ToMV ORF coat protein sequences supporting the group's hypothesis that ToMV could be retrieved (Castello et al. 1999). The group's results were challenged by Worobey et al. (2008) using phylogenetic analysis, which demonstrated that the ancient sequences were identical to modern ToMV sequences, a result that is highly unlikely for rapidly evolving RNA viruses. Worobey et al. (2008) further suggested that the sequences obtained from Castello et al. (1999) were either a contamination from the positive control used in the study or due to a field collection contamination (Worobey 2008).

5 Biosafety and Ethical Concerns

HTS and advances in molecular methods have enabled the full genome sequencing of several extinct viral pathogens such as the 1918 influenza ancient smallpox virus. Perhaps more surprising than identifying ancient viral genomic sequences, recently a 30,000-year-old replication-competent DNA virus and two 700-year-old intact RNA viruses were isolated from Siberian permafrost and from a permanent ice patch in Canada, respectively (Legendre et al. 2014; Ng et al. 2014). These results suggest that under certain conditions, viruses have the ability to survive for prolonged periods of time retaining their infectivity in the process (Holmes 2014). The isolation of such viruses along with the reconstruction of 1918 influenza virus raises biosafety and ethical dilemmas. Is it safe to work with samples such as permafrost, well-preserved animal mummies, or museum samples? Is it safe to either resurrect or reconstruct deadly or potentially deadly pathogens? Should the sequencing information be made generally available? The scientists that resurrected the 30,000-year-old *Pithovirus* suggested that global warming, intensive drilling, and mining operations in the Arctic could thaw and release potentially harmful human pathogens. However, even though such an occurrence is theoretically plausible, other scientists argue that such a risk is minimal, and thus far no zoonotic viruses have been revived directly from the environment (Holmes 2014; Reardon 2014). It remains to be seen how many different viruses can be cultured from ancient samples. This has never been tested systematically, and therefore the full extent of the risks remains unclear.

Critics of the reconstruction of the 1918 pandemic influenza virus claimed the study should have never been performed as it provides a guideline for the creation of weaponized influenza. The critics suggested that at the very least, the material and method sections of such studies remain unavailable to the public. The authors and the supporters of the 1918 influenza virus reconstruction argued that the medical benefits of such studies outweigh associated risks and further argued that the material and methods are crucial for replication and verification purposes (Selgelid 2009). Paleovirology is a young field, and many of the pathogens studied are not of zoonotic relevance. Nonetheless, as work continues on the pathogens involved in major historical infectious disease outbreaks further biosafety considerations, ethical discussions and potential risk and benefit analysis would be warranted (Miller and Selgelid 2007).

6 Future Perspectives and Conclusions

Paleovirology is a field that is still in its infancy with major limitations due to the degraded nature of historical and ancient DNA and RNA, the fragility of most viruses, and the relatively low number of virions relative to host or environmental cells in infected materials (Harkins and Stone 2015; Katzourakis 2013; van Regenmortel and Mahy 2010). Like modern virology, paleovirology also suffers from the lack of general markers that can be applied for pan viral assays in the way that microbiomes can be characterized by sequencing amplified 16S rDNA genes. Viruses have no such genes in common (Gasc et al. 2016). Even though the majority of viruses are not stable over time, we have summarized the successes and controversies in the emerging paleovirology field (also summarized in Fig. 2) (Asnicar et al. 2015). The isolation, sequence, and characterization from permafrost and paraffin samples of the 1918 influenza virus represent one of the earliest and most successful stories in paleovirology and paleogenomics in general. However, due to the unstable nature of viruses, paleovirology will likely remain primarily limited to certain kinds of viruses, e.g., DNA viruses and only under specific conditions of preservation. Advances in genomic enrichment techniques and sequencing technologies offer great potential to expand this challenging field. The development of hybridization enrichment techniques using synthetic oligonucleotides or PCR products as baits to select sequences of interest from the abundant noise that is present in ancient samples has enabled the discipline to move forward and pass the limitation of polymerase chain reaction (PCR) methods used until recently. Even though PCR still remains an integral method in paleogenomics research, the ability to perform solution hybridization, which can tolerate and enrich divergent sequences due to the longer bait sequences used, has considerably broadened the scope of research with ancient samples (Samorodnitsky et al. 2015; Gasc et al. 2016). Despite its potential, in-solution enrichment methods still generate a substantial amount of non-target-specific sequence, and as the sequences generated are very short, viral assembly or placing polymorphisms in phase, is challenging even at times with long read sequences. Nonetheless, combining enrichment methods with the sequencing depth that HTS offers, ancient viral genomes can be constructed from degraded short sequences that would otherwise take many years to obtain using classic molecular methods. Obtaining these sequences will allow for the identification of viral strains associated with historical outbreaks, determination of the origin of existing viral strains, and direct determination of evolutionary rates of viruses. As viruses do not leave behind fossils, such molecular information, if properly gathered and applied, could revolutionize virology.

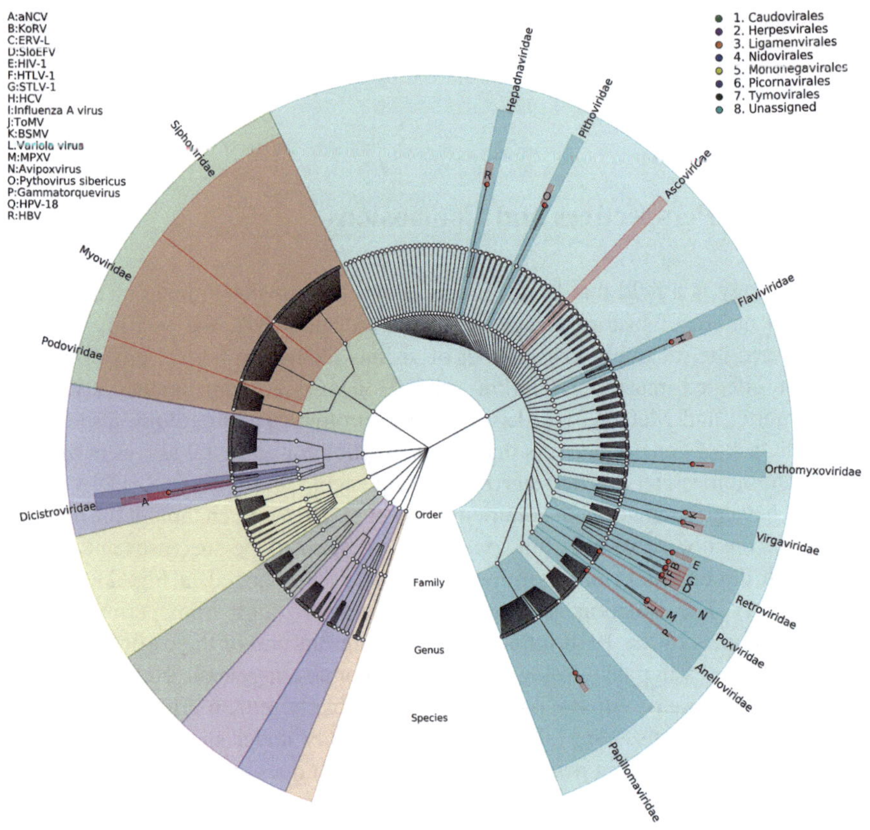

Fig. 2 Viral phylogeny illustration of ICTV taxonomic data plotted using GraPhlAn (Asnicar et al. 2015). Viral orders are illustrated with different colors in the circular phylogeny figure. Viral family, genus, and species that were identified in ancient DNA studies are illustrated using red background or red clade circles

References

Alter MJ. Epidemiology of viral hepatitis and HIV co-infection. J Hepatol. 2006;44(Suppl 1):S6–9. https://doi.org/10.1016/j.jhep.2005.11.004.

Appelt S, Fancello L, Le Bailly M, Raoult D, Drancourt M, Desnues C. Viruses in a 14th-century coprolite. Appl Environ Microbiol. 2014;80(9):2648–55. https://doi.org/10.1128/aem.03242-13.

Asnicar F, Weingart G, Tickle TL, Huttenhower C, Segata N. Compact graphical representation of phylogenetic data and metadata with GraPhlAn. PeerJ. 2015;3:e1029.

Aswad A, Katzourakis A. Paleovirology and virally derived immunity. Trends Ecol Evol. 2012;27 (11):627–36. https://doi.org/10.1016/j.tree.2012.07.007.

Avila-Arcos MC, Ho SY, Ishida Y, Nikolaidis N, Tsangaras K, Honig K, Medina R, Rasmussen M, Fordyce SL, Calvignac-Spencer S, Willerslev E, Gilbert MT, Helgen KM, Roca AL, Greenwood AD. One hundred twenty years of koala retrovirus evolution determined from museum skins. Mol Biol Evol. 2013;30(2):299–304. https://doi.org/10.1093/molbev/mss223.

Bedarida S, Dutour O, Buzhilova AP, de Micco P, Biagini P. Identification of viral DNA (Anelloviridae) in a 200-year-old dental pulp sample (Napoleon's Great Army, Kaliningrad, 1812). Infect Genet Evol. 2011;11(2):358–62. https://doi.org/10.1016/j.meegid.2010.11.007.

Biagini P, Thèves C, Balaresque P, Géraut A, Cannet C, Keyser C, Nikolaeva D, Gérard P, Duchesne S, Orlando L, Willerslev E, Alekseev AN, de Micco P, Ludes B, Crubézy E. Variola virus in a 300-year-old Siberian mummy. N Engl J Med. 2012;367(21):2057–9. https://doi.org/10.1056/NEJMc1208124.

Bolte AL, Meurer J, Kaleta EF. Avian host spectrum of avipoxviruses. Avian Pathol. 1999;28(5):415–32. https://doi.org/10.1080/03079459994434.

Bouvier NM, Palese P. The biology of influenza viruses. Vaccine. 2008;26:D49–53.

Calvignac S, Terme J-M, Hensley SM, Jalinot P, Greenwood AD, Hänni C. Ancient DNA identification of early 20th century simian T-cell leukemia virus type 1. Mol Biol Evol. 2008;25(6):1093–8. https://doi.org/10.1093/molbev/msn054.

Castello JD, Rogers SO, Starmer WT, Catranis CM, Ma L, Bachand GD, Zhao Y, Smith JE. Detection of tomato mosaic tobamovirus RNA in ancient glacial ice. Polar Biol. 1999;22(3):207–12. https://doi.org/10.1007/s003000050411.

Chen R, Holmes EC. Avian influenza virus exhibits rapid evolutionary dynamics. Mol Biol Evol. 2006;23(12):2336–41. https://doi.org/10.1093/molbev/msl102.

Cooper A, Poinar HN. Ancient DNA: do it right or not at all. Science. 2000;289(5482):1139. https://doi.org/10.1126/science.289.5482.1139b.

Drummond A, Pybus OG, Rambaut A. Inference of viral evolutionary rates from molecular sequences. Adv Parasitol. 2003;54:331–58.

Duchene S, Holmes EC, Ho SY. Analyses of evolutionary dynamics in viruses are hindered by a time-dependent bias in rate estimates. Proc Roy Soc Biol Sci. 2014;281(1786). https://doi.org/10.1098/rspb.2014.0732.

Duffy S, Shackelton LA, Holmes EC. Rates of evolutionary change in viruses: patterns and determinants. Nat Rev Genet. 2008;9(4):267–76. https://doi.org/10.1038/nrg2323.

Duggan AT, Perdomo MF, Piombino-Mascali D, Marciniak S, Poinar D, Emery MV, Buchmann JP, Duchêne S, Jankauskas R, Humphreys M, Golding GB, Southon J, Devault A, Rouillard J-M, Sahl JW, Dutour O, Hedman K, Sajantila A, Smith GL, Holmes EC, Poinar HN. 17th century variola virus reveals the recent history of smallpox. Curr Biol. 2016. https://doi.org/10.1016/j.cub.2016.10.061.

Fanning TG, Slemons RD, Reid AH, Janczewski TA, Dean J, Taubenberger JK. 1917 avian influenza virus sequences suggest that the 1918 pandemic virus did not acquire its Hemagglutinin directly from birds. J Virol. 2002;76(15):7860–2. https://doi.org/10.1128/JVI.76.15.7860-7862.2002.

Fornaciari G, Zavaglia K, Giusti L, Vultaggio C, Ciranni R. Human papillomavirus in a 16th century mummy. Lancet. 2003;362(9390):1160. https://doi.org/10.1016/S0140-6736(03)14487-X.

Franco E, Bagnato B, Marino MG, Meleleo C, Serino L, Zaratti L. Hepatitis B: epidemiology and prevention in developing countries. World J Hepatol. 2012;4(3):74–80. https://doi.org/10.4254/wjh.v4.i3.74.

Gasc C, Peyretaillade E, Peyret P. Sequence capture by hybridization to explore modern and ancient genomic diversity in model and nonmodel organisms. Nucleic Acids Res. 2016;44(10):4504–18. https://doi.org/10.1093/nar/gkw309.

Gessain A, Pecon-Slattery J, Meertens L, Mahieux R. Origins of HTLV-1 in South America (letter 1). Nat Med. 2000;6(3):232.

Gilbert MTP, Rambaut A, Wlasiuk G, Spira TJ, Pitchenik AE, Worobey M. The emergence of HIV/AIDS in the Americas and beyond. Proc Natl Acad Sci. 2007;104(47):18566–70. https://doi.org/10.1073/pnas.0705329104.

Gojobori T, Moriyama EN, Kimura M. Molecular clock of viral evolution, and the neutral theory. Proc Natl Acad Sci. 1990;87(24):10015–8.

Gray RR, Tanaka Y, Takebe Y, Magiorkinis G, Buskell Z, Seeff L, Alter HJ, Pybus OG. Evolutionary analysis of hepatitis C virus gene sequences from 1953. Philos Trans R Soc Biol Sci. 2013;368(1626):20130168. https://doi.org/10.1098/rstb.2013.0168.

Greenwood AD, Lee F, Capelli C, DeSalle R, Tikhonov A, Marx PA, MacPhee RDE. Evolution of endogenous retrovirus-like elements of the woolly mammoth (Mammuthus primigenius) and its relatives. Mol Biol Evol. 2001;18(5):840–7.

Harkins KM, Stone AC. Ancient pathogen genomics: insights into timing and adaptation. J Hum Evol. 2015;79:137–49. https://doi.org/10.1016/j.jhevol.2014.11.002.

Heintzman PD, Soares AER, Chang D, Shapiro B. Paleogenomics. In: Reviews in cell biology and molecular medicine. Weinheim: Wiley-VCH Verlag GmbH & Co. KGaA; 2006. https://doi.org/10.1002/3527600906.mcb.201500020.

Hofreiter M, Paijmans JL, Goodchild H, Speller CF, Barlow A, Fortes GG, Thomas JA, Ludwig A, Collins MJ. The future of ancient DNA: technical advances and conceptual shifts. Bioessays. 2015;37(3):284–93. https://doi.org/10.1002/bies.201400160.

Holman DM, Benard V, Roland KB, Watson M, Liddon N, Stokley S. Barriers to human papillomavirus vaccination among US adolescents: a systematic review of the literature. JAMA Pediatr. 2014;168(1):76–82. https://doi.org/10.1001/jamapediatrics.2013.2752.

Holmes EC. Freezing viruses in time. Proc Natl Acad Sci U S A. 2014;111(47):16643–4. https://doi.org/10.1073/pnas.1419827111.

Ishida Y, Zhao K, Greenwood AD, Roca AL. Proliferation of endogenous retroviruses in the early stages of a host germ line invasion. Mol Biol Evol. 2015;32(1):109–20.

Jern P, Coffin JM. Effects of retroviruses on host genome function. Annu Rev Genet. 2008;42:709–32.

Kahila Bar-Gal G, Kim MJ, Klein A, Shin DH, Oh CS, Kim JW, Kim TH, Kim SB, Grant PR, Pappo O, Spigelman M, Shouval D. Tracing hepatitis B virus to the 16th century in a Korean mummy. Hepatology. 2012;56(5):1671–80. https://doi.org/10.1002/hep.25852.

Kalish ML, Wolfe ND, Ndongmo CB, McNicholl J, Robbins KE, Aidoo M, Fonjungo PN, Alemnji G, Zeh C, Djoko CF, Mpoudi-Ngole E, Burke DS, Folks TM. Central African hunters exposed to simian immunodeficiency virus. Emerg Infect Dis. 2005;11(12):1928–30. https://doi.org/10.3201/eid1112.050394.

Katzourakis A. Paleovirology: inferring viral evolution from host genome sequence data. Phil Trans Roy Soc Biol Sci. 2013;368(1626):20120493. https://doi.org/10.1098/rstb.2012.0493.

Katzourakis A, Gifford RJ, Tristem M, Gilbert MTP, Pybus OG. Macroevolution of complex retroviruses. Science. 2009;325(5947):1512.

Lecuit M, Eloit M. The human virome: new tools and concepts. Trends Microbiol. 2013;21(10):510–5. https://doi.org/10.1016/j.tim.2013.07.001.

Legendre M, Bartoli J, Shmakova L, Jeudy S, Labadie K, Adrait A, Lescot M, Poirot O, Bertaux L, Bruley C, Coute Y, Rivkina E, Abergel C, Claverie JM. Thirty-thousand-year-old distant relative of giant icosahedral DNA viruses with a pandoravirus morphology. Proc Natl Acad Sci. 2014;111(11):4274–9. https://doi.org/10.1073/pnas.1320670111.

Li H-C, Fujiyoshi T, Lou H, Yashiki S, Sonoda S, Cartier L, Nunez L, Munoz I, Horai S, Tajima K. The presence of ancient human T-cell lymphotropic virus type I provirus DNA in an Andean mummy. Nat Med. 1999;5(12):1428–32.

Llamas B, Valverde G, Fehren-Schmitz L, Weyrich LS, Cooper A, Haak W. From the field to the laboratory: controlling DNA contamination in human ancient DNA research in the high-throughput sequencing era. Sci Technol Archaeol Res. 2017;3(1):1–14. https://doi.org/10.1080/20548923.2016.1258824.

Miller S, Selgelid MJ. Ethical and philosophical consideration of the dual-use dilemma in the biological sciences. Sci Eng Ethics. 2007;13(4):523–80. https://doi.org/10.1007/s11948-007-9043-4.

Moustafa A, Xie C, Kirkness E, Biggs W, Wong E, Turpaz Y, Bloom K, Delwart E, Nelson KE, Venter JC, Telenti A. The blood DNA virome in 8,000 humans. PLoS Pathog. 2017;13(3):e1006292. https://doi.org/10.1371/journal.ppat.1006292.

Ng TFF, Chen L-F, Zhou Y, Shapiro B, Stiller M, Heintzman PD, Varsani A, Kondov NO, Wong W, Deng X, Andrews TD, Moorman BJ, Meulendyk T, MacKay G, Gilbertson RL, Delwart E. Preservation of viral genomes in 700-y-old caribou feces from a subarctic ice patch. Proc Natl Acad Sci. 2014;111(47):16842–7. https://doi.org/10.1073/pnas.1410429111.

Noymer A, Garenne M. The 1918 influenza epidemic's effects on sex differentials in mortality in the United States. Popul Dev Rev. 2000;26(3):565–81.

Orlando L, Gilbert MT, Willerslev E. Reconstructing ancient genomes and epigenomes. Nat Rev Genet. 2015;16(7):395–408. https://doi.org/10.1038/nrg3935.

Pappas C, Aguilar PV, Basler CF, Solorzano A, Zeng H, Perrone LA, Palese P, Garcia-Sastre A, Katz JM, Tumpey TM. Single gene reassortants identify a critical role for PB1, HA, and NA in the high virulence of the 1918 pandemic influenza virus. Proc Natl Acad Sci. 2008;105 (8):3064–9. https://doi.org/10.1073/pnas.0711815105.

Parker PG, Buckles EL, Farrington H, Petren K, Whiteman NK, Ricklefs RE, Bollmer JL, Jiménez-Uzcátegui G. 110 Years of avipoxvirus in the Galapagos Islands. PLoS One. 2011;6 (1):e15989. https://doi.org/10.1371/journal.pone.0015989.

Patterson Ross Z, Klunk J, Fornaciari G, Giuffra V, Duchêne S, Duggan AT, Poinar D, Douglas MW, Eden J-S, Holmes EC, Poinar HN. The paradox of HBV evolution as revealed from a 16th century mummy. PLoS Pathog. 2018;14(1):e1006750. https://doi.org/10.1371/journal.ppat. 1006750.

Pybus OG, Markov PV, Wu A, Tatem AJ. Investigating the endemic transmission of the hepatitis C virus. Int J Parasitol. 2007;37(8–9):839–49. https://doi.org/10.1016/j.ijpara.2007.04.009.

Reardon S. Infectious diseases: smallpox watch. Nature. 2014;509(7498):22–4. https://doi.org/10. 1038/509022a.

Rector A, Lemey P, Tachezy R, Mostmans S, Ghim S-J, Van Doorslaer K, Roelke M, Bush M, Montali RJ, Joslin J, Burk RD, Jenson AB, Sundberg JP, Shapiro B, Van Ranst M. Ancient papillomavirus-host co-speciation in Felidae. Genome Biol. 2007;8(4):R57. https://doi.org/10. 1186/gb-2007-8-4-r57.

Reid AH, Fanning TG, Hultin JV, Taubenberger JK. Origin and evolution of the 1918 "Spanish" influenza virus hemagglutinin gene. Proc Natl Acad Sci. 1999;96(4):1651–6.

Rizzi E, Lari M, Gigli E, De Bellis G, Caramelli D. Ancient DNA studies: new perspectives on old samples. Genet Sel Evol. 2012;44:21. https://doi.org/10.1186/1297-9686-44-21.

Samorodnitsky E, Jewell BM, Hagopian R, Miya J, Wing MR, Lyon E, Damodaran S, Bhatt D, Reeser JW, Datta J, Roychowdhury S. Evaluation of hybridization capture versus amplicon-based methods for whole-Exome sequencing. Hum Mutat. 2015;36(9):903–14. https://doi.org/ 10.1002/humu.22825.

Selgelid MJ. Governance of dual-use research: an ethical dilemma. Bull World Health Organ. 2009;87(9):720–3.

Slater GJ, Cui P, Forasiepi AM, Lenz D, Tsangaras K, Voirin B, de Moraes-Barros N, MacPhee RDE, Greenwood AD. Evolutionary relationships among extinct and extant sloths: the evidence of mitogenomes and retroviruses. Genome Biol Evol. 2016;8(3):607–21. https:// doi.org/10.1093/gbe/evw023.

Smith O, Clapham A, Rose P, Liu Y, Wang J, Allaby RG. A complete ancient RNA genome: identification, reconstruction and evolutionary history of archaeological barley stripe mosaic virus. Sci Rep. 2014;4:4003. https://doi.org/10.1038/srep04003.

Tarlinton RE, Meers J, Young PR. Retroviral invasion of the koala genome. Nature. 2006;442(7098):79–81. http://www.nature.com/nature/journal/v442/n7098/suppinfo/ nature04841_S1.html.

Taubenberger JK, Morens DM. 1918 influenza: the mother of all pandemics. Emerg Infect Dis. 2006;12(1):15–22. https://doi.org/10.3201/eid1201.050979.

Taubenberger JK, Morens DM. The pathology of influenza virus infections. Annu Rev Pathol. 2008;3:499.

Taubenberger JK, Reid AH, Krafft AE, Bijwaard KE, Fanning TG. Initial genetic characterization of the 1918 "Spanish" influenza virus. Science. 1997;275(5307):1793–6. https://doi.org/10. 1126/science.275.5307.1793.

Taubenberger JK, Reid AH, Fanning TG. The 1918 influenza virus: a killer comes into view. Virology. 2000;274(2):241–5. https://doi.org/10.1006/viro.2000.0495.

Taubenberger JK, Reid AH, Lourens RM, Wang R, Jin G, Fanning TG. Characterization of the 1918 influenza virus polymerase genes. Nature. 2005;437(7060):889–93. https://doi.org/10.1038/nature04230.

Thèves C, Biagini P, Crubézy E. The rediscovery of smallpox. Clin Microbiol Infect. 2014;20(3):210–8. https://doi.org/10.1111/1469-0691.12536.

Tiee MS, Harrigan RJ, Thomassen HA, Smith TB. Ghosts of infections past: using archival samples to understand a century of monkeypox virus prevalence among host communities across space and time. R Soc Open Sci. 2018;5(1):171089.

Tsangaras K, Greenwood AD. Museums and disease: using tissue archive and museum samples to study pathogens. Ann Anat. 2012;194(1):58–73. https://doi.org/10.1016/j.aanat.2011.04.003.

Tsangaras K, Siracusa MC, Nikolaidis N, Ishida Y, Cui P, Vielgrader H, Helgen KM, Roca AL, Greenwood AD. Hybridization capture reveals evolution and conservation across the entire koala retrovirus genome. PLoS One. 2014;9(4):e95633. https://doi.org/10.1371/journal.pone.0095633.

Tumpey TM, Basler CF, Aguilar PV, Zeng H, Solorzano A, Swayne DE, Cox NJ, Katz JM, Taubenberger JK, Palese P, Garcia-Sastre A. Characterization of the reconstructed 1918 Spanish influenza pandemic virus. Science. 2005;310(5745):77–80. https://doi.org/10.1126/science.1119392.

Umemura T, Tanaka Y, Kiyosawa K, Alter HJ, Shih JW. Observation of positive selection within hypervariable regions of a newly identified DNA virus (SEN virus)(1). FEBS Lett. 2002;510(3):171–4.

van Regenmortel MHV, Mahy BWJ. Desk encyclopedia of general virology. Oxford: Academic Press; 2010.

Vandamme A-M, Hall WW, Lewis MJ, Goubau P, Salemi M. Origins of HTLV-1 in South America (letter 2). Nat Med. 2000;6(3):232–3.

VandeWoude S, Apetrei C. Going wild: lessons from naturally occurring T-lymphotropic lentiviruses. Clin Microbiol Rev. 2006;19(4):728–62. https://doi.org/10.1128/cmr.00009-06.

Weiss RA. The discovery of endogenous retroviruses. Retrovirology. 2006;3(1):67. https://doi.org/10.1186/1742-4690-3-67.

Willerslev E, Cooper A. Ancient DNA. Proc R Soc B Biol Sci. 2005;272(1558):3–16. https://doi.org/10.1098/rspb.2004.2813.

Wolfe ND, Heneine W, Carr JK, Garcia AD, Shanmugam V, Tamoufe U, Torimiro JN, Prosser AT, LeBreton M, Mpoudi-Ngole E, McCutchan FE, Birx DL, Folks TM, Burke DS, Switzer WM. Emergence of unique primate T-lymphotropic viruses among central African bushmeat hunters. Proc Natl Acad Sci. 2005;102(22):7994–9. https://doi.org/10.1073/pnas.0501734102.

Worobey M. Phylogenetic evidence against evolutionary stasis and natural abiotic reservoirs of influenza A virus. J Virol. 2008;82(7):3769–74. https://doi.org/10.1128/jvi.02207-07.

Worobey M, Gemmel M, Teuwen DE, Haselkorn T, Kunstman K, Bunce M, Muyembe J-J, Kabongo J-MM, Kalengayi RM, Van Marck E, Gilbert MTP, Wolinsky SM. Direct evidence of extensive diversity of HIV-1 in Kinshasa by 1960. Nature. 2008;455(7213):661–4. http://www.nature.com/nature/journal/v455/n7213/suppinfo/nature07390_S1.html.

Worobey M, Watts TD, McKay RA, Suchard MA, Granade T, Teuwen DE, Koblin BA, Heneine W, Lemey P, Jaffe HW. 1970s and "Patient 0" HIV-1 genomes illuminate early HIV/AIDS history in North America. Nature. 2016;539(7627):98–101. https://doi.org/10.1038/nature19827. http://www.nature.com/nature/journal/v539/n7627/abs/nature19827.html. Supplementary-information.

Wylie TN, Wylie KM, Herter BN, Storch GA. Enhanced virome sequencing using targeted sequence capture. Genome Res. 2015;25(12):1910–20. https://doi.org/10.1101/gr.191049.115.

Zhao F, Qi J, Schuster SC. Tracking the past: interspersed repeats in an extinct Afrotherian mammal, Mammuthus primigenius. Genome Res. 2009;19(8):1384–92. https://doi.org/10.1101/gr.091363.109.

Reconstructing Past Vegetation Communities Using Ancient DNA from Lake Sediments

Laura Parducci, Kevin Nota, and Jamie Wood

Abstract The field of ancient DNA has received much attention since the mid-1980s, when the first sequences of extinct species were obtained from museum and archaeological specimens. Early analyses focused on organellar DNA (mitochondrial in animals and chloroplast in plants) as these are present in multiple copies in the cells making isolation and analyses easier. Within the last decade, however, with considerable advances in high-throughput DNA sequencing technology and bioinformatics, it has become possible to analyse the more informative nuclear genome of a larger number of ancient samples and from a larger variety of substrates and environments. Here, we present recent progress made to reconstruct ancient vegetation communities from lake sediments and review recent key findings in the field. We synthesize and discuss the sources of plant DNA in sediment, the issues relating to DNA preservation after deposition, the criteria required for authentication and the technical advances recently made in the field for the analyses and the taxonomic identification of plant ancient DNA sequences obtained from these complex substrates. Together, these advances mean that we are on the way to an explosion of new information for the investigation of ancient plant environments.

Keywords Ancient DNA · High-throughput DNA sequencing · Lake sediments · Metabarcoding · Metagenomics · Pollen · Shotgun sequencing · Vegetation

L. Parducci (✉) · K. Nota
Department of Ecology and Genetics, Evolutionary Biology Centre, Uppsala University, Uppsala, Sweden
e-mail: laura.parducci@ebc.uu.se

J. Wood
Long-Term Ecology Laboratory, Landcare Research, Lincoln, Canterbury, New Zealand

Charlotte Lindqvist and Om P. Rajora (eds.), *Paleogenomics*,
Population Genomics [Om P. Rajora (Editor-in-Chief)],
https://doi.org/10.1007/13836_2018_38,

1 Introduction

Ancient DNA (aDNA) research may be broadly defined as the study of degraded DNA, such as that isolated from prehistoric plant or animal remains. As the rate of DNA degradation is dependent on local environmental conditions, the timeframe covered by aDNA studies is broad and may include the study of living individuals/ populations via hair, faeces or seeds, as well as the study of museum and archaeological specimens and subfossil remains from the Late Quaternary (the last 800,000 years). Ancient DNA research first began in the late 1980s following the development of the PCR technique, which allowed small quantities of DNA to be amplified and sequenced. Today, with the latest advances in high-throughput DNA sequencing (HTS) technology, the sequencing of DNA from specimens dating to thousands and even hundreds of thousands of years before present is almost commonplace. Nevertheless, strict laboratory protocols must be followed, and results replicated and carefully assessed, to help ensure the authenticity of aDNA sequences.

Ancient DNA research on plants from archaeological and palaeontological contexts has lagged behind equivalent work on animals, mainly because, compared to bones, it is often more difficult to find and obtain DNA from plant remains such as wood, seeds, fruits or pollen. For example, DNA from skeletal remains is usually well preserved, more abundant and easier to extract compared to DNA from plant remains like charred seeds (which represent 95% of the plant archaeological record) and wood megafossils. Extraction of DNA from other plant tissues, such as pollen or macrofossils, can be problematic and time-consuming, and moreover, the amount of endogenous DNA obtained is often limited.

In addition to tissue remains, plant aDNA molecules also occur in mixed substrates, such as middens and coprolites (environmental aDNA). The ability to identify plant species using DNA sequences obtained from such substrates is an important aspect of aDNA studies. For example, desiccated faecal material (coprolites) of herbivorous animals are frequently encountered in Late Quaternary and Holocene archaeological and palaeontological excavations and provide a source of plant aDNA for reconstructing the vegetation communities present at the time of deposition (Poinar et al. 2001; Hofreiter et al. 2003; Wood and Wilmshurst 2016). Many insights provided by aDNA analysis of coprolites are often unachievable through study of other fossil remains, including details about past diets, agricultural practices, seasonal migration, health and ecological interactions between species (e.g. Wood et al. 2012; Wood and Wilmshurst 2013). In a similar way, plant aDNA from rodent middens has been analysed to deduce the composition of Quaternary vegetation communities (Kuch et al. 2002). DNA from middens has also allowed identification of plant species not detected via macrofossil and palynological analyses of the same samples, some of which may be locally extinct or endemic (Murray et al. 2012). Recent applications of HTS to study middens and coprolites clearly demonstrate the potential of this approach for obtaining valuable information on the biology and ecology of extinct species and prehistoric humans, as well as allowing reconstruction of ancient vegetation communities and environments.

Ancient plant DNA can also be extracted from other environmental deposits, such as ice cores, permafrost, peat and lake sediments. In sediments, aDNA molecules are often adsorbed to mineral particles after release from their original cellular source, meaning that microscopic identification of plant remains is not always possible. The ability to sequence plant DNA from such substrates has transformed palaeovegetation research (see examples below), opening up new exciting ways of studying the plant fossil record, including the detection of taxa that do not leave traces in the fossil record ('fossil silent species') or the identification of 'cryptic' microrefugia at high latitudes increasing support from phylogeographic studies of a wide range of organisms (Stewart et al. 2010; Parducci et al. 2012).

In this chapter, we discuss recent progress in using HTS technologies to reconstruct ancient vegetation communities from lake sediments and review recent key findings in the field. We also synthesize and discuss the sources of plant DNA in lake sediments, issues relating to the preservation of DNA molecules and taphonomic processes occurring after deposition. The chapter also covers other important general and technical issues related to aDNA research, such as criteria required for authentication, details about the methodologies used in the research field (metabarcoding and shotgun sequencing analyses) and the challenges relating to the methodologies used for taxonomic identification of plant aDNA sequences. Finally, we make a critical evaluation of how aDNA can complement traditional proxies such as pollen or macrofossils for vegetation reconstruction studies and provide suggestions for future directions in the field.

2 Lake Sediments

Most recent studies reconstructing past floras have focused on lake sediment records, as these contain relatively continuous signals of both aquatic and terrestrial plant communities preserved within robust stratigraphic contexts and often anoxic conditions. Moreover, lakes can be found in a large number of environments around the world and are very abundant at high latitudes (Fig. 1). Sediment records from small lakes in particular should be better archives for molecular studies compared to larger lakes, as the effects of disturbances are low and the preserved records provide a good representation of the surrounding terrestrial environment. Because the temperature of lake waters depends principally upon geography and depth (Hutchinson 1957; Wetzel 2001), when lakes are small and deep enough, the water column can become thermally stratified, particularly in temperate and cold regions. In small lakes from such regions (Fig. 1a–c), therefore, the bottom water is normally colder than surface water in the summer, whereas it has a similar temperature in the winter when the whole lake is cold and may be ice covered. Sediments are thus constantly in contact with the coldest water and become insulated from the atmosphere, having greater temperature stability, favouring the development of anoxia and increasing the probability of DNA survival. On the contrary, sediments from larger lakes, more often found at lower latitudes (Fig. 1d–f), have less seasonal variation in temperature

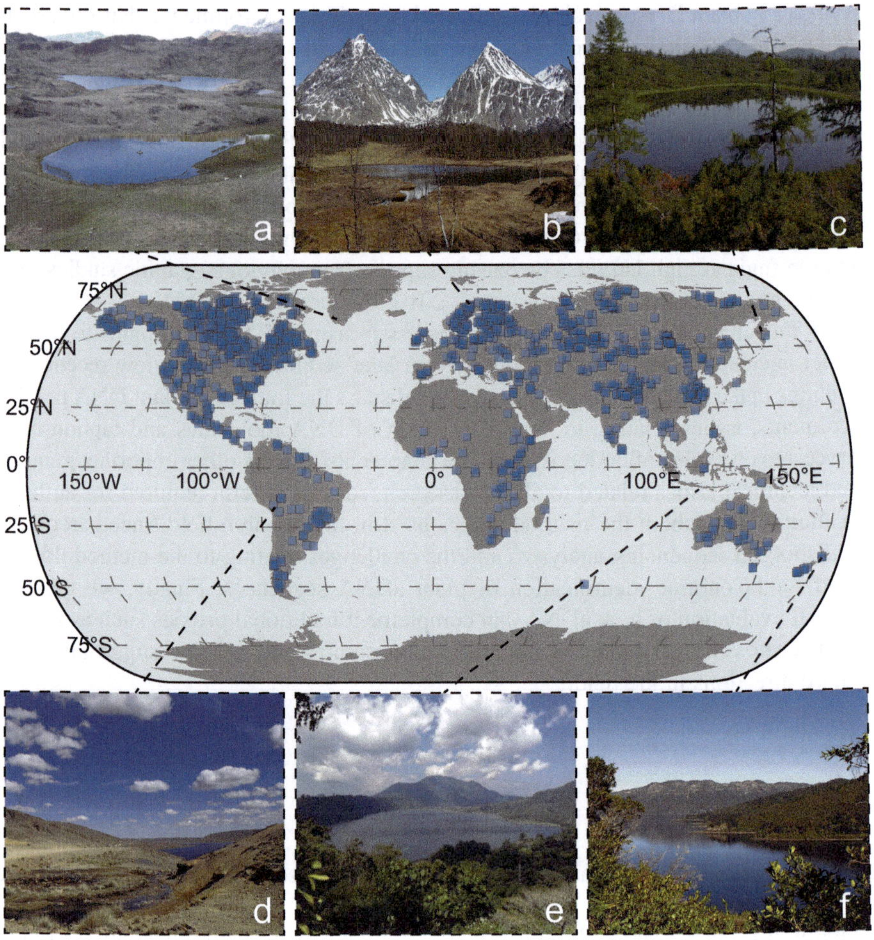

Fig. 1 Spatial distribution of larger lakes (blue squares) in the world (source: Natural Earth http://www.naturalearthdata.com). The map shows that lakes are widely distributed geographically and present in many different environments. Photos exemplify different types of lake environments. (**a**) Lake Comarum, South Greenland; (**b**) Fiskvatn, Troms, Norway; (**c**) Olive-backed lake, Kamchatka; (**d**) Lake Milluni, Bolivia; (**e**) Buyan, Bali; (**f**) Lake Waikaremoana, North Island, New Zealand. Source: Parducci et al. (2017)

and constant temperatures year-round and should in theory decrease the possibility of DNA preservation. However, an evaluation of the effect of the size and the geography of the lake and catchment area on the DNA signal retrieved in the sediments has not yet been studied experimentally. Lake sediments consist of variable proportions of autochthonous/allochthonous organic material and in-washed inorganic material. Microbial degradation of organic material frequently creates anoxic conditions in the benthic water zone and below 1–2 cm depth within the sediments (Sobek et al. 2009), which is also favourable for DNA preservation.

Anoxic conditions exclude also burrowing animals, thus minimizing bioturbation, water percolation and sediment reworking (Pansu et al. 2015), all of which are potential complications in palaeoecological and ancient molecular studies.

3 Recent Work on aDNA from Lake Sediments

Over the past few years, there have been an increasing number of studies that have used aDNA from lake sediment cores to reconstruct past ecosystems. Key findings from several of these studies have advanced the way in which this technique is used or have revealed novel ways in which the data can be used to understand the past. Since the first reports of aDNA from glacier ice, ice cores and permafrost (Willerslev et al. 1999, 2003, 2007), a number of studies have been published that investigate past biodiversity using so-called sedimentary aDNA (sedaDNA; Haile et al. 2009) from a large variety of palaeoenvironments. Most used an aDNA metabarcoding approach (see details below) in combination with classical palaeoecological analyses. Indeed, it has been through collaborative work between molecular ecologists and classical palaeoecologists that robust and reliable results have been produced to reconstruct past environments (Jørgensen et al. 2012; Parducci et al. 2012, 2013; Willerslev et al. 2014; Giguet-Covex et al. 2014; Pedersen et al. 2013, 2015, 2016; Alsos et al. 2016), and such collaboration is therefore strongly advocated (Hu et al. 2009; Anderson-Carpenter et al. 2011; Brown et al. 2014; Parducci et al. 2017). Merging molecular ecology (aDNA) and palaeoecology (e.g. pollen and macrofossils) techniques has shown clear benefits for reconstructing past vegetation. The different proxies can be used in combination, or singularly, depending on the scope of the study. For example, DNA and/or macrofossils are most suited to reconstructing the local vegetation community, whereas pollen analysis can also provide a signal of more regional vegetation. Where the aim of a study is to obtain a general and broad view of plant community change through time [e.g. investigating major vegetation types and plant introduction (Sjögren et al. 2016) or simply presence/absence of taxa (Niemeyer et al. 2017)], aDNA can be used as a stand-alone proxy, as the taxonomic resolution is sufficient for these general purposes and data generation is relatively fast and inexpensive compared with other techniques. On the other hand, if the aim is to monitor modern terrestrial plant biodiversity present around a lake, DNA alone is currently not good enough because rare species or species away from the lake shore are often not detected (Alsos et al. 2018).

As with any palaeoecological proxy, a detailed understanding of the biases involved is critical for ensuring robust interpretation of results. By comparing contemporary vegetation communities with those detected using environmental DNA extracted from surface lake sediments, Alsos et al. (2018) investigated the potential biases involved in the use of aDNA from lake sediments to reconstruct past vegetation communities. The study used metabarcoding to sequence plant aDNA from the upper 2 cm of the sediment column in 11 Norwegian lakes and compared the results to plant survey data from within and around the lakes. Forty-seven plant taxa were detected in the lake sediments. When plant taxa were detected using sedaDNA, it was found growing within 2 m from the lakeshores 73% of the time (the correlation was just 12% at 50 m from the lake edge). On average, 30% of the identifiable taxa recorded within the 2 m were detected with DNA, and the percentages varied largely between dominant and rare species (65% and 15%, respectively). This correspondence between DNA and plant communities and abundance declined with distance, suggesting that sedaDNA provides a localized signal. Interestingly, 16% of the plant records from DNA did not match taxa in the vegetation surveys, indicating that sedaDNA may be particularly useful for detecting rare species that were not observed in the field. Alsos et al. (2018) reported that detection rates generally varied in accordance with the abundance of a particular plant species in the local area but that some taxa appeared to be consistently under-represented in the DNA compared with their local abundance (e.g. Poaceae, Cyperaceae). Another interesting observation was that (similar to pollen) the taxonomic resolution varied markedly between groups, with species level resolution possible for some taxa (e.g. Ericaceae, Rosaceae), while in others only genus level resolution was possible (e.g. Betulaceae, Salicaceae).

In a previous study, Sjögren et al. (2016) used sedaDNA in two Scottish lakes to record vegetation changes accompanying recent afforestation experiments and found also that DNA worked as a temporally and floristically accurate proxy for major changes in local vegetation at both lakes. In addition, the proportion of reads seemed to relate well to the amount of DNA in the sediments, which in turn was related to the abundance and proximity of the taxa in the surrounding vegetation. In small catchments, there seems to be a direct coupling between slope conditions and sediment delivery into the lake, making small lakes highly sensitive to changes in snowmelt runoff, active layer instability and vegetation cover. While aDNA metabarcoding of plants from lake sediments has clear benefits in providing a new palaeoecological tool, additional baseline studies like that of Alsos et al. (2018) and Sjögren et al. (2016) are required to fully understand the benefits and limits of the technique and to assist with interpretation of the results.

Other studies have sequenced both plant and animal aDNA from lake sediments to explore different aspects of species movements through time. Pedersen et al. (2016) used shotgun metagenomics (see details below) in conjunction with radiocarbon dating, pollen and plant macrofossil evidence to reconstruct the age and biotic composition of the postglacial ice-free corridor between Beringia and southern North America. From 23 sediment samples taken from 8 lakes, they generated more than 1 billion DNA reads (~250 million of which passed quality control filters), from which 511,504 could be assigned to metazoan families and 2,596 to genera. Although there are advantages of shotgun sequencing, the unbiased sampling of taxa within a DNA sample and the very low percentages of reads that can be mapped and assigned to Eukaryotes compared to Bacteria are currently major drawbacks of this technique compared with metabarcoding (e.g. Stat et al. 2017; Ahmed et al. 2018; Parducci et al. 2018). Despite this, Pedersen et al. (2016) detected a range of plant and vertebrate taxa from the aDNA. The plant DNA concurred well with the pollen evidence of former vegetation communities, but some discrepancies relating to taphonomy and variability in pollen production levels were noted. The vertebrate fauna detected by the aDNA included representatives not recorded in local bone deposits and therefore supplemented existing knowledge about the palaeofauna of this region. Taken together, these lines of evidence were used to demonstrate that the corridor was only able to have begun supporting diverse biotic communities by around 12,600–12,500 years ago, too late to be provided a suitable route for earlier human migrations into the southern region.

More recent movements of species, and responses of ecosystems to those movements, have also been traced using aDNA approaches from lakes. Ficetola et al. (2018) used aDNA metabarcoding of lake sediments from the sub-Antarctic Kerguelen Islands in a multi-proxy study of ecosystem responses to invasive rabbits. DNA barcodes for both plants and mammals were amplified and sequenced. Rabbit DNA was first detected in sediments corresponding to the year range 1941–1948 AD. This matched the first occurrence of *Sporormiella* (a dung fungus), which was used as another proxy for rabbit presence. Although rabbits had been introduced to the Kerguelen Islands during the late nineteenth century, historic observations suggested they had not been present on the study island until sometime between 1932 and the 1960s, providing support for the reliability of the sedimentary proxies used. The aDNA analysis also revealed that the arrival of rabbits coincided with a change in dominant plant taxa. The cushion plant *Azorella selago*, for example, is susceptible to rabbit grazing and declines to very low levels. Other plants, perhaps more adapted to browse (e.g. *Acaena*), increased in relative abundance following the rabbit invasion.

Overall, sedaDNA signals in high latitude lakes appear to be a robust proxy for the plant communities growing immediately around the lake and usually include a good representation of aquatic and lakeshore plants with a background signal of more distant and regionally dominant taxa. Therefore, plant aDNA signals in lake sediments provide an accurate proxy for past changes in the local vegetation.

4 Sources of Plant DNA in Lake Sediments

The different sources of plant DNA and the factors influencing its transportation, deposition and preservation in lakes (i.e. the taphonomy of DNA in lakes) are important considerations when interpreting past ecology using aDNA from lakes. In this chapter, we focus primarily on Late Quaternary to Holocene lake sediments (i.e. the last 100,000 years), as lakes containing sediments from this time period are well represented at mid- and high latitudes of both hemispheres. These sediments are typically unconsolidated and consist of both inorganic (mineral particles of different sizes) and biogenic, or organic, components. The latter can be diverse and include tissues from a large variety of organisms, including plants, molluscs, fishes, insects, vertebrates, diatoms, algae, foraminifers, archaea and bacteria that lived around or in the lake.

Incorporation of remains into lake sediments can happen directly (plants and animals living in the lake, plants growing on or near the shore dropping leaves or fruits, animals defecating and urinating in the lake) or indirectly, through long-distance dispersal via wind or water. In most cases the remains will start to decay immediately following deposition and release their DNA into the sediment. Prokaryote DNA is deposited directly from microorganisms living in the sediments that release their DNA mainly through the secretion of plasmid and chromosomal DNA (Nielsen et al. 2007), while animal DNA can originate from skin flakes, fish scales, faeces, eggshells, hair, saliva, insect exuvia, regurgitation pellets and feathers imbedded in the sediment matrix. DNA from plants is deposited mainly from leaves, seeds, fruits, roots, wood remains and charcoal dropping into the lake. Several recent analyses of sedaDNA suggest that pollen does not contribute, at least not principally, to the plant DNA pool in lake sediments, as aDNA assemblages tend to be more similar to those of macrofossils than pollen, even where pollen grains are abundant (Parducci et al. 2015; Alsos et al. 2016, 2018; Bremond et al. 2017). It is possible, however, that with improved DNA extraction methods and DNA reference databases, the overlap between aDNA and conventional proxies will be improved.

An important issue to consider is the degree to which the vegetation community surrounding a lake is represented by the aDNA present in the sediments. This is not yet well understood. In particular, it is important to distinguish between aDNA signals of plant abundance around a lake and those of plant proximity. Alsos et al. (2016, 2018) and Sjögren et al. (2016) clearly showed that the type and the amount of sedaDNA (i.e. sequence abundance or number of reads) are tightly related to the distance of the local flora growing in and around lakes and that species can be detected even when these are not present as macrofossils. Other studies from modern soils have addressed quantitative questions to investigate relationships between plant biomass and DNA. Yoccoz et al. (2012) found that DNA from boreal soils is highly consistent with plant diversity estimated from conventional above-ground surveys and concluded that biomass and DNA sequence proportions are strongly related but the relationship differed among growth forms. For herbs, the relationship was approximately 1:1, while woody plants, which dominated the biomass, constituted a substantially lower proportion of the soil DNA. Conversely, forb representation in soil DNA was greater than in above-ground biomass.

A good understanding of these processes is not fully available at the moment, but we have at least a general picture of all processes involved and the factors that influence these processes. In Fig. 2, we show our current interpretation of how the biotic palaeoenvironmental plant proxies (pollen, macrofossils and DNA) originate, accumulate and develop through time in lake sediments. Some unknowns still remain regarding the molecular processes linking DNA and plant biomass in a lake, and further experiments are required to understand the underlying processes by which plant DNA information is originated, transported and preserved in sediments and to improve our ability to infer the presence and abundance of organisms from DNA signals.

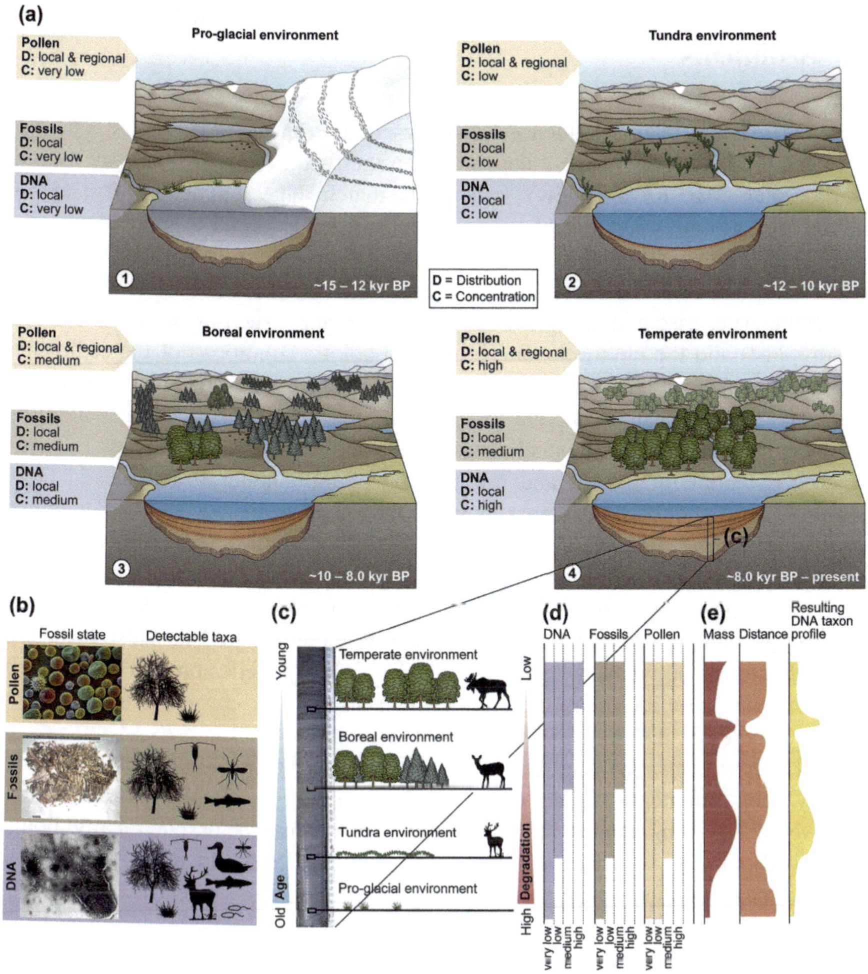

Fig. 2 Biotic palaeoenvironmental proxies in lake sediments. (**a**) Sequential environmental development for a temperate region, in which the lake sediments start accumulating as glacial ice retreats, incorporating glacially eroded debris and the sparse pioneering biota (1), which later is replaced by a tundra-steppe community (2) and then the boreal forest establishes (3) before eventually being replaced by a temperate forest (4). By identifying organisms detectable by DNA, macro- and microfossils accumulated and preserved in the lake sediments (**b**), it is possible to reconstruct the environments through time (**c**). It is important to notice that the rate of degradation is strongly correlated with the age of the sediments and that the input concentration (**d**) varies in different climatic environments from these three proxies. In addition, the resulting DNA profile (**e**) and macro- and microfossils are influenced by taphonomic processes such as differences in biomass production and the distance from source to deposit. This is why a combination of all these three proxies makes a more robust palaeoenvironmental reconstruction. Source: Parducci et al. (2017)

5 Factors Affecting Preservation of Plant DNA in Lake Sediments

Several different factors influence DNA preservation in the environment, and persistence times can vary widely depending on the type of environment (see Barnes and Turner 2016 for a review). It is likely that in lake sediments, plant aDNA represent a complex mixture of extracellular molecules bound to mineral particles and intact cells or aggregation of cells. The most important factors dictating preservation of DNA in lake sediments are the characteristics of the released DNA molecules (i.e. association with organellar membranes, such as mitochondria and chloroplast) and the environmental biotic and abiotic conditions of the sediments. Abiotic environmental conditions that can affect preservation include temperature, light, oxygen, pH, salinity and the composition of sediments, while biotic conditions include the presence of microbial communities and extracellular enzymes.

The potential for release and preservation of mitochondrial and plastid DNA differs from that of chromosomal (nuclear) DNA because the additional membranes around organellar contents provide greater protection against cytoplasmic nucleases (Nielsen et al. 2007; Allentoft et al. 2012). For this reason, most studies that reconstruct past floras focus on metabarcoding of chloroplast DNA (cpDNA) or analysing small regions (barcodes) present in this genome. CpDNA is also present in multiple copies within each organelle, with multiple organelles present in each cell, meaning that there are many more copies present than for chromosomal DNA, an ideal attribute for aDNA studies.

DNA from plant tissues in lake sediments can either be preserved intracellularly, i.e. in plant fragments visible under a microscope (Fig. 3b), or extracellularly, i.e. where DNA molecules are free within the sediments or chemically bound (adsorbed) to minerals such as clays or limes (Fig. 3c, d). In reality, total plant DNA in lake sediments likely represents a complex mixture of extracellular DNA molecules and whole plant cells. Evidence from studies of bacterial and plant DNA suggests that the majority of cells from plant and animal tissues are rapidly lysed, with their DNA immediately being released into the sediments (Nielsen et al. 2007). Extracellular DNA seems to therefore represent a relatively high proportion of the total DNA in sediments, where it forms an important nutrient source for heterotrophic organisms. Once in the sediments, DNA molecules can either be attacked by bacterial and fungal DNases, bind to organic and inorganic soil components like clay minerals or persist through natural transformation in bacterial and archaea cells (Nagler et al. 2018).

Despite the ubiquitous presence of DNase activity in soils and sediments (due to the presence of active microbial communities), high molecular weight extracellular DNA has been detected in significant amounts in soils (Blum et al. 1997; Nagler et al. 2018). Adsorption of DNA on mineral particles appears to be the main factor providing protection against DNases and the principal mechanism of DNA persistence in sediments (Pietramellara et al. 2009). Because microbial activity is high in surface sediments of lakes, we assume that the abundance and the quality of

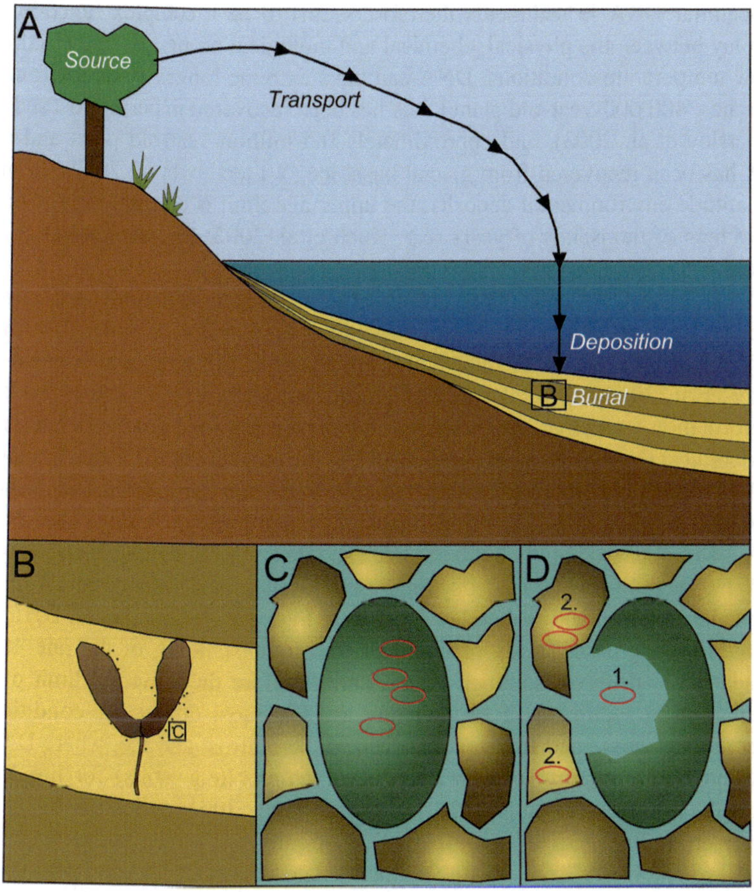

Fig. 3 Diagram illustrating different aspects of taphonomy and preservation of plant DNA in lake sediments; a seed from a tree (source of DNA) is transported via wind and deposited in the lake where it is buried in the sediments (**a**); DNA may be extracted from the seed itself, if preserved (**b**), but the tissues may begin to degrade in which case DNA will be present in cells or organelles (e.g. chloroplasts (**c**)) within the sediment. Over time these may also degrade, releasing DNA, which can be free within the sediment (1) or adsorbed to mineral grains within the sediment (2)

extracellular DNA in deeper sediments are functions of the adsorption capacity of the sediments. The rate of absorption on clay minerals like montmorillonite is extremely rapid due to their relatively large negatively charged surface area; c. 85–90% of total adsorption capacity can occur within 15 min. However, pH also plays a major role in determining the amount and type of adsorption, with minimal absorption above pH 5 and progressively more below this value (Greaves and Wilson 1969). Humic acids also bind DNA molecules, due to their negative surface charge, and therefore prolong DNA survival. Sands, however, have been found less effective in binding DNA compared with clays, due to the smaller surface area available (Pietramellara et al. 2009; Slon et al. 2017). The half-life of extracellular DNA in sediments therefore seems to be a complex function of the interplay between the physical, chemical and biological properties of the sediments. In low-temperature conditions, DNA can have extreme longevity in environmental substrates; 400,000-year-old plant DNA has been recovered in permafrost sediments (Willerslev et al. 2003), and approximately 0.5 million-year-old plant and animal DNA has been recovered from glacial basal ice (Willerslev et al. 2007). In mid- to low-latitude environmental deposits, the upper age limit is currently in the order of few to tens of thousands of years (e.g. Kuch et al. 2002; Boessenkool et al. 2013; Heinecke et al. 2017).

Extracellular DNA can also be preserved in sediments via natural transformation, a process through which some microorganisms take up DNA molecules from the environment and add them into their own genomes (Thomas and Nielsen 2005; Nagler et al. 2018). Several bacterial groups are known to be agents for natural transformation, as well as some archaea and a few eukaryotic groups (mainly micro-invertebrates) (Vries and Wackernagel 2005). The majority of microbial transformed DNA is rapidly metabolized within the cell, but some can persist and eventually recombine with the host genome. Normally, transformation is most efficient with DNA $\geq$1 kilobase in length, but it has been shown that even damaged and very short DNA fragments (down to 20 bp long) can be integrated into bacterial genomes (Overballe-Petersen and Willerslev 2014) (see Fig. 1 in Pedersen et al. 2016).

Based on a growing body of literature, our understanding of the rate of DNA degradation in different sediment types, and therefore the temporal limit of DNA preservation in sediments, is ever increasing. The most favourable conditions for preservation (anoxic and frozen/cold) occur in permafrost and ice, where DNA can persist in biotic remains and environmental samples (e.g. soils) for hundreds of thousands of years (Lindhal 1993; Hofreiter et al. 2001; Allentoft et al. 2012; Dabney et al. 2013; Orlando et al. 2014). Currently, the oldest authenticated plant aDNA sequences are from frozen sediments dated between 450 and 800 thousand years BP (Willerslev et al. 2007), while aDNA studies from lake sediments have so far been restricted to strata of much younger ages, likely limited by the age of lake formation (Heinecke et al. 2017). Such favourable conditions (cold and dry) are restricted to polar regions and high alpine environments. However, plant aDNA has also been extracted from arid and hot environments (Hofreiter et al. 2003; da Fonseca et al. 2015; Mascher et al. 2016) and from temperate middens and coprolites from arid environments (see Rawlence et al. 2014), suggesting that warm temperatures are not necessarily a barrier for the preservation of DNA molecules. Indeed, environmental conditions required for DNA preservation in sediments are not only related to temperatures and oxygen but also to pH range and to the presence of consistently dry or wet conditions during preservation. For example, taphonomic studies investigating conditions for optimizing access to authentic aDNA suggest that different environmental conditions (waterlogged and buried underground) are favourable for DNA preservation for wood megafossil remains dated to the early Holocene (Pollmann et al. 2005; Gómez-Zeledón et al. 2017; Wagner et al. 2018; Lendvay et al. 2018).

Several studies have investigated the post-mortem processes affecting DNA molecules in fossil tissue remains. The same enzymatic, hydrolytic and oxidative processes will also damage DNA molecules present in sediments. These processes are often reflected by misincorporation of C to T and G to A transitions, primarily towards the ends of the DNA molecules (Briggs et al. 2007; Jónsson et al. 2013), which leads to strand breakage and DNA fragmentation (Gates 2009; Dabney et al. 2013). For this reason, the highest success rates for plant aDNA isolation from sediments have been from frozen sediments (Willerslev et al. 2007) with limited bacterial abundance (and therefore limited presence of nucleases). It should be clear, however, that good preservation conditions only delay the fragmentation of DNA molecules, which even in ideal conditions (bones at −5°C) may only persist for several hundred thousand of years (Allentoft et al. 2012).

In Fig. 4, we summarize and compare our current understanding of the chain of processes that determine the transformation of the three main fossil assemblage types (pollen, macrofossils and aDNA) in lake sediments. Even if our understanding of taphonomical processes has improved in recent years, some challenging research questions still remain and need to be further investigated in the plant aDNA record (e.g. species quantification and abundances, absence of dominant terrestrial taxa, relative contribution of pollen DNA).

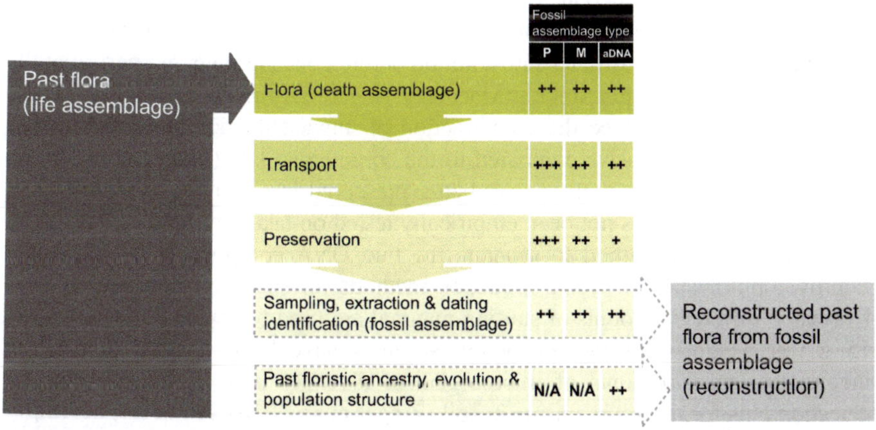

Fig. 4 Representation of the chain of processes involved in the transformation of plant information present in the three lake sediment assemblage types: pollen (P), macrofossils (M) and ancient DNA (aDNA). Current understanding of the processes is indicated as good (+++), reasonable (++) or poor (+). Figure redrawn from Birks and Birks (2016) and originally based on Jackson's (2012) general conceptual model for the representation of floristic material in palaeoecological assemblages

6 Extracting DNA from Lake Sediments

Different protocols for the extraction of aDNA have been proposed and tested on ancient materials such as the bones and teeth (e.g. Rohland and Hofreiter 2007), noncarbonized archaeobotanical remains (Wales et al. 2014) and cave sediments (e.g. Slon et al. 2017). Ancient DNA extraction methods from lake sediments have been aimed at either extracting total genomic DNA (e.g. extraction includes a lysis step) or extracting the extracellular fraction (e.g. extraction without a lysis step), before purifying the DNA to remove co-extracted PCR inhibitors (e.g. humic acids). Commonly used total genomic extraction protocols for lake sediments are based on a chemical lysis buffer including N-lauroylsarcosine, DDT or 2-mercaptoethanol in combination with proteinase K followed by a purification using phenol-chloroform (or octanol) and silica spin columns (Willerslev et al. 2003; Pedersen et al. 2016). Currently, protocols have been aiming at surpassing the use of organic solvents by using commercial kits (e.g. the Qiagen DNeasy PowerSoil Kit, formerly known as the PowerSoil DNA Isolation Kit from MoBio), which use a combination of chemical and physical lysis. The PowerSoil Kit has been developed for the extraction of total microbial DNA from soil, and modified lysis steps have been proposed for ancient plant material, e.g. adding DTT and proteinase K followed by overnight incubation (Epp et al. 2012; Zimmermann et al. 2017).

The extraction protocol for extracellular DNA from sediments has remained the same since it was published in 2012 (Taberlet et al. 2012; Giguet-Covex et al. 2014; Ficetola et al. 2018). This protocol uses a saturated phosphate buffer to release DNA bound to the sediment particles into solution before concentrating and purifying the DNA with the NucleoSpin Soil kit (Macherey-Nagel). Extracting extracellular DNA has been argued to reduce the co-extraction of unwanted microbial DNA and is significantly less expensive compared to the commercial kits (Zinger et al. 2016). Whether some taxonomic groups are better preserved in either the extracellular or intracellular fraction has not been empirically tested on lake sediments. Alawi et al. (2014) presented a method to separate the two DNA components, which might improve optimization of DNA extraction methods for specific taxonomic groups. Normally, different amounts of starting material are used for extraction (0.25 up to 15 g wet weight), but low extraction volumes have been argued to result in inconsistent measurements of microorganism diversity in soil and increase the detection bias for macroorganisms in soil (Zinger et al. 2016). However, no empirical studies have so far been published on this topic.

Currently, there is no generic extraction method that works well for all sediment types (Terrat et al. 2012) or in particular for lake sediments. Variations in chemistry, grain size and other sediment characteristics (organic content and pH) require protocol optimization for each specific material and setting (e.g. see Huang et al. 2016), as well as specific methodologies for downstream analyses like library preparation, sequencing and bioinformatic analyses. The most and widely used kit protocol, however, remains the DNeasy PowerSoil Kit that works well with all sediment types and produces large amounts of DNA, but it likely extracts too large fractions of bacterial DNA, which is not welcome when plants are the target of the study.

7 HTS Sequencing and Taxonomic Identification of Plant DNA from Lake Sediments

Today, HTS approaches are considered the most promising for future development in the aDNA field. There are two main HTS approaches for analysing aDNA from sediments: metabarcoding, which relies on PCR amplification of a single locus or 'barcode region' (chloroplast in plants), and metagenomics, which uses a shotgun sequencing approach, i.e. sequencing a non-discriminated pool of DNA.

Metabarcoding has recently been applied to sediments from a range of mid- to high latitude/altitude lakes (Coolen and Gibson 2009; Parducci et al. 2012; Pansu et al. 2015; Epp et al. 2015; Paus et al. 2015; Alsos et al. 2016), as well as tropical lakes (Boessenkool et al. 2013; Bremond et al. 2017). A key advantage of the metabarcoding approach is its ability to simultaneously amplify and sequence a large number of taxa for a limited cost. Plant metabarcoding data can readily be combined with other proxies of past environments to identify potential drivers of past vegetation change, using approaches derived from community ecology (Giguet-Covex et al. 2014). The performance of metabarcoding, however, is often limited by the often poor universality of the primers used (i.e. their inability to amplify a non-biased range of taxa), by the taxonomic resolution of the amplified region (i.e. capacity to differentiate and identify also closely related species) (Ficetola et al. 2010; Sønstebø et al. 2010) and by the need for extensive PCR replication experiments for the identification of rare taxa. Locus selection is particularly difficult for plant aDNA studies, as prerequisites, such as minimal bias in amplification of different, distantly related taxa and short sequence length, drastically reduce the ability to resolve plant taxa (Taberlet et al. 2007). Plant metabarcoding studies normally use a single marker but can also use a combination of primers to independently resolve different groups of taxa. For example, the *trn*L g/h primers (Taberlet et al. 2007) may be used first to obtain an overall assessment of plant diversity (they resolve most plant families) and then additional primers (e.g. ITS1-F/ITS1Poa-R for Poaceae and ITS1-F/TS1Ast-R for Asteraceae) could increase taxonomic resolution within selected families (Baamrane et al. 2012). However, different primers may preferentially amplify DNA of certain taxa, leading to biases in the final results (Yoccoz 2012). The power of metabarcoding for reconstructing ancient vegetation communities is strongly reliant on comprehensive taxonomic reference libraries for identifying sedaDNA sequences (Taberlet et al. 2012). Currently, many plant species (especially tropical) have no DNA sequences available in publicly available databases (such as GenBank). However, the number of reference sequences is now rapidly increasing, improving the utility of metabarcoding for plant aDNA studies. Custom-built local databases, containing loci of interest for the whole flora from the investigated regions, are also widely used and highly recommended in metabarcoding studies.

Shotgun sequencing (or metagenomic) analysis of ancient remains is increasingly being applied within the aDNA field (Orlando et al. 2015). This approach involves sequencing non-discriminated genomic DNA (i.e. no specific locus) from a sample.

Initially used for studying the fraction of uncultivable microbes in modern environmental samples (e.g. Vos et al. 2013), metagenomics has more recently been recognized as an important tool for circumventing the amplification biases inherent in metabarcoding (Ziesemer et al. 2015; Pedersen et al. 2016). So far, only four studies have published metagenomic data for palaeoenvironmental reconstruction (Smith et al. 2015; Pedersen et al. 2016; Ahmed et al. 2018; Parducci et al. 2018). While there is considerable potential in applying this technique for studying ancient palaeoenvironments, the lack of consensus approaches to data processing and inexperience with interpreting such large datasets can increase the chance of misinterpretations and false positives (Bennett 2015; Weiß et al. 2015). While different bioinformatic and statistical techniques exist for limiting false positives and confirming aDNA authenticity (Key et al. 2017) (see below), lack of reference genomes is a key issue. Indeed, a large part of the shotgun sequences (between 79 and 96%) remain unidentified (Pedersen et al. 2016; Slon et al. 2017; Ahmed et al. 2018; Parducci et al. 2018). Of the mapped reads, a large part is normally assigned to Bacteria and Archaea, while only few percentages (circa 5%) are assigned to Eukaryotes. Of the latter, only few reads (less than 0.01%) are assigned to plants (Viridiplantae). It is a promising trend, however, that genomic reference databases have rapidly improved over recent years and ongoing projects (e.g. PhyloAlps, https://www.france-genomique.org/spip/spip.php?article112&lang=fr, and NorBol, http://norbol.org/) are currently assembling the entire chloroplast genomes and nuclear ribosomal sequences of the floras of the Alps, Norway and parts of the Arctic, through genome skimming (Coissac et al. 2016). This will certainly promote improvement of taxonomical read assignment. Another way to circumvent the problem of poor assignments is to target-enrich the libraries prior to sequencing and to isolate specific DNA regions of interest (mitochondrial in animals and chloroplast in plants). The approach has been recently successfully used to sequence Neandertal and Denisovan DNA from Pleistocene cave sediments (Slon et al. 2017).

It is very likely that metagenomics will grow in the next years to become the primary HTS approach, allowing for efficient production of diversity metrics and resolution of population structure and helping bridge the gap between different scientific disciplines in palaeoecology.

8 Criteria Required for Authentication

Authentication of metabarcoding data, especially data originating from rare taxa or taxa located far away from lake shores (Alsos et al. 2018), requires extensive PCR replication experiments. Controlling false positives (contamination) and false negatives (missing detection of taxa that are actually present) is a major challenge in metabarcoding analyses from environmental DNA. Ficetola et al. (2015) discussed extensively the importance of controlling for false detection of taxa in all the steps of analyses, from those conducted in the laboratory to the bioinformatic ones. They found that multiple extractions and PCR amplifications of the same samples reduce the rate of false negatives, but the optimal level of replication varies among studies and is strongly dependent on the ability of taxa to be detected. As a rule of thumb, for aDNA studies where the level of detection is low, at least eight PCR replicates should be performed to reduce the presence of false negatives (Ficetola et al. 2015). On the other hand, with increased replication levels, the risk of false positives may also increase. False positives are another critical aspect of aDNA metabarcoding analyses where the low amount of template DNA generally requires multiple PCR cycles (>35). False positives may occur for different reasons, from contamination during sampling work and laboratory analyses to sequencing errors. To control false positives and to improve the quality of results, a number of approaches have been reported in the literature, depending on the type of study. In standard ecological analyses, a series of practices and control measures have been proposed to be adopted through all these steps, from using control blanks at all steps in the laboratory to controlling sequence quality during bioinformatic analyses (Ficetola et al. 2015).

In metagenomic analyses, the criteria for authentication are different, and there are a number of approaches that can be used to distinguish between endogenous and nonendogenous DNA sequences (Key et al. 2017). One of the major advantages of the shotgun sequencing approach is the large amounts of DNA data produced, which can be easily statistically investigated for post-mortem damage patterns and used as signal of authentication. *Post-mortem* DNA damage introduces specific nucleotide substitutions (C > T) that are normally present at the ends of the aDNA fragments, as well as specific fragmentation patterns (fragment normally shorter than 150 bp), which can be statically analysed by mapping HTS sequencing reads against reference genomes. mapDamage (Ginolhac et al. 2011) is the most and widely used package to identify such patterns from HTS sequence datasets (Briggs et al. 2007; Jónsson et al. 2013).

9 Molecular Versus Microscopic Methods

One of the major advantages offered by sedaDNA is that it can resolve more plant species and provide higher taxonomic resolution (depending on primers) than traditional morphological analyses of pollen and macrofossils. DNA can thus provide ecological and climatic insights that are otherwise difficult, or even impossible, to infer using traditional methods (Sønstebø et al. 2010; Parducci et al. 2015; Alsos et al. 2016). Despite improved identification keys for pollen, such analyses are still time demanding, and identifications are often restricted to genus or family level (rarely species). Several studies focusing on lakes have shown also that aDNA can be used to consistently identify certain taxonomic groups that are seldom, or not at all, resolved by microscopic techniques, such as Archaea (Ahmed et al. 2018), microbial eukaryote (Capo et al. 2016; Domaizon et al. 2013), algae (Stoof-Leichsenring et al. 2015), diatoms (Stoof-Leichsenring et al. 2012; Coolen and Gibson 2009), copepods (Bissett et al. 2005), bryophytes and many aquatic plants (Alsos et al. 2016). DNA analyses have also enabled potential drivers of vegetation change through time to be explored using approaches derived from community ecology (Giguet-Covex et al. 2014) and improved reconstruction of past biodiversity and palaeoenvironments (Willerslev et al. 2014; Pedersen et al. 2016). The latter aspect is an advantage compared with traditional morphological analyses of pollen, especially at high latitudes/altitudes where local pollen productivity is low and long-distance pollen dispersal is more common. In Svalbard, for example, sedaDNA from a lake allowed the detection of vascular, algal, aquatic and bryophyte taxa not present as macrofossils and demonstrated resilience of the local Arctic flora to climate change during the Holocene (Alsos et al. 2016).

Another important advantage offered by aDNA analyses of lake sediments is the possibility of working with the normally abundant pollen grains present in these deposits. Suyama et al. (1996) were the first to amplify DNA from ancient pollen extracted from peat. Successively, Parducci et al. (2005) used short cpDNA and mitochondrial DNA fragments from Holocene *Pinus* pollen to investigate population relationships through time, and the same technique was used to sequence cpDNA from angiosperm pollen from the Venice Lagoon (Paffetti et al. 2007) and conifer pollen from glaciers (Nakazawa et al. 2013). Using multiplex PCR on fresh pollen (Isagi and Suyama 2010) could also perform paternity analysis and infer the pattern and distance of pollen dispersal in modern plant populations (Matsuki et al. 2007, 2008; Hasegawa et al. 2009, 2015; Hirota et al. 2013). Despite these successes, the PCR success rate on ancient pollen grains is low, and the time required to analyse the single grains is high. With the advent of HTS technology and the possibility to sequence directly from single cells (single-cell sequencing technologies, SCS), it should now be possible to investigate more efficiently individual fossil pollen grains and hence conduct plant aDNA studies more effectively at the population level. Researchers are now testing the possibility of constructing HTS libraries directly from pollen employing fluorescence-based flow cytometry (FACS) for sorting single grains in combination with SCS technology. This approach allows the efficient examination of a large number of grains simultaneously, providing an improved alternative to the more time-consuming manual single-pollen genotyping technique. The FACS/SCS method offers the opportunity for analysing the genotype (or the whole genome) of a large number of plant individuals on millennial time scales and hence to conduct aDNA studies at the population level in ancient plants. Furthermore, prior to HTS library construction, individual pollen grains can be screened for DNA content and sorted in microwell plates using FACS on stained pollen suspensions with DAPI (4′,6-diamidino-2-phenylindole), so that the grains are selected based on DNA content using fluorescence-based flow cytometry. The successive steps include lysis of pollen walls with extraction buffers, extraction of DNA, whole-genome amplification using multiple displacement amplification (MDA) and successive downstream sequencing analyses on single grains (library preparation and pooled HTS sequencing). An alternative approach can be to sequence and work with individual nuclei extracted from pollen following bursting pollen through nylon meshes.

10 Conclusions and Future Perspectives

Despite early challenges, the field of aDNA has recently experienced significant advances in methods and technology and in understanding the taphonomy of plant DNA in sediments. These advances have resulted in improved application of aDNA techniques to lake sediments, allowing more defined reconstructions of past floras around lakes and, in general, a better understanding of certain palaeoecological issues. We expect that the application of shotgun and metabarcoding analyses of sedaDNA and SCS on pollen will rapidly grow in the coming years, as HTS methods are now becoming more accessible and less expensive and genomic reference databases are improving.

Below we present five key conclusions drawn from this chapter that we hope will be useful for plant aDNA researchers working with lake sediments:

1. Lake sediments provide continuous archives with fine temporal and spatial resolution, allowing the establishment of good molecular records for past vegetation history and the possibility for distinguishing the origin, dispersal and ancestry of plant species and populations through time.
2. With methodological and technological improvements achieved over the past decade and increased experience in applying these techniques, plant aDNA from lake sediments has now become an established tool for analysing past vegetation (presence and abundance of taxa). It also plays a key role in identifying 'fossil silent diversity', an important component for understanding past vegetation change and modelling vegetation response to future climate changes.
3. With improved techniques (e.g. target enrichment of cpDNA) and improved reference databases, plant aDNA from lakes will provide more precise reconstructions at the species level of past local vegetation than macrofossil and pollen analyses.
4. In coming years, SCS profiling of pollen from lake sediments will likely become a crucial tool for investigating histories and dynamics of plants at the population level.
5. Improved understanding of the taphonomy of DNA in lake sediments now allows a better understanding of the origin and fate of plant DNA molecules during and after deposition in lakes. However, further research on these processes is crucial, particularly those involved in DNA preservation (temperature, pH, adsorption onto mineral surfaces and oxygen availability), to improve our understanding of the power and limitations of aDNA reconstructions.

References

Ahmed E, Parducci L, Unneberg P, Ågren R, Schenk F, Rattray JE, Han L, Muschitiello F, Pedersen MW, Smittenberg RH, et al. Archaeal community changes in Lateglacial lake sediments: evidence from ancient DNA. Quat Sci Rev. 2018;181:19–29.

Alawi M, Schneider B, Kallmeyer J. A procedure for separate recovery of extra- and intracellular DNA from a single marine sediment sample. J Microbiol Methods. 2014;104:36–42.

Allentoft EA, Collins M, Harker D, Haile J, Oskam C, Hale M, et al. The half-life of DNA in bone: measuring decay kinetics in 158 dated fossils. Proc R Soc B. 2012;279:4724–33.

Alsos IG, Sjögren P, Edwards ME, Landvik JY, Gielly L, Forwick M, Coissac E, Brown AG, Jakobsen LV, Føreid MK, et al. Sedimentary ancient DNA from Lake Skartjørna, Svalbard: assessing the resilience of arctic flora to Holocene climate change. The Holocene. 2016;26:1–16.

Alsos IG, Lammers Y, Yoccoz NG, Jørgensen T, Sjögren P, Gielly L, Edwards ME. Plant DNA metabarcoding of lake sediments: how does it represent the contemporary vegetation. PLoS One. 2018;13:e0195403.

Anderson-Carpenter LL, McLachlan JS, Jackson ST, Kuch M, Lumibao CY, Poinar HN. Ancient DNA from lake sediments: bridging the gap between paleoecology and genetics. BMC Evol Biol. 2011;11:30–15.

Baamrane MAA, Shehzad W, Ouhammou A, Abbad A, Naimi M, Coissac E, Taberlet P, Znari M. Assessment of the food habits of the Moroccan dorcas gazelle in M'Sabih Talaa, west Central Morocco, using the *trn*L approach. PLoS One. 2012;7:e35643.

Barnes MA, Turner CR. The ecology of environmental DNA and implications for conservation genetics. Conserv Genet. 2016;17:1–17.

Bennett KD. Comment on 'sedimentary DNA from a submerged site reveals wheat in the British Isles 8,000 years ago'. Science. 2015;349:247.

Birks HJB, Birks HH. How have studies of ancient DNA from sediments contributed to the reconstruction of Quaternary floras? New Phytol. 2016;209:499–506.

Bissett A, Gibson JAE, Jarman SN, Swadling KM, Cromer L. Isolation, amplification, and identification of ancient copepod DNA from lake sediments. Limnol Oceanogr Methods. 2005;3:533–42.

Blum SAE, Lorenz MG, Wackernagel W. Mechanism of retarded DNA degradation and prokaryotic origin of DNases in nonsterile soils. Syst Appl Microbiol. 1997;20:513–21.

Boessenkool S, MCGlynn G, Epp LS, Taylor D, Pimentel M, Gizaw A, Memomissa S, Brochmann C, Popp M. Use of ancient sedimentary DNA as a novel conservation tool for high-altitude tropical biodiversity. Conserv Biol. 2013;28:446–55.

Bremond L, Favier C, Ficetola GF, Tossou MG, Akouégninou A, Gielly L, Giguet-Covex C, Oslisly R, Salzmann U. Five thousand years of tropical lake sediment DNA records from Benin. Quat Sci Rev. 2017;170:203–11.

Briggs AW, Stenzel U, Johnson PLF, Green RE, Kelso J, Prufer K, Meyer M, Krause J, Ronan MT, Lachmann M, et al. Patterns of damage in genomic DNA sequences from a Neandertal. Proc Natl Acad Sci U S A. 2007;104:14616–21.

Brown TA, Cappellini E, Kistler L, Lister DL, Oliveira HR, Wales N, Schlumbaum A. Recent advances in ancient DNA research and their implications for archaeobotany. Veg Hist Archaeobotany. 2014;24:207–14.

Capo E, Debroas D, Arnaud F, Guillemot T, Bichet V, Millet L, Gauthier E, Massa C, Develle AL, Pignol C, Lejzerowicz F, Domaizon I. Long-term dynamics in microbial eukaryotes communities: a palaeolimnological view based on sedimentary DNA. Mol Ecol. 2016;25:5925–43.

Coissac E, Hollingsworth PM, Lavergne S, Taberlet P. From barcodes to genomes: extending the concept of DNA barcoding. Mol Ecol. 2016;25:1423–8.

Coolen M, Gibson J. Ancient DNA in lake sediment records. PAGES News. 2009;17:104–6.

da Fonseca RR, Smith BD, Wales NA, Cappellini E, Skoglund P, Fumagalli M, Samaniego JA, Carøe C, Ávila-Arcos MAC, Hufnagel DE, et al. The origin and evolution of maize in the Southwestern United States. Nat Plants. 2015;1:1–5.

Dabney J, Knapp M, Glocke I, Gansauge M-T, Weihmann A, Nickel B, Valdiosera CE, García N, Pääbo S, Arsuaga JL, et al. Complete mitochondrial genome sequence of a middle Pleistocene cave bear reconstructed from ultrashort DNA fragments. Proc Natl Acad Sci U S A. 2013;110:15758–63.

Domaizon I, Savichtcheva O, Debroas D, Arnaud F, Villar C, Pignol C, Alric B, Perga ME. DNA from lake sediments reveals the long-term dynamics and diversity of Synechococcus assemblages. Biogeosciences. 2013;10:2515–64.

Epp LS, Boessenkool S, Bellemain EP, Haile J, Esposito A, Riaz T, Erseus C, Erséus C, Gusarov VI, Edwards ME, et al. New environmental metabarcodes for analysing soil DNA: potential for studying past and present ecosystems. Mol Ecol. 2012;21:1821–33.

Epp LS, Gussarova G, Boessenkool S, Olsen J, Haile J, Schrøder-Nielsen A, Ludikova A, Hassel K, Stenøien HK, Funder S, et al. Lake sediment multi-taxon DNA from North Greenland records early post-glacial appearance of vascular plants and accurately tracks environmental changes. Quat Sci Rev. 2015;117:152–63.

Ficetola GF, Coissac E, Zundel S, Riaz T, Shehzad W, Bessière J, Taberlet P, Pompanon F. An in silico approach for the evaluation of DNA barcodes. BMC Genomics. 2010;11:434–1572.

Ficetola GF, Pansu J, Bonin A, Coissac E, Giguet-Covex C, De Barba M, Gielly L, Lopes CM, Boyer F, Pompanon F, et al. Replication levels, false presences and the estimation of the presence/absence from eDNA metabarcoding data. Mol Ecol Resour. 2015;15:543–56.

Ficetola GF, Poulenard J, Sabatier P, Messager E, Gielly L, Leloup A, Etienne D, Bakke J, Malet E, Fanget B, et al. DNA from lake sediments reveals long-term ecosystem changes after a biological invasion. Sci Adv. 2018;4:eaar4292.

Gates KS. An overview of chemical processes that damage cellular DNA: spontaneous hydrolysis, alkylation, and reactions with radicals. Chem Res Toxicol. 2009;22(11):1747–60. https://doi.org/10.1021/tx900242k.

Giguet-Covex C, Pansu J, Arnaud F, Rey P-J, Griggo C, Gielly L, Domaizon I, Coissac E, David F, Choler P, et al. Long livestock farming history and human landscape shaping revealed by lake sediment DNA. Nat Commun. 2014;5:3211.

Ginolhac A, Rasmussen M, Gilbert MTP, Willerslev E, Orlando L. mapDamage: testing for damage patterns in ancient DNA sequences. Bioinformatics. 2011;27:2153–5.

Gómez-Zeledón J, Grasse W, Runge F, Land A. TaqMan qPCR pushes boundaries for the analysis of millennial wood. J Archeol Sci. 2017;79:53–61.

Greaves MP, Wilson MJ. The adsorption of nucleic acids by montmorillonite. Soil Biol Biochem. 1969;1:317–23.

Haile J, Froese DG, MacPhee RDE, Roberts RG, Arnold LJ, Reyes AV, Rasmussen M, Nielsen R, Brook BW, Robinson S, et al. Ancient DNA reveals late survival of mammoth and horse in interior Alaska. Proc Natl Acad Sci U S A. 2009;106:22352–7.

Hasegawa Y, Suyama Y, Seiwa K. Pollen donor composition during the early phases of reproduction revealed by DNA genotyping of pollen grains and seeds of *Castanea crenata*. New Phytol. 2009;182:994–1002.

Hasegawa Y, Suyama Y, Seiwa K. Variation in pollen-donor composition among pollinators in an entomophilous tree species, *Castanea crenata*, revealed by single-pollen genotyping. PLoS One. 2015;10:e0120393.

Heinecke L, Epp LS, Reschke M, Stoof Leichsenring KR, Mischke S, Plessen B, Herzschuh U. Macrophyte dynamics in Lake Karakul (Eastern Pamir) over the last 29 cal kyr BP. J Paleolimnol. 2017;58(3):403–17. https://doi.org/10.1007/s10933-017-9986-7.

Hirota SK, Nitta K, Suyama Y, Kawakubo N, Yasumoto AA, Yahara T. Pollinator-mediated selection on flower color, flower scent and flower morphology of *Hemerocallis*: evidence from genotyping individual pollen grains on the stigma. PLoS One. 2013;8:e85601.

Hofreiter M, Serre D, Poinar H, Kuch M, Pääbo S. Ancient DNA. Nat Rev Genet. 2001;2:353–9.

Hofreiter M, Betancourt JL, Sbriller AP, Markgraf V, McDonald HG. Phylogeny, diet, and habitat of an extinct ground sloth from Cuchillo Curá, Neuquén Province, Southwest Argentina. Quat Res. 2003;59:364–78.

Hu FS, Hampe A, Petit RJ. Paleoecology meets genetics: deciphering past vegetational dynamics. Front Ecol Environ. 2009;7:371–9.

Huang Y, Lowe DJ, Heng Z, Cursons R, Young JM, Churchman J, Schipper LA, Rawlence NJ, Wood JR, Cooper A. A new method to extract and purify DNA from allophanic soils and

paleosols, and potential for paleoenvironmental reconstruction and other applications. Geoderma. 2016;274:114–25.
Hutchinson GE. A treatise on limnology. In: Geography, physics and chemistry, vol. 1. New York: Wiley; 1957.
Isagi Y, Suyama Y. In: Isagi Y, Suyama Y, editors. Single-pollen genotyping. Tokyo: Springer; 2010.
Jackson ST. Representation of flora and vegetation in Quaternary fossil assemblages: known and unknown knowns and unknowns. Quat Sci Rev. 2012;49:1–15.
Jónsson H, Ginolhac A, Schubert M, Johnson PLF, Orlando L. mapDamage2.0: fast approximate Bayesian estimates of ancient DNA damage parameters. Bioinformatics. 2013;29:1682–4.
Jørgensen T, Haile J, Moller P, Andreev A, Boessenkool S, Rasmussen M, Kienast F, Coissac E, Taberlet P, Brochmann C, et al. A comparative study of ancient sedimentary DNA, pollen and macrofossils from permafrost sediments of northern Siberia reveals long-term vegetational stability. Mol Ecol. 2012;21:1989–2003.
Key FM, Posth C, Krause J, Herbig A, Bos KI. Mining metagenomic data sets for ancient DNA: recommended protocols for authentication. Trends Genet. 2017;33:508–20.
Kuch M, Rohland N, Betancourt J, Latorre C, Steppan S, Poinar H. Molecular analysis of a 11 700-year-old rodent midden from the Atacama Desert, Chile. Mol Ecol. 2002;11:913–24.
Lendvay B, Hartmann M, Brodbeck S, Nievergelt D, Reinig F, Zoller S, Parducci L, Gugerli F, Büntgen U, Sperisen C. Improved recovery of ancient DNA from subfossil wood – application to the world's oldest Late Glacial pine forest. New Phytol. 2018;217:1737–48.
Lindhal T. Instability and decay of the primary structure of DNA. Nature. 1993;362:709–15.
Mascher M, Schuenemann VJ, Davidovich U, Marom N, Himmelbach A, Hübner S, Korol A, David M, Reiter E, Riehl S, et al. Genomic analysis of 6,000-year-old cultivated grain illuminates the domestication history of barley. Nat Genet. 2016;48:1089–93.
Matsuki Y, Isagi Y, Suyama Y. The determination of multiple microsatellite genotypes and DNA sequences from a single pollen grain. Mol Ecol Notes. 2007;7:194–8.
Matsuki Y, Tateno R, Shibata M, Isagi Y. Pollination efficiencies of flower-visiting insects as determined by direct genetic analysis of pollen origin. Am J Bot. 2008;95:925–30.
Murray DC, Pearson SG, Fullagar R, Chase BM, Houston J, Atchison J, White NE, Bellgard MI, Clarke E, Macphail M, et al. High-throughput sequencing of ancient plant and mammal DNA preserved in herbivore middens. Quat Sci Rev. 2012;58:135–45.
Nagler M, Insam H, Pietramellara G, Ascher-Jenull J. Extracellular DNA in natural environments: features, relevance and applications. Appl Microbiol Biotechnol. 2018;116:67–14.
Nakazawa F, Uetake J, Suyama Y, Kaneko R, Takeuchi N, Fujita K, Motoyama H, Imura S, Kanda H. DNA analysis for section identification of individual Pinus pollen grains from Belukha glacier, Altai Mountains, Russia. Environ Res Lett. 2013;8:014032.
Nielsen KM, Johnsen PJ, Bensasson D, Daffonchio D. Release and persistence of extracellular DNA in the environment. Environ Biosaf Res. 2007;6:37–53.
Niemeyer B, Epp LS, Stoof-Leichsenring KR, Pestryakova LA, Herzschuh U. A comparison of sedimentary DNA and pollen from lake sediments in recording vegetation composition at the Siberian treeline. Mol Ecol Resour. 2017;26:41.
Orlando L, Ginolhac A, Zhang G, Froese D, Albrechtsen A, Stiller M, Schubert M, Cappellini E, Petersen B, Moltke I, et al. Recalibrating Equus evolution using the genome sequence of an early middle Pleistocene horse. Nature. 2014;498:74–8.
Orlando L, Gilbert MTP, Willerslev E. Reconstructing ancient genomes and epigenomes. Nat Rev Genet. 2015;16:395–408.
Overballe-Petersen S, Willerslev E. Horizontal transfer of short and degraded DNA has evolutionary implications for microbes and eukaryotic sexual reproduction. BioEssays. 2014;36:1005–10.
Paffetti D, Vettori C, Caramelli D, Vernesi C, Lari M, Paganelli A, Paule L, Giannini R. Unexpected presence of *Fagus orientalis* complex in Italy as inferred from 45,000-year-old DNA pollen samples from Venice lagoon. BMC Evol Biol. 2007;7:S6.

Pansu J, Giguet-Covex C, Ficetola GF, Gielly L, Boyer F, Zinger L, Arnaud F, Poulenard J, Taberlet P, Choler P. Reconstructing long-term human impacts on plant communities: an ecological approach based on lake sediment DNA. Mol Ecol. 2015;24:1485–98.

Parducci L, Suyama Y, Lascoux M, Bennett KD. Ancient DNA from pollen: a genetic record of population history in Scots pine. Mol Ecol. 2005;14:2873–82.

Parducci L, Jørgensen T, Tollefsrud MM, Elverland E, Alm T, Fontana SL, Bennett KD, Haile J, Matetovici I, Suyama Y, et al. Glacial survival of boreal trees in northern Scandinavia. Science. 2012;335:1083–6.

Parducci L, Matetovici I, Fontana SL, Bennett KD, Suyama Y, Haile J, Kjær KH, Larsen NK, Drouzas AD, Willerslev E. Molecular- and pollen-based vegetation analysis in lake sediments from Central Scandinavia. Mol Ecol. 2013;22:3511–24.

Parducci L, Väliranta M, Salonen JS, Ronkainen T, Matetovici I, Fontana SL, Eskola T, Sarala P, Suyama Y. Proxy comparison in ancient peat sediments: pollen, macrofossil and plant DNA. Philos Trans R Soc B. 2015;370:20130382.

Parducci L, Bennett KD, Ficetola GF, Alsos IG, Suyama Y, Wood JR, Pedersen MW. Ancient plant DNA in lake sediments. New Phytol. 2017;214:924–42.

Parducci L, Unneberg P, Pedersen MW, Han L, Lammers Y, Alsos Greve I, Salonen SJ, Väliranta M, Slotte T, Wohlfarth B. Shotgun sequencing Lateglacial-early Holocene lake sediment from Sweden to assess past plant diversity. 2018. Submitted.

Paus A, Boessenkool S, Brochmann C, Epp LS, Fabel D, Haflidason H, Linge H. Lake store Finnsjøen – a key for understanding Lateglacial/early Holocene vegetation and ice sheet dynamics in the central Scandes Mountains. Quat Sci Rev. 2015;121:36–51.

Pedersen MW, Ginolhac A, Orlando L, Olsen J, Andersen K, Holm J, Funder S, Willerslev E, Kjær KH. A comparative study of ancient environmental DNA to pollen and macrofossils from lake sediments reveals taxonomic overlap and additional plant taxa. Quat Sci Rev. 2013;75:161–8.

Pedersen MW, Overballe-Petersen S, Ermini L, Sarkissian CD, Haile J, Hellstrom M, Spens J, Thomsen PF, Bohmann K, Cappellini E, et al. Ancient and modern environmental DNA. Philos Trans R Soc B. 2015;370:20130383.

Pedersen MW, Ruter A, Schweger C, Friebe H, Staff RA, Kjeldsen KK, Mendoza MLZ, Beaudoin AB, Zutter C, Larsen NK, et al. Postglacial viability and colonization in North America's ice-free corridor. Nature. 2016;537:45–9.

Pietramellara G, Ascher J, Borgogni F, Ceccherini MT, Guerri G, Nannipieri P. Extracellular DNA in soil and sediment: fate and ecological relevance. Biol Fertil Soils. 2009;45:219–35.

Poinar H, Kuch M, Sobolik K, Barnes I, Stankiewicz A, Kuder T, Spaulding W, Bryant V, Cooper A, Pääbo S. A molecular analysis of dietary diversity for three archaic native Americans. Proc Natl Acad Sci U S A. 2001;98:4317–22.

Pollmann B, Jacomet S, Schlumbaum A. Morphological and genetic studies of waterlogged Prunus species from the Roman vicus Tasgetium (Eschenz, Switzerland). J Archaeol Sci. 2005;32:1471–80.

Rawlence NJ, Lowe DJ, Wood JR, Young JM, Churchman GJ, Huang Y-T, Cooper A. Using palaeoenvironmental DNA to reconstruct past environments: progress and prospects. J Quat Sci. 2014;29:610–26.

Rohland N, Hofreiter M. Comparison and optimization of ancient DNA extraction. BioTechniques. 2007;42:343–52.

Sjögren P, Edwards ME, Gielly L, Langdon CT, Croudace IW, Merkel MKF, Fonville T, Alsos IG. Lake sedimentary DNA accurately records twentieth century introductions of exotic conifers in Scotland. New Phytol. 2016;213:929–41.

Slon V, Hopfe C, Weiß CL, Mafessoni F, la Rasilla de M, Lalueza-Fox C, Rosas A, Soressi M, Knul MV, Miller R, et al. Neandertal and Denisovan DNA from Pleistocene sediments. Science. 2017;53:eaam9695.

Smith O, Momber G, Bates R, Garwood P, Fitch S, Pallen M, Gaffney V, Allaby RG. Sedimentary DNA from a submerged site reveals wheat in the British Isles 8,000 years ago. Science. 2015;347:998–1001.

Sobek S, Durisch-Kaiser E, Zurbrügg R. Organic carbon burial efficiency in lake sediments controlled by oxygen exposure time and sediment source. Limnol Oceanogr Methods. 2009;54:2243–54.
Sønstebø JH, Gielly L, Brysting AK, Elven R, Edwards M, Haile J, Willerslev E, Coissac E, Rioux D, Sannier J, et al. Using next-generation sequencing for molecular reconstruction of past Arctic vegetation and climate. Mol Ecol Resour. 2010;10:1009–18.
Stat M, Huggett MJ, Bernasconi R, DiBattista JD, Berry TE, Newman SJ, Harvey ES, Bunce M. Ecosystem biomonitoring with eDNA: metabarcoding across the tree of life in a tropical marine environment. Sci Rep. 2017;7:1–11.
Stewart JR, Lister AM, Barnes I, Dalén L. Refugia revisited: individualistic responses of species in space and time. Proc R Soc B Biol Sci. 2010;277:661–71.
Stoof-Leichsenring KR, Epp LS, Trauth MH, Tiedelman R. Hidden diversity in diatoms of Kenyan Lake Naivasha: a genetic approach detects temporal variation. Mol Ecol. 2012;21:1918–30.
Stoof-Leichsenring KR, Herzschuh U, Pestryakova LA, Klemm J, Epp LS, Tiedelman R. Genetic data from algae sedimentary DNA reflect the influence of environment over geography. Sci Rep. 2015;5:12924.
Suyama Y, Kawamuro K, Kinoshita I, Yoshimura K, Tsumura Y, Takahara H. DNA sequence from a fossil pollen of *Abies* spp. from Pleistocene peat. Genes Genet Syst. 1996;71:145–9.
Taberlet P, Coissac E, Pompanon F, Gielly L, Miquel C, Valentini A, Vermat T, Corthier G, Brochmann C, Willerslev E. Power and limitations of the chloroplast *trn*L (UAA) intron for plant DNA barcoding. Nucleic Acids Res. 2007;35:e14.
Taberlet P, Coissac E, Pompanon F, Brochmann C, Willerslev E. Towards next-generation biodiversity assessment using DNA metabarcoding. Mol Ecol. 2012;21:2045–50.
Terrat S, Christen R, Dequiedt S, Lelièvre M, Nowak V, Regnier T, Bachar D, Plassart P, Wincker P, Jolivet C, et al. Molecular biomass and MetaTaxogenomic assessment of soil microbial communities as influenced by soil DNA extraction procedure. Microb Biotechnol. 2012;5:135–41.
Thomas CM, Nielsen KM. Mechanisms of, and barriers to, horizontal gene transfer between bacteria. Nat Rev Microbiol. 2005;3:711–21.
Vos M, Wolf AB, Jennings SJ, Kowalchuk GA. Micro-scale determinants of bacterial diversity in soil. FEMS Microbiol Rev. 2013;37:936–54.
Vries J, Wackernagel W. Microbial horizontal gene transfer and the DNA release from transgenic crop plants. Plant Soil. 2005;266:91–104.
Wagner S, Lagane F, Seguin-Orlando A, Schbert M, Leroy T, Guichox E, Chancerel E, Bech-Hebelstrup I, Bernand V, Billard C, et al. High-Throughput DNA sequencing of ancient wood. Mol Ecol. 2018;27(5):1138–54.
Wales N, Andersen K, Cappellini E, Ávila-Arcos MC, Gilbert MTP. Optimization of DNA recovery and amplification from non-carbonized archaeobotanical remains. PLoS One. 2014;9:e86827–14.
Weiß CL, Dannemann M, Prufer K, Burbano HA, Pickrell JK. Contesting the presence of wheat in the British Isles 8,000 years ago by assessing ancient DNA authenticity from low-coverage data. elife. 2015;4:e10005.
Wetzel RG. Limnology. Lake and river ecosystems. 3rd ed. San Diego: Academic Press; 2001.
Willerslev E, Hansen AJ, Christensen B, Steffensen JP, Arctander P. Diversity of Holocene life forms in fossil glacier ice. Proc Natl Acad Sci U S A. 1999;96:8017–21.
Willerslev E, Hansen AJ, Binladen J, Brand TB, Gilbert MTP, Shapiro B, Bunce M, Wiuf C, Gilichinsky DA, Cooper A. Diverse plant and animal genetic records from Holocene and Pleistocene sediments. Science. 2003;300:791–5.
Willerslev E, Cappellini E, Boomsma W, Nielsen R, Hebsgaard MB, Brand TB, Hofreiter M, Bunce M, Poinar HN, Johnsen S, et al. Ancient biomolecules from deep ice cores reveal a forested southern Greenland. Science. 2007;317:111–4.

Willerslev E, Davison J, Moora M, Zobel M, Coissac E, Edwards ME, Lorenzen ED, Vestergård M, Gussarova G, Haile J, et al. Fifty thousand years of Arctic vegetation and megafaunal diet. Nature. 2014;506:47–51.

Wood J, Wilmshurst J. Pollen analysis of coprolites reveals dietary details of heavy-footed moa (*Pachyornis elephantopus*) and coastal moa (*Euryapteryx curtus*) from Central Otago. N Z J Ecol. 2013;37:151–5.

Wood JR, Wilmshurst JM. A protocol for subsampling Late Quaternary coprolites for multi-proxy analysis. Quat Sci Rev. 2016;138:1–5.

Wood JR, Wilmshurst JM, Wagstaff SJ, Worthy TH, Rawlence NJ, Cooper A. High-resolution coproecology: using coprolites to reconstruct the habits and habitats of New Zealand's extinct upland moa (*Megalapteryx didinus*). PLoS One. 2012;7:e40025.

Yoccoz NG. The future of environmental DNA in ecology. Mol Ecol. 2012;21:2031–8.

Yoccoz NG, Bråthen KA, Gielly L, Haile J, Edwards ME, Goslar T, Stedingk Von H, Brysting AK, Coissac E, Pompanon F, et al. DNA from soil mirrors plant taxonomic and growth form diversity. Mol Ecol. 2012;21:3647–55.

Ziesemer KA, Mann AE, Sankaranarayanan K, Schroeder H, Ozga AT, Brandt BW, Zaura E, Waters-Rist A, Hoogland M, Salazar-García DC, et al. Intrinsic challenges in ancient microbiome reconstruction using 16S rRNA gene amplification. Sci Rep. 2015;5:16498.

Zimmermann H, Raschke E, Epp L, Stoof-Leichsenring K, Schirrmeister L, Schwamborn G, Herzschuh U. The history of tree and Shrub Taxa on Bol'shoy Lyakhovsky Island (New Siberian Archipelago) since the last interglacial uncovered by sedimentary ancient DNA and pollen data. Genes. 2017;8:273–28.

Zinger L, Chave J, Coissac E, Iribar A, Louisanna E, Manzi S, Schilling V, Schimann H, Sommeria-Klein G, Taberlet P. Extracellular DNA extraction is a fast, cheap and reliable alternative for multi-taxa surveys based on soil DNA. Soil Biol Biochem. 2016;96:16–9.

Archaeogenomics and Crop Adaptation

Robin G. Allaby, Oliver Smith, and Logan Kistler

Abstract The genetic history of domestic plants is complex, protracted, and unique to often very specific factors including location, human intent, and the wider environment. In addition to well-addressed questions of domestication syndrome, and conscious versus unconscious selection, the issue of domestication poses a plethora of more nuanced questions, in particular regarding plants' abilities to adapt to new environments, and the genomic scars those forced changes leave behind. Ancient DNA from archaeobotanical remains offers a window through which we are now beginning to unravel these histories, in a large part through the technical advances in sequencing technologies and theoretical advances in genome evolution. In this chapter, we will explore how plant archaeogenomics is characterized in a large part by plasticity, of genome size, genome activity, and transposable elements, through specific mechanisms including introgression, mutation load, and stress response. We will also examine the various substrates from which invaluable information can be recovered, by no means limited to DNA from seeds.

Keywords Adaptation · Agriculture · Archaeobotany · Archaeogenomics · Domestication · Functionalization · Genome size · Heterozygosity · Introgression · Mutation load · Ploidy · Transposable elements

---

R. G. Allaby (✉)
School of Life Sciences, University of Warwick, Coventry, UK
e-mail: r.g.allaby@warwick.ac.uk

O. Smith
Natural History Museum of Denmark, Copenhagen, Denmark

L. Kistler
Department of Anthropology, Smithsonian Institution, National Museum of Natural History, Washington, DC, USA

Charlotte Lindqvist and Om P. Rajora (eds.), *Paleogenomics*, Population Genomics [Om P. Rajora (Editor-in-Chief)], https://doi.org/10.1007/13836_2018_51,

1 Introduction

The evolution of domesticated forms of plants involved their adaptation to the human environment. In response to human gathering from the wild and subsequent cultivation, plants being exploited for their food value, and in some cases their associated weeds, became modified in ways which enhanced their survival in this anthropogenic disturbed environment. These modifications increased their chances of becoming harvested, and so resown, to survive in increasingly dense stands, and the deeper burial associated with cultivation.

2 Plant Genome Evolution

Plants have some distinctive characteristics at the genome level relative to other eukaryotic groups that influence their ability to adapt. As we increasingly understand plant genome evolution both from modern and archaeogenomic (archaeological genomic) data, it is becoming increasingly clear that these distinctive characteristics are likely to be central to crop domestication and adaptation to a wide range of environments. The archaeogenomic perspective in particular, though, is beginning to highlight where the domestication trajectory is headed in terms of sustainability and our ability to adapt to the future.

Plant genomes have by far the widest range in both gene content and genome size of any eukaryotic group (for reviews see Schubert and Vu 2016; Pellicer et al. 2018). This variation is intrinsic to the evolution of the plant kingdom and underlies how plant genomes appear to be distinct from other eukaryotic groups in how they evolve and adapt. Episodes of speciation are correlated with changes in genome size (Puttick et al. 2015), and the rise of the angiosperms was itself associated with a large downsizing of genomes enabling a wider range of cell sizes and more efficient physiology leading to an explosion of speciation (Simonin and Roddy 2018). While the absolute size of a plant genome has a consequence on habitat and ecology (Pellicer et al. 2018), it is the instability and consequent plasticity of the genome that leads to changes in genome size that is linked to an adaptive quality (Pellicer et al. 2018; Schubert and Vu 2016; Oliver et al. 2013), with both increases and decreases in genome size being associated with speciation (Puttick et al. 2015).

Several processes drive dynamism in genome size. General trends to large genomes are often seen through transposable element expansion and whole genome duplication events, particularly prevalent in lower plants but also seen in some angiosperm groups, most notably in members of the Liliopsida and some asterids (Pellicer et al. 2018). Transposable element expansions can lead to changes in regulation of expression, while polyploidization leads to the adaptive opportunity for subfunctionalization and neofunctionalization. Genome expansion therefore provides one means of exploring an adaptive landscape for plants that is less available to the animal kingdom as evidenced by the stark difference in paleopolyploidy rates between the two (Blanc and Wolfe 2004), with most angiosperm lineages having a multiple polyploid history (Soltis et al. 2009). However, very large genomes lead to slower cell division cycles, larger cells and are often associated with stable environments (Cavalier-Smith 2005), which may explain why such plants are over-represented in endangered species lists (Vinogradov 2003). The counter mechanisms that reduce genome size include deletion-biased double-strand break repair (Schubert and Vu 2016) and illegitimate recombination events (Devos et al. 2002). While genome reduction gives rise to cellular features that underlie the weedy success of the angiosperms, such as small cells and rapid cycling, recombination itself is highly advantageous under conditions of adaptive evolution (Ziolkowski et al. 2017). This is because if multiple loci are under selective pressure, recombination alleviates the Hill-Robertson effect in which extensive linkage inhibits the successful assemblage of adaptive variants, which can prevent rapid adaptation (Hill and Robertson 1966).

3 Evolution of Domesticated Plant Genome Size and Regulation

The genomes of domesticated plants were plastic enough to be adaptable to human environments. Likely features at the genome level that made them adaptable include a large and active transposable element population, leading to a number of larger genomes involved (over 1,000 Mb), as well as a propensity to select for recombination. The principal plant domesticates from around the world (compiled from Larson et al. 2014) are broadly sampled from across the range of flowering plant genome size excluding only the extremes (Rensing 2017; Leitch et al. 2001; Fig. 1). Notably, the earliest domesticates, in SW Asia, were particularly enriched for large genomes. The number of polyploids in domesticates generally reflects the trend for angiosperms, although there is a notable enrichment for polyploids among the vegetative crops (typically tuber crops) of the tropical latitudes (Meyer et al. 2012).

Transposable element activity is often triggered by stress, such as in marginal or disturbed environments (Chénais et al. 2012). A number of crops have been identified to be associated with extensive recent and/or ongoing transposable element activity (Naito et al. 2006 [rice, *Oryza sativa*]; Zaki and Ghany 2004 [cotton, *Gossypium hirsutum*]; Mascagni et al. 2015 [sunflower, *Helianthus annuus*]; Middleton et al. 2013 [wheat, *Triticum aestivum*, and barley, *Hordeum vulgare*]; Diez et al. 2014 [maize, *Zea mays*]). Genomic changes associated with the activity of transposable elements in particular are likely to lead to changes in gene regulation. In general, TE movement is likely to be disruptive leading to reduction of expression, which under normal circumstances might be expected to be selectively disadvantageous. However, it can also create phenotypic variation upon which selection can act. It is interesting to note that many of the genetic loci associated with genetic domestication are also directly associated with transposable elements (Oliver et al.

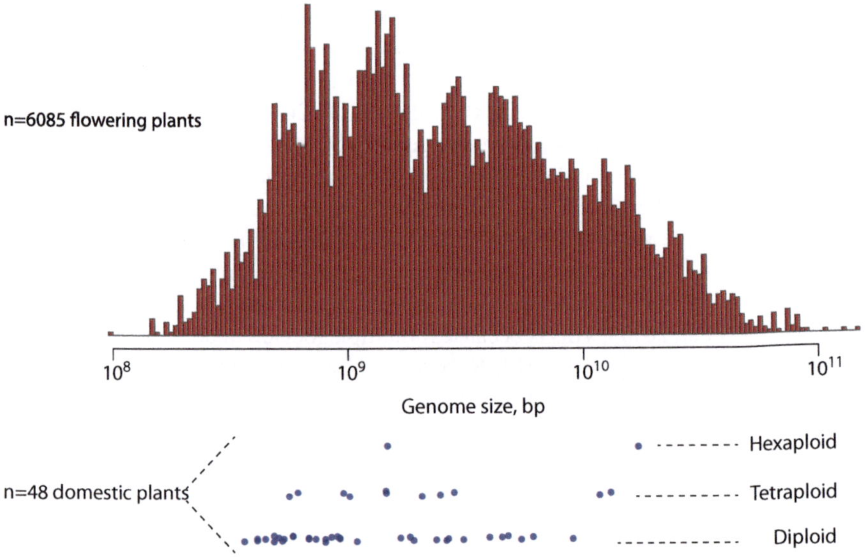

Fig. 1 Genome sizes of angiosperm genomes and domesticated genomes of plants based on C-values (Leitch et al. 2001)

2013), and more generally the majority of genetic loci that have been associated with domestication phenotypes have transpired to be regulatory in nature (Purugganan and Fuller 2009; Meyer and Purugganan 2013). The potentially disruptive nature of TE activity on expression levels suggests that the human environment in which plants undergo domestication is also one that allows for toleration of trait loss that is not directly adaptive for domestication, so-called relaxed negative selection (Fraser 2011). Therefore a range of phenotypic traits would have arisen through regulatory changes. Some of these were adaptive for the human environment, while others would have been neutral in gardens and crop fields – leading to loss of characteristics advantageous in the wild but unnecessary under domestication.

Recombination increases the efficacy of selection and leads to a reduction of genome size (Tiley and Burleigh 2015), suggesting that a reduction in genome size can be a marker of both recombination and adaption. Adaptive clines to altitude, for instance, have been found to involve reductions in genome size in maize (Bilinski et al. 2018). It is therefore likely that the adaptation involved in domestication will have a tendency to promote recombination, and reduce genome size, which has been suggested for maize (Sidhu et al. 2017). Data for very few domesticated species currently exist for both domesticate and known wild progenitor genome size, but the basic relationship between domestication, recombination, and genome size suggests that a physical contraction of the genome is likely a common outcome of domestication.

Much of the way in which plants could adapt to the human environment was tempered by plant genome architecture and their relative genomic plasticity. The typical regulatory changes that came about led to a characteristic number of traits that are seen repeated to various degrees across domestic plant species. This suite of traits is termed the 'domestication syndrome' (Harlan et al. 1973; Hammer 1984) and includes loss of seed dispersal mechanisms, changes in seed size, synchronicity of development, loss of photoperiod sensitivity, changes in architecture, and physiological changes in toxin levels (Fuller 2007). The syndrome is also associated with a loss of wild traits which are redundant in the domesticated environment, such as hooks and awns (Fuller 2007), and even seed production where vegetative propagation is the norm (Zerega et al. 2004; Perrier et al. 2011). The parallelism evident in the domestication syndrome is echoed in the underlying molecular parallelism where many instances of the same orthologs have been affected in different plant species (Lenser and Theißen 2013). This striking level of parallel evolution suggests that suitable genomic architecture for domestication may drive a canalization of changes in flowering plants in responses to the stressors of the human environment.

Conversely, this framework predicts that certain plant genome architectures would likely be unviable for domestication. This may be particularly true for plants with very small genomes (<300 Mb) and low TE content, less likely to produce the range of phenotypic variation through bursts of regulatory change. It could also be true of very large genomes with an inactive and consequently diverse TE complement (Oliver and Greene 2009, 2012).

4 Genetic Variation and the Pace of Adaptation to the Human Environment

The transition to the human environment involved numerous adaptive challenges, both initially in domestication centres and then during the subsequent agricultural expansions into new environments. Selection comes at a cost on several levels. Firstly, selection is associated with a reduction in population size termed the substitution load, first identified by Haldane (1957), which limits the amount of selection that can occur during any one period of time. As populations become smaller as a consequence of the substitution load, selection becomes less efficacious in the face of the increasing forces of drift, such that small populations are all but incapable of adaptive evolution. The consequent small population sizes that can result from bouts of selection are prone to a second cost, the mutation load, as drift in reduced populations pumps deleterious variation into the population (Henn et al. 2015). Modelling suggests that the number of loci that can be under selection simultaneously before the population size is eroded to sizes counteractive to agricultural production is limited (Allaby et al. 2015), despite the observation that some domestication traits can have many associated QTLs. However, the model demonstrates that loci offering adaptive solutions to the same selection pressure interfere with each other but behave neutrally with respect to one another since no single one has an adaptive advantage over the others, as previously suggested (Allaby 2010), resulting usually in only one or two of the adaptive variants becoming fixed. However, an unfeasibly large proportion of the total genomic gene population (>20%) would have to offer adaptive solutions before the number of selective challenges that can be accommodated increases (Allaby et al. 2015). It should be noted, however, that a positive interference can occur between variants of different loci to different specific selective pressures such that both become fixed more quickly together in outcrossing systems, or where there is enhanced recombination, but the Hill-Robertson effect in inbreeding systems causes the reverse effect (Allaby et al. 2015).

The ability of plants to adapt is therefore markedly limited, with a full suite of domestication syndrome traits possibly reflecting a maximum of adaptive change over a protracted period of time (Allaby et al. 2015). Archaeological measurements of selection on indicator traits for domestication – such as loss of seed dispersal or increases in seed size – show that the selection pressures involved were generally very weak for the cereals over a multimillennial timescale (Purugganan and Fuller 2009, 2011). Furthermore, those selection pressures changed over time. The onset of selection for Southwest Asian cereals was weak, possibly began over 20Kyr ago deep in the Pleistocene, and was most likely associated with gathering pressures, while much stronger pressures arrived later with the advent of stone technologies such as the sickle (Allaby et al. 2017).

The overall evidence points towards a protracted period of weak selection that in the case of the Near East may have occurred over tens of millennia. Such a long process would afford considerable opportunity for introgression between wild progenitor populations and cultivated populations, and indeed the selection signals imply a selection pressure on wild populations in situ. Some modern crops have been shown to have a genetic diversity that reflects a very wide wild gene pool input (Poets et al. 2015; Civáň et al. 2013; van Heerwaarden et al. 2011). Such widespread wild input will have introduced traits of local adaptation. Wild barley, for instance, occurs in multiple ecotome types possibly conferring a wide range of adaptive traits, which may have contributed to this becoming one of the most hardy and wide ranging plants of any agricultural expansion (Poets et al. 2015). Additionally, maize carried out of its tropical lowland domestication centre into nearby high-elevation environments benefited from extensive adaptive introgression with a wild sister subspecies endemic to the highlands (van Heerwaarden et al. 2011; Da Fonseca et al. 2015). This genetic assimilation of traits may also have occurred as crops spread into new environments. The archaeological record in Europe suggests that the spread of agriculture was tempered with frequent stalls and crashes (Shennan et al. 2013; Stephens and Fuller 2012), which may reflect the limits of the pace of adaptation possible for the plant assemblage of the agricultural package (Allaby et al. 2015; Banks et al. 2013; Colledge et al. 2005; Coward et al. 2008). In this case, introgression from locally adapted wild crops may have been an important source of adaptive variants, as is the case in flax for delayed flowering time (Gutaker et al. 2017) or in barley for photoperiod insensitivity (Jones et al. 2008, 2012).

The genetic diversity of domesticated species in general is reduced relative to their wild progenitors at a species wide level, typically by 20–60% (Gaut et al. 2015) using measures such as nucleotide diversity. This reduction in diversity has been widely thought to be due to a relatively small founding population of plants brought into cultivation coupled with the long-term effects of drift causing genetic erosion (Allaby et al. 2008; Eyre-Walker et al. 1998; Gaut et al. 2015; Meyer and Purugganan 2013; Zhu et al. 2007). The occurrence of such a bottleneck is thought to be behind the observation that almost all domesticate species observed so far have a larger mutation load than their respective wild progenitors (Liu et al. 2017; Makino et al. 2018; Marsden et al. 2016; Moyers et al. 2017; Renaut and Rieseberg 2015; Schubert et al. 2014; Wang et al. 2017). Furthermore, one of the features that makes plant genomes receptive to domestication, instability through TE activity, is now also recognized as a significant contributor to mutation load through the generation of deleterious structural variants (Gaut et al. 2018).

5 Untangling Domestication Through Archaeogenomics

Much of the genomic view discussed above has been built from data from modern samples. This has considerably refined our understanding of the dynamics of plant domestication and specifically the adaptation involved but also has thrown up new questions and paradoxes. In particular, there is a juxtaposition between (1) the strength of selection involved in domestication and (2) the demographic effects attributed to a domestication bottleneck leading to an elevated mutation load. The 'domestication bottleneck paradox' is that weak selection should not have been possible at the demographic levels required to generate mutation load in domesticate species (Allaby et al. 2018). Evolutionary models can be used to probe the past on the basis of modern diversity to attempt to reconstruct past demography (Beissinger et al. 2016; Meyer et al. 2016; Wang et al. 2017; Zhou et al. 2017) but can suffer from the limitations of population assumptions (Mazet et al. 2016). Ancient DNA (aDNA) studies have long held the promise of direct verification of evolutionary processes in the past that rely less on model assumptions. Principally, archaeogenomic contributions are aiding our understanding of demography over time, as well as the specific order of selection pressures on traits as well as providing time stamps on geographical spread. Together, these are beginning to reveal a more functional understanding of the evolutionary process of plant domestication and in some aspects to force quite fundamental reassessments of how domestication occurred.

Many studies in the past have focused on questions of origin of crops in terms of the number and locality of domestications (Brown et al. 2009; Palmer et al. 2012a, b; Schlumbaum et al. 2008). Increasingly, investigations are switching their emphasis from where and how many times domestication occurred for particular crops to how domestication came about in more functional terms, with regard to both the associated human behaviours that led to domestication (Fuller et al. 2010; Zeder 2015) and the genomic processes of adaptation to the human environment within plant genomes, with emphasis on those genes involved with the domestication syndrome

(Gutaker and Burbano 2017; Di Donato et al. 2018). A landmark study in 2003 demonstrated the potential of ancient DNA by showing the order in which three domestication trait genes governing architecture, storage proteins, and starch synthesis in maize appeared in archaeological maize (Jaenicke-Despres et al. 2003). In this case, two of the genes involved were shown to be under early selection, but the third, *su1* associated with starch synthesis, was shown to have come under selection in maize much later. This insight connected with observations in archaeology. It had been noted from the archaeobotanical record that the order in which domestication syndrome traits appear is staggered, with traits such as seed size increase occurring prior to rise of loss of seed shattering implying a complexity of different processes involved (Fuller 2007). Subsequent to this study, the advent of next-generation sequencing arrived in ancient DNA ushering in a new paleogenomic era (Poinar et al. 2006). It was a few years later before the first archaeogenomic level study of ancient plant remains appeared that showed the apparent speed with which genomes could change over time in regard to their TE content, with an implied consequence on genome size (Palmer et al. 2012a, b). These findings echo with later findings of a trend of association between adaptation in plant genomes and changing genome size largely driven by altered TE content (Bilinski et al. 2018). Subsequently, several archaeogenomes have been sequenced of maize (Da Fonseca et al. 2015; Ramos-Madrigal et al. 2016; Vallebueno-Estrada et al. 2016; Swarts et al. 2017), barley (Mascher et al. 2016), and sorghum (*Sorghum bicolor*; Smith et al. 2018).

The oldest plant archaeogenome to date is a 6,000-year-old barley from the Yoram caves in the Judean desert in Israel, reaching just over halfway back in time to the emergence of the first domesticates of barley (Mascher et al. 2016). This study established that the races here were most similar to modern barleys of the region, suggesting that little has changed in the past 6,000 years despite cultural turnover leading to the supposition that this lineage is locally adapted and for that reason favoured by sequential incoming cultures. This notion is in agreement with potential for locally adapted barley demonstrated by the differential input wild barley ecotypes from contrasting biomes (Poets et al. 2015). Further studies into the functional basis of such putative local adaptation may prove useful for modern crop breeding.

The earliest maize samples were only a little younger than the barley at around 5,300 years of age from two cave sites in the Tehuacan valley just outside the general area maize domestication is believed to have occurred (Ramos-Madrigal et al. 2016; Vallebueno-Estrada et al. 2016). This date represents ~60% of the time back to the rise of the first domesticates (Piperno et al. 2007, 2009). In both cases, the maize genomes were partially domesticated, with many domestication-associated genes showing the wild state. In the case of the San Marcos caves samples, the genomes were found to be highly inbred suggesting a small isolated population (Vallebueno-Estrada et al. 2016). Both archaeological samples were intermediate between wild teosinte and cultivated maize, either representing a basal lineage to modern maize or extinct domestication trajectory lineages. Both these studies support the notion of a gradual emergence of maize as has been shown of the cereals in the Southwest Asia (Tanno and Willcox 2006; Purugganan and Fuller 2009; Allaby et al. 2017).

A dissection of both selection processes and introgression during the early spread of maize was possible for samples between 2,000 and 750 years old in the SW United States (Da Fonseca et al. 2015). In this case, the older samples were shown to most likely originate from the Mexican highlands, because there was evidence for selection in architectural aspects of the domestication syndrome such as loss of shattering, thick glumes, and lateral branches. The younger samples showed an influx of introgression from lowland maize and selection at loci associated with drought stress and starch synthesis associated with the development of larger cobs. As such, this study was able to pull out the process of local adaptation as the crop moved to a new environment as well as inform on the changing phenotype to the modern form. This aspect of examining the extent of adaptation with latitude was refined considerably with maize samples from Turkey Pen that were ~1,900 years old (Swarts et al. 2017). In this case, the extensive genomic and phenotypic characterization of maize lines enabled the group to correlate genomic diversity to phenotype for several traits, including flowering time, plant height, and extent of tillering. The ancient plant was predicted to be short, bushy, and environmentally at the limit of its adaptive range given the latitudinal seasonal length and days it would have taken to mature. This work gives further insight into the temperance of agricultural spread by the time required for crops to adapt.

A time series of sorghum archaeogenomes provides a real-time picture of how domesticated genomes have evolved over time, leading to some surprising insights (Smith et al. 2018). As with previous archaeogenome studies on maize, we see the order of selection being unpicked with similar results. The earlier stages seem to be associated with architectural changes such as shattering and branching, while later stages see considerable selection on the sugar metabolism. In this study, though, it is the striking pattern of diversity loss over time that is particularly noted. In this case, the loss of genetic diversity associated with the crop domestication does not seem to be centred on the time of origin through an initial domestication bottleneck, but in fact dwindles over time in a linear way that is similar to previously observed serial founder effects in humans (de Giorgio et al. 2009), and here correlated with an agricultural cropping regime in which 25% of the harvest is set aside for sowing the following agricultural cycle. In this study, there is a second process of accumulating mutation load, which again is not associated with a domestication bottleneck but periods of selection over time.

The lack of an early loss of genetic diversity of the latter study prompted a survey of all the plant archaeogenomic data available for complete genomes at the time, which amounted to just three crops, barley, maize, and sorghum (Allaby et al. 2018). While much more data are needed across many genomes and species, the early indications are that both maize and barley also show no signs of an early loss of diversity.

The adaptive dynamic of crop plants has also been captured more indirectly with the increased temporal resolution of archaeogenomic levels of information providing biogeographical time stamps. For example, cucurbits plastid archaeogenomes have been used to establish the genus that was previously more widespread but suffered a contraction as the Late Pleistocene megafaunal extinctions removed their natural seed dispersers (Kistler et al. 2015). Moreover, in this case, extinct domestication lineages were discovered that had been domesticated from *Cucurbita fraterna* that only exists as an isolated wild population today, echoing previous characterization of lost domestication lineages through aDNA of chenopods (Kistler and Shapiro 2011). Conversely, the higher level of resolution provided by archaeogenomic information allows the fine dissection of diversity, which has helped to dramatically revise the dispersal route of bottle gourd to one of the oceanic current route across the Atlantic from Africa to South America rather than a Pacific or even Beringian route (Kistler et al. 2015). The extent and pace of crop spread is also becoming more detectable using NGS metagenomic techniques through sources such as sedaDNA (Smith et al. 2015). Here, the presence of the crop is suggested ahead of the main agricultural expansion but close to frontier settlements consistent with a pace of expansion tempered by constraints of latitudinal adaptation.

The adaptive challenges faced by crops as they spread from their homelands included not only the abiotic – the main focus of the studies above – but also biotic stressors. It is increasingly recognized that as alien species spread, their success or failure is often greatly influenced by the pathogenic arsenal they bring with them or meet (Anderson et al. 2004; Jones 2009; Stukenbrock and McDonald 2008). Archaeogenomic analysis of the *Phytophthora infestans* infection that caused the Irish potato famine showed it to be the result of an importation of a variant of the pathogen from outside its own natural range (Yoshida et al. 2013). Furthermore, the utility of RNA genomics combined with the paleomethylation state of the genome has revealed a dynamic interaction between a virus (Smith et al. 2014a) and its responding plant crop host (Smith et al. 2014b). Such studies give insight into the paleoepidemiology of past plant diseases and hold promise of information relevant to modern agriculture and food security.

6 Conclusions and Future Perspectives: Resurrection, Rescue, and Real-World Applications for Archaeogenomics

Over the next few decades, world populations are predicted to expand to nine billion, with an associated requirement of a 50% increase in food supply against a backdrop of a warming climate. There is therefore good reason to better understand the basis of local adaptation in crops, what has occurred in the past, and what the limits are for the future. Archaeogenomic records contain a long history of adaptation to changing conditions which have the potential to help us understand how plants coped with the new environments they reached and how long it took them to do so. Once we understand local adaptations of the past, can those adaptations be applied to modern crops? Studies are starting to produce examples of potential adaptations detected in past populations that are absent from modern crops (Smith et al. 2017), and more will surely follow. Crops that have become adapted to dry environments will be of particular interest, such as barley (Palmer et al. 2009; Allaby et al. 2014), as sources of material to drive genetic modification in modern crops.

Much has been made of the nutritional value of ancient crops, but biochemical evidence suggests that little is to be gained directly in terms of health and well-being (Shewry and Hey 2015). However, resurrection of lost crops may have other impacts. One promising avenue is through the exploitation of herbarium resources, from which it has become relatively easy to generate complete genomes. These can be used to identify crops that have fallen out of use for one reason or another where there are no living cultivated stands in existence. Efforts are currently under way in our labs to identify living populations of Bengal cotton (*Gossypium arboretum* var. *neglecta*), which fell from cultivated use some 150 years ago due its short fibre length, incompatibility with mechanization, and geopolitical pressure to utilize imported American cotton processed in British textile mills. However, examples of the crop do still exist in herbaria, and potential feral populations exist in South Asia. Matching the two has the potential to revive a lost industry.

The biodiversity that lies waiting in herbaria has highly impactful potential for restoration and rescue of past crops, while the archaeological record holds the promise of rescue for modern crops. As such, the archaeogenomic era, while still young, is becoming increasingly relevant to the present and future. It is only a matter of time before genetic botanical information that had been locked in the past will be reincorporated back into living nature.

References

Allaby RG. Integrating the processes in the evolutionary system of domestication. J Exp Bot. 2010;61:935–44.

Allaby RG, Fuller DQ, Brown TA. The genetic expectations of a protracted model for the origins of domesticated crops. Proc Natl Acad Sci U S A. 2008;105:13982–6.

Allaby RG, Gutaker R, Clarke AC, Pearson N, Ware R, Palmer SA, Kitchen JL, Smith O. Using archaeogenomics and computational approaches to unravel the history of local adaptation in crops. Philos Trans R Soc B. 2014;370:20130377.

Allaby RG, Kitchen JL, Fuller DQ. Surprisingly low limits of selection in plant domestication. Evol Bioinforma. 2015;11(S2):41–51.

Allaby RG, Stevens C, Lucas L, Maeda O, Fuller DQ. Geographic mosaics and changing rates of cereal domestication. Philos Trans R Soc B. 2017;372:20160429.

Allaby RG, Ware RL, Kistler L. A re-evaluation of the domestication bottleneck from archaeogenomic evidence. Evol Appl. 2018;1–9. https://doi.org/10.1111/eva.12680.

Anderson PK, Cunningham AA, Patel NG, Morales FJ, Epstein PR, Daszak P. Emerging infectious diseases of plants: pathogen pollution, climate change and agrotechnology drivers. Trends Ecol Evol. 2004;19:535–44.

Banks W, Antunes N, Rigaud S, d'Errico F. Ecological constraints on the first prehistoric farmers in Europe. J Archaeol Sci. 2013;40:2746–53.

Beissinger TM, et al. Recent demography drives changes in linked selection across the maize genome. Nat Plants. 2016;2:16084.

Bilinski P, Albert PS, Berg JJ, Birchler JA, Grote MN, Lorant A, Quezada J, Swarts K, Yang J, Ross-Ibarra J. Parallel altitudinal clines reveal trends in adaptive evolution of genome size in *Zea mays*. PLoS Genet. 2018;14(5):e1007162.

Blanc G, Wolfe KH. Widespread paleopolyploidy in mode plant species inferred from age distributions of duplicate genes. Plant Cell. 2004;16:1667–78.

Brown TA, Jones MK, Powell W, Allaby RG. The complex origins of domesticated crops. Trends Ecol Evol. 2009;24:103–9.

Cavalier-Smith T. Economy, sped and size matter: evolutionary forces driving nuclear genome miniaturization and expansion. Ann Bot. 2005;95:147–75.

Chénais B, Caruso A, Hiard S, Casse N. The impact of transposable elements on eukaryotic genomes: from genome size increase to adaptation to stressful environments. Gene. 2012;509:7–15.

Civáň P, Ivaničová Z, Brown TA. Reticulated origin of domesticated emmer wheat supports a dynamic model for the emergence of agriculture in the fertile crescent. PLoS One. 2013;8(11): e81955.

Colledge S, Conolly J, Shennan S. The evolution of early Neolithic farming from SW Asian origins to NW European limits. Eur J Archaeol. 2005;8:137–56.

Coward F, Shennan S, Colledge S, Conolly J, Collard M. The spread of Neolithic plant economies from the near East to northwest Europe: a phylogenetic analysis. J Archaeol Sci. 2008;35:42–56.

Da Fonseca R, et al. The origin and evolution of maize in the Southwestern United States. Nat Plants. 2015;1:14003.

De Giorgio M, Jakobsson M, Rosenberg NA. Explaining worldwide patterns of human genetic variation using a coalescent-based serial founder model of migration outward from Africa. Proc Natl Acad Sci U S A. 2009;106:16057–62.

Devos KM, Brown JKM, Bennetzen JL. Genome size reduction through illegitimate recombination counteracts genome expansion in Arabidopsis. Genome Res. 2002;12:1075–9.

Di Donato A, Filippone E, Ercolano MR, Frusciante L. Genome sequencing of ancient plant remains: findings, uses and potential applications for the study and improvement of modern crops. Front Plant Sci. 2018;9:441.

Diez CM, Meca E, Tenaillon MI, Gaut BS. Three groups of transposable elements with contrasting copy number dynamics and host responses in the maize (*Zea mays ssp. mays*) genome. PLoS Genet. 2014;10(4):e1004298.

Eyre-Walker A, Gaut RL, Hilton H, Feldman DL, Gaut BS. Investigation of the bottleneck leading to the domestication of maize. Proc Natl Acad Sci U S A. 1998;95:4441–6.

Fraser H. Genome-wide approaches to the study of adaptive gene expression evolution. BioEssays. 2011;33:469–77.

Fuller DQ. Contrasting patterns in crop domestication and domestication rates: recent archaeological insights from the Old World. Ann Bot. 2007;100:903–24.

Fuller DQ, Allaby RG, Stevens C. Domestication as Innovation: the entanglement of techniques, technology and chance in the domestication of cereal crops. World Archaeol. 2010;42:13–28.

Gaut BS, Diez CM, Morrell PL. Genomics and the contrasting dynamics of annual and perennial domestication. Trends Genet. 2015;31:709–19.

Gaut BS, Seymour DK, Liu Q, et al. Demography and its effects on genomic variation in crop domestication. Nat Plants. 2018;4:512–20.

Gutaker R, Burbano H. Reinforcing plant genomics using ancient DNA. Curr Opin Plant Biol. 2017;36:38–45.

Gutaker R, Zaidem M, Fu Y-B, Diederichsen A, Smith O, Ware R, Allaby RG. Adaptation to European latitudes through assimilation of wild diversity at the LuTFL1 locus altered architecture and promoted fiber production in flax. BioRxiv. 2017; https://doi.org/10.1101/178772.

Haldane JBS. The cost of selection. J Genet. 1957;55:511–24.

Hammer K. The domestication syndrome. Kulturpflanze. 1984;32:11–34.

Harlan J, de Wet JMJ, Price EG. Comparative evolution of cereals. Evolution. 1973;27:311–25.

Henn BM, Botigué LR, Bustamante CD, Clarke AG, Gravel S. Estimating the mutation load in humans. Nat Rev Genet. 2015;16:333–43.

Hill WG, Robertson A. The effect of linkage on limits to artificial selection. Genet Res Camb. 1966;8:269–94.

Jaenicke-Despres V, Buckler ES, Smith BD, Gilbert MTP, Cooper A, Doebley J, Paabo S. Early allelic selection in maize as revealed by ancient DNA. Science. 2003;302:1206–8.

Jones RAC. Plant virus emergence and evolution: origins, new encounter scenarios, factors driving emergence, effects of changing world conditions, and prospects for future control. Virus Res. 2009;141:113–30.

Jones H, Leigh FJ, Mackay I, Bower MA, Smith LMJ, Charles MP, Jones G, Jones MK, Brown TA, Powell W. Population-based resequencing reveals that the flowering time adaptation of cultivated barley originated east of the fertile crescent. Mol Biol Evol. 2008;25:2211e2219.

Jones G, Jones H, Charles MP, Jones MK, Colledge S, Leigh FJ, Lister DA, Smith LMJ, Powell W, Brown TA. Phylogeographic analysis of barley DNA as evidence for the spread of Neolithic agriculture through Europe. J Archaeol Sci. 2012;39:3230–8.

Kistler L, Shapiro B. Ancient DNA confirms a local origin of domesticated chenopod in eastern North America. J Archaeol Sci. 2011;38:3549–54.

Kistler L, Newsom LA, Ryan TM, Clarke AC, Smith BD, Perry GH. Gourds and squashes (*Cucurbita* spp.) adapted to megafaunal extinction and ecological anachronism through domestication. Proc Natl Acad Sci U S A. 2015;112:15107–12.

Larson G, et al. Current perspectives and the future of domestication studies. Proc Natl Acad Sci U S A. 2014;111:6139–46.

Leitch IJ, Hanson L, Winfield M, Parker J, Bennett MD. Nuclear DNA C-values complete familial representation in gymnosperms. Ann Bot. 2001;88:843–9.

Lenser T, Theißen G. Molecular mechanisms involved in convergent crop domestication. Trends Plant Sci. 2013;18(12):704–14.

Liu Q, Zhou Y, Morrell PL, Gaut BS. Deleterious variants in Asian rice and the potential cost of domestication. Mol Biol Evol. 2017;34:908–24.

Makino T, Rubin CJ, Carneiro M, Axelsson E, Andersson L, Webster MT. Elevated proportions of deleterious genetic variation in domestic animals and plants. Genome Biol Evol. 2018;10:276–90.

Marsden CD, et al. Bottlenecks and selective sweeps during domestication have increased deleterious genetic variation in dogs. Proc Natl Acad Sci U S A. 2016;113:152–7.

Mascagni F, Barghini E, Giordani T, Rieseberg L, Cavallini A, Natali L. Repetitive DNA and plant domestication: variation in copy number and proximity to genes of LTR-retrotransposons among wild and cultivated sunflower (Helianthus annuus) genotypes. Genome Biol Evol. 2015;7(12):3368–82.

Mascher M, et al. Genomic analysis of 6,000 year old cultivated grain illuminates the domestication history of barley. Nat Genet. 2016;48:1089–93.

Mazet O, Rodríguez W, Grusea S, Boitard S, Chikhi L. On the importance of being structured: instantaneous coalescence rates and human evolution – lessons for ancestral population size inference. Heredity. 2016;116:362–71.

Meyer RS, Purugganan MD. Evolution of crop species: genetics of domestication and diversification. Nat Rev Genet. 2013;14:840–52.

Meyer RS, et al. Domestication history and geographical adaptation inferred from a SNP map of African rice. Nat Genet. 2016;48:1083–8.

Meyer RS, DuVal AE, Jensen HR. Patterns and processes in crop domestication: an historical review and quantitative analysis of 203 global food crops. New Phytol. 2012; https://doi.org/10.1111/j.1469-8137.2012.04253.x.

Middleton CP, Stein N, Keller B, Kilian B, Wicker T. Comparative analysis of genome composition in Triticeae reveals strong variation in transposable element dynamics and nucleotide diversity. Plant J. 2013;73:347–56.

Moyers BT, Morrell PL, McKay JK. Genetic costs of domestication and improvement. Heredity. 2017;109:103–16.

Naito K, Cho E, Yang G, Campbell MA, Yano K, Okumoto Y, Tanisaka T, Wessler SR. Dramatic amplification of a rice transposable element during recent domestication. Proc Natl Acad Sci U S A. 2006;103:17620–5.

Oliver KR, Greene WK. Transposable elements: powerful facilitators of evolution. BioEssays. 2009;31:703–14.

Oliver KR, Greene WK. Transposable elements and viruses as factors in adaptation and evolution: an expansion and strengthening of the TE-Thrust hypothesis. Ecol Evol. 2012;2:2912–33.

Oliver KR, McComb JA, Greene WK. Transposable elements: powerful contributors to angiosperm evolution and diversity. Genome Biol Evol. 2013;5(10):1886–901.

Palmer SA, Moore JD, Clapham AJ, Rose P, Allaby RG. Archaeogenetic evidence of ancient Nubian barley evolution from six to two-row indicates local adaptation. PLoS One. 2009;4(7): e6301.

Palmer SA, Clapham AJ, Rose P, Freitas F, Owen BD, Beresford-Jones D, Moore JD, Kitchen JL, Allaby RG. Archaeogenomic evidence of punctuated genome evolution in Gossypium. Mol Biol Evol. 2012a;29(8):2031–8.

Palmer S, Smith O, Allaby RG. The blossoming of plant archaeogenetics. Ann Anat. 2012b;194:146–56.

Pellicer J, Hidalgo O, Dodsworth S, Leitch IJ. Genome size diversity and its impact on the evolution of land plants. Genes. 2018;9:88.

Perrier X, et al. Multidisciplinary perspectives on banana (Musa spp.) domestication. Proc Natl Acad Sci U S A. 2011;108:11311–8.

Piperno DR, Moreno JE, Iriarte J, Holst I, Lachniet M, Jones JG, Ranere AJ, Castanzo R. Lat Pleistocene and Holocene environmental history of the Iguala Valley, Central Balsas watershed of Mexico. Proc Natl Acad Sci U S A. 2007;104:11874–81.

Piperno DR, Ranere AJ, Holst I, Iriarte J, Dickau R. Starch grain and phytolith evidence from early ninth millennium B.P. maize from the Central Balsas River Valley, Mexico. Proc Natl Acad Sci U S A. 2009;106:5019–24.

Poets AM, Fang Z, Clegg MT, Morell PL. Barley landraces are characterized by geographically heterogeneous genomic origins. Genome Biol. 2015;16:173.

Poinar HN, Schwarz C, Qi J, Shapiro B, MacPhee RDE, Buigues B, Tikhonov A, Huson DH, Tomsho LP, Auch A, et al. Metagenomics to paleogenomics: large-scale sequencing of mammoth DNA. Science. 2006;311:392–4.

Purugganan MD, Fuller DQ. The nature of selection during plant domestication. Nature. 2009;457:843–8.

Purugganan MD, Fuller DQ. Archaeological data reveal slow rates of evolution during plant domestication. Evolution. 2011;65:171–83.

Puttick MN, Clark J, Donoghue PCJ. Size is not everything: rates of genome size evolution, not C-value, correlate with speciation in angiosperms. Proc R Soc B. 2015;282:20152289.

Ramos-Madrigal J, Smith BD, Moreno-Mayaer V, Gopalakrishnan S, Ross-Ibarra J, Gilbert MTP, Wales N. Genome sequence of a 5,310-year-old maize cob provides insights into the early stages of maize domestication. Curr Biol. 2016;26:3195–201.

Renaut S, Rieseberg LH. The accumulation of deleterious mutations as a consequence of domestication and improvement in sunflowers and other Compositae crops. Mol Biol Evol. 2015;32:2273–83.

Rensing SA. Why we need more non-seed plant models. New Phytol. 2017;216:355–60.

Schlumbaum A, Tensen M, Jaenicke-Després V. Ancient plant DNA in archaeobotany. Veg Hist Archaeobotany. 2008;17:233–44.

Schubert I, Vu TH. Genome stability and evolution: attempting a holistic view. Trends Plant Sci. 2016;21:749–57.

Schubert M, et al. Prehistoric genomes reveal the genetic foundation and cost of horse domestication. Proc Natl Acad Sci U S A. 2014;111:E5661–9.

Shennan S, Downey SS, Timpson A, et al. Regional population collapse followed initial agricultural booms in mid-Holocene Europe. Nat Commun. 2013;4:2486.

Shewry PR, Hey S. Do “ancient” wheat species differ from modern bread wheat in their contents of bioactive components? J Cereal Sci. 2015;65:236–43.

Sidhu GK, Warzecha T, Pawlowski WP. Evolution of meiotic recombination genes in maize and teosinte. BMC Genomics. 2017;18:106.
Simonin KA, Roddy AB. Genome downsizing, physiological novelty, and the global dominance of flowering plants. PLoS Biol. 2018;16(1):e2003706.
Smith O, Clapham A, Rose P, Liu Y, Wang J, Allaby RG. A complete ancient RNA genome: identification, reconstruction and evolutionary history of archaeological Barley Stripe Mosaic Virus. Sci Rep. 2014a;4:4003.
Smith O, Clapham A, Rose P, Liu Y, Wang J, Allaby RG. Genomic methylation patterns in archaeological barley show de-methylation as a time-dependent diagenetic process. Sci Rep. 2014b;4:5559.
Smith O, Momber G, Bates R, Garwood P, Fitch S, Pallen M, Gaffney V, Allaby RG. Sedimentary DNA from a submerged site reveals wheat in the British Isles 8000 years ago. Science. 2015;347:998–1001.
Smith O, Palmer SA, Clapham AJ, Rose P, Liu Y, Wang J, Allaby RG. Small RNA activity in archaeological barley shows novel germination inhibition in response to environment. Mol Biol Evol. 2017;34:2555–62.
Smith O, Nicholson W, Fuller D, Stephens C, Kistler L, Mace E, Jordan D, Barker G, Allaby RG. A domestication history of dynamic adaptation and genomic deterioration in sorghum. BioRxiv. 2018; https://doi.org/10.1101/336503.
Soltis DE, et al. Polyploidy and angiosperm diversification. Am J Bot. 2009;96:336–48.
Stephens CJ, Fuller DQ. Did Neolithic farming fail? The case for a Bronze Age agricultural revolution in the British Isles. Antiquity. 2012;86:707–22.
Stukenbrock EH, McDonald BA. The origins of plant pathogens in agro ecosystems. Annu Rev Phytopathol. 2008;46:75–100.
Swarts K, et al. Genomic estimation of complex traits reveals ancient maize adaptation to temperate North America. Science. 2017;357:512–5.
Tanno KI, Willcox G. How fast was wild wheat domesticated? Science. 2006;311:1886.
Tiley GP, Burleigh G. The relationship of recombination rate, genome structure, and patterns of molecular evolution across angiosperms. BMC Evol Biol. 2015;15:194.
Vallebueno-Estrada M, et al. The earliest maize from San Marcos Tehuacán is a partial domesticate with genomic evidence of inbreeding. Proc Natl Acad Sci U S A. 2016;113:14151–6.
van Heerwaarden J, Doebley J, Briggs WH, Glaubitz JC, Goodman MM, Jesus Sanchez Gonzalez J, Ross-Ibarra R. Genetic signals of origin, spread and introgression in a large sample of maize landraces. Proc Natl Acad Sci U S A. 2011;108:1088–92.
Vinogradov AE. Selfish DNA is maladaptive: evidence from the plant Red List. Trends Genet. 2003;19:609–14.
Wang L, Beissinger TM, Lorant A, Ross-Ibarra C, Ross-Ibarra J, Hufford MB. The interplay of demography and selection during maize domestication and expansion. Genome Biol. 2017;18:215.
Yoshida K, Schuenemann VJ, Cano LM, Pais M, Mishra B, Sharma R, Lanz C, Martin FN, Kamoun S, Krause J, et al. The rise and fall of the Phytophthora infestans lineage that triggered the Irish potato famine. elife. 2013;2:e00731.
Zaki EA, Ghany AAA. Ty3/gypsy retrotransposons in Egyptian cotton (G. barbadense). J Cotton Sci. 2004;8:179–85.
Zeder M. Core questions in domestication research. Proc Natl Acad Sci U S A. 2015;112:3191–8.
Zerega NJC, Ragone D, Motley T. Complex origins of breadfruit (*Artocarpus altilis*, Moraceae): implications for human migrations in Oceania. Am J Bot. 2004;91:760–6.
Zhou Y, Massonnet M, Sanjak JS, Cantu D, Gaut BS. Evolutionary genomics of grape (*Vitis vinifera* ssp. *vinifera*) domestication. Proc Natl Acad Sci U S A. 2017;114:11715–20.
Zhu Q, Zheng X, Luo J, Gaut B, Song G. Multilocus analysis of nucleotide variation of *Oryza sativa* and its wild relatives: severe bottleneck during domestication of rice. Mol Biol Evol. 2007;24:875–88.
Ziolkowski PA, et al. Natural variation and dosage of the HEI10 meiotic E3 ligase control *Arabidopsis* crossover recombination. Genes Dev. 2017;31:306–17.

Herbarium Genomics: Plant Archival DNA Explored

Freek T. Bakker

Abstract Herbarium genomics, allowing testing of historic biological hypotheses in plant science, is a promising field mainly driven by recent advances in next-generation sequencing (NGS) technology. Herbarium collections represent an enormous botanical repository of both specimens and of phenotypic observations and locality data, of sometimes long-extinct taxa. Herbarium specimens, a large part of which stem from the nineteenth and eighteenth century, are mostly pressed and mounted and were usually heat-treated and poisoned for preservation. Whereas the presence of post-mortem damage in herbarium DNA has been found to consist of mainly genome fragmentation (single- and double-stranded breaks), damage-derived miscoding lesions appear to be highly limited or even negligible. For organelle genomes and other repetitive genomic compartments, genome skimming appears effective in retrieving sequence data from plant herbarium specimens, whereas studies addressing herbarium nuclear-encoded genes and particularly whole genomes are still in minority. High levels of herbarium genomic fragmentation possibly lead to insert sizes being smaller than Illumina read lengths applied. Using a series of 93 herbarium DNA samples, representing 10 angiosperm families, near-complete plastomes were assembled for 80% of the specimens, some of which are 146 years old. Overlapping read pairs were found to occur in roughly 80% of all read pairs obtained. After merging such overlapping pairs, the resulting fragments and their distribution can be considered to reflect the ongoing process of genome fragmentation up to the moment of DNA extraction. Fragment length distributions appear to fit gamma distributions with either many small fragments present or an increasing number of longer fragments having accumulated. These distributions appear to differ from usually observed first-order genomic degradation kinetics, possibly due to the nonrepresentative nature of genome skimming samples.

Keywords Genomic fragmentation · Herbarium DNA · Plant aDNA · Plastomics

---

F. T. Bakker (✉)
Biosystematics Group, Wageningen University & Research, Wageningen, The Netherlands
e-mail: freek.bakker@wur.nl

Charlotte Lindqvist and Om P. Rajora (eds.), *Paleogenomics*,
Population Genomics [Om P. Rajora (Editor-in-Chief)],
https://doi.org/10.1007/13836_2018_40,

1 Introduction

Herbarium collections constitute an enormous repository of botanical (meta)data, i.e. both at specimen and collection locality data level, thereby documenting the botanical world (e.g. Bebber et al. 2010; Queenborough 2017; James et al. 2018). The estimated 350 million herbarium specimens deposited in 3,400 herbaria worldwide (Soltis 2007) collectively represent a huge past collection effort, sometimes in far from ideal conditions, and enable a time-series perspective in plant species' past ecology, phenotype, pathogens and demography (Bieker and Martin 2018). Including herbarium specimens with rare morphologies or from remote sampling locations, which would not easily be obtained through fieldwork, in genetic studies has become much more straightforward over the last decade (Bieker and Martin 2018; Olofsson et al. 2016). From a botanical perspective, paleogenomics takes a special place as most of the plant (and fungal) specimens deposited in herbarium collections are usually <300 years old. Although *herbarium genomics*, i.e. the recovery and analysis of genomic data from herbarium collections, may not be 'paleo' in terms of age, the term 'ancient DNA' can definitely be applied when it comes to the biochemical condition and degradation (Shapiro and Hofreiter 2014; Hofreiter et al. 2015) associated with herbarium plant DNA. On the other hand, for much older time spans, e.g. Neolithic, botanical remains, far fewer collections are available and with usually far inferior DNA quality (e.g. Mikić 2015). Nevertheless, in addition to the presence of a third genomic compartment (the plastid genome or 'plastome'), herbarium specimens do take a special place in paleogenomics (or *museomics*) as the possession of cell walls in plant (and fungal) material provides much better protection for DNA damage due to, for instance, oxidative stress than is the case in animal tissues (Mateiu and Rannala 2008; Roldán-Arjona and Ariza 2009). Indeed, herbarium genomics using next-generation sequencing (NGS) has already yielded valuable data and contributed importantly in testing historical biological hypotheses: for instance, genomes were sequenced from type specimens and rare or extinct species stored in herbaria. Zedane et al. (2015) retrieved a complete mitochondrial genome from a 1875 collection of *Hesperelaea* (Oleaceae), a species that is now extinct. Similarly, NGS and herbarium collections were used to sequence a complete plastome from an endemic, and now extinct, Hawaiian mint, *Stenogyne haliakalae*, a member of one of the largest plant lineages in the archipelago (Welch et al. 2016). Herbarium DNA was also used for finding previously unknown sister groups for important crops (Sebastian et al. 2010; Chomicki and Renner 2015), or in SNP analysis in genotyping by sequencing of species in *Solidago* (Asteraceae) (Beck and Semple 2015). Hart et al. (2016) described an approach to harvest hundreds of nuclear loci from herbarium DNA. To study historical pathogens, Yoshida et al. (2014, 2015) determined the genotype of the *Phytophthora infestans* (Mont.) de Bary strain that caused the great Irish potato famine in the nineteenth century. Likewise, herbarium DNA was crucial in discovering ancient alleles in *Alopecurus myosuroides* Huds. that are relevant to herbicide resistance but predating human influence (Délye et al. 2013). Reconstructing the shift to C4 photosynthesis in grasses could be conducted using DNA from a 100-year-old Malagasy herbarium specimen for which both its phylogenetic placement and the assessment of its 'genetic make-up' with regard to C4 photosynthesis could be assessed (Besnard et al. 2014; see also below). For taxonomy and DNA barcoding, herbaria collectively represent a potential treasure trove ready to be exploited (e.g. Xu et al. 2015), although only a few herbarium DNA barcoding projects have been realised to date (Osmundson et al. 2013; Costion et al. 2016; Enan et al. 2017). Bebber et al. (2010) estimated that around 70,000 new species are already in herbarium collections, 'waiting to be described', which further underlines the relevance of herbarium genomics, as it is expected to expedite archival DNA barcoding. Various published studies exist (Erkens et al. 2008; Särkinen et al. 2012; Drábková et al. 2002; Telle and Thines 2008; Gutaker et al. 2017) focussing explicitly on the efficiency of extraction and on the quality of herbarium DNA, mostly measured by PCR amplification. The expectation has been that heat treatment, but also the 'Schweinfurth method' (Schrenk 1888), will have been used in preparation of most herbarium specimens. The latter entails spraying specimens with ethanol in order to stop fungal growth, prior to heat fixation of the specimen. It was found that when extracting DNA from herbarium leaf material, most commercially available solutions are fine as long as some combination of CTAB protocols (Doyle and Dickson 1987; Doyle and Doyle 1987) and anion exchange purification is applied and that shorter PCR fragments amplify better (Särkinen et al. 2012). However, herbarium plant DNA yields are usually low, which can obviously be a problem when dealing with small, historic specimens, especially type specimens. Gutaker et al. (2017) report a DNA extraction protocol that is more suitable for recovering ultrashort herbarium DNA fragments than the commonly used CTAB protocol.

All in all, it is probably fair to say that we are currently at the dawn of a herbarium genomics era (Buerki and Baker 2015) and chances are high that a large body of plant archival genomic data will be generated in the years to come. This can only emphasise the vital importance of securing our herbarium collections for future molecular exploitation, a notion that is beginning to land. At the same time, it is good to realise that plant nuclear genomes are usually of much larger size than animal or fungal genomes (Gregory et al. 2007; and see below) and contain many repeats, which can hamper genome sequence assembly. Plant genomics is therefore challenging, be it from archival or fresh DNA. Plastomes on the other hand are characterised by both being of small size (around 160 kb) and exhibiting extensive structural conservation across land plants (Wicke and Schneeweiss 2015), enabling straightforward alignment and re-sequencing. In this chapter, I discuss recent findings on generating plastome sequences from herbarium angiosperm specimens and summarise the possible post-mortem damage incurred in the process of herbarium specimen fixation (see also Bakker 2015, 2017). In addition, future prospects for functional genomics in herbarium specimens are discussed.

2 Fragmentation in Herbarium DNA

Herbarium specimens are often dried with heat, which is known to have adverse effects on the immediate survival of DNA, most commonly through depurination (Lindahl and Andersson 1972). It is fairly well understood that applying heat to DNA when in a desiccating specimen is not favourable and can induce high levels of metabolic and cellular stress responses and ultimately cell death (Savolainen et al. 1995). The high temperatures (60–70°C), at which herbarium specimens are typically dried, cause cells to rupture quickly, releasing nucleases and other cellular enzymes (Gill and Tuteja 2010), as well as reactive oxygen species (ROS). Such physiological conditions resemble necrosis, and this cellular stress typically causes DNA to degrade randomly into smaller fragments, running as a smear on agarose gels (Reape et al. 2008; McCabe et al. 1997). Indeed, herbarium DNA is typically highly degraded into low molecular weight fragments (Doyle and Dickson 1987; Pyle and Adams 1989; Harris 1993), and this genomic fractionation causes the number of PCR-amplifiable template molecules to be reduced. To investigate this, studies were conducted involving next-generation sequencing of historic herbarium specimens (up to 65 years old) for which the actual individuals are still alive (Staats et al. 2011), in this case, trees growing in the Botanical Garden Leiden, The Netherlands. A new experimental herbarium was also prepared from the same individuals, allowing direct comparison of 'fresh plant material', 'young herbarium' and 'old herbarium' DNA. Some of the results are given in Fig. 1, which shows the extent of DNA fragmentation after specimen fixation (heating in a herbarium oven) and also that the subsequent time spent in a herbarium appears not to add significantly to overall fragmentation. The authors investigated this further by using qPCR assays directed to gene regions located at the three genomic compartments and

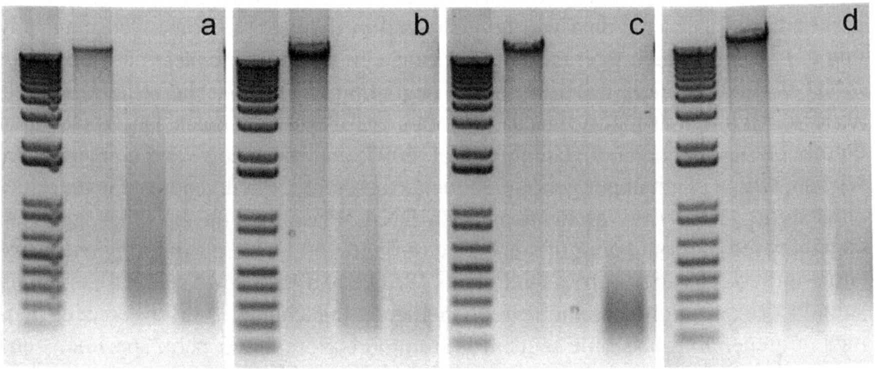

Fig. 1 Fresh plant material, young herbarium and historic herbarium DNA compared: agarose gels containing DNA extracts from *Lonicera maackii* (**a**), *Ginkgo biloba* (**b**), *Laburnum anagyroides* (**c**) and *Liriodendron tulipifera* (**d**). Fresh plants (left lane on the gels after the standard size marker) were collected; DNA is extracted immediately; young herbarium DNA (middle lane) is 3 weeks old; historic herbarium DNA (right lane) is 60, 107, 65 and 114 years old, respectively. DNA weight marker is 1 kb Plus DNA Ladder (Invitrogen). Adopted from Staats et al. (2011)

assessing how their copy numbers compare between ‘fresh’, ‘young herbarium’ and ‘old herbarium’ DNA extracts. Results indicated that 85–95% of the DNA was inaccessible to polymerases, probably due to double-stranded breaks directly after heat treatment or possibly due to blocking lesions, as observed in horse (*Equus ferus caballus*), mammoth (*Mammuthus primigenius*) and cave bear (*Ursus spelaeus*) ancient DNA by Heyn et al. (2010), and that significant difference in copy fold could not be detected when comparing young and old herbarium DNA. There was also no indication of preferential degradation of DNA in organellar compartments; this had been expected to some extent given the highly mutagenic environments of organelles with their ROS present at high concentrations.

After the PCR era, where fragmentation presented the main obstacle for successful amplification from herbarium DNA, everything changed in the NGS world, as fragmentation of the template genomic DNA is required for NGS library preparation, into which fragments are being incorporated directly and where the generally low yields sometimes are overcome by whole-genome amplification (WGA). The alternative to such low yields is obviously to use more starting herbarium material, but generally speaking one square centimetre of herbarium leaf tissue usually suffices for successful extraction, library preparation and superficial (Illumina) sequencing, which will be feasible for most specimens (including types). In any case, the DNA being fragmented no longer meant not being amenable to sequencing, which opened up great opportunities for herbarium DNA (e.g. Zedane et al. 2015; Hart et al. 2016; Staats et al. 2013).

Herbarium DNA fragmentation can sometimes be to such an extent that the efficiency of paired-end sequencing using Illumina HiSeq (and hence subsequent sequence assembly) can be affected. When template insert sizes are shorter than twice the Illumina read lengths applied, the actual sequencing reads will ‘meet in the middle’ of the insert and start to overlap (Fig. 2). However, when template insert sizes are smaller than the Illumina read length applied, this will result in the presence of adapter sequence at the end of the read (Turner 2014; see Fig. 2). In both scenarios of read overlap, the two reads can be merged into a single, longer read, which can be advantageous for subsequent assembly because (1) such reads can expedite finding overlaps and producing the De Bruyn graphs for the assembly process and (2) read errors can be detected and corrected where the reads overlap, which can result in a lower error rate than when using the raw reads. Another application of merging reads is that it enables assessing the distribution of fragment lengths in a herbarium DNA extract, as was carried out in Weiss et al. (2016). Here the authors inferred fragment length distributions for a series of herbarium specimens of up to 300 years old, by merging overlapping reads as outlined above. By assuming a log-normal fragment length distribution, the authors were able to deduce decay rates for their genomic extracts, based on the slope of the log (fragment length) plotted against specimen age. Weiss et al. (2016) conclude that the herbarium decay rate is ‘six times the rate of bone DNA decay’.

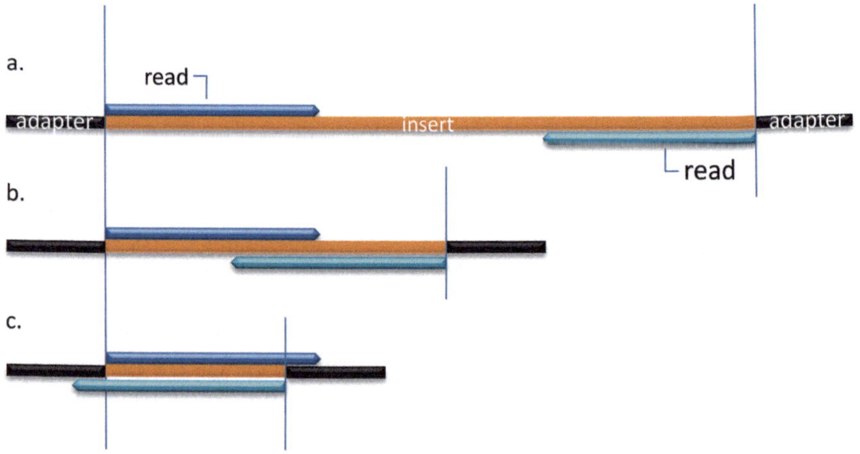

Fig. 2 Different scenarios for overlap in paired-end reads in combination with short insert sizes; (**a**) insert size is larger than 2× read length, and there is no overlap between reads; (**b**) insert size is between 1 and 2× read length, and reads overlap; (**c**) insert size is smaller than read length, and the reads include parts of the adapter. Vertical lines indicate ensuing fragment boundaries (as produced by BBMerge, see text)

3 Post-mortem Miscoding Lesions in Herbarium DNA

A point could be made about double- versus single-stranded DNA damage that the two are actually not mutually exclusive. Strand breaks (nicks, in the sugar-phosphate backbone) can occur in both strands, and when two nicks are more or less opposite from each other, a double-stranded break (and hence fragmentation) will result. For example, single-stranded library preparation methods (e.g. Gansauge et al. 2017) increase the number of molecules recovered compared with double-stranded library preparation methods, because of the presence of (single-stranded) nicks. On the other hand, miscoding lesions (i.e. damaged nucleotides, for instance, due to deamination) can occur in both double- and single-stranded DNA, be it at higher rates in the latter (Briggs et al. 2007). Damage-derived miscoding lesions (DDMLs; Brotherton et al. 2007) are damaged nucleotides that do permit polymerase extension but lead to 'wrong bases' in the newly amplified DNA and hence in the ensuing sequence data. Thus, if damaged nucleotides are present in herbarium DNA, they may result in damage-specific DDMLs by DNA polymerases during amplification (Hofreiter et al. 2001; Gilbert et al. 2003; Brotherton et al. 2007). This includes the occurrence of a-puric sites, deaminated cytosine residues and oxidised guanine residues, as found in studies in vivo and on ancient DNA (Lindahl 1993; Pääbo et al. 2004). Staats et al. (2011), based on their herbarium/fresh comparison panel described above, used the same nuclear, chloroplast and mitochondrial target genes as used in the qPCR assays, this time to generate PCR amplicons. These were sequenced using 454 sequencing technology, and comparison of the resulting reads was considered to reflect the occurrence of miscoding lesions in the template molecules (in this case the amplified molecules not the actual genomic DNA). Testing per nucleotide substitution type across all samples (and taking changes from both strands into account), it was found that only C → T/G → A transitions in plastid DNA from historic herbarium specimens occurred significantly more frequently than in plastid DNA from fresh and young herbarium and that this was not the case for the mitochondrial and nuclear compartments (Staats et al. 2011). The extra transitions were observed in only 0.03% of all nucleotides compared, allowing the assertion that herbarium DNA sequence data are most likely accurate. Based on these data, 'DNA repair protocols' such as those suggested by Yoshida et al. (2015) for herbarium DNA would therefore probably not have been necessary for these samples. In other herbarium DNA cases, the rate of extra transitions occurring could be higher and especially in the case of heterozygous SNP calling could confound base callers.

On the other hand, Weiss et al. (2016) in their study on herbarium DNA extracts found a pattern in herbarium DNA fragments that is typical of archival DNA, namely, an excess of C → T/G → A transitions primarily at the ends of the reads and declining exponentially inward (Briggs et al. 2007). The authors confirmed this in all historic herbarium samples analysed, but not in modern herbarium DNA that had not been heated. Therefore, the conclusion is probably fair that historic herbarium DNA from heated specimens looks rather similar to (non-heated) ancient DNA. To what extent the excess of C → T/G → A transitions at the ends of the reads drives post-mortem transitions in herbarium DNA sequences, the authors do not mention, but possibly the post-mortem transitions reported by Staats et al. (2011) correspond to these.

4 Herbarium Genome Skimming

In a follow-up study, Staats et al. (2013) demonstrated that by using Illumina HiSeq technology, herbarium DNA is perfectly amenable to plastome sequencing, in spite of the fact that mitochondrial, chloroplast and nuclear genomic copy numbers had decreased with 85–95% (as detected by qPCR; Staats et al. 2013). Between 81 and 100% plastome coverages were obtained from *Arabidopsis thaliana*, *Glycine max* and *Liriodendron tulipifera* herbarium specimens of up to 65 years old. Again, no increased nucleotide misincorporation rates were detected in all historic tissues, except for A → T/T → A transversions in chloroplast DNA which had an overall very low rate (0.025%; Staats et al. 2013) and according to the authors 'therefore appear to play little or no role in damage-derived miscoding lesions in herbarium DNA'.

In case of a 43-year-old *Arabidopsis thaliana* specimen, a full nuclear genome was sequenced as well, at 12× average coverage (Staats et al. 2013), which represented the first published herbarium angiosperm nuclear genome sequence. Herbarium-derived full nuclear genome sequencing is actually far from routine, and of course it is important to realise that *Arabidopsis thaliana* has a remarkably small nuclear genome size (119.7 Mb). This is atypical for angiosperms, for which genome size ranges from a minute 65 Mb (carnivorous *Genlisea*, Lentibulariaceae) up to a staggering 150,000 Mb (octaploid *Paris japonica*, Melianthaceae), with an average genome size considered to be 6,000 Mb long (Litt 2013). Well over half the angiosperm genomes estimated to date were found to be smaller than 5,000 Mb and about one-third to be under 1,000 Mb (Murray et al. 2010). Indeed, angiosperm genome sequence assembly represents a huge challenge (e.g. The Tomato Genome Consortium 2012) and is by far not as routine an undertaking as it is in animal and fungal genomics. Some parts of the angiosperm genome, however, are present in high copy number, notably the rDNA cistron repeats; the organellar genomes, i.e. the plastome and the chondrome (mitochondrial genome); and the different classes of highly repeated elements among which we distinguish microsatellite regions and long terminal repeats or transposable elements. Because of their repetitive nature, such regions will collectively be relatively well-represented, even in a limited or 'skimmed' second-generation sequencing sample that, by itself, would be too small to cover the entire nuclear genome. 'Genome skimming' has therefore been coined for the approach where superficial sequencing is performed and only genomic repeats or organellar genomes are represented with sufficient sequencing depth (Straub et al. 2012; Dodsworth et al. 2015). Usually this results in low costs compared with full/deep genome sequencing (although the cost for sequencing library preparation remains the same), and therefore it is an approach well suited for comparative studies involving many specimens. Another advantage of a skimming approach is that it prevents introducing rare variants and errors from various sources (Lonardi et al. 2015), whilst at the same time maintaining sufficient coverage for each repetitive genomic compartment. In a sense, it makes genome skimming comparable again with Sanger sequencing, in which 'rare variants' are marginalised in light of a main, average signal peak in Sanger trace files.

5 A Herbarium Plastomics Case Study

In a paper in a special issue on 'Collection-based research in the genome era' in the *Biological Journal of the Linnean Society*, we described an automated bioinformatics assembly pipeline for angiosperm organellar genomes, involving 'iterative organelle genome assembly' (IOGA) based on genome skimming data (Bakker et al. 2016). Our approach is similar to the 'baiting and iterative mapping' MitoBIM pipeline described by Hahn et al. (2013) for mitochondrial genomes, the difference being that IOGA does not require closely related reference organelle genome sequences. In addition, best assemblies in IOGA can be selected from multiple candidates using a maximum likelihood criterion (Clark et al. 2013). After scaffolding, i.e. correcting the relative orientation and order of contigs, final assemblies are then aligned to available 'nearest' reference plastome sequences using MUMmer plots (Fig. 3; Kurtz et al. 2004) in order to check accuracy of assembly.

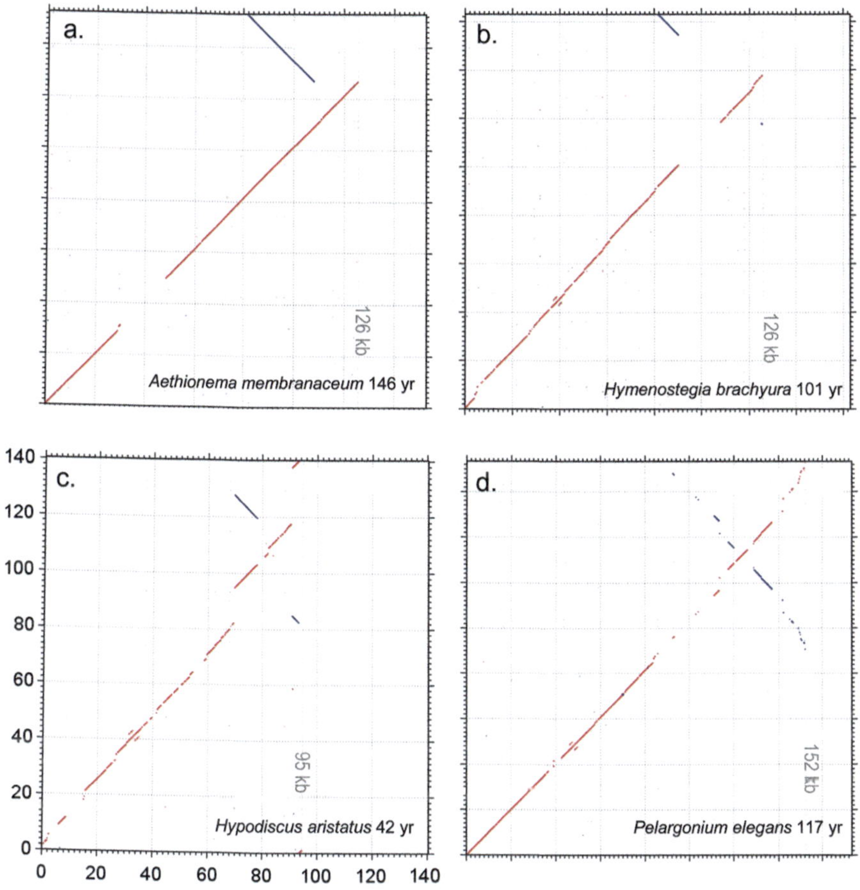

Fig. 3 Colinearity plots of IOGA-assembled plastomes from (**a**) a 146-year-old herbarium specimen of *Aethionema membranaceum* (Brassicaceae), 125,698 bp, compared with a reference plastome from *Brassica napus* (GenBank GQ861354); (**b**) a 101-year-old specimen of *Hymenostegia brachyura* (Fabaceae), 125,713 bp, compared with a reference plastome of *Prosopis* sp. (Fabaceae, GenBank KJ_68101, 163,040 bp); (**c**) a 42-year-old specimen of *Hypodiscus aristatus* (Restionaceae), 94,710 bp, compared with Zea mais plastome (Poaceae, GenBank X86563.2, 140,384 bp); (**d**) a 117-year-old specimen of *Pelargonium elegans* (152,457 bp) compared with reference plastome sequence of *Pelargonium alternans* (GenBank NC_023261, 173,374 bp). Repeat sequences are shown as off diagonals and reversals are shown in blue. Gridlines are 20 kb apart; herbarium plastome length is indicated by dashed lines

Using the IOGA pipeline, we compared 93 specimens from 12 angiosperm families, 73 of which were herbarium specimens up to 146 years old, to explore the feasibility of herbarium genomics (Bakker et al. 2016). After DNA extraction and quantification, carried out under standard conditions (i.e. not in an ancient DNA lab), sequence library preparation, index PCR and equimolar pooling of indexed libraries had been conducted; all libraries were then paired-end sequenced on four lanes on an Illumina HiSeq 2000 platform. For 84 out of our 93 specimens, sufficient numbers of paired-end reads were generated (at least 50,000), with all but two of the failed specimens being from historical herbarium material. A significant negative correlation was found between total reads per sample and specimen age, indicating that despite PCR enhancement of poor samples in the library preparation, older specimens still give fewer reads. The 84 successful samples were then subjected to IOGA (after subsampling to 1M reads in most cases) which yielded successful plastome assemblies for 74 specimens, or 80% of the specimens used. Nineteen of our 93 specimens did not yield plastome assemblies which could be due to the fact that plastome copy number was low in the leaf material used or that the equimolar mixing of specimens in the Illumina flow cell may have been unsuccessful somehow. Assembly lengths for the successful 74 accessions varied from 6 to220 kb with an overall average total assembly length of 136,167 bp, which is consistent with previously reported average angiosperm plastome lengths of 120–170 kb (e.g. Downie and Palmer 1992), including two inverted repeat (IR) regions of, on average, 25 kb each. In one case, *Pelargonium elegans*, a 117-year-old herbarium specimen, using only 24 ng of herbarium DNA, yielded a 167,770 bp assembly; from another, *Aethionema membranaceum*, a 146-year-old herbarium specimen, a complete plastome sequence was obtained. After checking pairwise alignments (MUMmer plots, Fig. 3) of best assemblies in selected samples, we found good colinearity with the published reference plastome sequence in cases for which reference and target were the same species, indicating accurate plastome sequence assembly. Reduced colinearity, and hence more off-diagonal elements in the MUMmer plots, was found in case of congenerics which probably reflects phylogenetic distance between target and reference plastome rather than misassembly.

When comparing fresh and herbarium specimens in terms of plastome assembly, it was found that differences were modest, with herbarium specimens yielding lower fractions of plastome-derived reads (4%) compared with those from fresh and silica gel dried specimens (13%; see Bakker et al. 2016). This would suggest that plastids may be lost preferentially in herbarium specimens, possibly after heat fixation. This would then contradict the studies by Staats et al. (2011), who could not find qPCR evidence for preferential degradation of organellar DNA in herbarium tissue. In any case, herbarium specimens appear to yield enough reads for effective plastome assembly. We found that total assembly length did not differ significantly between fresh and herbarium specimens, but that fresh samples on average yielded better N50s (Bakker et al. 2016). Total assembly length from herbarium DNA was the same, and herbarium assemblies just need slightly more editing and 'scaffolding'.

Specimen age per se did not seem to correlate with plastome assembly success. Of the 74 successful specimens in Bakker et al. (2016), there were 8 specimens older than 80 years, half of which gave plastome assemblies (>125 kb) that may be complete (or excluding one IR region). For all other specimens (i.e. younger than 80 years), this proportion was just over half (55%). Although there were more young than old specimens, which prevents making direct comparisons, it still appears that assembly success does not depend on specimen age. This is of course promising for the near-future further exploitation of herbarium collections worldwide, as many older (type) specimens are available. Herbarium specimens from wet-tropical conditions, of which there were 13 included in our study, appeared to 'behave' differently from all other accessions. Whereas 'dry collected' specimens sometimes may not even have been subjected to heat treatment (other than the sun) and usually do not get 'Schweinfürted' (Schrenk 1888), i.e. sprayed with ethanol in order to stop any fungi growing, for wet-tropical specimens this has been and sometimes still is common practice. It appears that preserving such specimens by immersion in ethanol prevents any DNA from being recovered later on (Mark Chase, personal communication). Bressan et al. (2014), however, found no difference in both quality and quantity of nuclear DNA recovered from tropical plant leaf tissue stored in liquid nitrogen versus 96% ethanol but also show how storage in ethanol causes cytoplasmic contents (including plastids) to be cleared from the leaf tissue cells. The Schweinfürth treatment in wet-tropical conditions nowadays usually entails keeping specimens inside a plastic bag under a saturated ethanol atmosphere, which can last for days before a drier is available. Alternatively, specimens are sometimes dried directly on a kerosene or gas stove (Jan Wieringa, personal communication).

When comparing the wet-tropical samples with the ones collected in drier conditions in the Bakker et al. (2016) study, generally, a higher number of contigs per assembly and lower N50 values were observed. When plotted against specimen age, it appeared as if the wet-tropical specimens seem to 'age' more quickly in terms of increased plastome assembly fragmentation when compared with dry habitat specimens. As exact preservation histories cannot usually be reconstructed for most herbarium specimens, no firm conclusions should be drawn here. Nevertheless, wet-tropical herbarium specimens may need some extra effort in plastome assembly and possibly require more additional Sanger sequencing-based confirmation in scaffolding.

6 Herbarium Genome Fragment Length Distributions

As mentioned above, fragmentation in herbarium DNA is a constant feature as reported in most relevant studies and can in principle be quantified by agarose gel electrophoresis (Fig. 1), automated electrophoresis or, given an Illumina HiSeq library, the in silico generation of fragments by merging overlapping reads (Fig. 2; Turner 2014). In order to explore the distribution of short-sized fragments in herbarium DNA and to what extent read overlap is occurring, I re-analysed a subset of 48 Illumina genome skimming herbarium samples from Bakker et al. (2016), listed below, using BBMerge from the BBTools package (http://jgi.doe.gov/data-and-tools/bbtools). This programme basically checks whether overlap exists and in case it finds an overlap, reads are subsequently merged. When insert size is shorter than the read length (in this case 100 bp), reads will have adapter sequence at the tail end, which is removed by BBMerge after merger (Brian Bushnell, personal communication). Using the default mode, proportion of overlapping reads, proportion of reads for which no overlap could be detected, as well as the average, SD and distribution of fragment lengths that resulted from merging the overlapping reads were recorded, and the results are given in Fig. 4. The 48 samples included species from *Lactuca*, *Karelinia* and *Nicolasia* (Asteraceae); *Polyscias* (Araliaceae); *Pelargonium* (Geraniaceae); *Aethionema* and *Tarenaya* (Brassicaceae); *Anthochortus*, *Dovea* and *Hypodiscus* (Restionaceae); *Anaxagorea*, *Desmopsis* and *Monanthotaxis* (Annonaceae); *Hymenostegia* and *Duparquetia* (Fabaceae); *Begonia* (Begoniaceae); *Paphiopedilum* (Orchidaceae); and *Rinorea* (Violaceae).

Across the 48 samples the average fragment length appears to be negatively correlated with specimen age ($R^2 = 0.29$), which implies that older herbarium DNA extracts contain smaller fragments (Fig. 4a). The standard deviation of the average fragment lengths tends to increase with longer fragments (polynomial regression, $R^2 = 0.94$), i.e. short fragments will probably have a narrow spread in size and occur in 'peaks' within a fragment length distribution. Longer fragments in contrast are then expected to occur across broader size ranges. This would be consistent with a scenario in which genomic fragments 'end up' in increasingly small sizes, and indeed the smaller sizes are correlated with higher specimen age as seen above. When looking at the number of overlapping reads, it is interesting to note that the percentage of read pairs that can be merged appears to be fairly independent of specimen age (Fig. 4c). In terms of actual numbers of reads, this relation is much more clear, with older specimens yielding lower amounts of reads in the first place and therefore also lower read pairs that could be merged.

When looking at the actual merged read fragment length distributions, I compared length distributions for two series of accessions from the Bakker et al. (2016) data: one for species of *Aethionema* (Brassicaceae), used and further described in Mohammadin et al. (2017) for phylogenetic analysis, and for *Lactuca* (Asteraceae) used and further described for the same purpose in Wei et al. (2017). The *Aethionema* series included both silica gel-dried and historic herbarium specimens of 23, 33, 40, 44, twice 50, 66 and 146 years old. The *Lactuca* series included silica gel-dried and historic herbarium specimens of 7, 36, 42, 43, twice 49, 54 and 64 years old. By comparing these congenerics, it can be assumed that genome size, GC contents, specimen tissue characteristics and specimen fixation histories (in most cases) are comparable too. Differences in fragment length distribution should therefore be due to specimen age, different specimen fixation (if applicable), herbarium collection locality or perhaps even stochasticity. Reads were merged using BBMerge as described above and fragment lengths between 26 and 184 bp plotted and their distributions compared (Fig. 5), after normalising each distribution to its total number of read pairs. Intuitively one would perhaps expect older specimens to be more fragmented than younger ones, given that more post-mortem time has been available. On the other hand, the results by Staats et al. (2011) indicated that this does not need to be the case (see above, and Fig. 1). Results show that for the *Lactuca* series, the oldest sample is indeed clearly the most highly fragmented (Fig. 5a). For the *Aethionema* series, however, results show that the older specimens do not have highest proportion of small fragments, but that specimens around 50 years do (Fig. 5b). For both series, we see that silica gel-dried samples show a gradual increase of longer fragment lengths that would probably have

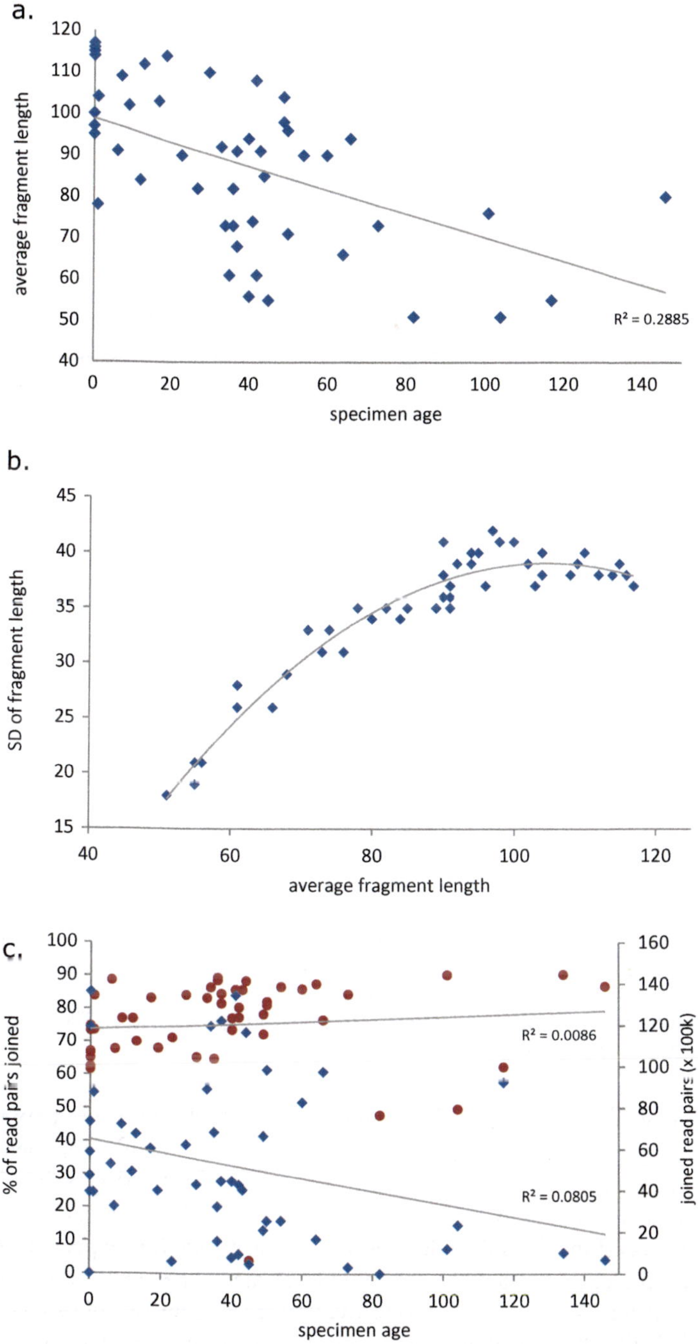

Fig. 4 Overlapping Illumina HiSeq reads from herbarium DNA extracts; the average fragment length after merging reads (**a**) plotted against specimen age; the SD of average fragment length (**b**) plotted against average fragment length; and (**c**) the percentage of total reads that could be merged (red dots) and the actual number of read pairs that could be joined (blue diamonds)

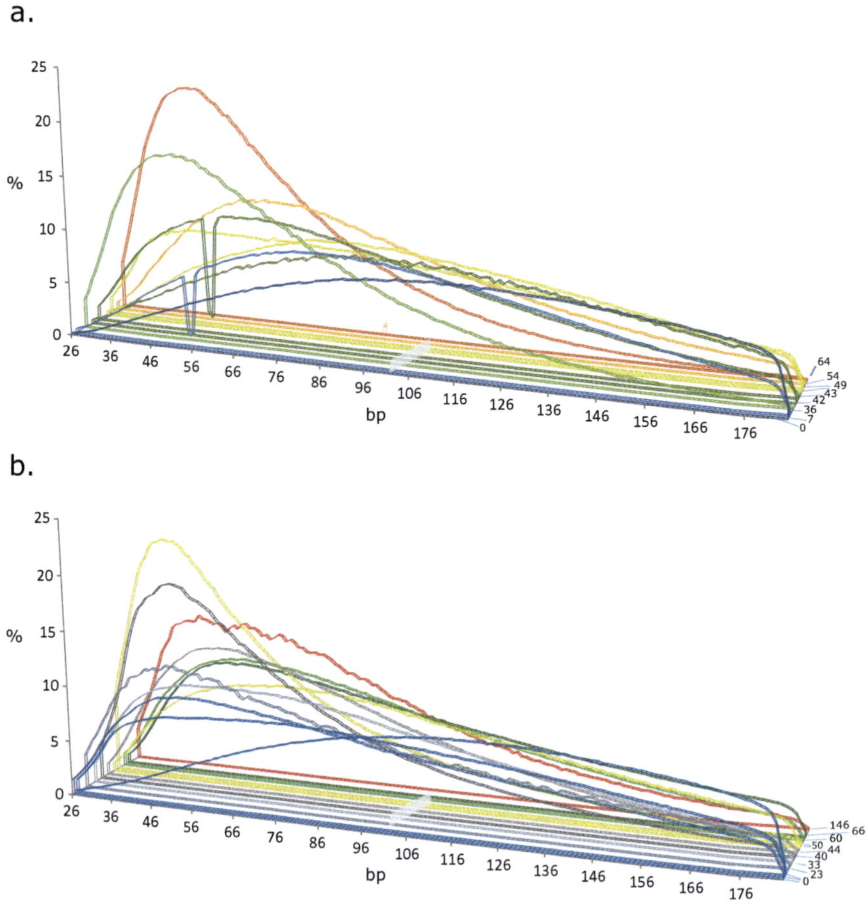

Fig. 5 Normalised distributions of fragment lengths (in bp) produced after merging 100 bp Illumina reads from fresh and herbarium specimens of different ages of *Lactuca* (**a**) and *Aethionema* (**b**). Reads up to 25 bp were discarded, and reads with length <100 were trimmed with regard to adapter sequences; distributions are sorted by (increased) specimen age. The transparent bar indicates the read length used (100 bp)

extended beyond 200 bp, had read lengths of say 150 bp been used. Weiss et al. (2016) found over-representation in A and G (purines) towards fragment ends, a pattern that reflects a similar signature in ancient DNA (Briggs et al. 2007). Depurination, or loss of A and G bases, is known to be a precursor to double-stranded breaks (Lindahl and Andersson 1972). Therefore, it is expected for purines to be overrepresented towards fragment ends. Some of the samples described here show indeed this fragment end purine over-representation (MapDamage data not shown). Whether the distribution of purines may indeed correlate with fragment length distributions as described here remains to be tested.

What is also interesting is that all distributions appear to fit well to a gamma-like distribution: either there are many short fragments and few longer ones, or there is a gradual increase in longer fragments. Weiss et al. (2016) and Allentoft et al. (2012) suggested log-normal distributions of fragment lengths in historic *Arabidopsis* and bone DNA, respectively. Yao et al. (2016) found the same for DNA degradation in human serum, urine and saliva DNA. These distributions would be consistent with a first-order kinetics at which DNA degrades, i.e. DNA has a half-life and the rate of degradation is constant (Allentoft et al. 2012). If the herbarium DNA degradation investigated here indeed fits a gamma rather than a log-normal distribution, this could indicate either a nonconstant rate of degradation or decay consistent with a higher-order kinetics and may reflect the complexity of these genomic samples. All reads in the analyses were derived from genome skimming sampling (see at Sect. 4), which means that of all the degraded DNA in the samples, repetitive compartments and sequences are overrepresented. Possibly, degradation of repetitive genomic compartments occurs at higher-order kinetics, i.e. a different half-life is present compared with non-repetitive DNA. However, this would need to be tested with (ancient) genomic samples that are deep-sequenced rather than genome-skimmed.

7 Future Perspectives and Conclusions

Herbarium genomics has seen great opportunities and development over the past decade, mainly driven by the ever-increasing availability of NGS technology. Especially when concerned with organelle genomes and other repetitive genomic compartments, approaches such as genome skimming appear effective in extracting DNA sequence data from large series of archival specimens. In combination with 'baiting and iterative mapping' assembly pipelines, routine assembly of nearly complete plastome sequences is feasible, with minimum specimen destruction and at reasonable costs. Overlapping read pairs are the result of template insert size used being smaller than twice the read length applied (or even smaller than the read length itself). Using a series of 93 herbarium DNA samples, representing 10 angiosperm families, overlapping read pairs were found to occur in roughly 80% of all read pairs obtained for most samples. Fragmentation is therefore confirmed as general feature of herbarium DNA, and insert sizes can be as small as <100 bp but still representing a majority of fragments.

Whereas obtaining organelle genomes or other high copy number regions (such as rDNA) will likely continue to prevail in studies of herbarium (and other archival) DNA, mostly benefitting phylogenetic and population genetic studies, studies addressing functional genomics questions using herbarium DNA are increasing in number. For instance, Besnard et al. (2014) were able to extract nuclear-encoded C4 photosynthesis genes from rare herbarium material (now extinct species) and place them in an evolutionary perspective, concluding that these genes were indeed absent from C3 plants. Beck and Semple (2015) successfully obtained herbarium SNP-based genotypes for comparison among *Solidago* species (Asteraceae) that are otherwise notorious for phylogenetic reconstruction, and Délye et al. (2013) relied on herbarium DNA for discovering ancient alleles in the grass *Alopecurus myosuroides* Huds. genome that are relevant to herbicide resistance (but predating human influence). Bieker and Martin (2018) list the main future applications of herbarium genomic, metagenomic and population genetic data to be (1) investigating how plant populations respond to environmental change, (2) infer temporal changes in genetic diversity, (3) identify genes under recent selection and (4) investigate past plant pathogen epidemics. For all four directions, we see exciting examples already being published (as listed above in Sect. 1).

With the advent of single-molecule long-read sequencing technologies (Ardui et al. 2018), such as PacBio and Oxford Nanopore, we may see a transition towards phasing out of 'short read technologies and its inherent limitations' (Ardui et al. 2018). Given the fragmented nature of herbarium DNA as described above, however, it is doubtful whether these 'third-generation' sequencing technologies will be highly useful for herbarium genomics, as they need DNA templates several dozen kilobase long. Hence, short-read technologies are expected to remain most relevant in archival DNA.

References

Allentoft ME, Collins M, Harker D, Haile J, Oskam CL, Hale ML, Campos PF, Samaniego JA, Gilbert MTP, Willerslev E, Zhang G, Scofield RP, Holdaway RN, Michael B. The half-life of DNA in bone: measuring decay kinetics in 158 dated fossils. Proc R Soc B. 2012;279 (1748):4724–33. https://doi.org/10.1098/rspb.2012.1745. Epub 2012 Oct 10.

Ardui S, Ameur A, Vermeesch JR, Hestand MS. Single molecule real-time (SMRT) sequencing comes of age: applications and utilities for medical diagnostics. Nucleic Acids Res. 2018;46:2159–68. https://doi.org/10.1093/nar/gky066.

Bakker FT. DNA sequences from plant herbarium tissue. In: Hörandl E, Appelhans M, editors. Next-generation sequencing in plant systematics. Bratislava: International Association for Plant Taxonomy (IAPT); 2015. p. 271–84.

Bakker FT. Herbarium genomics: skimming and plastomics from archival specimens. Webbia. 2017;72:35. https://doi.org/10.1080/00837792.2017.1313383.

Bakker FT, Lei D, Yu J, Mohammadin S, Wei Z, Van de Kerke S, Gravendeel B, Nieuwenhuis M, Staats M, Alquezar-Planas DE, Holmer R. Herbarium genomics: plastome sequence assembly from a range of herbarium specimens using an iterative organelle genome assembly (IOGA) pipeline. Biol J Linn Soc. 2016;117:33–43. https://doi.org/10.1111/bij.12642.

Bebber DP, Carine MA, Wood JRI, Wortley AH, Harris DJ, Prance GT, Davidse G, Paige J, Pennington TD, Robson NKB, Scotland RW. Herbaria are a major frontier for species discovery. PNAS. 2010;107:22169–71.

Beck JB, Semple JC. Next-generation sampling: pairing genomics with herbarium specimens provides species-level signal in *Solidago* (Asteraceae). Appl Plant Sci. 2015;3(6):1500014. https://doi.org/10.3732/apps.1500014.

Besnard G, Christin P-A, Malé P-JG, L'huillier E, Lauzeral C, Coissac E, Vorontsova MS. From museums to genomics: old herbarium specimens shed light on a C3 to C4 transition. J Exp Bot. 2014;65:6711. https://doi.org/10.1093/jxb/eru395.

Bieker VC, Martin MD. Implications and future prospects for evolutionary analyses of DNA in historical herbarium collections. Bot Lett. 2018; https://doi.org/10.1080/23818107.2018.1458651.

Bressan EA, Rossi ML, Gerald LT, Figueira A. Extraction of high-quality DNA from ethanol-preserved tropical plant tissues. BMC Res Notes. 2014;7:268. https://doi.org/10.1186/1756-0500-7-268.

Briggs AW, Stenzel U, Johnson PLF, Green RE, Kelso J, Prufer K, Meyer M, Krause J, Ronan MT, Lachmann M, Pääbo S. Patterns of damage in genomic DNA sequences from a Neandertal. PNAS. 2007;104:14616–21. https://doi.org/10.1073/pnas.0704665104.

Brotherton P, Endicott P, Sanchez JJ, Beaumont M, Barnett R, Austin J, Cooper A. Novel high-resolution characterization of ancient DNA reveals C > U-type base modification events as the sole cause of post mortem miscoding lesions. Nucleic Acids Res. 2007;35(17):5717–28. https://doi.org/10.1093/nar/gkm588.

Buerki S, Baker WJ. Collections-based research in the genomic era. Biol J Linn Soc. 2015;117:5. https://doi.org/10.1111/bij.12721.

Chomicki G, Renner SS. Watermelon origin solved with molecular phylogenetics including Linnaean material: another example of museomics. New Phytol. 2015;205:526–32.

Clark SC, Egan R, Frazier PI, Wang Z. ALE: a generic assembly likelihood evaluation framework for assessing the accuracy of genome and metagenome assemblies. Bioinformatics. 2013;29:435–43.

Costion CM, Lowe AJ, Rossetto M, Kooyman RM, Breed MF, Ford A, Crayn DM. Building a plant DNA barcode reference library for a diverse tropical flora: an example from Queensland, Australia. Divers Distrib. 2016;8:1–9. https://doi.org/10.3390/d8010005.

Délye C, Deulvot C, Chauvel B. DNA analysis of herbarium specimens of the grass weed *Alopecurus myosuroides* reveals herbicide resistance pre-dated herbicides. PLoS One. 2013;8(10):e75117.

Dodsworth S, Chase MW, Kelly LJ, Leitch IJ, Macas J, Novák P, et al. Genomic repeat abundances contain phylogenetic signal. Syst Biol. 2015;64(1):112–26. https://doi.org/10.1093/sysbio/syu080.

Downie SR, Palmer JD. Use of chloroplast DNA rearrangements in reconstruction plant phylogeny. In: Soltis PS, et al., editors. Molecular systematics of plants. New York: Chapman and Hall; 1992. p. 1–13.

Doyle JJ, Dickson EE. Preservation of plant species for DNA restriction endonuclease analysis. Taxon. 1987;36:715–22.

Doyle JJ, Doyle JL. A rapid DNA isolation procedure for small quantities of fresh leaf tissue. Phyt Bull. 1987;19:11–5.

Drábková L, Kirschner J, Vlcek C. Comparison of seven DNA extraction and amplification protocols in historic herbarium specimens of Juncaceae. Plant Mol Biol Rep. 2002;20:161–75.

Enan MR, Palakkott AR, Ksiksi TS. DNA barcoding of selected UAE medicinal plant species: a comparative assessment of herbarium and fresh samples. Phys Mol Biol Plants. 2017;23:221–7. https://doi.org/10.1007/s12298-016-0412-9.

Erkens RHJ, Cross H, Maas JW, Hoenselaar K, Chatrou LW. Age and greenness of herbarium specimens as predictors for successful extraction and amplification of DNA. Blumea. 2008;53:407–28.

Gansauge MT, Gerber T, Glocke I, Korlevic P, Lippik L, Nagel S, Riehl LM, Schmidt A, Meyer M. Single-stranded DNA library preparation from highly degraded DNA using T4 DNA ligase. Nucleic Acids Res. 2017;45(10):e79. https://doi.org/10.1093/nar/gkx033.

Gilbert MTP, Hansen AJ, Willerslev E, Rudbeck L, Barnes I, Lynnerup N, Cooper A. Distribution patterns of postmortem damage in human mitochondrial DNA. Am J Hum Genet. 2003;72:48–61.

Gill SS, Tuteja N. Reactive oxygen species and antioxidant machinery in abiotic stress tolerance in crop plants. Plant Physiol Biochem. 2010;48:909–30.

Gregory TR, Nicoll JA, Tamm H, Kullman B, Kullman K, Leitch IJ, Murray BG, Kapraun DF, Greilhuber J, Bennett MD. Eukaryotic genome size databases. Nucleic Acids Res. 2007;35 (Database issue):D332–D338. https://doi.org/10.1093/nar/gkl828.

Gutaker RM, Reiter E, Furtwängler A, Schuenemann VJ, Burbano HA. Extraction of ultrashort DNA molecules from herbarium specimens. BioTechniques. 2017;62:76–9. https://doi.org/10.2144/000114517.

Hahn C, Bachmann L, Chevreux B. Reconstructing mitochondrial genomes directly from genomic next-generation sequencing reads – a baiting and iterative mapping approach. Nucleic Acids Res. 2013;41(13):e129.

Harris SA. DNA analysis of tropical plant species: an assessment of different drying methods. Plant Syst Evol. 1993;188:57–64.

Hart ML, Forrest LL, Nicholls JA, Kidner CA. Retrieval of hundreds of nuclear loci from herbarium specimens. Taxon. 2016;65(5):1081–92.

Heyn P, Stenzel U, Briggs AW, Kircher M, Hofreiter M, Meyer M. Road blocks on paleogenomes – polymerase extension profiling reveals the frequency of blocking lesions in ancient DNA. Nucleic Acids Res. 2010;38(16):e161. https://doi.org/10.1093/nar/gkq572.

Hofreiter M, Jaenicke V, Serre D, von Haeseler A, Pääbo S. DNA sequences from multiple amplifications reveal artifacts induced by cytosine deamination in ancient DNA. Nucleic Acids Res. 2001;29:4793–9.

Hofreiter M, Paijmans JLA, Goodchild H, Speller CF, Barlow A, Fortes GG, Thomas JA, Ludwig A, Collins MJ. The future of ancient DNA: technical advances and conceptual shifts. BioEssays. 2015;37:284–93. https://doi.org/10.1002/bies.201400160.

James SA, Soltis PS, Belbin L, Chapman AD, Nelson G, Paul DL, Collins M. Herbarium data: global biodiversity and societal botanical needs for novel research. Appl Plant Sci. 2018;6: e1024. https://doi.org/10.1002/aps3.1024.

Kurtz S, Phillippy A, Delcher AL, Smoot M, Shumway M, Antonescu C, Salzberg SL. Versatile and open software for comparing large genomes. Genome Biol. 2004;5:R12.

Lindahl T. Instability and decay of the primary structure of DNA. Nature. 1993;362:709–15.

Lindahl T, Andersson A. Rate of chain breakage at apurinic sites in double-stranded deoxyribonucleic acid. Biochemistry. 1972;11:3618–23.

Litt A. Comparative evolutionary genomics of land plants. Ann Plant Rev. 2013;45:227–76. https://doi.org/10.1002/9781118305881.ch8.

Lonardi S, Mirebrahim H, Wanamaker S, Alpert M, Ciardo G, Duma D, Close TJ. When less is more: 'slicing' sequencing data improves read decoding accuracy and de novo assembly quality. Bioinformatics. 2015;31:2972–80.

Mateiu LM, Rannala BH. Bayesian inference of errors in ancient DNA caused by postmortem degradation. Mol Biol Evol. 2008;25(7):1503–11. https://doi.org/10.1093/molbev/msn095.

McCabe PF, Levine A, Meijer PJ, Tapon NA, Pennell RI. A programmed cell death pathway activated in carrot cells cultured at low cell density. Plant J. 1997;12:267–80.

Mikić AM. The first attested extraction of ancient DNA in legumes (Fabaceae). Front Plant Sci. 2015;6:1006. https://doi.org/10.3389/fpls.2015.01006.

Mohammadin S, Peterse K, van de Kerke SJ, Chatrou LW, Dönmez AA, Mummenhoff K, Pires JC, Edger PP, Al-Shehbaz IA, Schranz ME. Anatolian origins and diversification of *Aethionema*, the sister lineage of the core Brassicaceae. Am J Bot. 2017;104:1042–54.

Murray BG, Leitch IJ, Bennett MD. Gymnosperm DNA C-values database. Release 4.0, Dec 2010. http://data.kew.org/cvalues.

Olofsson JK, Bianconi M, Besnard G, Dunning LT, Lundgren MR, Holota H, Vorontsova MS, et al. Genome biogeography reveals the intraspecific spread of adaptive mutations for a complex trait. Mol Ecol. 2016;25(24):6107–123.

Osmundson TW, Robert VA, Schoch CL, Baker LJ, Smith A, Robich G, Mizzan L, Garbelotto M. Filling gaps in biodiversity knowledge for macrofungi: contributions and assessment of an herbarium collection DNA barcode sequencing project. PLoS One. 2013;8:1–8.

Pääbo S, Poinar H, Serre D, Jaenicke-Despres V, Hebler J, Rohland N, Kuch M, Krause J, Vigilant L, Hofreiter M. Genetic analyses from ancient DNA. Annu Rev Genet. 2004;38:645–79.

Pyle MM, Adams RP. In situ preservation of DNA in plant specimens. Taxon. 1989;38:576–81.

Queenborough S. Collections-based studies of plant functional traits. In: Friis I, Balslev H, editors. Tropical plant collections: legacies from the past? Essential tools for the future? Scientia Danica B (Biologica). Vol 6. 2017. p. 15–38, 223–36.

Reape TJ, Molony EM, McCabe PF. Programmed cell death in plants: distinguishing between different modes. J Exp Bot. 2008;59:435–44.

Roldán-Arjona T, Ariza RR. Repair and tolerance of oxidative DNA damage in plants. Mutat Res. 2009;681:169–79.

Särkinen T, Staats M, Richardson JE, Cowan RS, Bakker FT. How to open the treasure chest? Optimising DNA extraction from herbarium specimens. PLoS One. 2012;7:e43808. https://doi.org/10.1371/journal.pone.0043808.

Savolainen V, Cuénoud P, Spichiger R, Martinez MDP, Crèvecoeur M, Manen J-F. The use of hebarium specimens in DNA phylogenetics: evaluation and improvement. Plant Syst Evol. 1995;197:87–98.

Schrenk J. Schweinfurth's method of preserving plants for herbaria. Bull Torrey Bot Club. 1888;15:292–3.

Sebastian P, Schaefer H, Telford IRH, Renner SS. Cucumber (*Cucumis sativus*) and melon (*C. melo*) have numerous wild relatives in Asia and Australia, and the sister species of melon is from Australia. PNAS. 2010;107:14269–73.

Shapiro B, Hofreiter M. A Paleogenomic perspective on evolution and gene function: new insights from ancient DNA. Science. 2014;343:1236573. https://doi.org/10.1126/science.1236573.

Soltis PS. Digitization of herbaria enables novel research. Am J Bot. 2017;104:1–4.

Staats M, Cuence A, Richardson JE, Vrielink-van Ginkel R, Petersen G, Seberg O, Bakker FT. DNA damage in plant herbarium tissue. PLoS One. 2011;6:e28448.

Staats M, Erkens RHJ, van de Vossenberg B, Wieringa JJ, Kraaijeveld K, Stielow B, Geml J, Richardson JE, Bakker FT. Genomic treasure troves: complete genome sequencing of herbarium and insect museum specimens. PLoS One. 2013;8(7):e69189. https://doi.org/10.1371/journal.pone.0069189.

Straub SCK, Parks M, Weitemeir K, Fishbein M, Cronn R, et al. Navigating the tip of the genomic iceberg: next-generation sequencing for plant systematics. Am J Bot. 2012;99:349–64.

Telle S, Thines M. Amplification of *cox2* (~620 bp) from 2 mg of up to 129 years old herbarium specimens, comparing 19 extraction methods and 15 polymerases. PLoS One. 2008;3:e3584.

The Tomato Genome Consortium. The tomato genome sequence provides insights into fleshy fruit evolution. Nature. 2012;485:635. https://doi.org/10.1038/nature11119.

Turner FS. Assessment of insert sizes and adapter content in fastq data from NexteraXT libraries. Front Genet. 2014;5:1–7. https://doi.org/10.3389/fgene.2014.00005.

Wei Z, Zhu SX, van den Berg RG, Bakker FT, Schranz ME. Phylogenetic relationships within *Lactuca* L. (Asteraceae), including African species, based on chloroplast DNA sequence comparisons. Genet Resour Crop Evol. 2017;64:55–71.

Weiss CL, Schuenemann VJ, Devos J, Shirsekar G, Reiter E, Gould BA, Stinchcombe JR, Krause J, Burbano HA. Temporal patterns of damage and decay kinetics of DNA retrieved from plant herbarium specimens. R Soc Open Sci. 2016;3:160239

Welch AJ, Collins K, Ratan A, Drautz-Moses DI, Schuster SC, Lindqvist C. The quest to resolve recent radiations: plastid phylogenomics of extinct and endangered Hawaiian endemic mints (Lamiaceae). Mol Phylogenet Evol. 2016;99:16–33.

Wicke S, Schneeweiss GM. Next-generation organellar genomics: potentials and pitfalls of high-throughput technologies for molecular evolutionary studies and plant systematics. In: Hörandl E, Appelhans MS, editors. Next generation sequencing in plant systematics. Bratislava: International Association for Plant Taxonomy (IAPT); 2015. p. 9–50.

Xu C, Dong W, Shi S, Cheng T, Li C, Liu Y, Wu P, Wu H, Gao P, Zhou S. Accelerating plant DNA barcode reference library construction using herbarium specimens: improved experimental techniques. Mol Ecol Resour. 2015;15:1366–74. https://doi.org/10.1111/1755-0998.12413.

Yao W, Mei C, Nan X, Hui L. Evaluation and comparison of in vitro degradation kinetics of DNA in serum, urine and saliva: a qualitative study. Gene. 2016;590(1):142–8. https://doi.org/10.1016/j.gene.2016.06.033. Epub 2016 June 16.

Yoshida K, Burbano HA, Krause J, Thines M, Weigel D, et al. Mining herbaria for plant pathogen genomes: back to the future. PLoS Pathog. 2014;10(4):e1004028. https://doi.org/10.1371/journal.ppat.1004028.

Yoshida K, Sasaki E, Kamoun S. Computational analyses of ancient pathogen DNA from herbarium samples: challenges and prospects. Front Plant Sci. 2015;6:771.

Zedane L, Hong-Wa C, Murienne J, Jeziorsky C, Baldwin BG, Besnard G. Museomics illuminate the history of an extinct, paleoendemic plant lineage (*Hesperelaea*, Oleaceae) known from an 1875 collection from Guadalupe Island, Mexico. Biol J Linn Soc. 2015;117:44–57.

Paleogenomics of Animal Domestication

Evan K. Irving-Pease, Hannah Ryan, Alexandra Jamieson, Evangelos A. Dimopoulos, Greger Larson, and Laurent A. F. Frantz

Abstract Starting with dogs, over 15,000 years ago, the domestication of animals has been central in the development of modern societies. Because of its importance for a range of disciplines – including archaeology, biology and the humanities – domestication has been studied extensively. This chapter reviews how the field of paleogenomics has revolutionised, and will continue to revolutionise, our understanding of animal domestication. We discuss how the recovery of ancient DNA from archaeological remains is allowing researchers to overcome inherent shortcomings arising from the analysis of modern DNA alone. In particular, we show how DNA, extracted from ancient substrates, has proven to be a crucial source of information to reconstruct the geographic and temporal origin of domestic species. We also discuss how ancient DNA is being used by geneticists and archaeologists to directly observe evolutionary changes linked to artificial and natural selection to generate a richer understanding of this fascinating process.

Keywords Ancient DNA · Archaeology · Domestication · Entomology · Evolution · Genomics · Zoology

E. K. Irving-Pease (✉) · H. Ryan · A. Jamieson · E. A. Dimopoulos · G. Larson
The Palaeogenomics and Bio-Archaeology Research Network, Research Laboratory for Archaeology and History of Art, University of Oxford, Oxford, UK
e-mail: evan.irving-pease@arch.ox.ac.uk

L. A. F. Frantz (✉)
The Palaeogenomics and Bio-Archaeology Research Network, Research Laboratory for Archaeology and History of Art, University of Oxford, Oxford, UK

School of Biological and Chemical Sciences, Queen Mary University of London, London, UK
e-mail: laurent.frantz@qmul.ac.uk

Charlotte Lindqvist and Om P. Rajora (eds.), *Paleogenomics*,
Population Genomics [Om P. Rajora (Editor-in-Chief)],
https://doi.org/10.1007/13836_2018_55,

1 Introduction

The domestication of plants and animals was one of the most significant transformations in human history. Domestication was central to the emergence of settled agricultural communities (Larson et al. 2014). The advent of farming and pastoralism, during the Neolithic transition, lead to massive social, economic, religious and demographic changes (Zeder 2012a). It supported vastly increased human population sizes (Bocquet-Appel 2011) and laid the foundation for the development of complex civilisations (Larson and Burger 2013). Ultimately these changes transformed the biosphere and ushered in the age of the Anthropocene (Smith and Zeder 2013).

The study of animal domestication is a broad endeavour, which draws in expertise from archaeology, genetics, ecology and physical sciences (Zeder et al. 2006; Vigne 2011; Larson et al. 2012, 2014; Zeder 2016; MacHugh et al. 2017). This multidisciplinary approach has provided the power to address the critical questions of when, where and how animal domestications took place (Larson et al. 2014), as well as to help elucidate the biological basis for animal domestication (Jensen 2014). More recently, the study of animal domestication has been transformed by the revolution in modern and ancient genome sequencing (Larson and Burger 2013; Larson and Bradley 2014; Gerbault et al. 2014).

This chapter will review how paleogenomics has informed, and will continue to inform, our understanding of animal domestication. We will discuss how paleogenomic approaches applied to domestic species have been used to resolve their geographic and temporal origin, to track human migration, and to understand how animal genomes have been shaped by changes in human culture and technology.

2 Sequencing Ancient DNA

Early ancient DNA (aDNA) studies were constrained by the high cost and low yield of the sequencing technology which was available at the time. The first aDNA study, which recovered DNA from an extinct quagga (Higuchi et al. 1984), used molecular cloning to amplify target DNA molecules, by ligating them into plasmids and replicating them within bacteria (Maniatis et al. 1982). This approach was rapidly superseded by the discovery of the polymerase chain reaction (PCR) (Saiki et al. 1985; Mullis and Faloona 1987), which allowed researchers to efficiently amplify predetermined genomic loci, for sequencing using the Sanger chain-terminating method (Sanger et al. 1977). In vitro amplification (PCR) also had its limitations as it required a priori knowledge of the loci being targeted, which restricted analyses to species and genes which had already been sequenced in modern populations. PCR targets the intended locus using a pair of primers (forward and reverse) which flank the target region. As aDNA is highly fragmented – mostly less than 100 base pairs (bp) (Sawyer et al. 2012) – the loci targeted by the PCR primers need to be shorter than the average length of endogenous molecules in an ancient sample, or the experiment might fail.

These early PCR-based studies focused primarily on the recovery of a single gene locus from the mitochondrial DNA (mtDNA). The most commonly targeted regions were highly variable loci, such as cytochrome *b* and the mtDNA control region, which were used extensively for resolving molecular phylogenies (Irwin et al. 1991; Meyer 1994). Unlike the nuclear genome, which has only two copies in each cell, there can be many thousands of copies of the mitochondrial genome in each cell (Reynier et al. 2001). This greater relative abundance of mtDNA improves the likelihood of retrieving any particular locus via PCR amplification. Whilst mtDNA is easier to recover, its information content is more limited than nuclear autosomal DNA. Autosomal DNA is inherited equally from both parents, in contrast to mtDNA which is uniparentally inherited, along the maternal line only. Consequently, mtDNA may not reflect the broader evolutionary history of the species as a whole (reviewed in Ballard and Whitlock 2004). Discrepancies between mtDNA and nuclear DNA analysis can be particularly acute when there are sex-biased processes, population replacement or gene flow occurring at the population level, such as those documented in horses (Vilà et al. 2001; Lippold et al. 2011b), pigs (Frantz et al. 2013b) and cattle (Hanotte et al. 2002).

The advent of high-throughput or 'next-generation sequencing' (NGS) platforms in the mid-2000s (Margulies et al. 2005; Bentley et al. 2008) dramatically reduced the cost of sequencing and massively increased the volume of throughput (reviewed in Goodwin et al. 2016). For paleogenomics, NGS technology was instrumental in the sequencing of the first ancient whole genomes, beginning with a ~40,000-year-old woolly mammoth (*Mammuthus primigenius*) (Miller et al. 2008) and shortly followed by a similarly aged Neanderthal (*Homo neanderthalensis*) (Green et al. 2010). In the years since then, ancient whole genomes have been published for several non-human mammalian taxa, including the horse (Orlando et al. 2013; Schubert et al. 2014; Librado et al. 2015), Przewalski's horse (Der Sarkissian et al. 2015; Gaunitz et al. 2018), quagga (Jónsson et al. 2014), auroch (Park et al. 2015), mammoth (Palkopoulou et al. 2015; Lynch et al. 2015), wolf (Skoglund et al. 2015), dog (Frantz et al. 2016b; Botigué et al. 2017; Ní Leathlobhair et al. 2018) and goat (Daly et al. 2018). Whilst ancient whole genomes have yet to be published for domestic pigs, cattle, sheep or chicken, sequences for these taxa will likely be forthcoming in the near future.

Sequencing ancient genomes, however, even with NGS technologies remains challenging – the primary constraint being the poor preservation of endogenous aDNA in subfossil remains recovered from archaeological sites. It is not uncommon for the endogenous DNA fraction of an NGS sequencing run to be below 1% (Carpenter et al. 2013). This problem is particularly acute in geographic regions with warm climates (Hofreiter et al. 2015), where most domestic animals originated. Many factors contribute to the degradation of ancient DNA, including time, temperature, humidity, soil pH and microbial action. Despite decades of research, however, the decay kinetics of DNA degradation are still not well understood (Allentoft et al.

2012). In practice, the heterogeneity of DNA degradation makes preservation infeasible to accurately predict.

Recent studies have shown that aDNA preservation is also highly variable across different archaeological samples – awareness of which has led to dramatic improvements in aDNA recovery by focusing research on samples with higher endogenous yields. The petrous portion of the temporal bone can contain up to 183 times the concentration of endogenous DNA found in less dense bone (Gamba et al. 2014; Pinhasi et al. 2015). Tooth cementum has also been shown to contain comparably high levels of endogenous DNA content (Adler et al. 2011; Higgins et al. 2013; Damgaard et al. 2015). In experiments comparing petrous bones and tooth cementum, recovered from corresponding skeletons, the petrous bone was found to contain higher endogenous yields in only one tested assemblage, with the majority showing no systematic difference in yield (Hansen et al. 2017). As teeth are often over-represented in archaeological assemblages (Lam et al. 1999), they are an ideal target for aDNA recovery. In addition, teeth are great markers of domestication in multiple species, including pigs (Evin et al. 2013), horses (Cucchi et al. 2017) and dogs (Ameen et al. 2017).

Even with these strong constraints, genome-wide datasets have recently been published for early Neolithic farmers from sites across the Near East, including Anatolia, the Levant and Zagros Mountains (Broushaki et al. 2016; Gallego-Llorente et al. 2016; Lazaridis et al. 2016; Kılınç et al. 2016). Comparable sequences for domestic animals from the region have so far been limited to goats (Daly et al. 2018). Given the great importance of the Near East as a centre for domestication, it is likely that more genome-wide sequences from ancient domestic animals will be forthcoming in the near future. Recovery of nuclear aDNA will be crucial to our understanding of the underlying process of domestication.

3 Pathways to Animal Domestication

When, where and how animals were domesticated are central questions to our understanding of human civilisation. The current consensus amongst archaeologists and geneticists is that most domestic animals originated in a small number of 'core' zones, from whence they were dispersed across the globe (Larson and Fuller 2014). As such, animal domestication is thought to be a rare process. Ancient DNA has been key to establish (as well as to challenge) our perception of the geographical and temporal origin of many species and to test the idea that domestication is a rare phenomenon.

The idea that domestication is rare is also based on current theoretical perspective that depict domestication as a non-linear, diffuse and long-term process that requires specific conditions to occur (Conolly et al. 2011; Vigne et al. 2011). The complexity and nuance of these processes have informed the development of two new theoretical models of animal domestication, by Vigne (2011) and Zeder (2012b), which have cast off the anthropocentrism of many previous models.

Vigne's (2011) model described animal domestication as the ultimate phase of intensification in the relationship between animal and human populations. This multistage model proposes a continuum of intensification, progressing through phases of (1) anthropophily, (2) commensalism, (3) control in the wild, (4) control of captive animals, (5) extensive breeding, (6) intensive breeding and ultimately (7) pet keeping (Vigne 2011). Not all domestic animals, however, progressed through each of these stages. By focusing on the shared phases of intensification between different groups of domestic taxa, Zeder's (2012b) model has proposed three main pathways to domestication. This model describes animal domestication as a mutualistic process, with progressive intensification of animal-human relationships; however, it further distinguishes between three distinct evolutionary trajectories: a (1) commensal pathway, a (2) prey pathway and a (3) directed pathway (Zeder 2012b) (Fig. 1).

3.1 *The Commensal Pathway*

Under the commensal pathway, wild animals were firstly attracted to, and then entangled by, elements of the human constructed niche (Zeder 2012b). The attraction occurred as wild animals were drawn to food sources available on the margins of human occupation – such as refuse scavenging (e.g. wolves and wild boars), food stores (e.g. mice or chickens) or increased prey availability (e.g. cats). These commensal animals would have been subjected to subtle selection favouring individuals who were more adapted to exploit the human niche. Over time this human-animal relationship intensified, ultimately leading to a full domestic partnership. This commensal pathway implies no intentionality or forethought on the part of the human partners during most of the process but rather describes a slowly evolving beneficial relationship (Zeder 2012b).

3.2 *The Prey Pathway*

Under the prey pathway, wild animals were firstly exploited for their meat and hides before demographic pressures led humans to take an ever-greater role in herd management (Zeder 2012b). Where hunting pressures may have changed the size and composition of prey herds, humans responded by adjusting their hunting strategies to maintain sufficient prey availability – such as preferential targeting of young males (Zeder 2006). Over time, these hunting strategies developed progressively through more advanced systems of herd management and captive breeding through to directed breeding for favourable behavioural and phenotypic traits. In this way, the early stages of the prey pathway can be seen as just as unintentional as the commensal pathway. In contrast, however, the latter stages of the prey pathway are characterised by an intensification of human intervention, in an attempt to maintain supply of a diminishing resource (Zeder 2012b).

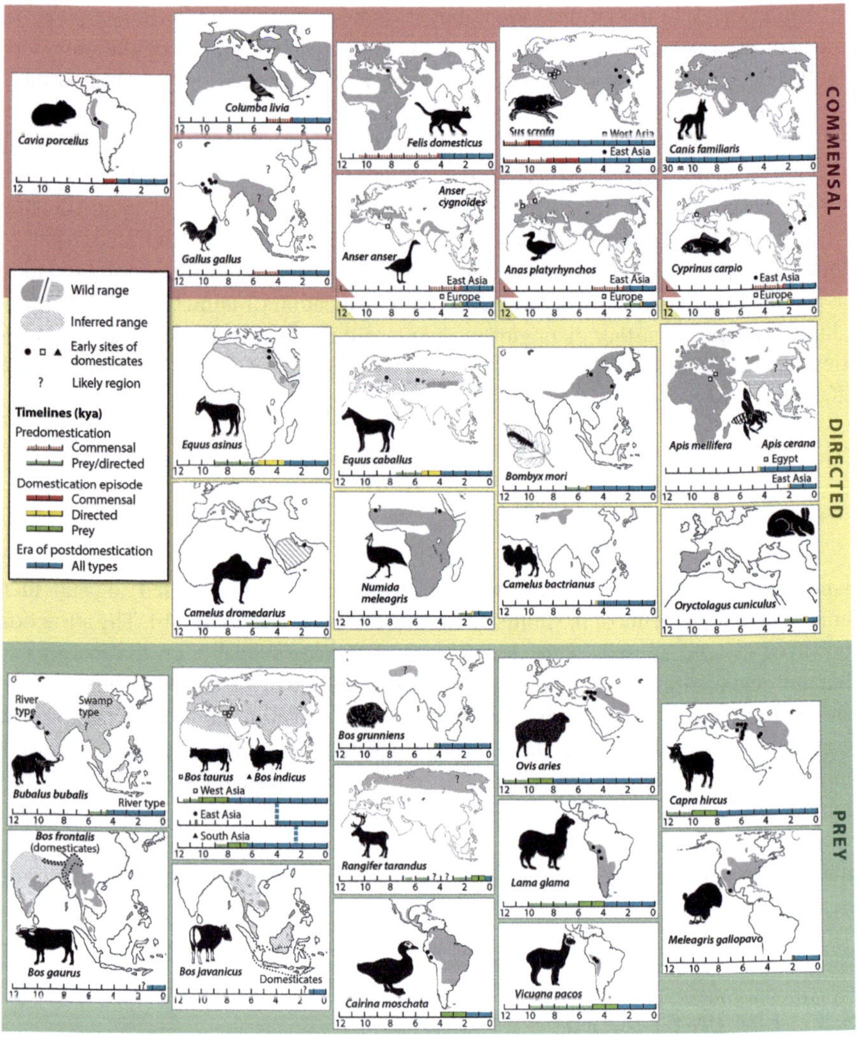

Fig. 1 Geographical/chronological time frame of domestication and potential pathways for major domestic animals. The timelines are in ky (1,000 years) increment. Adapted after Larson and Fuller (2014)

3.3 *The Directed Pathway*

Under the directed pathway, humans leveraged their prior experience with domestic animals, and their emergent understanding of directed breeding, to capture wild animals and intentionally bring them under increasing levels of human control (Zeder 2012b). The directed pathway describes the route taken for almost all recently domesticated taxa – particularly the exponential increase in aquatic species – but was of much lesser importance in the distant past. A recent meta-analysis found that 97% of all aquatic domesticates have been domesticated in the past hundred years, including more than 100 species in the preceding decade alone (Duarte et al. 2007). This recent prevalence of the directed pathway, coupled with modern intensive breeding practices, has been formative in the minds of many researchers and obscured a clearer understanding of early animal domestications. The idea of the directed pathway as the preeminent mode of domestication is typified by the theories of Galton (1865) and Clutton-Brock (1994), amongst many others, in which domestication is seen as the logical outcome of the intentional taming of wild animals.

4 When, Where and Which Pathway

The domestication of animals began more than 15,000 years ago, with the domestication of the grey wolf (*Canis lupus*) by nomadic hunter-gatherers (Larson et al. 2012). It was not until much later (beginning around 11,000 years ago) that people in the Near East intensified their relationships with wild populations of sheep, goat, aurochs and boar, such that incipient domestication processes began to emerge (Conolly et al. 2011). By 10,000 years ago, these four elements of the so-called Neolithic package had spread extensively throughout Southwest Asia and the Eastern Mediterranean (Vigne 2008). Despite the later ubiquity of these domesticates across the region, detailed zooarchaeological studies have revealed the complex non-linear nature of these domestication processes, complete with ebbs and flows in tempo in response to the local environment and conditions (Conolly et al. 2011; Vigne et al. 2011).

In the following section, we will briefly review what is known about the domestication of a range of key mammalian, avian and insect species – with a particular focus on paleogenomic contributions to our understanding of these domestications. The species profiles are grouped by the pathways they each took towards domestication, to highlight shared elements of the underlying process.

4.1 Commensal Domesticates

4.1.1 Dogs

The first animal likely to have followed the commensal pathway to domestication was the grey wolf (*Canis lupus*) (reviewed in Thalmann and Perri 2018). It has been theorised that wolves which were naturally less wary of people would have been drawn to human encampments to scavenge refuse left by hunters (Thalmann and Perri 2018). Where, when and how many times wolves were domesticated remains a contentious issue, due to the sparsity of evidence and conflicting interpretations of both the archaeology and genetics (Germonpré et al. 2009; Larson and Bradley

2014; Skoglund et al. 2015; Frantz et al. 2016b; Botigué et al. 2017; Ní Leathlobhair et al. 2018).

The earliest widely accepted archaeological dog remains date to about 15,000 years ago (Thalmann and Perri 2018). Earlier canid remains, dating back to over 30,000 years ago (Germonpré et al. 2009), were recently described as dogs, but their status (as dogs or wolves) remains highly controversial (Perri 2016). Paleogenomic data has provided additional information about the potential time frame for dog domestication. In particular, analyses of genome-wide data from an ancient Siberian wolf (Skoglund et al. 2015) and an ancient Irish dog (Frantz et al. 2016b) together with radiocarbon dates have provided the means to estimate a reliable mutation rate for canids and to obtain an estimate of the divergence time between extant wolves and dogs of 20,000–40,000 years ago. This timing, which represents an upper bound for dog domestication, needs to be interpreted with caution as the ancestor of dogs may have become extinct (Thalmann et al. 2013; Freedman et al. 2014; Frantz et al. 2016b). This would mean this time instead represents the time of divergence between extant wolves and the ancestor of dogs, rather than the time at which dogs were domesticated.

Over the years, genomic (including paleogenomic) studies have provided conflicting information about the geographical origin of dogs, with papers suggesting that dogs originated in East Asia (Pang et al. 2009; Wang et al. 2015), Central Asia (Shannon et al. 2015), the Middle East (von Holdt et al. 2010) and Europe (Thalmann et al. 2013). Additional genome-wide paleogenomic studies, however, have provided novel clues on the geographical origin of dogs. For example, studies based on ancient genomes from European dogs have suggested that modern Western Eurasian populations (including Africa, Europe and Middle East) were most likely imported from Asia, over 7,000 years ago (Frantz et al. 2016b; Botigué et al. 2017). Based on additional archaeological data and multiple ancient mtDNA sequences, the authors of one of these studies suggested that populations that inhabited Europe and the Middle East, prior to the arrival of dogs from East Asia, had been domesticated independently (Frantz et al. 2015). This hypothesis, which implies that the dogs that were present prior to the arrival of East Asian dogs are now extinct, remains to be tested.

4.1.2 Pigs

Although they were hunted like other ungulate species (sheep, goat, cattle, etc.), the omnivorous lifestyle of pigs provided them with the ability to consume human waste, suggesting that they were potentially domesticated via a commensal pathway (Larson and Fuller 2014). Interestingly, pigs are the only animals for which we have unequivocal, genetic and archaeological evidence for two independent domestication processes, from two different subspecies of *Sus scrofa*, in China and Anatolia, respectively (Larson et al. 2005). Ancient DNA studies have played a key role in unravelling a complex domestication history marked by frequent population replacements.

The Western Eurasian domestic pigs were most likely first domesticated in Anatolia, over 10,000 years ago, as suggested by zooarchaeological evidence of selection and culling from long-term occupation sites such as the Çayönü Tepesi (Hongo and Meadow 1998; Ervynck et al. 2001). Ancient DNA evidence suggests that they were then transported, from the Near East into Europe as part of the Neolithic package (Larson et al. 2007a), around 9,000 years ago (Conolly et al. 2011). Evidence for such an early human-mediated dispersal of pigs, from the Near East into Europe, however, is absent from modern DNA sequences (Larson et al. 2005). Lack of Near Eastern ancestry in modern domestic breeds is most likely the result of a population turnover resulting from long-term gene flow between European wild boars and domestic pigs (Frantz et al. 2015), a process that likely started as soon as pigs were introduced in Europe (Larson et al. 2007a).

Further ancient DNA evidence suggests that European domestic pigs, lacking Near Eastern ancestry, were later introduced back into the Near East (Anatolia), during the Iron Age where they replaced pigs with Near Eastern ancestry (Ottoni et al. 2013). More recently, Chinese pigs, which were domesticated from a highly divergent subspecies (Frantz et al. 2013b, 2016a), were imported in Europe to improve production traits during the industrial revolution (White 2011; Bosse et al. 2014a). This process dramatically affected the genetic (Bosse et al. 2014b) and phenotypic make-up (Bosse et al. 2014a) of European populations.

In East Asia, the first unequivocal evidence of pig domestication dates back to ~8,600 years ago at the site of Jiahu near the Yellow River (China) (Cucchi et al. 2011). Similar to the process seen in Europe, ancient mtDNA evidence suggests that East Asian domestic pigs were transported from their domestication centre to Island Southeast Asia, Papua and Polynesia where they were later replaced by pigs of European decent (Larson et al. 2010; Linderholm et al. 2016). During their human-mediated dispersal throughout Island Southeast Asia, domestic pigs encountered a high diversity of wild suid species (and subspecies) which readily interbreed with domestic pigs (Frantz et al. 2013b, 2014; Ai et al. 2015). Future ancient nuclear DNA will be able to assess whether deliberate interbreeding with wild stock may have allowed for adaptation of domestic pigs to the wide range of habitat they encountered in Europe, Asia and Polynesia (Frantz et al. 2016a).

4.1.3 Cats

Cats (*Felis catus*) also became domesticated via the commensal pathway; however, despite their worldwide popularity, relatively little is known about the origins of the domestic cat (reviewed in Geigl and Grange 2018). The archaeological and genetic evidence points to both the Near East and Egypt as important regions for the domestication of the cat (Vigne et al. 2004; Driscoll et al. 2007; Ottoni et al. 2017). The wild progenitor of the domestic cat (*Felis catus*) is the Near Eastern wildcat (*Felis silvestris lybica*), which has a natural range spanning North Africa and the Near East (Driscoll et al. 2007). Archaeological remains of wildcats in the Near East point to a long history of commensal relationship with early farming communities, where they are thought to have predated on invasive rodent populations (Vigne et al. 2004, 2012). This relationship persisted for thousands of years before the appearance of any classic domestication traits – such as reduction in overall body size and the emergence of novel coat colours (Vigne et al. 2016). The dispersal of domestic cats around the world was aided by their role on ships and trade vessels as protection against rodents. This is reflected in their patterns of dispersal, which mirror major trade routes (Lipinski et al. 2008; Ottoni et al. 2017).

A worldwide phylogenetic study of modern cats, using a fragment of the mitochondrial genome (mtDNA) and microsatellite markers, has shown that the Near Eastern wildcat (*F. s. lybica*) is more closely related to the domestic cat than other subspecies of wildcat (Driscoll et al. 2007). Based on the current distribution of wildcats, the authors concluded that cats were most likely domesticated in the Near East. Of the 979 analysed samples, they identified 15 wildcats from Israel, the United Arab Emirates, Bahrain and Saudi Arabia with mtDNA and microsatellite markers consistent with those found in modern domestic cats (Driscoll et al. 2007). Further research, using microsatellite markers to analyse the phylogeographical structure of modern domestic cats, also found support for a Mediterranean basin origin for their dispersal (Lipinski et al. 2008).

A recent ancient mtDNA study of 209 archaeological cat remains has shown that mitochondrial lineages from both the Near East and Egypt contributed to worldwide domestic cat populations at different times (Ottoni et al. 2017). Their analysis showed that domestic cats are drawn from five deeply divergent mtDNA subclades (IV-A to IV-E) of *F. s. lybica* and that the relative proportions of domestic cat haplogroups have shifted over time. The IV-A and IV-B subclades were identified as originating in the Near East and represent the first wave of domestic cats which spread across the Old World. The IV-C subclade originated in Egypt and was found in the majority of Egyptian cat mummies. Despite a supposed ban on the export of Egyptian cats (Zeuner 1963), the Egyptian subclade increased in frequency outside Egypt, such that during the first millennium AD in western Anatolia, it had expanded to twice the frequency of the local Near Eastern subclade (Ottoni et al. 2017). The authors speculated that the cause of this increase might have been due to more desirable behavioural characteristics of the Egyptian cats.

The same study also looked at the history of the tabby coat trait, one of the most widely used markers for identifying domestic cats (Ottoni et al. 2017). Their analysis found that the coat-colour variant responsible for the derived blotched tabby marking only reached high frequency after the Middle Ages, around the time that cat pelts were being traded for clothing. Coupled with the relatively small changes in overall size of domestic cats, this suggests that directed breeding of cats for morphological novelty was a very late phenomenon (Ottoni et al. 2017).

4.1.4 Chickens

Genetic data from modern domestic chickens and wild junglefowl have established that the red junglefowl (*Gallus gallus*) is the primary wild ancestor of the domestic chicken (Liu et al. 2006; Miao et al. 2013). Studies of nuclear genetic data have demonstrated, however, that the yellow skin allele found in domestic chickens was not inherited from red junglefowl but was instead inherited from the grey junglefowl (*Gallus sonneratii*) demonstrating that the genome of modern domestic chickens combines elements of at least two junglefowl species (Eriksson et al. 2008).

An initial review of the archaeological evidence argued that chicken domestication had begun by the third millennium BC, since the first robust evidence for poultry farming has been recovered in the Indus Valley ~2,600–1,900 BC, before chickens were then translocated to the Near East, Africa and Europe during the first millennium BC (Zeuner 1963). Based on an analysis of osteological evidence that attested to the presence of chickens during the Middle Neolithic (~6,000–4,000 BC) in the Yellow River basin, West and Zhou (1988) concluded that chickens were domesticated in the Southeast Asian native range of red junglefowl prior to 6,000 BC before being dispersed westwards along a northern route through Central and Western Eurasia. Two subsequent studies (Berke 1995; Peters 1997) questioned whether the Chinese specimens actually belonged to domestic chickens since they possessed morphological features typical of other galliform birds including pheasants. Despite these critiques, a mid-Holocene origin of domestic chickens has been frequently claimed in the literature.

A recent ancient DNA analysis of galliform bone specimens from Early and Middle Neolithic sites in the Yellow River basin reinforced the claims for an early domestication of chickens (Xiang et al. 2014). This study suggested that red junglefowl dispersed naturally to Northern China following the Younger Dryas where they were then domesticated during the early Neolithic. This assertion has since been questioned. An independent morphological re-evaluation of galliform bones from Northern Chinese Neolithic sites concluded that the bones in question belonged primarily to the common pheasant (*Phasianus colchicus*) (Peters et al. 2016; Eda et al. 2016). In addition, several lines of evidence including an assessment of associated wild mammalian faunas and high-resolution climate and precipitation records from temperate Holocene East Asia suggested that the (sub-)tropical forest habitat conducive to thermophilic red junglefowl did not extend into Northern China during the mid-Holocene climatic optimum (Peters et al. 2016). Lastly, multiple studies of modern domestic chickens have suggested that red junglefowl from peninsular Southeast Asia is the likely initial population from which domestic chickens were derived, and a recent genetic study of complete mitochondrial genomes has cast doubt on the likelihood that chickens were domesticated in Northern China (Huang et al. 2018). Future archaeological and genomic studies of modern and ancient chickens are necessary to reveal not only the spatial and temporal pattern of chicken domestication but also the process which led to the close association between chickens and people.

4.2 Prey Domesticates

4.2.1 Goats

Goats (*Capra hircus*), along with sheep (*Ovis aries*) and cattle (*Bos taurus*) all followed the prey pathway to domestication in the Fertile Crescent region of the Near East (Zeder 2012b). Archaeological and genetic evidence has established that goats were domesticated from the bezoar ibex (*Capra aegagrus*), a species of wild goat inhabiting the mountainous region spanning southwestern Turkey to central Afghanistan and southern Pakistan (Zeder and Hesse 2000; Naderi et al. 2008).

Detailed zooarchaeological studies of wild goat assemblages have allowed researchers to reconstruct the age and sex-specific harvest profiles employed by hunters prior to domestication. These harvest profiles reveal incipient herd management strategies, in which hunters transitioned from targeting of prime age males, which maximised short-term meat return, towards selective culling of subadult males and older adult females, to promote growth in herd sizes (Zeder and Hesse 2000; Zeder 2006, 2008). These management strategies of wild ranging goats gradually intensified from herding towards a fully domestic relationship, and domestic phenotypes appear in archaeological goat assemblages around 10,500 years ago at multiple sites across Southeastern Anatolia, the Zagros Mountains and Cyprus (Conolly et al. 2011).

Domestic goats were subsequently brought into Europe as part of the Neolithic package; however, unlike with pigs and cattle, there were no extant wild populations for the incoming domestic population to admix with (Scheu et al. 2012). Studies of modern mitochondrial DNA in domestic goat populations have revealed unusually high levels of genetic diversity coupled with low levels of geographical structuring (Luikart et al. 2001; Naderi et al. 2007, 2008). This diversity has been attributed to population structure in the region of domestication, followed by extensive trade and transport of domestic goats. Modern goat populations comprise six maternal haplogroups (A, B, C, D, F and G), with most domestic goats belonging to haplogroup A (Naderi et al. 2007). The first ancient DNA study of goats established that haplogroups A and C were both present in the Early Neolithic in France, with moderately high genetic diversity, a result that the authors interpreted as potential evidence for two independent domestications with subsequent gene flow between populations (Fernández et al. 2006).

Recently, the diverse origins of domestic goats were further investigated in the first genome-wide study of ancient caprids (Daly et al. 2018). The authors selectively targeted petrous bones to retrieve genome-wide data from 51 ancient goats and used mtDNA capture to retrieve complete mtDNA genomes for 83 ancient goats. Their analyses of nuclear genomes provided evidence for variable proportions of ancestry shared between pre-domestic wild goats and early domestic goat populations, which suggested local recruitment of divergent wild populations during domestication (Daly et al. 2018). This was mirrored by the mtDNA data, which showed that multiple highly divergent haplogroups were involved in the domestication process and have differentially contributed to the genetic make-up of modern populations. This study also revealed that, in contrast to modern populations (Naderi et al. 2008), mtDNA haplogroups were highly structured in ancient populations (Daly et al. 2018). Interestingly, the collapse in haplogroup structure happened relatively early in their evolutionary history (~7,000 years ago), when haplogroup A replaced most others to become the dominant haplogroup across the region (Daly et al. 2018).

4.2.2 Sheep

Sheep (*Ovis aries*) also followed a prey pathway to domestication in the Fertile Crescent, around 10,500 years ago, with the Asiatic mouflon (*Ovis orientalis*) as the most likely wild progenitor (Conolly et al. 2011). Both the urial (*Ovis vignei*) and the argali (*Ovis ammon*) have also been suggested as potential ancestors; however, no mitochondrial lineages from either species have been observed in domestic sheep populations (Meadows et al. 2011). The European mouflon (*Ovis aries musimon*) is a feral descendent of a primitive domestic population (Bruford and Townsend 2006), and a recent genome-wide analysis revealed widespread bidirectional admixture between European mouflon and modern domestic sheep (Barbato et al. 2017).

Domestic sheep populations comprise five maternal haplogroups (A, B, C, D and E), with most modern sheep belonging to haplogroups A, B and C (Meadows et al. 2011). Two major Y-chromosome patrilines have also been identified, showing limited geographic structure (Meadows and Kijas 2009). Similar to the pattern seen in domestic goats, the maternal haplogroups diverged long before domestication, suggesting that multiple divergent lineages were involved in the domestication process (Pedrosa et al. 2005; Meadows et al. 2011). The relative abundance of these haplogroups has changed over time, with haplogroups A and B dominating the initial expansion into Europe, followed by haplogroup C around 3,000 years ago (Tapio et al. 2006). This first wave of domestic sheep, bred primarily for meat production, was replaced by a second wave of domestic stock carrying improved production traits for wool and milk (Chessa et al. 2009; Demars et al. 2017).

Recently, a genome-wide study of selection in modern sheep and goats found 90 selective sweep regions which segregated between domestic and wild populations of *Capra* and *Ovis* (Alberto et al. 2018). A gene ontology enrichment analysis (reviewed in Huang et al. 2009) identified significant enrichment for genes involved in nervous system, immunity and productivity traits (Alberto et al. 2018). Interestingly, this analysis identified only 20 regions under selection which were common to both *Capra* and *Ovis*, suggesting that convergent phenotypes in goats and sheep were primarily established by selection on non-homologous gene regions.

4.2.3 Cattle

Cattle (*Bos taurus* and *Bos indicus*) also followed the prey pathway to domestication; however, there is ongoing uncertainty about how many times cattle were domesticated (Loftus et al. 1994; Troy et al. 2001; Hanotte et al. 2002; Beja-Pereira et al. 2006; Chen et al. 2010; Pitt et al. 2018). Large genome-wide studies of modern domestic cattle have shown that they form three deeply divergent groups: (1) Eurasian and (2) African taurine cattle (*Bos taurus*) and (3) Asian indicine cattle, or zebu (*Bos indicus*) (Gibbs et al. 2009; Decker et al. 2014).

The earliest cattle domestication occurred in the Fertile Crescent, approximately 10,500–10,000 years ago, where Eurasian taurine cattle were domesticated from wild Eurasian aurochs (*Bos primigenius*) (Hanotte et al. 2002; Helmer et al. 2005; Hongo et al. 2009; Conolly et al. 2011). The domestication of Asian indicine cattle occurred in South Asia, approximately 8,000–7,500 years ago, and was the product of either an independent domestication process or admixture between domestic taurine cattle and Asian aurochs (*Bos primigenius namadicus*) (Meadow 1983; Loftus et al. 1994; Chen et al. 2010; Larson and Burger 2013). Current archaeological and genetic evidence is consistent with an independent domestication process; however, without ancient genome-wide data, admixture between Asian aurochs and domestic taurine cattle cannot be ruled out as the potential source of indicine cattle domestication (Larson and Burger 2013). Uncertainty around a hypothesised independent domestication of African aurochs (*Bos primigenius africanus*) (Bradley et al. 1996; Hanotte et al. 2002; Wendorf and Schild 2005; Stock and Gifford-Gonzalez 2013), in the Western Desert of Egypt, has been largely resolved following reanalysis of the archaeological material (Brass 2018) and explicit model-based testing of the genetic data (Pitt et al. 2018), which found no evidence for an independent African domestication.

A recent study published the first whole-genome sequence of an extinct Eurasian aurochs (*Bos primigenius*), recovered from a 6,750-year-old British specimen (Park et al. 2015). Analysis of the genome-wide data revealed localised nuclear gene flow into the ancestors of British and Irish taurine cattle, contrary to previous mtDNA studies, which found no evidence of introgression (Edwards et al. 2007). Model-based testing of ancient genetic data suggests that the matrilineal founding population of taurine cattle may have been as low as just 80 individuals (Bollongino et al. 2012; Scheu et al. 2015). As taurine cattle migrated from the Near East into Europe, their mtDNA genetic diversity decreased along the axis of migration, and intercontinental migration continued up until ~7,000 years ago (Scheu et al. 2015). When whole-genome sequences of early domestic cattle become available, we will be able to better resolve the role of introgression between Eurasian domestic cattle and wild aurochs.

Within Asia, the evolutionary history of the *Bos* genus is characterised by reticulate admixture between domestic cattle populations and other *Bos* species (Wu et al. 2018). East Asian cattle populations show a mosaic of ancestry components, including an ancestral East Asian taurine component, a later Eurasian taurine component and a deeply divergent Chinese indicine component (Chen et al. 2018). Cattle populations from Tibet also show signs of adaptive introgression of yak (*Bos grunniens*) genes, in the response-to-hypoxia pathway, likely supporting an adaptation to high altitude (Chen et al. 2018; Wu et al. 2018) – similar to the adaptive introgression from Denisovans into Tibetans (Huerta-Sánchez et al. 2014) and Tibetan wolves into Tibetan mastiffs (Miao et al. 2017).

4.2.4 New World Camelids

In South America, llamas (*Lama glama*) and alpacas (*Vicugna pacos*) likely also followed a prey pathway. Archaeological evidence suggests the domestication of llamas and alpacas from their potential wild progenitors, vicuñas (*Vicugna vicugna*) and guanacos (*Lama guanicoe*), began ~6,000 years ago (Diaz-Lameiro 2016) within their overlapping native ranges in the mountainous regions of Bolivia, Chile and Peru and the central Andes Mountains (Barreta et al. 2013). There are two current hypotheses for how the domestication of these two species took place. The first is that both llamas and alpacas are domesticated forms of guanacos. Alternatively, alpacas may be a domesticated form of vicuñas, while llamas were derived from guanacos.

Both these hypotheses have support from genetic data. Ancient mitochondrial DNA sequenced from llama and alpaca remains from pre-Columbian South American sites (Cerro Narrio, Ecuador and Iwawi, Bolivia) demonstrated that the ancient alpacas and llamas clustered together within a well-supported monophyletic group more closely related to guanacos than to vicuñas, thus suggesting that both species were domesticated from guanacos in the northern South American Andes (Diaz-Lameiro 2016). The second hypothesis is supported by a study using modern nuclear data, which suggested that alpacas and llamas are more closely related to vicuñas and guanacos, respectively (Kadwell et al. 2001; Wheeler et al. 2006). Though this second study based upon a larger number of nuclear and mitochondrial loci has more weight, the large observed differences between the wild species may be partly due to a strong bottlenecking in the recent past. For instance, the vicuña population in the 1960s had a population size of only 2,000 across South America (Barreta et al. 2013), and guanaco populations have been small over the past century. The biases associated with these recent demographic shifts may have had an effect on the interpretation of these datasets. Understanding the origins and domestication history of these two species will be much more clearly understood through the generation and interpretation of ancient nuclear DNA datasets derived from archaeological material across the spatio-temporal range of the wild and domestic species.

4.3 *Directed Domesticates*

4.3.1 Horses

The earliest suggested case of an animal following the directed pathway to domestication is that of the horse (*Equus ferus caballus*) (Zeder 2012b), which may have been domesticated to assist steppe pastoralists in hunting wild horses (Levine 1999; Olsen 2006a). Identifying horse domestication in the archaeological record is difficult because many of the classic markers of domestication show no discernible variation between early wild and domestic populations – e.g. morphological changes (Eisenmann and Mashkour 2005) and mortality profiles (Olsen 2006a, b). The earliest evidence for horse domestication (reviewed in Orlando 2018) comes from Central Asia, around ~5,500 years ago, where skeletal pathologies indicate horses were bridled and probably ridden and stable isotope analysis of lipid residues in pottery indicate processing of mare's milk (Outram et al. 2009).

Modern horse populations comprise 18 major maternal haplogroups (A–R), 17 of which are found in domestic horses and 1 of which (haplogroup F) is found only in Przewalski's horses (*Equus ferus przewalskii*) (Achilli et al. 2012). This high number of mtDNA haplogroups, which diverged long before the start of domestication, has been interpreted as evidence of extensive restocking of wild mares during the domestication process (Vilà et al. 2001; Lippold et al. 2011b). In contrast, modern Y-chromosome patrilines have an extreme lack of diversity (Lindgren et al. 2004), likely caused by a strong bottleneck in male horses. The timing of this bottleneck is not clear, however, as aDNA studies have revealed that ancient domestic horses had greater Y-chromosome diversity than modern horses (Lippold et al. 2011a; Librado et al. 2017). The recent publication of the first complete assembly of the horse Y-chromosome should assist in future aDNA studies of male-biased processes in horse domestication (Janečka et al. 2018).

A recent genome-wide aDNA study of 14 ancient domestic horses has also challenged the traditional view that the high rate of deleterious mutations found in modern horses can be attributed to a male population bottleneck during domestication (Librado et al. 2017). The 'cost of domestication' hypothesis (reviewed in Moyers et al. 2018) argues that the process of domestication leads to increased levels of deleterious mutations in domestic animals – principally via population bottlenecks and strong artificial selection. In the case of horses, however, aDNA has revealed that ancient domestic horses had high rates of genetic diversity, and an analysis of the fitness consequences of that diversity found that the mutational load of ancient horses was less than that of both modern horses and pre-domestic horses (Librado et al. 2017). This implies that current levels of deleterious mutations are most likely a product of subsequent breeding practices, rather than a consequence of the domestication process itself.

Przewalski's horses are often described as the only extant wild horses (e.g. Der Sarkissian et al. 2015), after they were rescued from extinction in the wild following a captive breeding programme involving 12 wild-caught individuals (Volf et al. 1991). A recent genome-wide aDNA study of ancient domestic and Przewalski's horses, from the domestication centre in Central Asia, however, showed that Przewalski's horses are not truly wild but are instead the feral descendants of the first domestic horses (Gaunitz et al. 2018). This study revealed that it was the ancestors of modern Przewalski's horses which were first domesticated ~5,500 years ago and that by ~4,000 years ago, there had been a nearly complete genetic turnover amongst domestic horses, coinciding with the dramatic population expansion associated with the Yamnaya culture during the Early Bronze Age (Allentoft et al. 2015; Gaunitz et al. 2018). The exact timing of this turnover, and the geographic origin of the population, which gave rise to all modern domestic horses, remains unknown. Whilst there is still much to discover about the evolutionary history of horses, this study highlights the incredible insights that paleogenomics can bring to our understanding of the history of domestication.

4.3.2 Rabbits

The European rabbit (*Oryctolagus cuniculus*) is often reported to have been domesticated via the directed pathway. In the most widely cited historical account, rabbits were supposedly domesticated by Catholic monks in France, circa AD 600, when they were granted a dispensation to eat foetal rabbits during Lent (Zeuner 1963; Clutton-Brock 1981). The practice of eating *laurices* – newborn or foetal rabbits – goes back to at least the first century AD, when Pliny the Elder describes the Spanish delicacy of cutting foetal rabbits from the belly of their mother and eating them whole and uneviscerated (*Naturalis Historia*, 8.55). It follows that by granting permission to consume *laurices* during the many fasting days of the mediaeval calendar, French monks were suddenly motivated to move the breeding of rabbits above ground to obtain a reliable supply of newborn rabbits. First put forward by Nachtsheim (1936), this account has its origins in a widely miscited text from the late sixth century by St Gregory of Tours (Gregory 1969). Through successive retellings, the account became incrementally embellished, such that consumption of *laurices* became especially popular amongst the monks during Lent (Nachtsheim 1936), then permitted by the Church because they were not considered meat (Zeuner 1963), and ultimately that the dispensation was granted by Pope Gregory the Great (Carneiro et al. 2011), an unrelated contemporary of St Gregory of Tours. In fact, there is no evidence that eating *laurices* was ever commonplace nor that they were not considered meat, and the timing and nature of rabbit domestication remains unknown (Irving-Pease et al. 2018).

Despite this, European rabbits have a well-resolved geographic origin, in Southwest France, and the presence of an extant wild progenitor makes it comparatively easy to obtain modern genomic samples from which to model the process of selection during domestication (Carneiro et al. 2011, 2014, 2015). A recent study compared genome-wide data from six breeds of domestic rabbits and wild rabbits from across their native range, to scan for segregating signatures of selection (Carneiro et al. 2014). The authors found more than 100 selective sweep regions distinct to domestic rabbits, and a gene ontology enrichment analysis identified significant enrichment for genes involved in brain and neuronal development (Carneiro et al. 2014). Interestingly, the authors found very few fixed derived alleles in the domestic breeds, suggesting that domestication was achieved via changes in allele frequencies at hundreds of loci, each with low effect size. When ancient genome-wide data becomes available for European rabbits, it should be possible to test the timing of selection at these loci, to better elucidate the process of rabbit domestication.

4.3.3 Old World Camels

The progenitor of modern Old World camels reached Eurasia ~3 million years ago (Gauthier-Pilters and Dagg 1981; Köhler 1981; Peters 1997). By the Middle

Pleistocene, Old World camels ranged from China and Mongolia over Central Asia to the Arabian Peninsula, including parts of North Africa and Eastern Europe (Köhler 1981; Titov 2008). By the end of the Pleistocene, the range of wild camelids had contracted dramatically (Gaulthiers-Pilters and Dagg 1981; Kozhamkulova 1986, Titov 2008), and several wild camel species became extinct leaving only the species *Camelus ferus*.

The distribution of the small extant wild population is restricted to China and Mongolia (Bannikov 1976; Hare 1997; Reading et al. 1999; Mix et al. 1997, 2002), though the domestic form, *Camelus bactrianus*, has spread throughout Central Asia and is now found from Northeast China, Mongolia, South Russia and Central Asia. In Asia Minor, its distribution overlaps with that of the dromedary. The one-humped dromedary camel, probably once found as a wild animal throughout the Arabian region but known with certainty only in the domestic or feral state, is now widespread in the hot deserts of North Africa and Arabia (Walker 1964).

Archaeological records show evidence for a relationship between people and the Bactrian camel ~5,000 years ago (Bulliet 1975; Benecke 1994), and the earliest records of camel bones come from sites Turkmenistan and Iran (Kuzmina 2008). Given the presence of camel bones in Bronze Age strata from sites in Iran and southern Turkmenistan, it has been hypothesised that the inhabitants of the Iranian Plateau and the Kopet Dagh foothills area played a major role in the domestication of the two-humped camel (Benecke 1994).

Due to their use as pack animals, the modern populations of dromedary camels do not possess significant phylogeographic structure (Almathen et al. 2016). A recent study of dromedary camels (Almathen et al. 2016) successfully recovered DNA from ancient dromedary remains. The authors concluded that the founders of the modern domestic dromedary camels were likely a population of wild camels present in the southeastern corner of the Arabian Peninsula and that domestic populations were routinely hybridised with wild individuals with novel mtDNA haplotypes.

4.3.4 Insects

Two domesticated insect species likely followed the directed pathway: silkworms (*Bombyx mori*) and honeybees (*Apis mellifera*). People probably began selectively breeding moths for silk production ~5,000 years ago (Bisch-Knaden et al. 2014). The extreme changes in morphology and their reliance on humans for survival and reproduction have led to the recognition of the domestic form as a unique species, *B. mori.* Recent genetic analyses of complete mitochondrial sequences from different geographic regions (Li et al. 2010) and a mixture of mitochondrial and nuclear loci (Sun et al. 2012) now suggests that silkworm domestication began in China, in line with fossil, historical and archaeological lines of evidence (Sun et al. 2012).

Though there is a clear genetic distinction between wild and domestic silkworm lineages, *B. mori* retain ~83% of the genetic variance of its wild relatives.

Xia et al. (2009) interpreted this observation as evidence for a short domestication process with a large starting population. Yang et al. (2014) used coalescence simulations and approximate Bayesian computation (ABC) methods on 29 nuclear loci to suggest that domestication began ~7,500 years ago with a subsequent bottleneck ~4,000 years ago. Though the genetic architecture of domestication remains uncertain, several studies have identified genes and phenotypes that have been selected during domestication, including loci related to the olfactory system (Xiang et al. 2013), orphan genes (Sun et al. 2015) (reviewed in Tautz and Domazet-Lošo 2011) and epigenetic changes (Xiang et al. 2013).

The genus *Apis* has ten distinct species, nine of which are confined to Asia which suggests that the domesticated species, *A. mellifera*, also originated in Asia. This is supported by the fact that the closest species to *A. mellifera*, *Apis cerana,* is found in Western and Central Asia. Unlike domestic silkworms however, there is a range of subspecies of domesticated honeybee, and these are phenotypically distinct in different geographic regions. Because these species are adapted to their environment of origin, the basis for this phenotypic variation is largely unknown (Wallberg et al. 2014). These subspecies fall into four categories supported by morphometric and genetic studies: A are subspecies found throughout Africa, M from Western and Northern Europe, and C Eastern Europe and O includes species from Turkey and the Middle East (Han et al. 2012). Despite the parsimonious explanation of an Asian origin, an early paper using 1,136 nuclear SNPs suggested Africa as the origin of *A. mellifera* due to the unexpected rooting of their phylogenetic tree in the African clade (Whitfield et al. 2006). More recent studies have questioned this conclusion. One study demonstrated that some of the analysed subspecies were actually recent hybrids, and by removing these species from the analyses, the root of phylogenetic trees did not fall unequivocally into the A clade (Han et al. 2012). A similarly ambiguous conclusion was drawn when trees were built using 8.3 million SNPs (Wallberg et al. 2014). As a result, Asia remains the most likely origin of *A. mellifera*.

Harpur et al. (2012) found that honeybees exhibit unusually high levels of genetic diversity, as domestic bees are more genetically diverse than wild populations in Europe. This high level of diversity is believed to be maintained by the crossing of queens from diverse locations to produce more diverse hives. De la Rúa et al. (2013) pointed out that backcrossing with the local populations may be reducing the overall variation in the global honeybee population. Interbreeding between wild and domestics may be reducing the number of individuals with local adaptations that may be advantageous in a changing environment.

Relative to domestic mammal species, it is far more difficult to identify domestic insects in the archaeological record. As a result, investigations into the early process of domestication will have to rely upon genetic and morphological insights derived from museum specimens of silkworm and honeybees (e.g. Cridland et al. 2018).

5 The Biological Architecture of Domestication

Given its importance for our understanding of evolution, domestication has been extensively studied by experimental biologists and geneticists. These studies have focused on characterising the nature of the specific biological changes underlying the differences between domestic and wild species, as well as the interspecific similarities amongst domestic animals (known as the 'domestication syndrome'; Fig. 2). Paleogenomics has an enormous potential to address many questions regarding the biological underpinning of domestication by, for example, providing time-series data that can help detect artificial selection in the genome. Here we review how studies have, and will continue to, leverage the power of paleogenomics to answer fundamental questions in domestication.

5.1 *Theories and Experiments*

The evolutionary basis of animal domestication is one of the most enduring questions in evolutionary biology. Shortly after Charles Darwin (1859) published the theory of evolution by natural selection, he turned his attention to the study of domestication. Darwin's (1868) seminal work on the topic, *The Variation of Animals and Plants under Domestication*, examined in extensive detail the remarkable phenotypic similarity shown by a diverse range of domestic animals. Darwin's observations on the role of selection during domestication distinguished between two phases of artificial selection: termed 'unconscious' and 'methodical'. Darwin argued that the initial phase of domestication would have involved people unknowingly selecting for domestication traits by, for example, choosing the more productive cattle to breed and the less productive to eat (Darwin 1868). Over time, these unconscious selective pressures formed the many regional landraces of animals. More recently, he theorised people began practising conscious or methodical selection, in which animals were bred with a specific phenotypic outcome in mind – a view largely informed by the animal husbandry practices of the nineteenth century (Marshall et al. 2014). This perspective on animal domestication placed central focus on the role of human intent in the development of domestication traits and reproductive isolation from wild populations to preserve them.

These ideas were further developed by Francis Galton (1865, 1883), based on ethnographic observation of pet keeping in hunter-gatherer communities. Galton argued that the domestication of animals was a direct consequence of the human desire to capture and tame wild animals. All animals would be exposed to this process, but only those with a natural predisposition towards domestication would be permanently tamed. These anthropocentric views of domestication proved very influential, placing human intent at the heart of many contemporary definitions of domestication (Bökönyi 1989; Ducos 1989; Clutton-Brock 1994).

	occur in all individuals/have occurred in early domesticated forms of a species								occur in some varieties/breeds of a species								
	increased tameness	decreased brain size	decreased heart weight	shorter muzzle	reduced tooth size	increased variability of vertebrae count*	change in caudal vertebrae count**	more frequent oestrus cycles	floppy ears	curly tail	supernumerary toes	disproportionate dwarfism***	depigmentation	increased skin area; skin folds	hairlessness	wool	curly hair
dog	x	x	x	x	x	x	x	x	x	x	x	x	x	x	x	x	x
silver fox	x	x (?)		x				x	x	x			x				
ferret	x	x	x										x				
mink	x	x											x				
cat	x	x		x			x		x	x		x	x		x	x	
donkey	x	x							x				x (?)				
horse	x	x				x			x		x	x	x		x		
buffalo	x												x				
cattle	x	x		x					x		x	x	x	x	x		x
zebu	x								x				x (?)	x			
yak	x	x											x				
goat	x	x		x				x	x			x	x		x		x
sheep	x	x		x			x		x			x	x	x	x	x	x
reindeer	x												x				
pig	x	x	x	x	x	x			x	x	x	x	x	x	x	x	x
camel	x	x											x				
dromedary	x	x											x				
llama	x	x	x								x		x (?)			x	
alpaca	x	x	x									x	x	x		x	
rabbit	x	x	x						x				x	x	x	x	x
guinea pig	x	x	x						x (?)				x				x
chinchilla	x												x (?)				
hamster	x												x				
mouse	x		x	x	x			x					x		x		x
rat	x	x	x					x					x				
gerbil	x	x						x					x (?)				

Fig. 2 Multiple traits, commonly referred to as 'domestication syndrome', and their occurrence in different mammalian species. Adapted after Sánchez-Villagra et al. (2016)

Darwin's (1868) study of domestication identified a series of behavioural, physiological and morphological traits shared by domestic animals, but not by their wild progenitors. These shared traits subsequently became known as the 'domestication syndrome' (Hammer 1984). Amongst domestic animals, these traits are now considered to include increased docility and tameness, reduction in body mass and brain size, novel coat colours and patterns, altered tails and floppy ears, smaller teeth and shorter snouts, prolonged physical and behavioural neoteny, more frequent and nonseasonal reproductive cycles as well as changes in hormonal and neurotransmitter expression (Darwin 1868; Hammer 1984; Wilkins et al. 2014). The prevalence of these traits amongst domestic animals, including birds, fish and mammals, suggests that domestic animals respond similarly to artificial selection. The resultant domestication syndrome (Fig. 2) has been hypothesised to result from a disruption in developmental process of the neural crest (Wilkins et al. 2014; Sánchez-Villagra et al. 2016).

Experimental studies of animal domestication have played a key role in our understanding of how the domestication syndrome develops. The earliest experiments involved domestication of the brown rat (*Rattus norvegicus*) (King and Donaldson 1929; Castle 1947); however, the most informative experiments involved the silver fox – a melanistic form of the red fox (*Vulpes vulpes*). Beginning in 1959 at the Institute of Cytology and Genetics in Novosibirsk, Dmitry Belyaev established an experimental breeding programme which selectively bred silver foxes, brown rats and European mink (*Mustela lutreola*) for tameness (Belyaev 1969; Trut et al. 2004, 2009). Captive silver foxes were sourced from fur farms, where they had been selectively bred for their unique coat pigmentation (Belyaev 1969). Their level of aggression towards humans was tested by attempting to hand feed, stroke or handle the foxes, and those which exhibited the least aggressive response were chosen for subsequent breeding (Trut et al. 2004). The selective pressures in each generation were very high, with only 3% of males and 8–10% of females permitted to breed (Trut et al. 2004). Within 30 generations, almost half of the experimental foxes had behavioural relationships with humans that were analogous to domestic dogs. Interestingly, they also exhibited classic symptoms of the domestication syndrome – changes in coat colour and snout length, floppy ears and altered developmental timing (Trut et al. 2004).

Whole-genome sequences for tame, aggressive and conventional foxes raised under these experimental conditions have recently been published (Kukekova et al. 2018). Analysis of this data identified more than 100 regions showing signatures of selection in one or more of the experimental populations, and the *SorCS1* gene was identified as a strong candidate gene for tame behaviour.

5.2 Genetic Changes During Domestication

Many researchers have investigated the genetic basis for the phenotypic and behavioural changes seen in the domestication syndrome (Dobney and Larson 2006; Trut et al. 2009; Albert et al. 2009; Driscoll et al. 2009; Axelsson et al. 2013; Jensen 2014; Wilkins et al. 2014; Carneiro et al. 2014). With regard to plant domestication, good progress has been made in identifying genes linked to domestication and crop improvement (reviewed in Doebley et al. 2006; Olsen and Wendel 2013); however, the identification of similar genes linked to animal domestication has been more elusive. Increasingly, research has suggested that the phenotypic diversity found in domestic animal populations is based on complex genetic architectures involving hundreds of genes and regulatory regions, each with small effect sizes (Larson et al. 2014; Wilkins et al. 2014; Carneiro et al. 2014).

Evidence drawn from across the range of domestic taxa, and phenotypic traits, suggests complex pleiotropic, polygenic and epistatic effects (Reissmann and Ludwig 2013; Wilkins et al. 2014; Wright 2015). For example, pleiotropy – in which single genes affect multiple discrete phenotypic traits – has been putatively identified in behavioural, morphological, life history and sexual ornament traits in domestic chickens (*Gallus gallus*) (Wright et al. 2010; Johnsson et al. 2012). Polygenic traits – in which single phenotypic traits are controlled by multiple genes – are most clearly evident in pigmentation traits for hair, skin and eyes, where more than 125 causal genes have been identified in domestic mice (*Mus musculus*) (Bennett and Lamoreux 2003). Epistasis – in which the expression of a genetic variant is dependent on the effect of one or more variants in modifier regions (Cordell 2002) – has been putatively identified in more than a dozen epistatic pairs effecting tameness, flight and startle responses, body weight and other traits, in experimentally domesticated brown rats (Albert et al. 2009). Amongst domesticated crops, where the architecture of domestication traits is better understood, epistasis is thought to play a key role in phenotypic expression (reviewed in Doust et al. 2014).

The identification of genes involved in animal domestication and their mapping to complex traits has been achieved via two main approaches: (1) quantitative trait locus (QTL) mapping (reviewed in Mackay et al. 2009) and (2) genome-wide association studies (GWAS) (reviewed in McCarthy et al. 2008). Both techniques have been critical in identifying candidate genes associated with traits that differentiate domestic populations (Goddard and Hayes 2009). This work has been aided by the development of online databases, cataloguing known gene associations. The Animal QTLdb now contains more than 57,000 trait mappings (Hu et al. 2007, 2016), and the Online Mendelian Inheritance in Animals (OMIA) database (Nicholas 2003; Lenffer et al. 2006) catalogues thousands of monogenic traits in domestic animals. QTL mapping and GWAS studies, however, often focus on traits that are important for economic productivity rather than traits that differentiate wild and domestic populations.

Population genomic studies that focus on identifying signatures of selection in genome-wide sequence data from wild and domestic populations have also identified numerous candidate genes putatively involved in animal domestication. This approach has recently been used to identify putative selection in polygenic loci involved in brain and neuronal development traits in domestic rabbits (*Oryctolagus cuniculus*) (Carneiro et al. 2014) and digestion and nervous system development traits in dogs (Axelsson et al. 2013). Genome-wide sequencing data has also been used to test the hypothesis of gene loss as a driver of rapid evolutionary change (Olson 1999), which has been discounted as an important process in the domestication of dogs (Freedman et al. 2016), chickens (Rubin et al. 2010), pigs (Rubin et al. 2012) and rabbits (Carneiro et al. 2014).

5.3 *Temporal Pattern of Genetic and Morphological Changes*

Identifying the genetic basis of animal domestication based solely on modern DNA, however, can be problematic (Larson and Burger 2013). In order to identify the genetic basis of traits that are associated with the early stages of the domestication process, it is necessary to dissociate these traits from changes that happened during later stage of the process (Vigne 2011). This can be problematic as domestic animals bear little direct resemblance to their early forbears, due to thousands of years of artificial selection, divergent environmental conditions and introgression with populations unrelated to the initial domestication. More recently, this has been further complicated by intensive breeding practices which have made reconstructing the early stages of domestication much harder (Larson and Burger 2013).

Recently, a genome-wide selection scan identified a putative domestication locus in the thyroid-stimulating hormone receptor (*TSHR*) gene in domestic chickens (Rubin et al. 2010). Thyroid hormone metabolism has previously been suggested as a key factor in animal domestication (Crockford 2002; Dobney and Larson 2006), and the *TSHR* gene has been shown to play an important role in metabolic regulation and control of seasonal reproduction in birds (Nakao et al. 2008) and mammals (Hanon et al. 2008). Single nucleotide polymorphisms (SNPs) from the *TSHR* sweep region, including a candidate causal missense mutation, were genotyped in hundreds of domestic chickens, from dozens of geographically dispersed populations. The missense mutation was found to be almost completely fixed in the domestic population, with an allele frequency of 0.987 (Rubin et al. 2010). The same SNPs were typed in more than 50 red junglefowl (*Gallus gallus*) – thought to be the primary wild ancestor of the domestic chicken (Eriksson et al. 2008). The missense mutation was found with an allele frequency of 0.35, which the authors attributed to introgression from domestic chickens into zoo populations of red junglefowl (Rubin et al. 2010).

The identification of the *TSHR* gene as a domestication locus relies on the assumption that selective pressure on the allele that is now almost fixed in domestic chickens took place during the early stage of the domestication process. This assumption was directly tested when another research group recovered aDNA from 80 domestic chickens, from a dozen sites across Europe, with a temporal range of approximately 2,000 years (Girdland Flink et al. 2014). The authors were able to genotype the SNP from the *TSHR* sweep region in 44 ancient samples. The missense mutation was found with an allele frequency of just 0.432, and only 18% of the samples were homozygous for the derived allele (Girdland Flink et al. 2014). This analysis clearly demonstrated that the fixation of the *TSHR* missense mutation was associated with later trait improvements rather than the initial domestication process (Fig. 3).

In a subsequent study, selection on the *TSHR* locus was revisited, with an expanded ancient DNA dataset and the application of a novel Bayesian statistical framework for modelling the strength of selection over time (Loog et al. 2017). The authors concluded that selection on the derived allele began around AD 920,

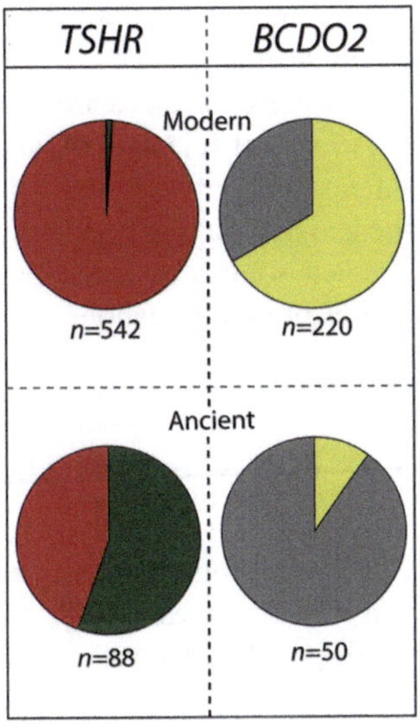

Fig. 3 Pie charts representing the allele frequency of variants at *TSHR* and *BCDO2* (affecting skin colour) in ancient chickens. This figure demonstrates that the variants at those genes, which are thought to influence traits in modern domestic chickens, were not found at high frequency in ancient chickens. Their rise in frequency is thought to be associated with breeding during the Middle Ages, but not with the domestication process (Loog et al. 2017). Adapted after Girdland Flink et al. (2014)

coinciding with mediaeval religious dietary reforms, which may have increased demand for both chicken and eggs (Loog et al. 2017). These findings are supported by zooarchaeological assemblages from England and Germany, spanning the mediaeval period, which show an increase in both the overall frequency of chickens and the relative proportion of adult hens – interpreted as sign of increased egg production (Serjeantson 2006; Sykes 2007; Holmes 2014). Functional genetic investigation of the pleiotropic effects of the *TSHR*-derived allele in chickens has shown that it is associated with increased egg production (Karlsson et al. 2016), decreased aggression and less fearful behaviours (Karlsson et al. 2015) – consistent with artificial selection for intensified egg production during the mediaeval period. Evidently, *TSHR* has played an important role in the evolutionary history of domestic chickens; however, its identification as a domestication locus is erroneous, and it can better be described as an improvement trait.

Similar cases, involving misidentified domestication genes, have been reported in domestic dogs and wheat. In the latter, a derived allele, fixed in modern populations of wheat, was identified as a putative domestication locus in the *NAM-B1* gene (Uauy et al. 2006). Ancient DNA recovered from herbarium seeds, however, established that the ancestral allele was still commonly found in cultivated populations as recently as 150 years ago (Asplund et al. 2010). Similarly, a recent genome-wide study of dogs demonstrated that most modern populations harbour a high number of copies of the *AMY2B* genes (Axelsson et al. 2013). This high copy number is almost fixed in modern dogs (Freedman et al. 2014) and allows them to better process starch (Axelsson et al. 2013). Ancient DNA studies, however, showed that these genetic variations only started to occur following the onset of farming, more than 7,000 years after dogs were domesticated (Arendt et al. 2016; Ollivier et al. 2016). More recent aDNA analysis further suggests that selection on *AMY2B* copy-number variation did not begin until well after the advent of agriculture (Botigué et al. 2017).

These examples clearly demonstrate the importance of ancient DNA in verifying the timing of selection during the domestication process and the pitfalls inherent in inference based solely on modern DNA. As the number of aDNA studies increases, the geographic range and temporal resolution of these datasets will allow ever more detailed studies to investigate which loci were under selection during the early phases of domestication.

5.4 *Genes as Domestic Markers*

Genetic markers can potentially be used to evaluate whether animal remains belong to a wild or domestic individual; however, the use of genetics is controversial due to the disputed importance of genetic changes during the early phases of domestication (Zeder 2012a; Vigne 2015). These controversies stem from a general lack of consensus regarding the definition of domestication, particularly one which unifies both plants and animals. This lack of clear definition has recently been identified as one of the key challenges in domestication research (Zeder 2015). There are, however, some clear examples of genetic (and phenotypic) changes that are highly diagnostic of the domestication status of an animal. For example, multiple non-synonymous (protein changing) mutations have been found in the melanocortin 1 receptor (*MC1R*) gene of pigs which leads to a black coat colour, or black spotted colour, and loss of their wild-type camouflage coat pattern (Fang et al. 2009). At least three independent mutations, resulting in similar phenotypes, exist in pigs, one in European pigs, one in East Asian pigs and one in Hawaiian feral pigs (introduced during the Polynesian expansion; Fig. 4) (Linderholm et al. 2016). In modern European domestic pigs, this dominant allele, which leads to loss of camouflage, is found at very high frequency, while it is almost absent from wild populations (Koutsogiannouli et al. 2010; Frantz et al. 2013a). This suggests a strong negative selection in wild boars.

This European dominant black allele was recently found in four ~6,500-year-old pig remains from the site of Ertebølle (Mesolithic of northern Germany) (Krause-Kyora et al. 2013). These animals also had a mtDNA haplogroup originating in Near Eastern domestic populations, and geometric morphometric (GMM) analysis revealed they had molars with domestic shape characteristics and pathologies (Krause-Kyora et al. 2013). Their domestic status, however, conflicted with the cultural context in which they were found – Mesolithic hunter-gatherer rather than Neolithic farmers.

This sparked a controversy and led to several published replies (Evin et al. 2014; Rowley-Conwy and Zeder 2014a, b). The principle critique centred on the lack of evidence that humans at Ertebølle had a special relationship with these animals, distinct from that of wild boar (Rowley-Conwy and Zeder 2014a). The authors of the reply argued that domestication involves more than the phenotypic expression of

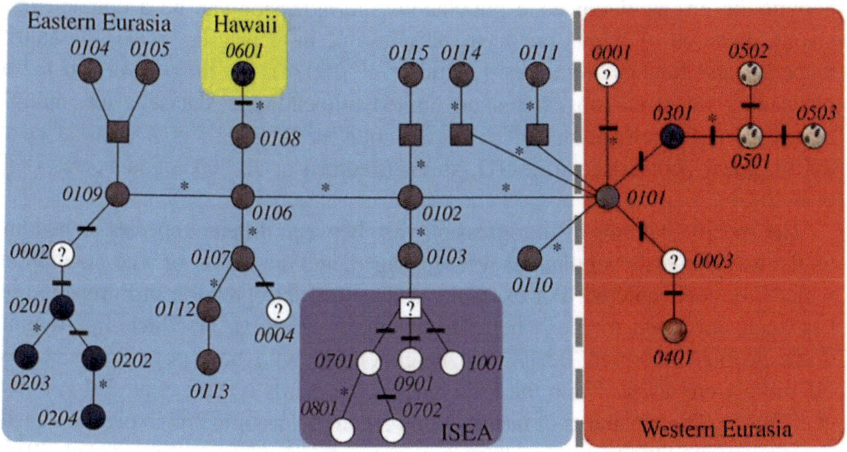

Fig. 4 Haplotype network for the MC1R gene coding region. This figure demonstrates the existence of three lineages of black pigs, in Hawaii, Europe and East Asia. Adapted after Linderholm et al. (2016)

genetic traits and requires a mutualistic relationship between the domestic and the domesticator. Therefore, even unambiguously domestic animals – with the complete set of behavioural and phenotypic traits – identified in this undifferentiated context would shed no light on the process of domestication or the adoption of agriculture in the region (Rowley-Conwy and Zeder 2014b). All together this highlights the fact that biological markers, even highly discriminative as those described above, cannot on their own provide the sole basis for a definition of domestication.

5.5 *Introgression in Domestication*

Animal domestication is often thought to be defined, not solely by genetic and phenotypic characteristics but also by population processes such as a strong bottlenecks, reproductive isolation from wild populations and directed breeding (Marshall et al. 2014). More recently, modern and ancient genomic datasets have revealed that these conditions were much less common than previously thought – revealing complex and varied patterns of introgression between wild and domestic pigs (Frantz et al. 2015), goats (Daly et al. 2018), cattle (Park et al. 2015), horses (Schubert et al. 2014), dromedary camels (Almathen et al. 2016), cats (Ottoni et al. 2017) and many other species (Marshall et al. 2014). As early farmers spread outwards from the major centres of domestication, the domestic animals that accompanied them frequently interbred with wild populations encountered along their routes of dispersal. Successive waves of gene flow over thousands of years have resulted in modern genomes which are complex palimpsests, containing traces of many different ancestral populations. For researchers that use DNA to re-trace the temporal and geographic origin of domestic populations, introgression can be a double-edged sword. Patterns of admixture have been useful in untangling routes of animal dispersal and human migration (Larson et al. 2007b), but they have also led to misleading interpretations, based on limited mitochondrial datasets, for multiple independent domestications of cattle (Hanotte et al. 2002), pigs (Larson et al. 2005), goats (Luikart et al. 2001), sheep (Pedrosa et al. 2005) and horses (Vilà et al. 2001).

The specific patterns of introgression vary between different species, depending on the way domestic populations were managed, and the variety of wild populations which were encountered. For example, widespread introgression in European pigs has been attributed to loose herd management practices, in which free-ranging domestic pigs interbred freely with neighbouring wild boar populations, whose offspring were adopted into the loosely managed herds (Ottoni et al. 2013; Frantz et al. 2015). These patterns of introgression are highly asymmetric in pigs, with wild boars receiving little to no gene flow from domestic populations (Frantz et al. 2015). In general, the directionality of admixture is biased towards gene flow from local populations into migrant groups, especially with increasing distance from the source of the migration (Currat et al. 2008). Notable exceptions do occur, however, such as the *K*-locus variant introgressed from dogs into wolves (Schweizer et al. 2018) and *MITF* gene variants introgressed from cattle into yaks (Wu et al. 2018).

African cattle, which early genetic evidence suggested might have been independently domesticated (Hanotte et al. 2002), are now better explained by introgression between Near Eastern domestic cattle (*Bos taurus*), wild North African aurochs (*Bos primigenius africanus*) and successive waves of domestic Asian zebus (*Bos indicus*) (Mwai et al. 2015; Brass 2018; Pitt et al. 2018). In domestic chickens, the now ubiquitous yellow leg trait was acquired via introgression from the wild grey junglefowl (*Gallus sonneratii*) (Eriksson et al. 2008). For some species, introgression with wild populations continues to be an active process – particularly amongst reindeer (*Rangifer tarandus*) (Røed et al. 2008) and honeybees (*Apis mellifera*) (Harpur et al. 2012), which exhibit very high levels of genetic diversity.

An important, but limited, approach to investigating these complex histories is to use large genome-wide datasets to characterise the patterns of diversity and admixture seen in modern domestic populations – like cattle (Gibbs et al. 2009; Decker et al. 2014), sheep (Kijas et al. 2012), goats (Wang et al. 2016; Brito et al. 2017), pigs (Ai et al. 2013), horses (McCue et al. 2012; Petersen et al. 2013; Schaefer et al. 2017), chickens (Muir et al. 2008; Stainton et al. 2017), dogs (von Holdt et al. 2010; Shannon et al. 2015) and mice (Yang et al. 2011; Staubach et al. 2012). These large modern datasets benefit from the relative ease of sampling and low cost of data generation, compared to aDNA. The recent development of novel computational methods using phased haplotypes (Lawson et al. 2012; Hellenthal et al. 2014) has increased the precision with which the timing, direction and fraction of admixture can be resolved in these high-quality modern datasets.

The inferences which can be made from modern DNA alone, however, are limited by the use of modern genetic variation as a proxy for ancestral populations. Modern DNA can be blind to population replacement (e.g. Haak et al. 2015; Gaunitz et al. 2018), because the extirpated populations made little contribution to modern genetic variation. Nor can modern DNA effectively detect or measure admixture from unsampled extinct species (e.g. Prüfer et al. 2014; Park et al. 2015), because the ancestral states of both species are unknown – although statistical methods have been developed to infer admixture from unsampled archaic populations (e.g. Plagnol and Wall 2006; Vernot and Akey 2014). The best approaches are those that combine both ancient and modern DNA with explicit testable models of evolutionarily processes (Gerbault et al. 2014). The recent development of novel Bayesian techniques for modelling serially sampled DNA holds particular promise to reveal important insights into the evolutionary process of domestication (Schraiber et al. 2016; Ferrer-Admetlla et al. 2016; Loog et al. 2017).

6 Future Perspectives

6.1 Ancient Epigenomes

The role of epigenetics in the domestication process, and in regulating domestic phenotypes, is a promising area of new research. For example, researchers working on the experimental domestication of the silver fox have suggested that observed differences in hormonal expression, associated with the domestication syndrome, may be linked to epigenetic modifications (Trut et al. 2009). A recent study comparing methylation patterns between dogs and wolves found 68 significantly differentially methylated sites across the two species, which included sites linked to the *GABRB1* and *SLC17A8* neurotransmitter genes, associated with a range of cognitive functions (Janowitz Koch et al. 2016). The role of epigenetics in a wide range of livestock phenotypes has also recently garnered a lot of attention (Feeney et al. 2014; Ibeagha-Awemu and Zhao 2015; Triantaphyllopoulos et al. 2016).

As our understanding of epigenetics improves, the ability to retrieve epigenetic information from ancient DNA will become increasingly important (reviewed in Hanghøj and Orlando 2018). Technical advances have recently made the recovery of ancient methylation maps possible (Briggs et al. 2010), which has resulted in the publication of the first genome-wide methylation maps for an ancient human (Pedersen et al. 2014), a Neanderthal and a Denisovan (Gokhman et al. 2014). Specialist computational tools for performing these analyses have also recently become available (Hanghøj et al. 2016). Presently, an equivalent ancient genome-wide methylation map has yet to be produced for domestic animals; however, as the number of ancient whole-genome sequences increase, it is only a matter of time before these become available.

6.2 Technical Advances

On the technical front, paleogenomics has benefited greatly from the development of increasingly cheaper and higher-throughput sequencing platforms. As development of these machines continues apace, we can expect the cost of DNA sequencing to continue to reduce. In some experimental designs, the limiting factor is no longer the cost of sequencing, but the costs of reagents and skilled labour for sample preparation (Rohland and Reich 2012). Protocols and laboratory equipment for automated library preparation, using liquid handling robots, are already available (Farias-Hesson et al. 2010; Lundin et al. 2010), and such approaches will likely become more commonplace in the future. As the cost of sequencing and sample preparation continues to drop, the number of samples and range of taxa which can be sequenced will increase concomitantly. Domestic animals are well represented in many archaeological sites, providing the potential for aDNA studies with fine-grained transects through time.

As paleogenomic studies scale up, increasingly sophisticated population genetic models will be necessary to interpret the process of animal domestication (Gerbault et al. 2014). Current methods for inferring patterns of admixture will need to be extended and improved to deal with more complicated models and larger datasets. Model-based clustering techniques, like *STRUCTURE* (Pritchard et al. 2000) and *ADMIXTURE* (Alexander et al. 2009), are very popular but widely over-interpreted (Lawson et al. 2018). Graph fitting approaches, like *TreeMix* (Pickrell and Pritchard 2012) and *MixMapper* (Lipson et al. 2013), are useful for inferring models of admixture but lack a formal statistical test of fit (Patterson et al. 2012). Formal models of admixture can be tested with *f*-statistics (Reich et al. 2009; Patterson et al. 2012) and *D*-statistics (Green et al. 2010; Durand et al. 2011), but these methods cannot resolve complex admixture topologies. Haplotype-based methods (Lawson et al. 2012; Hellenthal et al. 2014) work well on high-quality phased data, but are not suitable for low-coverage ancient data. Bayesian techniques, like *admixturegraph* (Leppälä et al. 2017), can test goodness of fit between models using Bayes factors, but computing these factors is computationally expensive, making automated model exploration very slow. As datasets continue to increase in size, the main constraint on genome analysis will be scaling computation to contend with the growth in sequence data (Muir et al. 2016).

6.3 Novel Substrates for aDNA

Paleogenomics is branching out into the recovery of aDNA from a range of novel substrates (reviewed in Green and Speller 2017). For example, the recent demonstration that aDNA can be successfully retrieved from historic parchments has opened up a whole new avenue for the study of domestic animals (Teasdale et al. 2015). Large numbers of historical parchments exist in archival and private collections across Europe. These parchments represent an exceptionally well-dated source of aDNA for reconstructing the evolutionary history of regional landraces of sheep, goat and cattle (Teasdale et al. 2015). Ancient coprolites from domestic animals have also recently been shown to be a suitable substrate for the recovery of aDNA. Using a combination of microscopy and aDNA sequencing, a recent study of domestic dog coprolites was able to establish the major diet components of ancient Polynesian dogs (Wood et al. 2016). Additionally, ancient latrines have been shown to contain retrievable quantities of parasite aDNA, the host specificity of which can be used to infer the presence of domestic animal species (Søe et al. 2018).

Calcified dental plaque, known as dental calculus, has also recently been established as an important new substrate for aDNA recovery (Adler et al. 2013; Warinner et al. 2014, 2015; Weyrich et al. 2015). Archaeological studies of dental calculus in domestic animals have a long history, the earliest of which used light microscopy to study phytoliths trapped in dental calculus from cattle, sheep and horse teeth (Armitage 1975). Other early studies identified a broad range of organic substances in dental calculus (Dobney and Brothwell 1986) and developed a system for quantifying dental calculus in human, cattle and sheep teeth (Dobney and Brothwell 1987). More recently, paleogenomic studies of dental calculus have focused on changes in human health and diet. For example, a recent study used aDNA from dental calculus to establish that Mesolithic foragers in the Balkans were consuming domesticated plant foods (Cristiani et al. 2016). As paleogenomics broadens its focus away from human-centred studies, similar studies of animal diet and oral health will no doubt be applied to domestic taxa and their wild progenitors.

Environmental and sediment DNA are also showing strong potential for reconstructing the movement of domestic animals and their environmental impacts. A recent study used DNA metabarcoding of alpine lake sediments to build a high-resolution picture of agricultural land use since the Neolithic (Giguet-Covex et al. 2014). The authors were able to identify ancient sediment DNA from cattle, goats, sheep, horses and chickens and to correlate their abundance with changes in plant cover and erosion. The potential of environmental DNA, however, is moderated by the risk of vertical DNA movement through sediment stratigraphy. For example, one study identified sheep DNA in a New Zealand cave site from layers which predated European contact, demonstrating that DNA leaching can be problematic under some soil conditions (Haile et al. 2007). The inability to directly date environmental DNA from sediments which lack macrofossils is also a significant concern and has caused some to question the identification of the earliest domestic wheat in Britain, from an 8,000-year-old layer of a sediment core (Smith et al. 2015a, b; Bennett 2015).

7 Conclusion

The future of paleogenomics and its application to the study of animal domestication looks bright. Ten years ago, the retrieval of a single gene locus from few ancient samples was cause for celebration. Now, studies involving genome-wide data from dozens (Haak et al. 2015; Fu et al. 2016; Lazaridis et al. 2016) or even hundreds (Mathieson et al. 2015; Lipson et al. 2017) of ancient samples are increasingly commonplace. So far, large paleogenomic studies have favoured retrieval of ancient human DNA, but similarly sized studies of domestic animals are certainly on the horizon. As our understanding of aDNA preservation (Hansen et al. 2017) and decay kinetics (Kistler et al. 2017) improves, more informed choice of skeletal elements and sampling locations will also permit the retrieval of aDNA from older time depths and warmer climates. We anticipate that the trend will be towards larger studies with many more samples and much older and finer temporal resolution.

References

Achilli A, Olivieri A, Soares P, et al. Mitochondrial genomes from modern horses reveal the major haplogroups that underwent domestication. Proc Natl Acad Sci U S A. 2012;109:2449–54. https://doi.org/10.1073/pnas.1111637109.

Adler CJ, Haak W, Donlon D, Cooper A. Survival and recovery of DNA from ancient teeth and bones. J Archaeol Sci. 2011;38:956–64. https://doi.org/10.1016/j.jas.2010.11.010.

Adler CJ, Dobney K, Weyrich LS, et al. Sequencing ancient calcified dental plaque shows changes in oral microbiota with dietary shifts of the Neolithic and industrial revolutions. Nat Genet. 2013;45:ng.2536. https://doi.org/10.1038/ng.2536.

Ai H, Huang L, Ren J. Genetic diversity, linkage disequilibrium and selection signatures in Chinese and Western pigs revealed by genome-wide SNP markers. PLoS One. 2013;8:e56001. https://doi.org/10.1371/journal.pone.0056001.

Ai H, Fang X, Yang B, et al. Adaptation and possible ancient interspecies introgression in pigs identified by whole-genome sequencing. Nat Genet. 2015;47:217–25. https://doi.org/10.1038/ng.3199.

Albert FW, Carlborg Ö, Plyusnina I, et al. Genetic architecture of tameness in a rat model of animal domestication. Genetics. 2009;182:541–54. https://doi.org/10.1534/genetics.109.102186.

Alberto FJ, Boyer F, Orozco-terWengel P, et al. Convergent genomic signatures of domestication in sheep and goats. Nat Commun. 2018;9:813. https://doi.org/10.1038/s41467-018-03206-y.

Alexander DH, Novembre J, Lange K. Fast model-based estimation of ancestry in unrelated individuals. Genome Res. 2009;19:1655–64. https://doi.org/10.1101/gr.094052.109.

Allentoft ME, Collins M, Harker D, et al. The half-life of DNA in bone: measuring decay kinetics in 158 dated fossils. Proc Biol Sci. 2012;279:4724–33. https://doi.org/10.1098/rspb.2012.1745.

Allentoft ME, Sikora M, Sjögren K-G, et al. Population genomics of Bronze Age Eurasia. Nature. 2015;522:167–72. https://doi.org/10.1038/nature14507.

Almathen F, Charruau P, Mohandesan E, et al. Ancient and modern DNA reveal dynamics of domestication and cross-continental dispersal of the dromedary. Proc Natl Acad Sci U S A. 2016;113:6707–12. https://doi.org/10.1073/pnas.1519508113.

Ameen C, Hulme-Beaman A, Evin A, et al. A landmark-based approach for assessing the reliability of mandibular tooth crowding as a marker of dog domestication. J Archaeol Sci. 2017;85:41–50. https://doi.org/10.1016/j.jas.2017.06.014.

Arendt M, Cairns KM, Ballard JWO, et al. Diet adaptation in dog reflects spread of prehistoric agriculture. Heredity. 2016;117:301–6. https://doi.org/10.1038/hdy.2016.48.

Armitage PL. The extraction and identification of opal phytoliths from the teeth of ungulates. J Archaeol Sci. 1975;2:187–97. https://doi.org/10.1016/0305-4403(75)90056-4.

Asplund L, Hagenblad J, Leino MW. Re-evaluating the history of the wheat domestication gene NAM-B1 using historical plant material. J Archaeol Sci. 2010;37:2303–7. https://doi.org/10.1016/j.jas.2010.04.003.

Axelsson E, Ratnakumar A, Arendt M-L, et al. The genomic signature of dog domestication reveals adaptation to a starch-rich diet. Nature. 2013;495:360–4. https://doi.org/10.1038/nature11837.

Ballard JWO, Whitlock MC. The incomplete natural history of mitochondria. Mol Ecol. 2004;13:729–44. https://doi.org/10.1046/j.1365-294X.2003.02063.x.

Bannikov A. Wild camels of the Gobi. Wildlife. 1976;18:2.

Barbato M, Hailer F, Orozco-terWengel P, et al. Genomic signatures of adaptive introgression from European mouflon into domestic sheep. Sci Rep. 2017;7:7623. https://doi.org/10.1038/s41598-017-07382-7.

Barreta J, Gutiérrez-Gil B, Iñiguez V, et al. Analysis of mitochondrial DNA in Bolivian llama, alpaca and vicuna populations: a contribution to the phylogeny of the South American camelids. Anim Genet. 2013;44:158–68. https://doi.org/10.1111/j.1365-2052.2012.02376.x.

Beja-Pereira A, Caramelli D, Lalueza-Fox C, et al. The origin of European cattle: evidence from modern and ancient DNA. Proc Natl Acad Sci U S A. 2006;103:8113–8. https://doi.org/10.1073/pnas.0509210103.

Belyaev DK. Domestication of animals. Sci J. 1969;5:47–52.

Benecke N. Mensch und seine Haustiere. Thesis. 1994.

Bennett KD. Comment on "Sedimentary DNA from a submerged site reveals wheat in the British Isles 8000 years ago". Science. 2015;349:247. https://doi.org/10.1126/science.aab1886.

Bennett DC, Lamoreux ML. The color loci of mice – a genetic century. Pigment Cell Res. 2003;16:333–44. https://doi.org/10.1034/j.1600-0749.2003.00067.x.

Bentley DR, Balasubramanian S, Swerdlow HP, et al. Accurate whole human genome sequencing using reversible terminator chemistry. Nature. 2008;456:53–9. https://doi.org/10.1038/nature07517.

Berke H. Der Mensch und seine Haustiere Die Geschichte einer jahrtausendealten Beziehung. 1995.

Bisch-Knaden S, Daimon T, Shimada T, et al. Anatomical and functional analysis of domestication effects on the olfactory system of the silkmoth Bombyx mori. Proc R Soc Lond B Biol Sci. 2014;281:20132582. https://doi.org/10.1098/rspb.2013.2582.

Bocquet-Appel J-P. When the world's population took off: the springboard of the Neolithic Demographic Transition. Science. 2011;333:560–1. https://doi.org/10.1126/science.1208880.

Bökönyi S. Definitions of animal domestication. In: Clutton-Brock J, editor. The Walking larder: patterns of domestication, pastoralism, and predation. London: Unwin Hyman; 1989. p. 22–7.

Bollongino R, Burger J, Powell A, et al. Modern taurine cattle descended from small number of near-eastern founders. Mol Biol Evol. 2012;29:2101–4. https://doi.org/10.1093/molbev/mss092.

Bosse M, Megens H-J, Frantz LAF, et al. Genomic analysis reveals selection for Asian genes in European pigs following human-mediated introgression. Nat Commun. 2014a;5:4392. https://doi.org/10.1038/ncomms5392.

Bosse M, Megens H-J, Madsen O, et al. Untangling the hybrid nature of modern pig genomes: a mosaic derived from biogeographically distinct and highly divergent Sus scrofa populations. Mol Ecol. 2014b;23:4089–102. https://doi.org/10.1111/mec.12807.

Botigué LR, Song S, Scheu A, et al. Ancient European dog genomes reveal continuity since the early Neolithic. Nat Commun. 2017;8:16082. https://doi.org/10.1038/ncomms16082.

Bradley DG, MacHugh DE, Cunningham P, Loftus RT. Mitochondrial diversity and the origins of African and European cattle. Proc Natl Acad Sci U S A. 1996;93:5131–5. https://doi.org/10.1073/pnas.93.10.5131.

Brass M. Early North African Cattle domestication and its ecological setting: a reassessment. J World Prehist. 2018;31:81–115. https://doi.org/10.1007/s10963-017-9112-9.

Briggs AW, Stenzel U, Meyer M, et al. Removal of deaminated cytosines and detection of in vivo methylation in ancient DNA. Nucleic Acids Res. 2010;38:e87. https://doi.org/10.1093/nar/gkp1163.

Brito LF, Kijas JW, Ventura RV, et al. Genetic diversity and signatures of selection in various goat breeds revealed by genome-wide SNP markers. BMC Genomics. 2017;18:229. https://doi.org/10.1186/s12864-017-3610-0.

Broushaki F, Thomas MG, Link V, et al. Early Neolithic genomes from the eastern Fertile Crescent. Science. 2016:aaf7943. https://doi.org/10.1126/science.aaf7943.

Bruford MW, Townsend SJ. Mitochondrial DNA diversity in modern sheep: implications for domestication. In: Documenting domestication: new genetic and archaeological paradigms. Oakland: University of California Press; 2006. p. 306–16.

Bulliet RW. The camel and the wheel. Cambridge: Harvard University Press; 1975.

Carneiro M, Afonso S, Geraldes A, et al. The genetic structure of domestic rabbits. Mol Biol Evol. 2011;28:1801–16. https://doi.org/10.1093/molbev/msr003.

Carneiro M, Rubin C-J, Di Palma F, et al. Rabbit genome analysis reveals a polygenic basis for phenotypic change during domestication. Science. 2014;345:1074–9. https://doi.org/10.1126/science.1253714.

Carneiro M, Piorno V, Rubin C-J, et al. Candidate genes underlying heritable differences in reproductive seasonality between wild and domestic rabbits. Anim Genet. 2015;46:418–25. https://doi.org/10.1111/age.12299.

Carpenter ML, Buenrostro JD, Valdiosera C, et al. Pulling out the 1%: whole-genome capture for the targeted enrichment of ancient DNA sequencing libraries. Am J Hum Genet. 2013;93:852–64. https://doi.org/10.1016/j.ajhg.2013.10.002.

Castle WE. The domestication of the rat. Proc Natl Acad Sci U S A. 1947;33:109–17. https://doi.org/10.1073/pnas.33.5.109.

Chen S, Lin B-Z, Baig M, et al. Zebu cattle are an exclusive legacy of the South Asia Neolithic. Mol Biol Evol. 2010;27:1–6. https://doi.org/10.1093/molbev/msp213.

Chen N, Cai Y, Chen Q, et al. Whole-genome resequencing reveals world-wide ancestry and adaptive introgression events of domesticated cattle in East Asia. Nat Commun. 2018;9:2337. https://doi.org/10.1038/s41467-018-04737-0.

Chessa B, Pereira F, Arnaud F, et al. Revealing the history of sheep domestication using retrovirus integrations. Science. 2009;324:532–6. https://doi.org/10.1126/science.1170587.

Clutton-Brock J. Domesticated animals from early times. London: British Museum (Natural History) and William Heinemann Ltd.; 1981.

Clutton-Brock J. The unnatural world: behavioural aspects of humans and animals in the process of domestication. In: Manning A, Serpell J, editors. Animals and human society: changing perspectives. New York: Routledge; 1994. p. 23–35.

Conolly J, Colledge S, Dobney K, et al. Meta-analysis of zooarchaeological data from SW Asia and SE Europe provides insight into the origins and spread of animal husbandry. J Archaeol Sci. 2011;38:538–45. https://doi.org/10.1016/j.jas.2010.10.008.

Cordell HJ. Epistasis: what it means, what it doesn't mean, and statistical methods to detect it in humans. Hum Mol Genet. 2002;11:2463–8. https://doi.org/10.1093/hmg/11.20.2463.

Cridland JM, Ramirez SR, Dean CA, et al. Genome sequencing of museum specimens reveals rapid changes in the genetic composition of honey bees in California. Genome Biol Evol. 2018;10:458–72. https://doi.org/10.1093/gbe/evy007.

Cristiani E, Radini A, Edinborough M, Borić D. Dental calculus reveals Mesolithic foragers in the Balkans consumed domesticated plant foods. Proc Natl Acad Sci U S A. 2016;113:10298–303. https://doi.org/10.1073/pnas.1603477113.

Crockford SJ. Animal domestication and heterochronic speciation. In: Minugh-Purvis N, McNamara KJ, editors. Human evolution through developmental change. Baltimore: JHU Press; 2002. p. 122–53.

Cucchi T, Hulme-Beaman A, Yuan J, Dobney K. Early Neolithic pig domestication at Jiahu, Henan Province, China: clues from molar shape analyses using geometric morphometric approaches. J Archaeol Sci. 2011;38:11–22. https://doi.org/10.1016/j.jas.2010.07.024.

Cucchi T, Mohaseb A, Peigné S, et al. Detecting taxonomic and phylogenetic signals in equid cheek teeth: towards new palaeontological and archaeological proxies. R Soc Open Sci. 2017;4:160997. https://doi.org/10.1098/rsos.160997.

Currat M, Ruedi M, Petit RJ, et al. The hidden side of invasions: massive introgression by local genes. Evolution. 2008;62:1908–20. https://doi.org/10.1111/j.1558-5646.2008.00413.x.

Daly KG, Delser PM, Mullin VE, et al. Ancient goat genomes reveal mosaic domestication in the Fertile Crescent. Science. 2018;361:85–8. https://doi.org/10.1126/science.aas9411.
Damgaard PB, Margaryan A, Schroeder H, et al. Improving access to endogenous DNA in ancient bones and teeth. Sci Rep. 2015;5:11184. https://doi.org/10.1038/srep11184.
Darwin C. On the origin of species by means of natural selection. London: John Murray; 1859.
Darwin C. The variation of animals and plants under domestication. New York: O. Judd & Company; 1868.
De la Rúa P, Jaffé R, Muñoz I, et al. Conserving genetic diversity in the honeybee: comments on Harpur et al. (2012). Mol Ecol. 2013;22:3208–10. https://doi.org/10.1111/mec.12333.
Decker JE, McKay SD, Rolf MM, et al. Worldwide patterns of ancestry, divergence, and admixture in domesticated cattle. PLoS Genet. 2014;10:e1004254. https://doi.org/10.1371/journal.pgen.1004254.
Demars J, Cano M, Drouilhet L, et al. Genome-wide identification of the mutation underlying fleece variation and discriminating ancestral hairy species from modern woolly sheep. Mol Biol Evol. 2017;34:1722–9. https://doi.org/10.1093/molbev/msx114.
Der Sarkissian C, Ermini L, Schubert M, et al. Evolutionary genomics and conservation of the endangered Przewalski's horse. Curr Biol. 2015;25:2577–83. https://doi.org/10.1016/j.cub.2015.08.032.
Diaz-Lameiro AM. Evolutionary origins and domestication of South American camelids, the alpaca (Vicugna pacos) and the llama (Lama glama) explained through molecular DNA methods. Binghamton: State University of New York at Binghamton; 2016.
Dobney K, Brothwell D. Dental calculus: its relevance to ancient diet and oral ecology. In: Cruwys E, Foley R, editors. Teeth and anthropology. Oxford: BAR; 1986.
Dobney K, Brothwell D. A method for evaluating the amount of dental calculus on teeth from archaeological sites. J Archaeol Sci. 1987;14:343–51. https://doi.org/10.1016/0305-4403(87)90024-0.
Dobney K, Larson G. Genetics and animal domestication: new windows on an elusive process. J Zool. 2006;269:261–71. https://doi.org/10.1111/j.1469-7998.2006.00042.x.
Doebley JF, Gaut BS, Smith BD. The molecular genetics of crop domestication. Cell. 2006;127:1309–21. https://doi.org/10.1016/j.cell.2006.12.006.
Doust AN, Lukens L, Olsen KM, et al. Beyond the single gene: how epistasis and gene-by-environment effects influence crop domestication. Proc Natl Acad Sci U S A. 2014;111:6178–83. https://doi.org/10.1073/pnas.1308940110.
Driscoll CA, Menotti-Raymond M, Roca AL, et al. The Near Eastern origin of cat domestication. Science. 2007;317:519–23. https://doi.org/10.1126/science.1139518.
Driscoll CA, Macdonald DW, O'Brien SJ. From wild animals to domestic pets, an evolutionary view of domestication. Proc Natl Acad Sci U S A. 2009;106:9971–8. https://doi.org/10.1073/pnas.0901586106.
Duarte CM, Marbá N, Holmer M. Rapid domestication of marine species. Science. 2007;316:382–3. https://doi.org/10.1126/science.1138042.
Ducos P. Defining domestication: a clarification. In: Clutton-Brock J, editor. The Walking larder: patterns of domestication, pastoralism, and predation. London: Unwin Hyman; 1989. p. 28–30.
Durand EY, Patterson N, Reich D, Slatkin M. Testing for ancient admixture between closely related populations. Mol Biol Evol. 2011;28:2239–52. https://doi.org/10.1093/molbev/msr048.
Eda M, Lu P, Kikuchi H, et al. Reevaluation of early Holocene chicken domestication in northern China. J Archaeol Sci. 2016;67:25–31. https://doi.org/10.1016/j.jas.2016.01.012.
Edwards CJ, Bollongino R, Scheu A, et al. Mitochondrial DNA analysis shows a Near Eastern Neolithic origin for domestic cattle and no indication of domestication of European aurochs. Proc R Soc Lond B Biol Sci. 2007;274:1377–85. https://doi.org/10.1098/rspb.2007.0020.
Eisenmann V, Mashkour M. Chevaux de Botaï, chevaux récents et autres souches de la domestication. In: Gardeisen A, editor. Les équidés dans le monde méditerranéen antique. Lattes: Edition de l'Association pour le développement de l'archéologie en Languedoc-Roussillon; 2005. p. 41–9.

Eriksson J, Larson G, Gunnarsson U, et al. Identification of the yellow skin gene reveals a hybrid origin of the domestic chicken. PLoS Genet. 2008;4:e1000010. https://doi.org/10.1371/journal.pgen.1000010.

Ervynck A, Dobney K, Hongo H, Meadow R. Born Free ? New Evidence for the Status of "Sus scrofa" at Neolithic Çayönü Tepesi (Southeastern Anatolia, Turkey). Paléorient. 2001;27:47–73.

Evin A, Cucchi T, Cardini A, et al. The long and winding road: identifying pig domestication through molar size and shape. J Archaeol Sci. 2013;40:735–43. https://doi.org/10.1016/j.jas.2012.08.005.

Evin A, Flink LG, Krause-Kyora B, et al. Exploring the complexity of domestication: a response to Rowley-Conwy and Zeder. World Archaeol. 2014;46:825–34. https://doi.org/10.1080/00438243.2014.953711.

Fang M, Larson G, Ribeiro HS, et al. Contrasting mode of evolution at a coat color locus in wild and domestic pigs. PLoS Genet. 2009;5:e1000341. https://doi.org/10.1371/journal.pgen.1000341.

Farias-Hesson E, Erikson J, Atkins A, et al. Semi-automated library preparation for high-throughput DNA sequencing platforms. J Biomed Biotechnol. 2010;2010:617469.

Feeney A, Nilsson E, Skinner MK. Epigenetics and transgenerational inheritance in domesticated farm animals. J Anim Sci Biotechnol. 2014;5:48. https://doi.org/10.1186/2049-1891-5-48.

Fernández H, Hughes S, Vigne J-D, et al. Divergent mtDNA lineages of goats in an early Neolithic site, far from the initial domestication areas. Proc Natl Acad Sci U S A. 2006;103:15375–9. https://doi.org/10.1073/pnas.0602753103.

Ferrer-Admetlla A, Leuenberger C, Jensen JD, Wegmann D. An approximate Markov model for the wright-fisher diffusion and its application to time series data. Genetics. 2016;203:831–46. https://doi.org/10.1534/genetics.115.184598.

Frantz AC, Zachos FE, Kirschning J, et al. Genetic evidence for introgression between domestic pigs and wild boars (Sus scrofa) in Belgium and Luxembourg: a comparative approach with multiple marker systems: introgression between pigs and boars. Biol J Linn Soc Lond. 2013a;110:104–15. https://doi.org/10.1111/bij.12111.

Frantz LAF, Schraiber JG, Madsen O, et al. Genome sequencing reveals fine scale diversification and reticulation history during speciation in Sus. Genome Biol. 2013b;14:R107. https://doi.org/10.1186/gb-2013-14-9-r107.

Frantz LAF, Madsen O, Megens H-J, et al. Testing models of speciation from genome sequences: divergence and asymmetric admixture in Island South-East Asian Sus species during the Plio-Pleistocene climatic fluctuations. Mol Ecol. 2014;23:5566–74. https://doi.org/10.1111/mec.12958.

Frantz LAF, Schraiber JG, Madsen O, et al. Evidence of long-term gene flow and selection during domestication from analyses of Eurasian wild and domestic pig genomes. Nat Genet. 2015;47:1141–8. https://doi.org/10.1038/ng.3394.

Frantz LAF, Meijaard E, Gongora J, et al. The evolution of Suidae. Annu Rev Anim Biosci. 2016a;4:61–85. https://doi.org/10.1146/annurev-animal-021815-111155.

Frantz LAF, Mullin VE, Pionnier-Capitan M, et al. Genomic and archaeological evidence suggest a dual origin of domestic dogs. Science. 2016b;352:1228–31. https://doi.org/10.1126/science.aaf3161.

Freedman AH, Gronau I, Schweizer RM, et al. Genome sequencing highlights the dynamic early history of dogs. PLoS Genet. 2014;10:e1004016. https://doi.org/10.1371/journal.pgen.1004016.

Freedman AH, Schweizer RM, Ortega-Del Vecchyo D, et al. Demographically-based evaluation of genomic regions under selection in domestic dogs. PLoS Genet. 2016;12:e1005851. https://doi.org/10.1371/journal.pgen.1005851.

Fu Q, Posth C, Hajdinjak M, et al. The genetic history of Ice Age Europe. Nature. 2016;534:200–5. https://doi.org/10.1038/nature17993.

Gallego-Llorente M, Connell S, Jones ER, et al. The genetics of an early Neolithic pastoralist from the Zagros, Iran. Sci Rep. 2016;6:31326. https://doi.org/10.1038/srep31326.

Galton F. The first steps towards the domestication of animals. Trans Ethnol Soc Lond. 1865;3:122. https://doi.org/10.2307/3014161.

Galton F. Inquiries into human faculty and its development. New York: Macmillan and Company; 1883.

Gamba C, Jones ER, Teasdale MD, et al. Genome flux and stasis in a five millennium transect of European prehistory. Nat Commun. 2014;5:5257. https://doi.org/10.1038/ncomms6257.

Gaunitz C, Fages A, Hanghøj K, et al. Ancient genomes revisit the ancestry of domestic and Przewalski's horses. Science. 2018;360:111–4. https://doi.org/10.1126/science.aao3297.

Gauthier-Pilters H, Dagg AI. The camel. Its evolution, ecology, behavior, and relationship to man. Chicago: The University of Chicago Press; 1981.

Geigl E-M, Grange T. Of cats and men: ancient DNA reveals how the cat conquered the ancient world. In: Lindqvist C, Rajora OP, editors. Paleogenomics. Cham: Springer; 2018. p. 1–18.

Gerbault P, Allaby RG, Boivin N, et al. Storytelling and story testing in domestication. Proc Natl Acad Sci U S A. 2014;111:6159–64. https://doi.org/10.1073/pnas.1400425111.

Germonpré M, Sablin MV, Stevens RE, et al. Fossil dogs and wolves from Palaeolithic sites in Belgium, the Ukraine and Russia: osteometry, ancient DNA and stable isotopes. J Archaeol Sci. 2009;36:473–90. https://doi.org/10.1016/j.jas.2008.09.033.

Gibbs RA, Taylor JF, Van Tassell CP, et al. Genome-wide survey of SNP variation uncovers the genetic structure of cattle breeds. Science. 2009;324:528–32. https://doi.org/10.1126/science.1167936.

Giguet-Covex C, Pansu J, Arnaud F, et al. Long livestock farming history and human landscape shaping revealed by lake sediment DNA. Nat Commun. 2014;5:3211. https://doi.org/10.1038/ncomms4211.

Girdland Flink L, Allen R, Barnett R, et al. Establishing the validity of domestication genes using DNA from ancient chickens. Proc Natl Acad Sci U S A. 2014;111:6184–9. https://doi.org/10.1073/pnas.1308939110.

Goddard ME, Hayes BJ. Mapping genes for complex traits in domestic animals and their use in breeding programmes. Nat Rev Genet. 2009;10:381–91. https://doi.org/10.1038/nrg2575.

Gokhman D, Lavi E, Prüfer K, et al. Reconstructing the DNA methylation maps of the Neanderthal and the Denisovan. Science. 2014;344:523–7. https://doi.org/10.1126/science.1250368.

Goodwin S, McPherson JD, McCombie WR. Coming of age: ten years of next-generation sequencing technologies. Nat Rev Genet. 2016;17:333–51. https://doi.org/10.1038/nrg.2016.49.

Green EJ, Speller CF. Novel substrates as sources of ancient DNA: prospects and hurdles. Genes. 2017;8:180. https://doi.org/10.3390/genes8070180.

Green RE, Krause J, Briggs AW, et al. A draft sequence of the Neanderthal genome. Science. 2010;328:710–22. https://doi.org/10.1126/science.1188021.

Gregory, Saint, Bishop of Tours. History of the Franks. New York: Norton; 1969.

Haak W, Lazaridis I, Patterson N, et al. Massive migration from the steppe was a source for Indo-European languages in Europe. Nature. 2015;522:207–11. https://doi.org/10.1038/nature14317.

Haile J, Holdaway R, Oliver K, et al. Ancient DNA chronology within sediment deposits: are paleobiological reconstructions possible and is DNA leaching a factor? Mol Biol Evol. 2007;24:982–9. https://doi.org/10.1093/molbev/msm016.

Hammer K. Das Domestikationssyndrom. Kulturpflanze. 1984;32:11–34. https://doi.org/10.1007/BF02098682.

Han F, Wallberg A, Webster MT. From where did the Western honeybee (Apis mellifera) originate? Ecol Evol. 2012;2:1949–57. https://doi.org/10.1002/ece3.312.

Hanghøj K, Orlando L. Ancient epigenomics. In: Lindqvist C, Rajora OP, editors. Paleogenomics. Cham: Springer; 2018. p. 1–37.

Hanghøj K, Seguin-Orlando A, Schubert M, et al. Fast, accurate and automatic ancient nucleosome and methylation maps with epiPALEOMIX. Mol Biol Evol. 2016;33:3284–98. https://doi.org/10.1093/molbev/msw184.

Hanon EA, Lincoln GA, Fustin J-M, et al. Ancestral TSH mechanism signals summer in a photoperiodic mammal. Curr Biol. 2008;18:1147–52. https://doi.org/10.1016/j.cub.2008.06.076.
Hanotte O, Bradley DG, Ochieng JW, et al. African pastoralism: genetic imprints of origins and migrations. Science. 2002;296:336–9. https://doi.org/10.1126/science.1069878.
Hansen HB, Damgaard PB, Margaryan A, et al. Comparing ancient DNA preservation in petrous bone and tooth cementum. PLoS One. 2017;12:e0170940. https://doi.org/10.1371/journal.pone.0170940.
Hare J. The wild Bactrian camel Camelus bactrianus ferus in China: the need for urgent action. Oryx. 1997;31:45–8. https://doi.org/10.1046/j.1365-3008.1997.d01-2.x.
Harpur BA, Minaei S, Kent CF, Zayed A. Management increases genetic diversity of honey bees via admixture. Mol Ecol. 2012;21:4414–21. https://doi.org/10.1111/j.1365-294X.2012.05614.x.
Hellenthal G, Busby GBJ, Band G, et al. A genetic atlas of human admixture history. Science. 2014;343:747–51. https://doi.org/10.1126/science.1243518.
Helmer D, Gourichon L, Monchot H, et al. Identifying early domestic cattle from Pre-Pottery Neolithic sites on the Middle Euphrates using sexual dimorphism. In: Vigne J-D, Peters J, Helmer D, editors. The first steps of animal domestication: new archaeozoological approaches. Oxford: Oxbow; 2005. p. 86–95.
Higgins D, Kaidonis J, Townsend G, et al. Targeted sampling of cementum for recovery of nuclear DNA from human teeth and the impact of common decontamination measures. Investigative Genet. 2013;4:18. https://doi.org/10.1186/2041-2223-4-18.
Higuchi R, Bowman B, Freiberger M, et al. DNA sequences from the quagga, an extinct member of the horse family. Nature. 1984;312:282–4. https://doi.org/10.1038/312282a0.
Hofreiter M, Paijmans JLA, Goodchild H, et al. The future of ancient DNA: technical advances and conceptual shifts. Bioessays. 2015;37:284–93. https://doi.org/10.1002/bies.201400160.
Holmes M. Animals in Saxon and Scandinavian England: backbones of economy and society. Leiden: Sidestone Press; 2014.
Hongo H, Meadow RH. Pig exploitation at Neolithic Cayonu Tepesi (Southeastern Anatolia). MASCA Res Papers Sci. 1998;15:77–98.
Hongo H, Pearson J, Öksüz B, Ilgezdi G. The process of ungulate domestication at Çayönü, Southeastern Turkey: a multidisciplinary approach focusing on Bos sp. and Cervus elaphus. Anthropozoologica. 2009;44:63–78. https://doi.org/10.5252/az2009n1a3.
Hu Z-L, Fritz ER, Reecy JM. AnimalQTLdb: a livestock QTL database tool set for positional QTL information mining and beyond. Nucleic Acids Res. 2007;35:D604–9. https://doi.org/10.1093/nar/gkl946.
Hu Z-L, Park CA, Reecy JM. Developmental progress and current status of the Animal QTLdb. Nucleic Acids Res. 2016;44:D827–33. https://doi.org/10.1093/nar/gkv1233.
Huang DW, Sherman BT, Lempicki RA. Bioinformatics enrichment tools: paths toward the comprehensive functional analysis of large gene lists. Nucleic Acids Res. 2009;37:1–13. https://doi.org/10.1093/nar/gkn923.
Huang X-H, Wu Y-J, Miao Y-W, et al. Was chicken domesticated in northern China? New evidence from mitochondrial genomes. Sci Bull Fac Agric Kyushu Univ. 2018;63:743–6. https://doi.org/10.1016/j.scib.2017.12.004.
Huerta-Sánchez E, Jin X, Asan, et al. Altitude adaptation in Tibetans caused by introgression of Denisovan-like DNA. Nature. 2014;512:194–7. https://doi.org/10.1038/nature13408.
Ibeagha-Awemu EM, Zhao X. Epigenetic marks: regulators of livestock phenotypes and conceivable sources of missing variation in livestock improvement programs. Front Genet. 2015;6:302. https://doi.org/10.3389/fgene.2015.00302.
Irving-Pease EK, Frantz LAF, Sykes N, et al. Rabbits and the specious origins of domestication. Trends Ecol Evol. 2018;33:149–52. https://doi.org/10.1016/j.tree.2017.12.009.
Irwin DM, Kocher TD, Wilson AC. Evolution of the cytochrome b gene of mammals. J Mol Evol. 1991;32:128–44. https://doi.org/10.1007/BF02515385.

Janečka JE, Davis BW, Ghosh S, et al. Horse Y chromosome assembly displays unique evolutionary features and putative stallion fertility genes. Nat Commun. 2018;9:2945. https://doi.org/10.1038/s41467-018-05290-6.

Janowitz Koch I, Clark MM, Thompson MJ, et al. The concerted impact of domestication and transposon insertions on methylation patterns between dogs and grey wolves. Mol Ecol. 2016;25:1838–55. https://doi.org/10.1111/mec.13480.

Jensen P. Behavior genetics and the domestication of animals. Annu Rev Anim Biosci. 2014;2:85–104. https://doi.org/10.1146/annurev-animal-022513-114135.

Johnsson M, Gustafson I, Rubin C-J, et al. A sexual ornament in chickens is affected by pleiotropic alleles at HAO1 and BMP2, selected during domestication. PLoS Genet. 2012;8:e1002914. https://doi.org/10.1371/journal.pgen.1002914.

Jónsson H, Schubert M, Seguin-Orlando A, et al. Speciation with gene flow in equids despite extensive chromosomal plasticity. Proc Natl Acad Sci U S A. 2014;111:18655–60. https://doi.org/10.1073/pnas.1412627111.

Kadwell M, Fernandez M, Stanley HF, et al. Genetic analysis reveals the wild ancestors of the llama and the alpaca. Proc R Soc Lond B Biol Sci. 2001;268:2575–84. https://doi.org/10.1098/rspb.2001.1774.

Karlsson A-C, Svemer F, Eriksson J, et al. The effect of a mutation in the thyroid stimulating hormone receptor (TSHR) on development, behaviour and TH levels in domesticated chickens. PLoS One. 2015;10:e0129040. https://doi.org/10.1371/journal.pone.0129040.

Karlsson A-C, Fallahshahroudi A, Johnsen H, et al. A domestication related mutation in the thyroid stimulating hormone receptor gene (TSHR) modulates photoperiodic response and reproduction in chickens. Gen Comp Endocrinol. 2016;228:69–78. https://doi.org/10.1016/j.ygcen.2016.02.010.

Kijas JW, Lenstra JA, Hayes B, et al. Genome-wide analysis of the World's sheep breeds reveals high levels of historic mixture and strong recent selection. PLoS Biol. 2012;10:e1001258. https://doi.org/10.1371/journal.pbio.1001258.

Kılınç GM, Omrak A, Özer F, et al. The demographic development of the first farmers in anatolia. Curr Biol. 2016;26:2659–66. https://doi.org/10.1016/j.cub.2016.07.057.

King HD, Donaldson HH. Life processes and size of the body and organs of the gray Norway rat during ten generations in captivity. Am Anat Mem. 1929;14:106.

Kistler L, Ware R, Smith O, et al. A new model for ancient DNA decay based on paleogenomic meta-analysis. Nucleic Acids Res. 2017;45:6310–20. https://doi.org/10.1093/nar/gkx361.

Köhler I. Zur Domestikation Des Kamels. Institut Fur Zoologie, Tierarztliche Hochschule. 1981.

Koutsogiannouli EA, Moutou KA, Sarafidou T, et al. Detection of hybrids between wild boars (Sus scrofa scrofa) and domestic pigs (Sus scrofa f. domestica) in Greece, using the PCR-RFLP method on melanocortin-1 receptor (MC1R) mutations. Mamm Biol – Zeitschrift für Säugetierkunde. 2010;75:69–73. https://doi.org/10.1016/j.mambio.2008.08.001.

Kozhamkulova BS. The late Cenozoic two-humped (Bactrian) camels of Asia. Quartärpläontolgie (Abh Ber Inst Quartärpläontolgie Weimar). 1986;6:93–7.

Krause-Kyora B, Makarewicz C, Evin A, et al. Use of domesticated pigs by Mesolithic hunter-gatherers in northwestern Europe. Nat Commun. 2013;4:2348. https://doi.org/10.1038/ncomms3348.

Kukekova AV, Johnson JL, Xiang X, et al. Red fox genome assembly identifies genomic regions associated with tame and aggressive behaviours. Nat Ecol Evol. 2018;2:1479–91. https://doi.org/10.1038/s41559-018-0611-6.

Kuzmina EE. The prehistory of the silk road. Philadelphia: University of Pennsylvania Press; 2008.

Lam YM, Chen X, Pearson OM. Intertaxonomic variability in patterns of bone density and the differential representation of bovid, cervid, and equid elements in the archaeological record. Am Antiq. 1999;64:343–62. https://doi.org/10.2307/2694283.

Larson G, Bradley DG. How much is that in dog years? The advent of canine population genomics. PLoS Genet. 2014;10:e1004093. https://doi.org/10.1371/journal.pgen.1004093.

Larson G, Burger J. A population genetics view of animal domestication. Trends Genet. 2013;29:197–205. https://doi.org/10.1016/j.tig.2013.01.003.

Larson G, Fuller DQ. The evolution of animal domestication. Annu Rev Ecol Evol Syst. 2014;45:115–36. https://doi.org/10.1146/annurev-ecolsys-110512-135813.

Larson G, Dobney K, Albarella U, et al. Worldwide phylogeography of wild boar reveals multiple centers of pig domestication. Science. 2005;307:1618–21. https://doi.org/10.1126/science.1106927.

Larson G, Albarella U, Dobney K, et al. Ancient DNA, pig domestication, and the spread of the Neolithic into Europe. Proc Natl Acad Sci U S A. 2007a;104:15276–81. https://doi.org/10.1073/pnas.0703411104.

Larson G, Cucchi T, Fujita M, et al. Phylogeny and ancient DNA of Sus provides insights into neolithic expansion in Island Southeast Asia and Oceania. Proc Natl Acad Sci U S A. 2007b;104:4834–9. https://doi.org/10.1073/pnas.0607753104.

Larson G, Liu R, Zhao X, et al. Patterns of East Asian pig domestication, migration, and turnover revealed by modern and ancient DNA. Proc Natl Acad Sci U S A. 2010;107:7686–91. https://doi.org/10.1073/pnas.0912264107.

Larson G, Karlsson EK, Perri A, et al. Rethinking dog domestication by integrating genetics, archeology, and biogeography. Proc Natl Acad Sci U S A. 2012;109:8878–83. https://doi.org/10.1073/pnas.1203005109.

Larson G, Piperno DR, Allaby RG, et al. Current perspectives and the future of domestication studies. Proc Natl Acad Sci U S A. 2014;111:6139–46. https://doi.org/10.1073/pnas.1323964111.

Lawson DJ, Hellenthal G, Myers S, Falush D. Inference of population structure using dense haplotype data. PLoS Genet. 2012;8:e1002453. https://doi.org/10.1371/journal.pgen.1002453.

Lawson DJ, van Dorp L, Falush D. A tutorial on how not to over-interpret STRUCTURE and ADMIXTURE bar plots. Nat Commun. 2018;9:3258. https://doi.org/10.1038/s41467-018-05257-7.

Lazaridis I, Nadel D, Rollefson G, et al. Genomic insights into the origin of farming in the ancient Near East. Nature. 2016;536:419–24. https://doi.org/10.1038/nature19310.

Lenffer J, Nicholas FW, Castle K, et al. OMIA (Online Mendelian Inheritance in Animals): an enhanced platform and integration into the Entrez search interface at NCBI. Nucleic Acids Res. 2006;34:D599–601. https://doi.org/10.1093/nar/gkj152.

Leppälä K, Nielsen SV, Mailund T. admixturegraph: an R package for admixture graph manipulation and fitting. Bioinformatics. 2017;33:1738–40. https://doi.org/10.1093/bioinformatics/btx048.

Levine MA. Botai and the origins of horse domestication. J Anthropol Archaeol. 1999;18:29–78. https://doi.org/10.1006/jaar.1998.0332.

Li D, Guo Y, Shao H, et al. Genetic diversity, molecular phylogeny and selection evidence of the silkworm mitochondria implicated by complete resequencing of 41 genomes. BMC Evol Biol. 2010;10:81. https://doi.org/10.1186/1471-2148-10-81.

Librado P, Sarkissian CD, Ermini L, et al. Tracking the origins of Yakutian horses and the genetic basis for their fast adaptation to subarctic environments. Proc Natl Acad Sci U S A. 2015;112: E6889–97. https://doi.org/10.1073/pnas.1513696112.

Librado P, Gamba C, Gaunitz C, et al. Ancient genomic changes associated with domestication of the horse. Science. 2017;356:442–5. https://doi.org/10.1126/science.aam5298.

Linderholm A, Spencer D, Battista V, et al. A novel MC1R allele for black coat colour reveals the Polynesian ancestry and hybridization patterns of Hawaiian feral pigs. R Soc Open Sci. 2016;3:160304. https://doi.org/10.1098/rsos.160304.

Lindgren G, Backström N, Swinburne J, et al. Limited number of patrilines in horse domestication. Nat Genet. 2004;36:335–6. https://doi.org/10.1038/ng1326.

Lipinski MJ, Froenicke L, Baysac KC, et al. The ascent of cat breeds: genetic evaluations of breeds and worldwide random-bred populations. Genomics. 2008;91:12–21. https://doi.org/10.1016/j.ygeno.2007.10.009.

Lippold S, Knapp M, Kuznetsova T, et al. Discovery of lost diversity of paternal horse lineages using ancient DNA. Nat Commun. 2011a;2:450. https://doi.org/10.1038/ncomms1447.

Lippold S, Matzke NJ, Reissmann M, Hofreiter M. Whole mitochondrial genome sequencing of domestic horses reveals incorporation of extensive wild horse diversity during domestication. BMC Evol Biol. 2011b;11:328. https://doi.org/10.1186/1471-2148-11-328.

Lipson M, Loh P-R, Levin A, et al. Efficient moment-based inference of admixture parameters and sources of gene flow. Mol Biol Evol. 2013;30:1788–802. https://doi.org/10.1093/molbev/mst099.

Lipson M, Szécsényi-Nagy A, Mallick S, et al. Parallel palaeogenomic transects reveal complex genetic history of early European farmers. Nature. 2017;551:368–72. https://doi.org/10.1038/nature24476.

Liu Y-P, Wu G-S, Yao Y-G, et al. Multiple maternal origins of chickens: out of the Asian jungles. Mol Phylogenet Evol. 2006;38:12–9. https://doi.org/10.1016/j.ympev.2005.09.014.

Loftus RT, MacHugh DE, Bradley DG, et al. Evidence for two independent domestications of cattle. Proc Natl Acad Sci U S A. 1994;91:2757–61. https://doi.org/10.1073/pnas.91.7.2757.

Loog L, Thomas MG, Barnett R, et al. Inferring allele frequency trajectories from ancient DNA indicates that selection on a chicken gene coincided with changes in medieval husbandry practices. Mol Biol Evol. 2017;34:1981–90. https://doi.org/10.1093/molbev/msx142.

Luikart G, Gielly L, Excoffier L, et al. Multiple maternal origins and weak phylogeographic structure in domestic goats. Proc Natl Acad Sci U S A. 2001;98:5927–32. https://doi.org/10.1073/pnas.091591198.

Lundin S, Stranneheim H, Pettersson E, et al. Increased throughput by parallelization of library preparation for massive sequencing. PLoS One. 2010;5:e10029. https://doi.org/10.1371/journal.pone.0010029.

Lynch VJ, Bedoya-Reina OC, Ratan A, et al. Elephantid genomes reveal the molecular bases of Woolly Mammoth adaptations to the arctic. Cell Rep. 2015;12:217–28. https://doi.org/10.1016/j.celrep.2015.06.027.

MacHugh DE, Larson G, Orlando L. Taming the past: ancient DNA and the study of animal domestication. Annu Rev Anim Biosci. 2017;5:329–51. https://doi.org/10.1146/annurev-animal-022516-022747.

Mackay TFC, Stone EA, Ayroles JF. The genetics of quantitative traits: challenges and prospects. Nat Rev Genet. 2009;10:565–77. https://doi.org/10.1038/nrg2612.

Maniatis T, Fritsch EF, Sambrook J, et al. Molecular cloning: a laboratory manual. New York: Cold Spring Harbor Laboratory; 1982.

Margulies M, Egholm M, Altman WE, et al. Genome sequencing in microfabricated high-density picolitre reactors. Nature. 2005;437:376–80. https://doi.org/10.1038/nature03959.

Marshall FB, Dobney K, Denham T, Capriles JM. Evaluating the roles of directed breeding and gene flow in animal domestication. Proc Natl Acad Sci U S A. 2014;111:6153–8. https://doi.org/10.1073/pnas.1312984110.

Mathieson I, Lazaridis I, Rohland N, et al. Genome-wide patterns of selection in 230 ancient Eurasians. Nature. 2015;528:499–503. https://doi.org/10.1038/nature16152.

McCarthy MI, Abecasis GR, Cardon LR, et al. Genome-wide association studies for complex traits: consensus, uncertainty and challenges. Nat Rev Genet. 2008;9:356–69. https://doi.org/10.1038/nrg2344.

McCue ME, Bannasch DL, Petersen JL, et al. A high density SNP array for the domestic horse and extant Perissodactyla: utility for association mapping, genetic diversity, and phylogeny studies. PLoS Genet. 2012;8:e1002451. https://doi.org/10.1371/journal.pgen.1002451.

Meadow RH. Animal domestication in the Middle East: a view from the eastern margin. In: Clutton-Brock J, Grigson C, editors. Animals and archaeology. Oxford: BAR; 1983. p. 309–37.

Meadows JRS, Kijas JW. Re-sequencing regions of the ovine Y chromosome in domestic and wild sheep reveals novel paternal haplotypes. Anim Genet. 2009;40:119–23. https://doi.org/10.1111/j.1365-2052.2008.01799.x.

Meadows JRS, Hiendleder S, Kijas JW. Haplogroup relationships between domestic and wild sheep resolved using a mitogenome panel. Heredity. 2011;106:700–6. https://doi.org/10.1038/hdy.2010.122.

Meyer A. Shortcomings of the cytochrome b gene as a molecular marker. Trends Ecol Evol. 1994;9:278–80. https://doi.org/10.1016/0169-5347(94)90028-0.

Miao Y-W, Peng M-S, Wu G-S, et al. Chicken domestication: an updated perspective based on mitochondrial genomes. Heredity. 2013;110:277–82. https://doi.org/10.1038/hdy.2012.83.

Miao B, Wang Z, Li Y. Genomic analysis reveals hypoxia adaptation in the Tibetan Mastiff by introgression of the gray wolf from the Tibetan plateau. Mol Biol Evol. 2017;34:734–43. https://doi.org/10.1093/molbev/msw274.

Miller W, Drautz DI, Ratan A, et al. Sequencing the nuclear genome of the extinct woolly mammoth. Nature. 2008;456:387–90. https://doi.org/10.1038/nature07446.

Mix H, Reading RP, Lhagvasuren B. Zum Status des wilden Trampeltieres (Camelus bactrianus ferus) in der Mongolei. Zool Ges Arten-u Pop schutz. 1997;13:1–3.

Mix H, Reading RP, Blumer ES, Badamjaviin L. Status and distribution of wild bactrian camels in Mongolia. Ecol Conserv Wild Bact Camel. 2002:39–48.

Moyers BT, Morrell PL, McKay JK. Genetic costs of domestication and improvement. J Hered. 2018;109:103–16. https://doi.org/10.1093/jhered/esx069.

Muir WM, Wong GK-S, Zhang Y, et al. Genome-wide assessment of worldwide chicken SNP genetic diversity indicates significant absence of rare alleles in commercial breeds. Proc Natl Acad Sci U S A. 2008;105:17312–7. https://doi.org/10.1073/pnas.0806569105.

Muir P, Li S, Lou S, et al. The real cost of sequencing: scaling computation to keep pace with data generation. Genome Biol. 2016;17:53. https://doi.org/10.1186/s13059-016-0917-0.

Mullis KB, Faloona FA. Specific synthesis of DNA in vitro via a polymerase-catalyzed chain reaction. In: Methods in enzymology. New York: Academic Press; 1987. p. 335–50.

Mwai O, Hanotte O, Kwon Y-J, Cho S. African indigenous cattle: unique genetic resources in a rapidly changing world. Asian-Australas J Anim Sci. 2015;28:911–21. https://doi.org/10.5713/ajas.15.0002R.

Nachtsheim H. Vom Wildtier zum Haustier. Berlin: Alfred Metzner; 1936.

Naderi S, Rezaei H-R, Taberlet P, et al. Large-scale mitochondrial DNA analysis of the domestic goat reveals six haplogroups with high diversity. PLoS One. 2007;2:e1012. https://doi.org/10.1371/journal.pone.0001012.

Naderi S, Rezaei H-R, Pompanon F, et al. The goat domestication process inferred from large-scale mitochondrial DNA analysis of wild and domestic individuals. Proc Natl Acad Sci U S A. 2008;105:17659–64. https://doi.org/10.1073/pnas.0804782105.

Nakao N, Ono H, Yamamura T, et al. Thyrotrophin in the pars tuberalis triggers photoperiodic response. Nature. 2008;452:317–22. https://doi.org/10.1038/nature06738.

Ní Leathlobhair M, Perri AR, Irving-Pease EK, et al. The evolutionary history of dogs in the Americas. Science. 2018;361:81–5. https://doi.org/10.1126/science.aao4776.

Nicholas FW. Online Mendelian Inheritance in Animals (OMIA): a comparative knowledgebase of genetic disorders and other familial traits in non-laboratory animals. Nucleic Acids Res. 2003;31:275–7. https://doi.org/10.1093/nar/gkg074.

Ollivier M, Tresset A, Bastian F, et al. Amy2B copy number variation reveals starch diet adaptations in ancient European dogs. R Soc Open Sci. 2016;3:160449. https://doi.org/10.1098/rsos.160449.

Olsen SL. Early horse domestication on the Eurasian steppe. In: Zeder MA, Bradley DG, Smith BD, Emshwiller E, editors. Documenting domestication: new genetic and archaeological paradigms. Berkeley: University of California Press; 2006a. p. 245–69.

Olsen SL. Early horse domestication: weighing the evidence. In: Olsen SL, Grant S, Choyke AM, Bartosiewicz L, editors. Horses and humans: the evolution of human-equine relationships. Oxford: Archaeopress; 2006b. p. 81–113.

Olsen KM, Wendel JF. A bountiful harvest: genomic insights into crop domestication phenotypes. Annu Rev Plant Biol. 2013;64:47–70. https://doi.org/10.1146/annurev-arplant-050312-120048.

Olson MV. When less is more: gene loss as an engine of evolutionary change. Am J Hum Genet. 1999;64:18–23. https://doi.org/10.1086/302219.

Orlando L. An ancient DNA perspective on horse evolution. In: Lindqvist C, Rajora OP, editors. Paleogenomics. Cham: Springer; 2018. p. 1–27.

Orlando L, Ginolhac A, Zhang G, et al. Recalibrating Equus evolution using the genome sequence of an early Middle Pleistocene horse. Nature. 2013;499:74–8. https://doi.org/10.1038/nature12323.

Ottoni C, Flink LG, Evin A, et al. Pig domestication and human-mediated dispersal in western Eurasia revealed through ancient DNA and geometric morphometrics. Mol Biol Evol. 2013;30:824–32. https://doi.org/10.1093/molbev/mss261.

Ottoni C, Van Neer W, Cupere BD, et al. The palaeogenetics of cat dispersal in the ancient world. Nat Ecol Evol. 2017;1:0139. https://doi.org/10.1038/s41559-017-0139.

Outram AK, Stear NA, Bendrey R, et al. The earliest horse harnessing and milking. Science. 2009;323:1332–5. https://doi.org/10.1126/science.1168594.

Palkopoulou E, Mallick S, Skoglund P, et al. Complete genomes reveal signatures of demographic and genetic declines in the woolly mammoth. Curr Biol. 2015;25:1395–400. https://doi.org/10.1016/j.cub.2015.04.007.

Pang J-F, Kluetsch C, Zou X-J, et al. mtDNA data indicate a single origin for dogs south of Yangtze River, less than 16,300 years ago, from numerous wolves. Mol Biol Evol. 2009;26:2849–64. https://doi.org/10.1093/molbev/msp195.

Park SDE, Magee DA, McGettigan PA, et al. Genome sequencing of the extinct Eurasian wild aurochs, Bos primigenius, illuminates the phylogeography and evolution of cattle. Genome Biol. 2015;16:234. https://doi.org/10.1186/s13059-015-0790-2.

Patterson N, Moorjani P, Luo Y, et al. Ancient admixture in human history. Genetics. 2012;192:1065–93. https://doi.org/10.1534/genetics.112.145037.

Pedersen JS, Valen E, Velazquez AMV, et al. Genome-wide nucleosome map and cytosine methylation levels of an ancient human genome. Genome Res. 2014;24:454–66. https://doi.org/10.1101/gr.163592.113.

Pedrosa S, Uzun M, Arranz J-J, et al. Evidence of three maternal lineages in near eastern sheep supporting multiple domestication events. Proc R Soc Lond B Biol Sci. 2005;272:2211–7. https://doi.org/10.1098/rspb.2005.3204.

Perri A. A wolf in dog's clothing: initial dog domestication and Pleistocene wolf variation. J Archaeol Sci. 2016;68:1–4. https://doi.org/10.1016/j.jas.2016.02.003.

Peters J. Hahn oder Kapaun? Zur Kastration von Hühnen in der Antike. Archiv für Geflügelkunde. 1997;61:1–8.

Peters J, Lebrasseur O, Deng H, Larson G. Holocene cultural history of Red jungle fowl (Gallus gallus) and its domestic descendant in East Asia. Quat Sci Rev. 2016;142:102–19. https://doi.org/10.1016/j.quascirev.2016.04.004.

Petersen JL, Mickelson JR, Cothran EG, et al. Genetic diversity in the modern horse illustrated from genome-wide SNP data. PLoS One. 2013;8:e54997. https://doi.org/10.1371/journal.pone.0054997.

Pickrell JK, Pritchard JK. Inference of population splits and mixtures from genome-wide allele frequency data. PLoS Genet. 2012;8:e1002967. https://doi.org/10.1371/journal.pgen.1002967.

Pinhasi R, Fernandes D, Sirak K, et al. Optimal ancient DNA yields from the inner ear part of the human petrous bone. PLoS One. 2015;10:e0129102. https://doi.org/10.1371/journal.pone.0129102.

Pitt D, Sevane N, Nicolazzi EL, et al. Domestication of cattle: two or three events? Evol Appl. 2018;18:R157. https://doi.org/10.1111/eva.12674.

Plagnol V, Wall JD. Possible ancestral structure in human populations. PLoS Genet. 2006;2:e105. https://doi.org/10.1371/journal.pgen.0020105.

Pritchard JK, Stephens M, Donnelly P. Inference of population structure using multilocus genotype data. Genetics. 2000;155:945–59. https://doi.org/10.1111/j.1471-8286.2007.01758.x.

Prüfer K, Racimo F, Patterson N, et al. The complete genome sequence of a Neanderthal from the Altai Mountains. Nature. 2014;505:43–9. https://doi.org/10.1038/nature12886.

Reading RP, Mix H, Lhagvasuren B, Blumer ES. Status of wild Bactrian camels and other large ungulates in south-western Mongolia. Oryx. 1999;33:247–55. https://doi.org/10.1046/j.1365-3008.1999.00064.x.

Reich D, Thangaraj K, Patterson N, et al. Reconstructing Indian population history. Nature. 2009;461:489–94. https://doi.org/10.1038/nature08365.

Reissmann M, Ludwig A. Pleiotropic effects of coat colour-associated mutations in humans, mice and other mammals. Semin Cell Dev Biol. 2013;24:576–86. https://doi.org/10.1016/j.semcdb.2013.03.014.

Reynier P, May-Panloup P, Chrétien M-F, et al. Mitochondrial DNA content affects the fertilizability of human oocytes. Mol Hum Reprod. 2001;7:425–9. https://doi.org/10.1093/molehr/7.5.425.

Røed KH, Flagstad Ø, Nieminen M, et al. Genetic analyses reveal independent domestication origins of Eurasian reindeer. Proc R Soc Lond B Biol Sci. 2008;275:1849–55. https://doi.org/10.1098/rspb.2008.0332.

Rohland N, Reich D. Cost-effective, high-throughput DNA sequencing libraries for multiplexed target capture. Genome Res. 2012;22:939–46. https://doi.org/10.1101/gr.128124.111.

Rowley-Conwy P, Zeder M. Mesolithic domestic pigs at Rosenhof – or wild boar? A critical re-appraisal of ancient DNA and geometric morphometrics. World Archaeol. 2014a;46:813–24. https://doi.org/10.1080/00438243.2014.953704.

Rowley-Conwy P, Zeder M. Wild boar or domestic pigs? Response to Evin et al. World Archaeol. 2014b;46:835–40. https://doi.org/10.1080/00438243.2014.953712.

Rubin C-J, Zody MC, Eriksson J, et al. Whole-genome resequencing reveals loci under selection during chicken domestication. Nature. 2010;464:587–91. https://doi.org/10.1038/nature08832.

Rubin C-J, Megens H-J, Barrio AM, et al. Strong signatures of selection in the domestic pig genome. Proc Natl Acad Sci U S A. 2012;109:19529–36. https://doi.org/10.1073/pnas.1217149109.

Saiki RK, Scharf S, Faloona F, et al. Enzymatic amplification of beta-globin genomic sequences and restriction site analysis for diagnosis of sickle cell anemia. Science. 1985;230:1350–4. https://doi.org/10.1126/science.2999980.

Sánchez-Villagra MR, Geiger M, Schneider RA. The taming of the neural crest: a developmental perspective on the origins of morphological covariation in domesticated mammals. R Soc Open Sci. 2016;3:160107. https://doi.org/10.1098/rsos.160107.

Sanger F, Nicklen S, Coulson AR. DNA sequencing with chain-terminating inhibitors. Proc Natl Acad Sci U S A. 1977;74:5463–7.

Sawyer S, Krause J, Guschanski K, et al. Temporal patterns of nucleotide misincorporations and DNA fragmentation in ancient DNA. PLoS One. 2012;7:e34131. https://doi.org/10.1371/journal.pone.0034131.

Schaefer RJ, Schubert M, Bailey E, et al. Developing a 670k genotyping array to tag ~2M SNPs across 24 horse breeds. BMC Genomics. 2017;18:565. https://doi.org/10.1186/s12864-017-3943-8.

Scheu A, Geörg C, Schulz A, et al. The arrival of domesticated animals in South-Eastern Europe as seen from ancient DNA. In: Kaiser E, Burger J, Schier W, editors. Population dynamics in prehistory and early history. Berlin: De Gruyter; 2012. p. 45–54.

Scheu A, Powell A, Bollongino R, et al. The genetic prehistory of domesticated cattle from their origin to the spread across Europe. BMC Genet. 2015;16:54. https://doi.org/10.1186/s12863-015-0203-2.

Schraiber JG, Evans SN, Slatkin M. Bayesian inference of natural selection from allele frequency time series. Genetics. 2016;203:493–511. https://doi.org/10.1534/genetics.116.187278.

Schubert M, Jónsson H, Chang D, et al. Prehistoric genomes reveal the genetic foundation and cost of horse domestication. Proc Natl Acad Sci U S A. 2014;111:E5661–9. https://doi.org/10.1073/pnas.1416991111.

Schweizer RM, Durvasula A, Smith J, et al. Natural selection and origin of a melanistic allele in North American gray wolves. Mol Biol Evol. 2018;35:1190–209. https://doi.org/10.1093/molbev/msy031.

Serjeantson D. Birds: food and a mark of status. In: Woolgar CM, Serjeantson D, Waldron T, editors. Food in medieval England: diet and nutrition. Oxford: Oxford University Press; 2006. p. 131–47.

Shannon LM, Boyko RH, Castelhano M, et al. Genetic structure in village dogs reveals a Central Asian domestication origin. Proc Natl Acad Sci U S A. 2015;112:13639–44. https://doi.org/10.1073/pnas.1516215112.

Skoglund P, Ersmark E, Palkopoulou E, Dalén L. Ancient wolf genome reveals an early divergence of domestic dog ancestors and admixture into high-latitude breeds. Curr Biol. 2015;25:1515–9. https://doi.org/10.1016/j.cub.2015.04.019.

Smith BD, Zeder MA. The onset of the Anthropocene. Anthropocene. 2013;4:8–13. https://doi.org/10.1016/j.ancene.2013.05.001.

Smith O, Momber G, Bates R, et al. Sedimentary DNA from a submerged site reveals wheat in the British Isles 8000 years ago. Science. 2015a;347:998–1001. https://doi.org/10.1126/science.1261278.

Smith O, Momber G, Bates R, et al. Response to comment on "Sedimentary DNA from a submerged site reveals wheat in the British Isles 8000 years ago". Science. 2015b;349:247. https://doi.org/10.1126/science.aab2062.

Søe MJ, Nejsum P, Seersholm FV, et al. Ancient DNA from latrines in Northern Europe and the Middle East (500 BC-1700 AD) reveals past parasites and diet. PLoS One. 2018;13:e0195481. https://doi.org/10.1371/journal.pone.0195481.

Stainton JJ, Charlesworth B, Haley CS, et al. Use of high-density SNP data to identify patterns of diversity and signatures of selection in broiler chickens. J Anim Breed Genet. 2017;134:87–97. https://doi.org/10.1111/jbg.12228.

Staubach F, Lorenc A, Messer PW, et al. Genome patterns of selection and introgression of haplotypes in natural populations of the house mouse (Mus musculus). PLoS Genet. 2012;8:e1002891. https://doi.org/10.1371/journal.pgen.1002891.

Stock F, Gifford-Gonzalez D. Genetics and African cattle domestication. Afr Archaeol Rev. 2013;30:51–72. https://doi.org/10.1007/s10437-013-9131-6.

Sun W, Yu H, Shen Y, et al. Phylogeny and evolutionary history of the silkworm. Sci China Life Sci. 2012;55:483–96. https://doi.org/10.1007/s11427-012-4334-7.

Sun W, Zhao X-W, Zhang Z. Identification and evolution of the orphan genes in the domestic silkworm, Bombyx mori. FEBS Lett. 2015;589:2731–8. https://doi.org/10.1016/j.febslet.2015.08.008.

Sykes NJ. The Norman conquest: a zoological perspective. Oxford: Archaeopress; 2007.

Tapio M, Marzanov N, Ozerov M, et al. Sheep mitochondrial DNA variation in European, Caucasian, and Central Asian areas. Mol Biol Evol. 2006;23:1776–83. https://doi.org/10.1093/molbev/msl043.

Tautz D, Domazet-Lošo T. The evolutionary origin of orphan genes. Nat Rev Genet. 2011,12:692–702. https://doi.org/10.1038/nrg3053.

Teasdale MD, van Doorn NL, Fiddyment S, et al. Paging through history: parchment as a reservoir of ancient DNA for next generation sequencing. Philos Trans R Soc Lond B Biol Sci. 2015;370:20130379. https://doi.org/10.1098/rstb.2013.0379.

Thalmann O, Perri AR. Paleogenomic inferences of dog domestication. In: Lindqvist C, Rajora OP, editors. Paleogenomics. Cham: Springer; 2018. p. 1–34.

Thalmann O, Shapiro B, Cui P, et al. Complete mitochondrial genomes of ancient canids suggest a European origin of domestic dogs. Science. 2013;342:871–4. https://doi.org/10.1126/science.1243650.

Titov VV. Habitat conditions for Camelus knoblochi and factors in its extinction. Quat Int. 2008;179:120–5. https://doi.org/10.1016/j.quaint.2007.10.022.

Triantaphyllopoulos KA, Ikonomopoulos I, Bannister AJ. Epigenetics and inheritance of phenotype variation in livestock. Epigenetics Chromatin. 2016;9:31. https://doi.org/10.1186/s13072-016-0081-5.

Troy CS, MacHugh DE, Bailey JF, et al. Genetic evidence for Near-Eastern origins of European cattle. Nature. 2001;410:1088–91. https://doi.org/10.1038/35074088.

Trut LN, Plyusnina IZ, Oskina IN. An experiment on fox domestication and debatable issues of evolution of the dog. Russ J Genet. 2004;40:644–55. https://doi.org/10.1023/B:RUGE.0000033312.92773.c1.

Trut L, Oskina I, Kharlamova A. Animal evolution during domestication: the domesticated fox as a model. Bioessays. 2009;31:349–60. https://doi.org/10.1002/bies.200800070.

Uauy C, Distelfeld A, Fahima T, et al. A NAC gene regulating senescence improves grain protein, zinc, and iron content in wheat. Science. 2006;314:1298–301. https://doi.org/10.1126/science.1133649.

Vernot B, Akey JM. Resurrecting surviving Neanderthal lineages from modern human genomes. Science. 2014;343:1017–21. https://doi.org/10.1126/science.1245938.

Vigne J-D. Zooarchaeological aspects of the Neolithic diet transition in the Near East and Europe, and their putative relationships with the Neolithic demographic transition. In: The Neolithic demographic transition and its consequences. Dordrecht: Springer; 2008. p. 179–205.

Vigne J-D. The origins of animal domestication and husbandry: a major change in the history of humanity and the biosphere. C R Biol. 2011;334:171–81. https://doi.org/10.1016/j.crvi.2010.12.009.

Vigne J-D. Early domestication and farming: what should we know or do for a better understanding? Anthropozoologica. 2015;50:123–50. https://doi.org/10.5252/az2015n2a5.

Vigne J-D, Guilaine J, Debue K, et al. Early taming of the cat in Cyprus. Science. 2004;304:259. https://doi.org/10.1126/science.1095335.

Vigne J-D, Carrère I, Briois F, Guilaine J. The early process of mammal domestication in the Near East: new evidence from the Pre-Neolithic and Pre-Pottery Neolithic in Cyprus. Curr Anthropol. 2011;52:S255–71. https://doi.org/10.1086/659306.

Vigne J-D, Briois F, Zazzo A, et al. First wave of cultivators spread to Cyprus at least 10,600 y ago. Proc Natl Acad Sci U S A. 2012;109:8445–9. https://doi.org/10.1073/pnas.1201693109.

Vigne J-D, Evin A, Cucchi T, et al. Earliest "Domestic" cats in China identified as leopard cat (Prionailurus bengalensis). PLoS One. 2016;11:e0147295. https://doi.org/10.1371/journal.pone.0147295.

Vilà C, Leonard JA, Götherström A, et al. Widespread origins of domestic horse lineages. Science. 2001;291:474–7. https://doi.org/10.1126/science.291.5503.474.

Volf J, Kus E, Prokopova L. General studbook of the Przewalski horse. Prague: Zoological Garden Prague; 1991.

von Holdt BM, Pollinger JP, Lohmueller KE, et al. Genome-wide SNP and haplotype analyses reveal a rich history underlying dog domestication. Nature. 2010;464:898–902. https://doi.org/10.1038/nature08837.

Walker EP. Mammals of the world, vol. 3. Baltimore: John Hopkins University Press; 1964.

Wallberg A, Han F, Wellhagen G, et al. A worldwide survey of genome sequence variation provides insight into the evolutionary history of the honeybee Apis mellifera. Nat Genet. 2014;46:ng.3077. https://doi.org/10.1038/ng.3077.

Wang G-D, Zhai W, Yang H-C, et al. Out of southern East Asia: the natural history of domestic dogs across the world. Cell Res. 2015;26:21. https://doi.org/10.1038/cr.2015.147.

Wang X, Liu J, Zhou G, et al. Whole-genome sequencing of eight goat populations for the detection of selection signatures underlying production and adaptive traits. Sci Rep. 2016;6. https://doi.org/10.1038/srep38932.

Warinner C, Rodrigues JFM, Vyas R, et al. Pathogens and host immunity in the ancient human oral cavity. Nat Genet. 2014;46:336–44. https://doi.org/10.1038/ng.2906.

Warinner C, Speller C, Collins MJ. A new era in palaeomicrobiology: prospects for ancient dental calculus as a long-term record of the human oral microbiome. Philos Trans R Soc Lond B Biol Sci. 2015;370:20130376. https://doi.org/10.1098/rstb.2013.0376.

Wendorf F, Schild R. Are the early holocene cattle in the eastern sahara domestic or wild? Evol Anthropol. 2005;3:118–28. https://doi.org/10.1002/evan.1360030406.

West B, Zhou B-X. Did chickens go North? New evidence for domestication. J Archaeol Sci. 1988;15:515–33. https://doi.org/10.1016/0305-4403(88)90080-5.

Weyrich LS, Dobney K, Cooper A. Ancient DNA analysis of dental calculus. J Hum Evol. 2015;79:119–24. https://doi.org/10.1016/j.jhevol.2014.06.018.

Wheeler JC, Chikhi L, Bruford MW. Genetic analysis of the origins of domestic South American camelids. In: Archaeology and animal domestication: new genetic and archaeological paradigms. Berkeley: University of California Press; 2006. p. 329–41.

White S. From globalized pig breeds to capitalist pigs: a study in animal cultures and evolutionary history. Environ Hist Durh N C. 2011;16:94–120. https://doi.org/10.1093/envhis/emq143.

Whitfield CW, Behura SK, Berlocher SH, et al. Thrice out of Africa: ancient and recent expansions of the honey bee, Apis mellifera. Science. 2006;314:642–5. https://doi.org/10.1126/science.1132772.

Wilkins AS, Wrangham RW, Fitch WT. The "domestication syndrome" in mammals: a unified explanation based on neural crest cell behavior and genetics. Genetics. 2014;197:795–808. https://doi.org/10.1534/genetics.114.165423.

Wood JR, Crown A, Cole TL, Wilmshurst JM. Microscopic and ancient DNA profiling of Polynesian dog (kurī) coprolites from northern New Zealand. J Archaeol Sci Rep. 2016;6:496–505. https://doi.org/10.1016/j.jasrep.2016.03.020.

Wright D. The genetic architecture of domestication in animals. Bioinform Biol Insights. 2015;9:11–20. https://doi.org/10.4137/BBI.S28902.

Wright D, Rubin C-J, Martinez Barrio A, et al. The genetic architecture of domestication in the chicken: effects of pleiotropy and linkage. Mol Ecol. 2010;19:5140–56. https://doi.org/10.1111/j.1365-294X.2010.04882.x.

Wu D-D, Ding X-D, Wang S, et al. Pervasive introgression facilitated domestication and adaptation in the Bos species complex. Nat Ecol Evol. 2018;2:1139–45. https://doi.org/10.1038/s41559-018-0562-y.

Xia Q, Guo Y, Zhang Z, et al. Complete resequencing of 40 genomes reveals domestication events and genes in silkworm (Bombyx). Science. 2009;326:433–6. https://doi.org/10.1126/science.1176620.

Xiang H, Li X, Dai F, et al. Comparative methylomics between domesticated and wild silkworms implies possible epigenetic influences on silkworm domestication. BMC Genomics. 2013;14:646. https://doi.org/10.1186/1471-2164-14-646.

Xiang H, Gao J, Yu B, et al. Early Holocene chicken domestication in northern China. Proc Natl Acad Sci U S A. 2014;111:17564–9. https://doi.org/10.1073/pnas.1411882111.

Yang H, Wang JR, Didion JP, et al. Subspecific origin and haplotype diversity in the laboratory mouse. Nat Genet. 2011;43:648–55. https://doi.org/10.1038/ng.847.

Yang S-Y, Han M-J, Kang L-F, et al. Demographic history and gene flow during silkworm domestication. BMC Evol Biol. 2014;14:185. https://doi.org/10.1186/s12862-014-0185-0.

Zeder MA. Archaeological approaches to documenting animal domestication. In: Zeder MA, Bradley DG, Smith BD, Emshwiller E, editors. Documenting domestication: new genetic and archaeological paradigms. Berkeley: University of California Press; 2006. p. 171–80.

Zeder MA. Animal domestication in the Zagros: an update and directions for future research. Publications de la Maison de l'Orient et de la Méditerranée. 2008;49:243–77.

Zeder MA. The domestication of animals. J Anthropol Res. 2012a;68:161–90. https://doi.org/10.3998/jar.0521004.0068.201.

Zeder MA. Pathways to animal domestication. In: Gepts PL, editor. Biodiversity in agriculture: domestication, evolution, and sustainability. New York: Cambridge University Press; 2012b. p. 227–59.

Zeder MA. Core questions in domestication research. Proc Natl Acad Sci U S A. 2015;112:3191–8. https://doi.org/10.1073/pnas.1501711112.

Zeder MA. Domestication as a model system for niche construction theory. Evol Ecol. 2016;30:325–48. https://doi.org/10.1007/s10682-015-9801-8.

Zeder MA, Hesse B. The Initial Domestication of Goats (Capra hircus) in the Zagros Mountains 10,000 Years Ago. Science. 2000;287:2254–7. https://doi.org/10.1126/science.287.5461.2254.

Zeder MA, Emshwiller E, Smith BD, Bradley DG. Documenting domestication: the intersection of genetics and archaeology. Trends Genet. 2006;22:139–55. https://doi.org/10.1016/j.tig.2006.01.007.

Zeuner FE. A history of domesticated animals. London: Hutchinson & Co. (Publishers) Ltd.; 1963.

Paleogenomic Inferences of Dog Domestication

Olaf Thalmann and Angela R. Perri

Abstract Domestication is the result of a complex interplay of both biological and cultural processes. The mechanism underlying today's variety of domesticates has long sparked the interest of researchers but has always been difficult to define. While dogs (*Canis familiaris*) are now firmly established as the earliest domesticated animal, most questions about their domestication are still unresolved, including the location, the timing, and the potential drivers of domestication. Genetic evidence accumulated over the last 20 years unequivocally identified *Canis lupus*—a gray wolf—as the species giving rise to all modern dogs and suggested locations including Eastern Asia, Central Asia, the Middle East, and Europe as potential domestication origins. Inferences about the timing of dog domestication are equally controversial. A date in the latest Upper Paleolithic, between 15,000 and 12,000 years ago, has long been the accepted timing of domestication due to clear archaeological evidence of morphologically distinct modern dogs by this time, and more recent genetic findings have confirmed the onset of dog domestication in the Late Pleistocene. In order to disentangle the complexity and thus derive a comprehensive understanding of dog domestication, we need to develop evolutionary models that include all available evidence from archaeology, morphology, and genetics. While time travel is still fiction and studying domestication at the moment of action impossible, paleogenomic approaches provide intriguing prospects and necessary

O. Thalmann (✉)
Department of Pediatric Gastroenterology and Metabolic Diseases, Poznan University of Medical Sciences, Poznan, Poland
e-mail: othalmann@ump.edu.pl

A. R. Perri
Department of Human Evolution, Max Planck Institute for Evolutionary Anthropology, Leipzig, Germany

Department of Archaeology, Durham University, Durham, UK
e-mail: angela.r.perri@durham.ac.uk

Charlotte Lindqvist and Om P. Rajora (eds.), *Paleogenomics*, Population Genomics [Om P. Rajora (Editor-in-Chief)], https://doi.org/10.1007/13836_2018_27,

means to decipher the many facets of domestication and deliver such evidence. Consequently, recent paleogenomic work has proposed a dual domestication process in Europe and Eastern Asia, which might be a first step toward reconciling some of the previous divergent conclusions.

Keywords Dogs · Domestication · Fossils · Genetics · Paleogenomics · Wolves

1 Introduction

Some 160 years ago, Charles Darwin published an essay on the origin of man's best friend, the dog (*Canis familiaris*). In his paramount work, *On the Origin of Species* (Darwin 1859), he proposed that the phenotypic variation of domesticated dogs was not explained by a sole, single ancestor and required other diversifying influxes. Prior to Darwin there was much discussion regarding the ancestral lineage of modern dogs, specifically whether dogs descended from wolves (*Canis lupus*) or the golden jackal (*Canis aureus*). Hunter and Mears (1787) proposed that the dog, wolf, and jackal should be considered a single species given that they produced fertile offspring, but Linnaeus (1758) believed the unique upturned tail of some dogs distinguished it as a separate species.

While Darwin proposed a multi-species ancestry of dogs (Darwin 1859), some of his peers were convinced that all modern dogs originated from a single species. Thomas Bell (1837, pp. 197–198) suggested that based on the identical gestation periods of dogs and wolves, dog domestication from a wolf ancestor was the most parsimonious explanation. This hypothesis was later communicated to Darwin by his colleague Charles Lyell, who was perturbed by Darwin concluding a single-species origin for the phenotypically varied pigeon but argued against a single-species origin with regard to the dog (University of Cambridge, The Darwin correspondence project; letters by C. Lyell (22.10. and 21.11.1859). Following his *Origin of Species*, Darwin concluded that we would probably never be able to ascertain dogs' true origins (Darwin 1859).

Konrad Lorenz (1954) later suggested that some dogs were descended from the wolf, while others were descended from the jackal, but changed this view upon studying the complex vocalizations of the jackal, which are distinctly unlike those in wolves or dogs (Lorenz 1975). Thus, the debate about a multi- or single-species origin of modern dogs became a central question in domestication research, being approached from a behavioral (Scott and Fuller 1965; Zimen 1981), morphological (Wayne 1986), archaeological (Davis and Valla 1978; Morey 1994; Clutton-Brock 1995; Koler-Matznick 2002; Raisor 2005; Perri 2016), genetic (Savolainen et al. 2002; Lindblad-Toh et al. 2005; Larson et al. 2012), and paleogenomic (Frantz et al. 2016; Botigué et al. 2017) standpoint, with each unequivocally supporting a single ancestral species hypothesis.

Fig. 1 Example depicting the phenotypic variation in dogs overlaying its ancestor. Photo credits: M. Katarzyńska, the Poznan animal shelter, and D. Schwochow

With the advent of and developments in genetic research, in particular the vast progress in paleogenomics, it became undisputable that a gray wolf is the sole ancestor of modern dogs (Fig. 1). In the following, we will review this evidence and further discuss the various hypotheses on the temporal and geographic origin of dogs resulting from recent paleontological, archaeological, genetic, and paleogenomic research.

2 Paleontological and Archaeological Background

2.1 Wolves in the Pleistocene

Given their gray wolf ancestry, the origin of domesticated dogs is couched in the paleobiogeography of Pleistocene wolf populations. An eastern Beringian origin of *Canis lupus* is, most likely, based on the earliest evidence of the species coming from the sites of Cripple Creek Sump (Alaska, United States) and Old Crow (Yukon, Canada) (Tedford et al. 2009). Though the geological attribution and dating is controversial (Repenning and Brouwers 1992; Tedford et al. 2009), this material may date back up to one million years ago (Ma). In Eurasia, *Canis lupus* appeared nearly simultaneously during the late Middle Pleistocene, including in Siberia [500–300 thousand years ago [ka] BP; (Sotnikova and Rook 2010)], France [400–350 ka BP; (Bonifay 1971; Brugal and Boudadi-Maligne 2011)], and Italy [340–320 ka BP; (Anzidel et al. 2012)], probably representing the origin of true modern gray wolves (Sardella et al. 2014). By the end of the Middle Pleistocene, gray wolves are found across all of Eurasia [e.g., (Kahlke 1994; Boeskorov and Baryshnikov 2013)].

By the Late Pleistocene, there was considerable morphological diversity among gray wolves, but they are generally considered more cranio-dentally robust than modern gray wolves, often with some specialized adaptations (e.g., shortened rostrum, pronounced development of the temporalis muscle, robust premolars) for carcass and bone processing (Kuzmina and Sablin 1993; Leonard et al. 2007; Baryshnikov et al. 2009) associated with megafaunal hunting and scavenging. Compared to modern wolves, some Pleistocene wolves show increased tooth breakage (Leonard et al. 2007), similar to that seen in extinct dire wolves (Binder et al. 2002; Binder and Van Valkenburgh 2010), suggesting they regularly processed carcasses and/or they increasingly competed with other carnivores within the same niche (Leonard et al. 2007; Meachen et al. 2016). Similarities in the frequency and location of tooth fractures in these wolves and spotted hyenas (*Crocuta crocuta*) suggest that they were habitual bone crackers (Leonard et al. 2007).

2.2 Transition to Domesticated Dog

More recently a number of specimens have been proposed as incipient domesticated dogs, predating the 15,000-year-old benchmark (Table 1), though these proposals have been met with significant skepticism [e.g., (Crockford and Kuzmin 2012; Drake et al. 2015; Morey and Jeger 2015; Perri 2016)]. These sites are all within Europe or southern Siberia and date from c. 40,000 to 17,000 years ago. They include a site in Germany [Hohle Fels; (Camarós et al. 2016)], a site in Belgium [Goyet Cave; (Germonpré et al. 2009)], a site in the Czech Republic [Predmosti; (Germonpré et al. 2012, 2015a)], and four sites in Russia (Sablin and Khlopachev

Table 1 Proposed incipient Paleolithic dogs

Site	Date (B.P.)	Location	Reference
Hohle Fels	40,000–35,000	Germany	(Camarós et al. 2016)
Goyet Cave	36,500	Belgium	(Germonpré et al. 2009)
Razboinichya Cave	33,500	Russia	(Ovodov et al. 2011)
Kostenki 8	33,500–26,500	Russia	(Germonpré et al. 2015b)
Predmosti	31,000	Czech Republic	(Germonpré et al. 2012, 2015a)
Ulakhan Sular	17,200	Russia	(Germonpré et al. 2017)
Eliseevichi 1	17,000–16,000	Russia	(Sablin and Khlopachev 2002)

2002; Ovodov et al. 2011; Germonpré et al. 2015b, 2017). There has also been suggestion of domesticated dog paw prints from Chauvet Cave in France [c. 26,000 years ago, (Garcia 2005); but see (Ledoux and Boudadi-Maligne 2015)].

While the taxonomy of these much earlier proposed Paleolithic dogs is contentious, there are also a number of later Paleolithic proposed dogs whose taxonomic status has not been satisfactorily confirmed since their initial identification. These include a number of specimens from Germany [Kniegrotte, Oelknitz, Teufelsbrucke; (Morey 2014; Musil 2000; Benecke 1987; Housley et al. 1997; Napierala and Uerpmann 2012)], Switzerland [Monruz, Kesslerloch, Champre-veyres-Hauterive; (Napierala and Uerpmann 2012; Müller 2004, 2006, 2013; Morey 2014)], and Ukraine [Mezin, Mezhirich; (Napierala and Uerpmann 2012; Pidoplichko 1998; Germonpré et al. 2009; Morey 2014; Morey and Jeger 2015)]. A set of specimens dating to between 15,000 and 13,500 years ago from France [Montespan, Le Morin, Le Closeau, Pont d'Ambon; (Pionnier-Capitan et al. 2011; Morey 2014; Boudadi-Maligne et al. 2012)], Spain [Erralla; (Vigne 2005)], and Germany [Bonn-Oberkassel; (Benecke 1987; Street et al. 2015; Street 2002)] are among the first confidently identified domesticated dogs, based on their distinct morphology and archaeological contexts (coburied with humans in the case of Bonn-Oberkassel). From this period onward, the remains of domesticated dogs are identified regularly at archaeological sites from across Eurasia [e.g., (Dikov 1996; Turnbull and Reed 1974; Lawrence and Reed 1983; Tchernov and Valla 1997; Losey et al. 2013)].

2.3 *Limitations of Morphological Inferences*

As seen from a number of proposed "Paleolithic dog" examples (Table 1) and the resulting debate over these specimens, the time period between 40,000 and 15,000 years ago has been difficult for domestication researchers to illuminate and remains a gray area in dog domestication. In part, this is due to the intrinsic morphological similarities between *Canis lupus* and *Canis familiaris* and a general morphological flexibility in the *Canis* genus, as demonstrated by the morphological range of modern dog breeds. This is also due, in part, to an underrepresentation of Pleistocene wolf specimens in morphological analyses, leaving the range of Pleistocene wolf variation largely unexplored (Perri 2016). Without a baseline of morphological variation in the source wolf population, it is difficult for zooarchaeologists to differentiate between varied Pleistocene wolf ecomorphs and the initial indicators of dog domestication, leading to potentially incorrect identifications of both dogs and wolves in the early record. Ecological factors such as habitat type, climate, and prey specialization greatly affect the morphological plasticity of gray wolves (Geffen et al. 2004; Musiani et al. 2007; Flower and Schreve 2014), resulting in a range of morphologically, genetically, and ecologically distinct wolf morphotypes within Pleistocene gray wolf populations. In addition to this, a potentially prolonged process of domestication and ongoing prehistoric admixture with local wolves may have led to canids, which were behaviorally domesticated but morphologically wolflike in their skeletal appearance. Identifying these earliest iterations of tamed wolves, wolf dogs, or proto-dogs through the morphological analysis of canid faunal remains at archaeological sites may ultimately prove impossible without the incorporation of genetic analyses with morphological techniques.

2.4 The How and Why of Dog Domestication

Further complicating the issues surrounding dog domestication is the uncertainty about the events leading to or drivers of domestication. While much research effort has been spent on investigating the question of the *where* and *when* in dog domestication (see below), relatively little attention has been paid to the *how* and *why*. Extant hypotheses of dog domestication largely fall under three models: a hunting partnership, wolf pup adoption, and trash pile scavenging.

A hunting partnership between wolves and humans as a catalyst for dog domestication has long been suggested (Scott and Fuller 1965; Downs 1960). The basis of this hypothesis is most often cited as a similar social and ecological framework between humans and wolves—communal living and hunting, shared care of offspring, and daylight hunters who work together to target prey larger than themselves. This pathway to domestication was recently renewed by Shipman, who proposed wolves may have been domesticated by Paleolithic hunter-gatherers as megafaunal hunting tools (Shipman 2015). She proposes that co-hunting led to mutualistic benefits to both humans and wolves, including increased nutrition and population growth for humans and increased meat intake and decreased risk of injury for wolves (Shipman 2015).

The habit of recent hunter-gatherers regularly capturing and taming pet animals likely resulted in the widespread concept that this relationship led to the domestication of some animals (Cuvier 1817; Darwin 1868; Galton 1883), including the dog (Reed 1959; Zeuner 1963; Clutton-Brock 1981; Morey 1994). Proponents of this domestication hypothesis suggest that prehistoric hunter-gatherers must have come across wolf pups regularly enough to not only capture them but to raise them to adulthood and breed them, thus initiating the process of taming and domestication. Yet, while the adoption of wild pets, be it the orphaned offspring of hunting game or those taken directly from dens or nests, has long been a practice of many hunter-gatherer groups (Serpell 1989), none of these have resulted in domestication.

Finally, the most recent proposal for dog domestication comes in the form of the trash pile or town dump hypothesis. Originally proposed by Coppinger and Coppinger (2001), this model suggests that dog domestication is the result of an unintentional commensal relationship between wolves and humans, whereby wolves fed off the disposed subsistence remains of prehistoric populations, leading to an eventual domestic relationship between the two (Coppinger and Coppinger 2001). While Coppinger and Coppinger (2001) grounded their model on a relationship between wolves and settled agriculturalists, the domestication of dogs predating agriculture (Frantz et al. 2016) necessitates this model to be adjusted to accommodate wolves and prehistoric hunter-gatherer populations [e.g., (Morey and Jeger 2015; Zeder 2015)].

3 Genetic Evidence

Rapid advances in molecular technologies allowed us to develop an intimate view on the genetic composition of dogs, and early microscopical observations have now been replaced by comprehensive reconstructions of full genomes including those from fossils.

While the karyotypes of all *Canis* species are identical (Wurster-Hill and Centerwall 1982) and consist of 78 chromosomes with 2 × 38 acrocentric autosomes and two metacentric sex chromosomes, early comparisons of proteins, i.e., allozymes, identified the highest similarity between dogs and wolves among all of *Canis* (Wayne and O'Brien 1987). With the discovery of more informative, highly variable markers, such as short tandem repeats or microsatellites (Ellegren 2004), the close relationship between dogs and wolves became more apparent, a fact manifested by higher sharing of genetic variants between them compared to any other *Canis* pair (Roy et al. 1994; Garcia-Moreno et al. 1996).

3.1 *Mitochondrial DNA Studies*

Further development in DNA sequence analyses led to the first investigation of mitochondrial genome variation. One milestone in dog domestication research was a publication by Vilà et al. (1997). The authors investigated the genetic variation in the mitochondrial control region, the nonprotein-coding segment of the molecule, of 162 wolves sampled worldwide and 140 modern dogs of various breeds. Aside from the confirmation of a gray wolf ancestry, this study revealed other intriguing insights into dog domestication. The phylogenetic arrangement of the mitochondrial sequences generated in the study [Fig. 2a in Vilà et al. 1997)] shows that sequences derived from dogs cluster into four major clades without any breed specificity. The

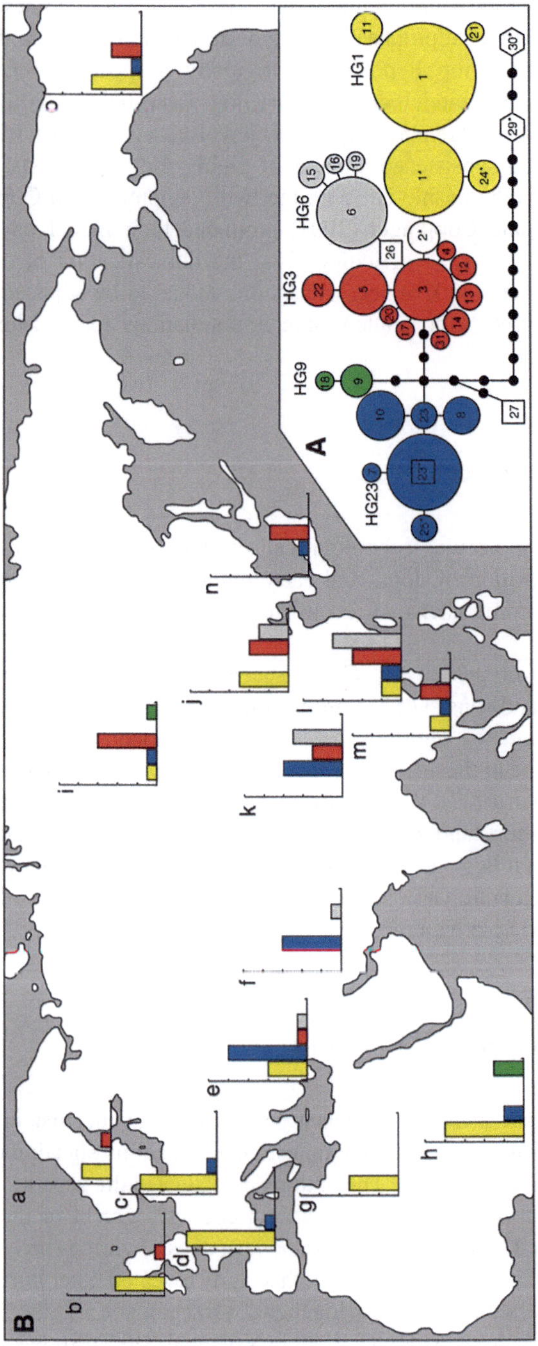

Fig. 2 Frequency and distribution of Y-chromosome variants throughout the world as illustrated in Ding et al. (2012). Different colors indicate major haplogroups. (**a**) Relationship of the major five clusters of Y-chromosome variation. (**b**) Global distribution of the respective variants

biggest clade [clade I in Fig. 2a in (Vilà et al. 1997)] exclusively contained dog sequences, and the divergence within this clade was estimated to over ~100,000 years ago. It needs to be pointed out that divergence times do not reflect population split times and should thus be treated with caution whenever inferring the timings of population separations [e.g., (Hudson 1992; Nichols 2001; Ruvolo et al. 1994)]. However, it was the first time that scientists had put a potential timing on dog domestication and was overwhelmingly welcomed. This date marked the dog as the oldest domestic animal species, but subsequent research (see below) found the estimate to be too old and suggested a much later timing, albeit still predating the domestication of other animals (Larson and Burger 2013). The agglomeration of dog and wolf sequences in two other clades could have been attributed to a phenomenon called incomplete lineage sorting (ILS) or might have indicated occurrences of admixtures between the two taxa after the initial domestication. Both aspects have hampered genetic investigations and will be discussed in more detail below.

While the spatially limited sampling of wolves in Vilà et al. (1997) did not allow phylogenetic inferences on the geographical origin of dogs, a subsequent study (Savolainen et al. 2002) addressed this issue in more depth. The authors concentrated on mitochondrial DNA variation in a large sampling of dogs ($n = 654$) and wolves ($n = 38$). Since the genetic variation of dogs worldwide reflected a subset of that prevalent in Southeast Asia today, Savolainen et al. (2002) placed the origin of modern dogs in this region. The authors further concluded that dogs derived from a single gene pool, consistent of at least five divergent female wolf lineages and that the size of population expanded following its initial domestication and starting at approximately 15,000 years ago (Savolainen et al. 2002). The timing in conjunction with the proposed geographic origin of modern dogs posed some conflicts with the fossil record, however, as some putative proto-dogs predating this suggested onset of dog domestication were found at several archaeological sites throughout Eurasia (Germonpré et al. 2009; Sablin and Khlopachev 2002; Nobis 1979). Consequently, a more complex scenario of dog domestication was needed in order to account for and incorporate all available evidence (Larson et al. 2012).

Nonetheless, these first genetic studies provided evidence for an ancestry of all domesticated dogs from *Canis lupus* (Fig. 1), placing the onset of its domestication within the late Pleistocene and potentially in Southeast Asia. Given the Holarctic distribution of the dog's ancestor, it seems logical to hypothesize that regional dog breeds might have originated from local variants of gray wolves, but genetic investigations have often revealed alternative scenarios. Examples include the dogs of the New World, Australia, and Madagascar.

America When investigating the origin of dogs from the Americas, Leonard et al. (2002) tested whether they derived from an independent domestication of a local gray wolf population or if they migrated there accompanying humans (Leonard et al. 2002). Bearing in mind the issue of recent admixture between and potential replacement of native dogs by Old World dogs brought by Europeans, the authors, for the first time, concentrated their research on the remains of pre-Columbian dogs and sequenced a small fragment of the mitochondrial genome. The majority of the ancient sequences fall within the most diverse dog clade, and no American wolf sequence was closely related to any of the ancient sequences, leading to the conclusion that New and Old World dogs share the same ancestry and that the first dogs followed humans into the Americas via Beringia (Leonard et al. 2002; Nielsen et al. 2017). This result was further confirmed by analyses of an elaborate sample of modern dogs (van Asch et al. 2013) and also by a study investigating the variation prevalent in ancient canids (Thalmann et al. 2013).

Australia The Australian dingo is considered to be a basal dog breed, and its history has sparked some controversy. Genetic investigations targeting the mitochondrial genome have revealed that the dingo derived from East Asian dogs arriving in Australia with the expansion of the Austronesian culture ~6000 years ago (Savolainen et al. 2004). A follow-up study confirmed a Southeast Asian origin but further refined the timing and the source of Australian dingoes and Polynesian dogs (Oskarsson et al. 2011). Based on the relationships between mitochondrial sequences, the authors concluded that contrary to traditional beliefs, these indigenous dogs derived from an introduction of dogs from Indonesia through the mainland Southeast Asia, which was estimated to have happened before the Neolithic period, over 18,000 years ago.

Madagascar The dogs of Madagascar were believed to mirror the Indonesian ancestry of the first settlers of the island [e.g., (Hurles et al. 2005)]. However, since 90% of the dogs in Madagascar carried mitochondrial sequences prevalent in sub-Saharan Africa and only 26% of the dogs shared variation of Indonesian ancestry but none carried the sequence typical for 40% of all Polynesian and Indonesian dogs, Ardalan et al. (2015) concluded that the ancestry of Madagascan dogs is African.

The mitochondrial DNA studies exemplified the usefulness of genetic markers located on the mitochondrial genome to infer the evolutionary history of dogs. However, its inheritance mode, being exclusively transmitted through the maternal line, limits the scope of the interpretation of the results and exclusively provides the female side of the history. In addition, owing to the mutation rate, the mitochondrial DNA only offers limited resolution for inferring recent demographic events.

3.2 *The Y-Chromosome*

While the mitochondrial genome reflects the history of female lineages in a population, the Y-chromosome provides the population history of the male lineage. As such they can complement each other for a more comprehensive picture of a species' evolutionary history.

The first analyses of Y-chromosome variation in worldwide samples of dogs supported a Southeast Asian ancestry of modern dogs (Ding et al. 2012). The authors analyzed a ~15,000 bp fragment of the Y-chromosome in 151 dogs, 12 wolves, and 2 coyotes. The results confirmed the ancestry of dogs from wolves rather than coyotes as evident by a closer genetic distance of dog and wolf Y-chromosome sequences. Furthermore, since all dog sequences assembled into five groups (Fig. 2a), it can be assumed that modern dogs descended from multiple patrilines of wolves. More precisely, considering factors such as mutational biases, the authors concluded that 13–24 different Y-chromosome lineages gave rise to today's variation in dogs. While most of the Y-chromosome variation was globally shared (Fig. 2b), only East and Southwestern Asia harbored the full panel of Y-chromosome lineages. Ding et al. (2012) argued that the region south of the Yangtze River contained more genetic variation compared to other potential regions of dog origin, such as Europe and Southwest Asia (13, 6.5, and 9.58 out of 23 sequences, respectively), and provided further evidence in support of an Asian ancestry of modern dogs [see also (Pang et al. 2009)].

Despite the specific inheritance modes unique to the mitochondrial and Y-chromosome markers, a congruency in the results has been demonstrated whenever inferring the history of dogs. For instance, Brown et al. (2011) investigated both markers in ~600 modern dogs with a high proportion originating from villages in the Middle East and Southeast Asia. Contrary to purebred dogs, village dogs are thought to resemble some of the original features of the first dogs, being unaffected by strong artificial selection (closed breeding programs), and do therefore represent a more ancestral state. The study by Brown et al. (2011) confirmed higher diversity in village dogs from Southeast Asia compared to the Middle East, independently supporting the hypothesis of a dog origin in this region [see (Savolainen et al. 2002; Pang et al. 2009)] and also providing some intriguing insights into the ancestry of purebred dogs. While both Middle Eastern and East Asian dog breeds clustered with Middle Eastern and Southeast Asian village dogs, respectively, European and American dog breeds did not cluster with geographically proximate village dogs but instead with Southeast Asian ones. The authors concluded that this reflects an extensive influence of an "exotic" stock in modern European and American purebred dogs (Brown et al. 2011).

In a third study investigating Y-chromosome variation in modern dogs, Sacks et al. (2013) proposed an alternative hypothesis of dog domestication involving a dramatic population expansion of dogs out of Southeast Asia into the North and West and thereby replacing indigenous lineages. As an advance over the two previously discussed studies (Ding et al. 2012; Brown et al. 2011), Sacks et al. (2013) applied a more precise mutation rate model and thereby assigned a date to specific demographic events in dog history. This led the authors to reject the hypothesis of a sole origin of dogs in Southeast Asia. Their logic was the following: the estimated age of the European sequence group was some 8400–5800 years old (depending on the calibration point used), and thus connection between pre-Victorian European and Southeast Asian dogs falls within the Neolithic. This young age contradicts a single origin in Southeast Asia some 15,000 years ago and rather supports the massive Neolithic expansion from Southeast Asia. Furthermore, the authors proposed a scenario in which dogs underwent significant diversification in Southeast Asia, and such changes might have transformed them and allowed them to accompany humans on their migrations as beasts of burden or valuable trade objects (Sacks et al. 2013; Germonpré et al. 2009; Coppinger and Schneider 1995).

In summary, investigations of the Y-chromosome confirmed gray wolves as the sole ancestor of dogs and their East Asian ancestry but also raised concerns and provided an alternative scenario involving intensive diversification and expansion of dogs from Southeast Asia, thereby replacing indigenous forms worldwide during the onset of the Neolithic.

3.3 The Dog Genome Project and Insights into Dog Domestication from Whole-Genome Studies

3.3.1 The Dog Genome Project and Genome-Wide Analyses

As molecular methods became more sophisticated, comprehensive genomic information was made available (Kirkness et al. 2003), cumulating in the release of the first draft of the dog genome (Lindblad-Toh et al. 2005). This genetic resource opened new avenues for research aiming at both the understanding of dog domestication and also the establishment of the dog as a model species.

The actual genome paper was centered on an in-depth description of the genome's peculiarities, especially in relation to other mammalian genomes, and the analyses revealed many intriguing features.

- *Functionally related genes show similar patterns in humans and in dogs*—a fact giving insights into the coevolution of gene function.
- *The linkage of genes within breeds can extend several megabases, whereas between breeds linkage breaks down after tens of kilobases*—indicating two principle bottlenecks in the history of dogs.
- *Long haploblocks are shared between breeds*—hence, genetic risk factors might overcome breed boundaries.
- *Generation of a catalogue of >2 million variants (SNP) made available to the research community*—a valuable resource for future genetic studies.

The pool of genetic variants generated during the dog genome project served as a basis for investigating the potential domestication history of the dog in unprecedented depth. One influential paper surveying the genome-wide distribution of ~50,000 variants (SNP) in modern dogs and wolves proposed yet another geographical origin of dogs—the Middle East (von Holdt et al. 2010). The authors observed a higher similarity of genetic variants constituting unique haplotypes (variants located at the same chromosome) between dogs and wolves from the Middle East compared to wolves from other geographical regions (Fig. 3) previously considered the cradle of dog domestication. Almost all breeds shared significantly more haplotypes with Middle Eastern wolves in the shorter 5-SNP window (five consecutive variants), and surprisingly, wolves from Europe and not from Asia (China) shared the second

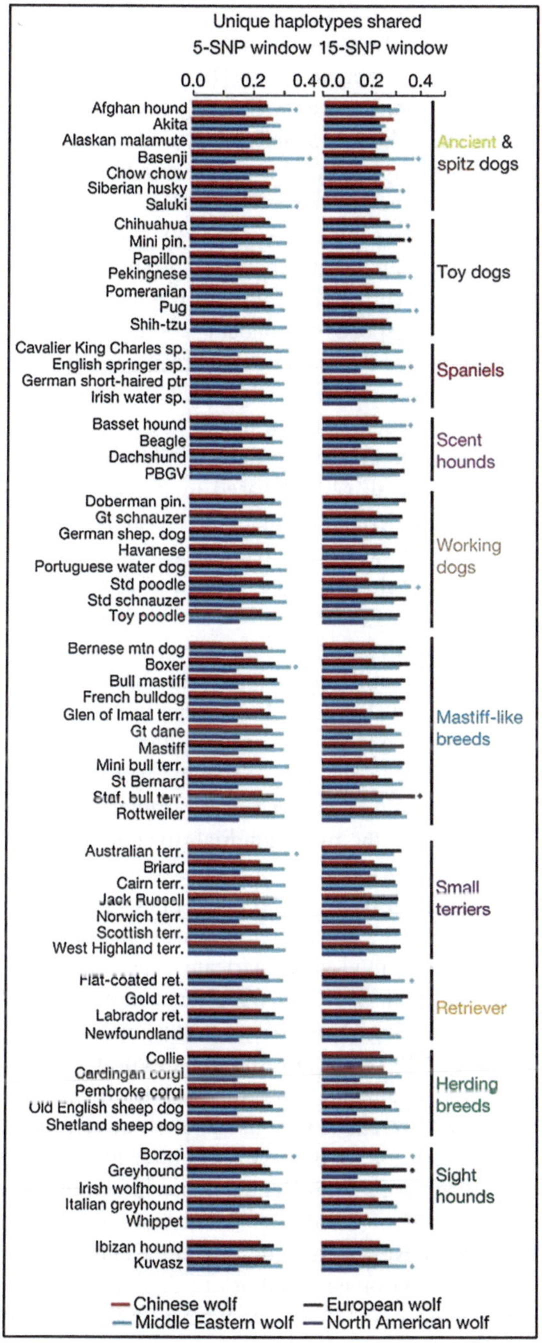

Fig. 3 Haplotype sharing between modern dogs and wolves from different regions of the world (after von Holdt et al. 2010). The longer the bar, the more variation is shared between each dog breed and the respective wolf population (as labelled in the legend). Two genomic windows were considered: the smaller five consecutive variant (5-SNP) regions and the larger one spanning 15 consecutive variants (15-SNP window). Modern dogs are also clustered according to breed groups

highest number of haplotypes with modern dogs, challenging previous conclusions derived from the work on mitochondrial (Savolainen et al. 2002; Pang et al. 2009) and Y-chromosome markers (Ding et al. 2012; Brown et al. 2011). An exception to this pattern are some ancient Asian dog breeds exhibiting higher affinity to Chinese wolves in the larger 15-SNP window (15 consecutive variants) and suggesting either an ancestry of those in this region or admixture after initial domestication. Furthermore, the large number of SNPs analyzed in the study allowed von Holdt et al. (2010) to assess the breed specificity of variants [see also (Parker et al. 2004)], and the authors found that almost all investigated dogs could be genetically assigned to their respective breed. In light of dog domestication, this result points to a limited number of respective founders, continuous inbreeding, and thus small effective population size of the many dog breeds. Lastly, divergent lineages of dogs identified based on the genome-wide haplotype composition may represent old variants persisting in today's breed pool.

Despite incorporating "ancient" dog breeds, such as the basenji and the dingo, von Holdt et al. (2010) based their conclusions mainly on data generated from purebred, modern dogs, a fact often criticized (see Sect. 3.4). Furthermore, evidence exists pointing to an admixed ancestry of Middle Eastern wolves and dogs (Fan et al. 2016). The inclusion of such specimens would camouflage the genetic composition of these wolves as being closely related to dogs and hence mislead implications on the origin of dogs.

In a study utilizing genome-wide SNP, microsatellite, as well as mitochondrial data, Boyko et al. (2009) shed further light on dog domestication by investigating the genetic variation in African village dogs (Boyko et al. 2009). The study revealed that African village dogs are genetically distinct from nonnative and mixed-breed dogs, hence suggesting an older ancestry rather than an admixed version of modern dogs. By comparing the mitochondrial diversity of the African village dogs to estimates from East Asia (Savolainen et al. 2002), Boyko et al. (2009) found a similar amount of variation in these two regions indicating a global distribution of mitochondrial variation and thereby questioning a single origin of modern dogs in East Asia. While this study was focused on African village dogs, a follow-up study by the same group (Shannon et al. 2015) enlarged the sampling (4676 modern dogs from 161 breeds, 549 village dogs from 38 countries) and extended the genetic dataset to almost 200,000 SNPs on the autosomes, sex chromosomes, and the mitochondrial genome, thereby providing an unprecedented depth of analyses. Given the nature of the data, the authors were able to consider alternative genetic estimators, such as the decay of linkage between markers (LD) to infer the evolutionary history of dogs. It appears that village dogs from Central Asia show a more rapid decay of LD than any other village dog population (Fig. 4), suggesting a likely origin in this region. In support of this hypothesis, village dogs from Central Asia show minimal admixture with European dog breeds contrary to dogs from the Neotropics and the South Pacific, which almost exclusively derived from European breeds.

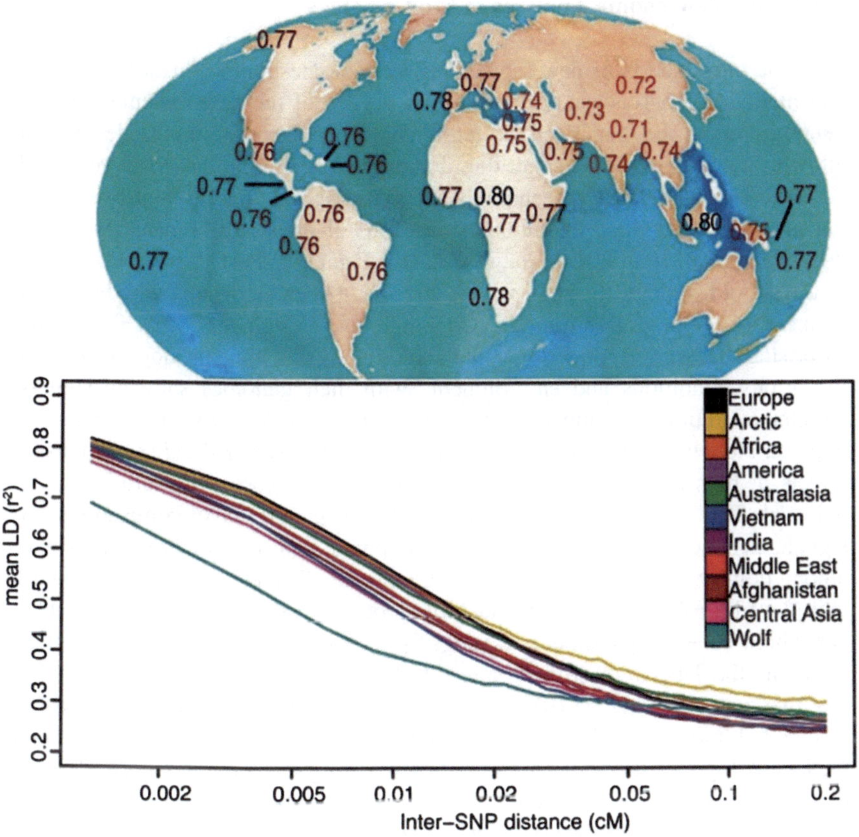

Fig. 4 Global distribution of linkage disequilibrium values (LD) and its decay between markers spaced throughout the genome. Figure adapted from Shannon et al. (2015). The world panel shows exact measures of LD in accordance to geographic location of dogs and should be interpreted with lower numbers indicating higher variation in the respective region. The lower panel highlights the genomic variation as measured with LD over a specific genomic distance and indicates higher variation whenever mean LD is lower

In summary, evidence from pre-genomic studies have suggested an origin of modern dogs in Southeast Asia (Savolainen et al. 2002; Pang et al. 2009; Ding et al. 2012; Brown et al. 2011), the Middle East (von Holdt et al. 2010), and Central Asia (Shannon et al. 2015) and place the timing to either the Middle (Vilà et al. 1997) or Late (Savolainen et al. 2002) Pleistocene or possibly as recent as the Neolithic in conjunction with the advent of farming in modern human societies. The latter timing has been hypothesized since dogs possess a higher copy number of the amylase gene and might thus be better adapted to a starch-rich diet [but see below, (Axelsson et al. 2013)].

3.3.2 Whole-Genome Analyses of Modern Dogs

With the introduction of new sequencing technologies and sophisticated analytical and modelling approaches during the first decade of the new millennia, complete genomes could be generated at reasonable costs and at unprecedented speed allowing us to better understand dog domestication. Within the last 5 years, influential genome papers have been published shedding light on this controversial topic but also the intertwined evolutionary histories of dogs and humans (Wang et al. 2013; Freedman et al. 2014; Gou et al. 2014; Fan et al. 2016).

As the oldest domesticated species, dogs have been associated with humans and their respective migration routines (Coppinger and Coppinger 2002). Thus, it is not surprising that while conquering new habitats alongside humans, dogs have adapted to various conditions and environments with their genomes showing signals of parallel evolution. In humans, one such signal is left at loci encoding for the hypoxia-inducible factor pathway with its major player *EPAS1*, which exhibits effects in high-altitude adapted populations [e.g., (Alkorta-Aranburu et al. 2012; Xu et al. 2011)]. Similarly, the genomes of their canine travel companions reveal signatures of selection in accordance with hypoxia. In a survey of genomic variation among dog breeds adapted to life at high altitudes, Gou et al. (2014) identified loci that played a role in shaping the adaptive processes in such habitats. They investigated high-coverage genomes of 60 dogs from six breeds living at different altitudes, including the Tibetan mastiff. As a result of these genome scans, the authors found that two genes (*EPAS1*, *HBB*) showed high differentiation among the breeds from different altitudes as well as reduced genetic diversity and increased LD surrounding the loci. Altogether this indicates the action of strong selection pressure and suggests that these genomic areas played a vital role in adaptation to hypoxic environments paralleled in the two species (Gou et al. 2014).

Similarly, Wang et al. (2013) have provided an extensive list of genes that showed signatures of parallel evolution in dogs and humans in a survey performed on genomes of four wolves, three indigenous Chinese dogs, and three representatives of modern breeds. The authors detected a suite of 311 genes under positive selection in dogs with a large number of overlapping loci showing the same patterns in humans. Among those candidates are genes playing a crucial role in metabolism and digestion, neurological processes, and some being involved in cancer (Wang et al. 2013). This finding in particular highlights the usefulness of the dog as a model organism whenever inferring the evolutionary trajectory of diseases or behavioral traits, such as the reduction of aggressive behavior while living in crowded environments, as indicated by the selection acting upon genes of the serotonin system. The authors further drew conclusions on the demographic history of dogs and wolves. Contrary to previous studies (Lindblad-Toh et al. 2005; von Holdt et al. 2010), they described almost the same effective population size of extant wolves compared to ancestral ones (94%) and detected only a mild bottleneck within dogs and more specifically Chinese indigenous dogs. An estimated split time between dogs and wolves of 32,000 years ago proposed an onset of dog domestication in the

Late Pleistocene. However, the authors also found signatures of moderate gene flow in both directions (Fig. 5a) following this split. While the data from Wang et al. (2013) supported an East Asian origin of dogs, Freedman et al. (2014) derived to another conclusion in their genome comparison of dogs and wolves (Fig. 5b). The authors employed high-quality genomes newly generated from three wolves originating from the putative geographical centers of dog domestication, Asia, the Middle East, and Europe, as well as representatives of "ancient" dog lineages (dingo and basenji) and the publicly available dog genome (Lindblad-Toh et al. 2005). Based on this data, they tested different domestication scenarios, and the model with the best support involved population bottlenecks in both lineages as well as subsequent gene flow (Fig. 5b). In contrast to the shallow bottleneck suggested by Wang et al. (2013), Freedman's group found evidence for a 16-fold population size reduction in association with the domestication bottleneck (Freedman et al. 2014). Moreover, the two studies deviated with respect to their inferences about the timing of domestication and the putative geographic origin of dogs, with Freedman et al. (2014) estimating the former to have occurred between 16,000 and 11,000 years ago and not finding any of the wolves from the three regions being more closely associated with dogs. The authors concluded that the wolf population potentially being ancestral to all modern dogs might have gone extinct [see also (Thalmann et al. 2013)]. Freedman et al. (2014) further tested the hypothesis that dogs might have originated as a result of the agricultural revolution (Axelsson et al. 2013). Axelsson et al. (2013) based their hypothesis on the finding that dogs have an increased copy number of the amylase gene (*AMY2B*) and are thus better adapted to a starch-rich diet. Whereas no doubt exists about the evolutionary advantage of an ability to digest starch for dogs living in an agriculturally dominated environment, the question remains, if this was a causative effect. Freedman et al. (2014) investigated the copy number variation of *AMY2B* in the analyzed genomes as well as additional breeds and wolves. The authors found that contrary to Axelsson's study, wolves were polymorphic with 16 of 40 tested wolves having more than two copies of *AMY2B* and dingoes only carrying two copies, thus questioning the conclusions drawn by Axelsson et al. (2013). As this pattern could have arisen from admixture between dogs and wolves, further evidence against the hypothesis of Axelsson et al. (2013) was presented by investigations of Neolithic dogs (Botigué et al. 2017). Two of the analyzed ancient dogs showed only two copies of the gene, whereas a third had an additional copy, most likely generated by a large-scale, segmental duplication rather than an increase in copy number. This means that low copy number pattern of *AMY2B* persisted in dogs through time and questions its role as a driving force in dog domestication.

These genomic studies together with the archaeological record put the origin of dogs into a hunter-gatherer context sometime between 34,000 and 9000 years ago (with mutation rate uncertainty). Yet, the quest for the geographic origin of dogs remains unsolved. Surprisingly, not even a comprehensive study combining all canid genomes available at the time (34 in total) resulted in an unequivocal answer and led the authors to conclude, again, that dogs derived from an extinct wolf population (Fan et al. 2016).

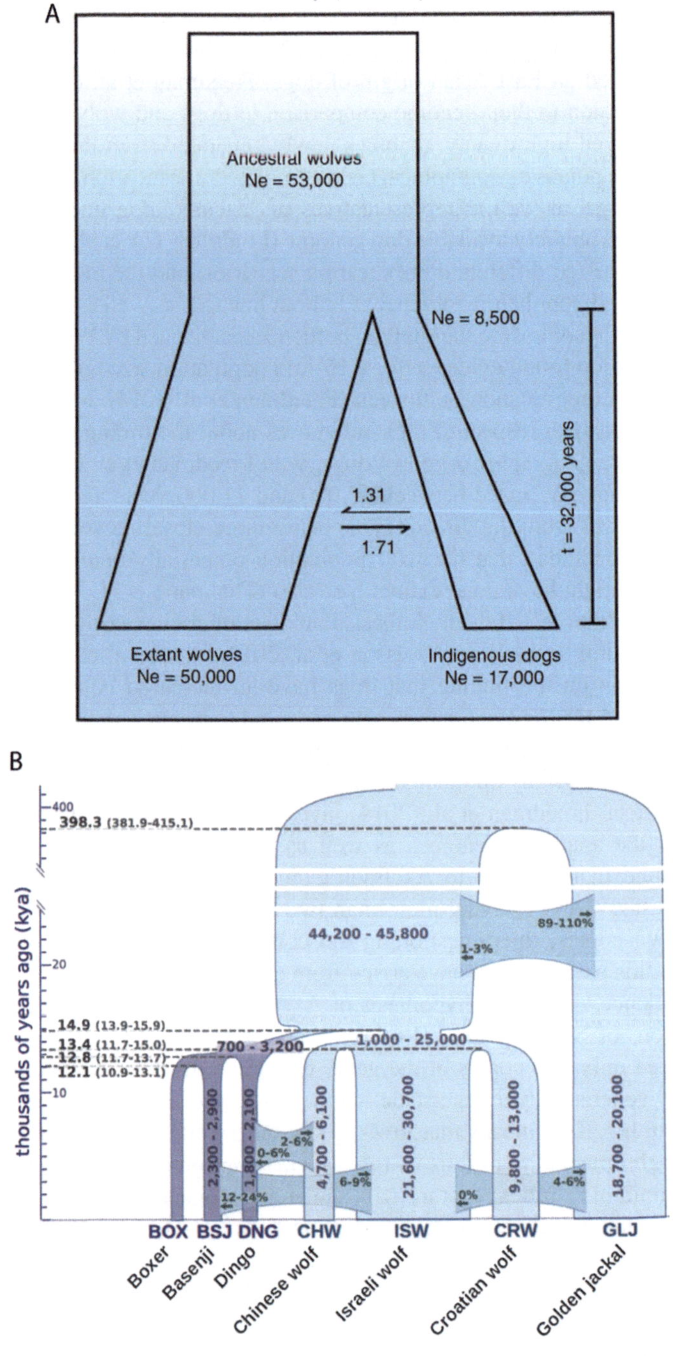

Fig. 5 Two models depicting the demographic histories of dogs and wolves and highlighting potential domestication scenarios. Adapted from Wang et al. (2013) and Freedman et al. (2014). Panel (**a**) shows a more simplified model of dog evolution, whereas panel (**b**) depicts many parameters and specifies breeds as well as wolf populations. Numbers within bars indicate the effective size of the respective populations

3.4 Critical Remarks on Genetic Inferences from Modern Dogs

The majority of genetic studies performed over the last decades included contemporary dog breeds and extant wolf populations. Despite the intriguing results and important insights of such studies, we need to keep in mind some of the limitations with this approach, when inferring the origin and evolutionary history of domesticated dogs (Freedman and Wayne 2016). For instance, analyses of genetic data are based on simplified assumptions (i.e., random mating, non-overlapping generations, etc.) that allow us to compare the patterns of genetic variation observed in natural populations with those of an idealized one and derive conclusions regarding the respective evolutionary forces that shaped the observed patterns. The closer these assumptions are met, the more robust the inferences become. However, often these assumptions can't be fulfilled, and we ought to be cautious of the interpretations, implications, and derived conclusions.

3.4.1 Genetic Diversity and Phylogenetic Argument

In natural populations it often holds true that the population with the highest genetic variation constitutes the ancestral one, such as sub-Saharan Africans in comparison to other human populations (Tishkoff et al. 2009). This is simply owed to the fact that more genetic variation has been accumulated in older populations as they had more time to differentiate.

In the case of dogs, researchers have identified the geographic origin of the most diversified dog population and acknowledged this as the ancestral one (Savolainen et al. 2002; Ding et al. 2012). A priori there is nothing wrong with this, but unlike other wild animals (including our ancestors), dogs do not behave like ideal, natural populations. Dogs have been domesticated by humans and living primarily in their vicinity ever since (Coppinger and Coppinger 2002). This means that their dispersal patterns, and thus gene flow, have been heavily impacted by intense anthropogenic pressure ultimately leading to strict breeding schemes. While artificial selection caused by humans has tremendously diversified dogs, at the same time, it has blurred the genetic signatures of their ancestry. For instance, early dogs likely followed humans on their migration routes, thereby admixing and hence diluting patterns of local variation (Malmström et al. 2008), which is an unlikely scenario deriving from looking exclusively at modern dogs. Consequently, other mechanisms than mere ancestry (such as introgression of breeds, human migration routes, and trade) have to be considered whenever interpreting higher genetic diversity in specific regions and ascertaining ancestry. Being the first domesticated animal has created a strong bond between dogs and humans thus intertwining their histories. Southeast Asia, for instance, was notably a center of active trading over the millennia, which has impacted the genetic composition of dogs originating from this region (Freedman et al. 2014). While there is certainly no doubt about the importance of Southeast Asia in the history of dogs, the observed patterns in modern dogs require alternative explanations, for instance, that given above by Sacks et al. (2013).

Yet another obstacle arises whenever putting modern dogs into context with contemporary wolves and employing a phylogenetic argument. We tend to interpret phylogenetic arrangements as follows: the geographic origin of the wild population being the closest in a phylogenetic tree relative to the domesticate most likely points toward its geographic origin. This argument is built upon two main assumptions: first, the population of wild progenitors still exists, and second, its demographic history did not undergo population size changes and can simply be assumed to be constant over time. In case of the dog's ancestor, the gray wolf, we now know that both assumptions are incorrect. Thalmann et al. (2013) and Freedman et al. (2014) independently came to the conclusion that the wolf population that gave rise to modern dogs went extinct, and thus the extant wolf population most closely related to modern dogs is likely the one being admixed, but not the ancestral one. With regard to the second assumption, evidence exists (Fan et al. 2016) that the demographic history of wolves was characterized by episodes of admixture and population size changes, which is in stark contrast to the simplified constant population size model that has been the basis of most demographic models inferring dog domestication. Furthermore, in agreement with the hypothesis that the original source population went extinct, a scenario of replacement needs to be invoked in such models. Consequently, in order to fully comprehend all the details of dog domestication, we need to employ models that account for the ancestor's history as well (Perri 2016).

3.4.2 Admixture and Incomplete Lineage Sorting

Short divergence time and continuous hybridization/admixture left traces in the genomes of modern dogs and wolves to an extent that up to 20% of the genome of East Asian wolves and 7–25% of that of European and Middle Eastern wolves show dog contribution (Fan et al. 2016). Recent admixture is a worrisome obstacle having severe effects on the outcome of genetic studies utilizing modern specimens and, if ignored, might complicate attempts to decipher the particulars of the evolutionary histories of modern dogs and wolves [see reviews (Freedman et al. 2016; Freedman and Wayne 2016)]. The picture is even more complicated as both canids share a very recent common ancestor. Cumulative evidence suggests an onset of dog domestication could be placed within the last 30,000 years (see above). While some of the genome reflects this short divergence time, other parts do not mirror this time span due to older coalescence of these loci, a phenomenon termed incomplete lineage sorting (ILS). Whereas ILS could have remained undetected with single-locus approaches in the past, recent technological developments allow for the generation of multi-loci genealogies (in theory, for each single nucleotide in the genome). As an effect of ignored ILS, genealogies of single loci might be falsely interpreted as evidence for gene flow thus impacting attempts to date population splits. However, contemporary genome-wide analyses deliver a multitude of coalescence events and hence can diminish the effects of ILS.

4 Combining Fossils and Genetics: The Paleogenomic Era

> "The inference of complex patterns of gene flow is challenging, or even impossible, when only modern samples are studied. Therefore, the acquisition of a broader set of ancient samples including ancient representatives from Central and Southeast Asia, and the Middle east will be crucial to further clarify the details of dog domestication and evolution" (Botigué et al. 2017).

One way to circumvent the problems outlined above is to go back in time and assess the genetic variation at the cradle of domestication (Larson et al. 2012). As this is obviously impossible, we ought to consider alternative approaches. One solution presents itself in the technological advances over the last decade and the availability of fossil remains that still contain useful DNA. Paleogenomic studies employing fossil materials and the latest molecular technologies are now being regularly published and promise intriguing insights (see various examples in this book). Almost annually the limits of such studies are challenged, and while a decade ago the Neanderthal genome was rather fiction than science, we now witness a wealth of analyses based on its initial publication in 2010 (Green et al. 2010). The prospects of paleogenomic studies are intriguing and are opening new horizons for our understanding of evolution and filling gaps otherwise not approachable.

4.1 Mitochondrial Genomes of Ancient Canids

In an attempt to utilize novel molecular technologies on a set of canid fossils dating back some 36,000 years ago, Thalmann et al. (2013) captured and sequenced complete mitochondrial genomes of 18 ancient canids and put these sequences into context with mitochondrial genomes from modern dogs and wolves. The authors described more genetic variation prevalent in the fossils compared to all modern dogs and wolves (Fig. 6). The analyses included the oldest fossils proposed as putative dogs (Germonpré et al. 2009; Ovodov et al. 2011) but could not unambiguously solve the debate about their respective classification (Crockford and Kuzmin 2012), as they constitute an ancient sister group to all modern dogs and wolves. Such a pattern could be explained by either an early domestication attempt that did not leave any direct descendants in modern dogs or alternatively that the morphological features that set those specimens apart from modern wolves could be part of the Pleistocene diversity within wolves and simply represent a specialized ecomorph (Leonard et al. 2007; Perri 2016). While these specimens were excluded as direct ancestors of modern dogs, other candidates did come to light. The phylogenetic analyses revealed that three of the four major dog clades [labelled A–D in

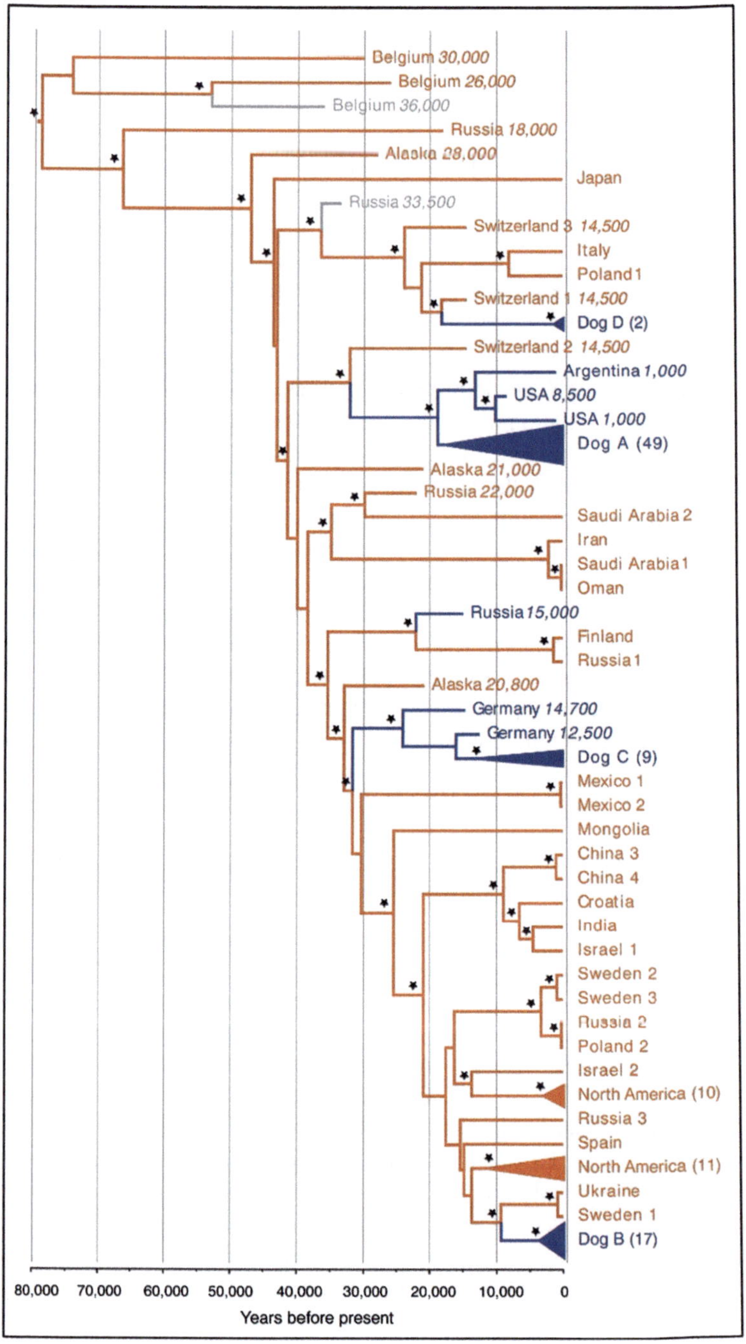

Fig. 6 Phylogenetic tree combining ancient and modern canids. Mitochondrial genomes derived from wolves are shown in orange and those from dogs are in blue. Adapted after Thalmann et al. (2013). Modern dogs are summarized into clades (A–D) according to the number of haplotypes they contain with clade A comprising the largest variation and hence depicting the oldest dog clade

Fig. 6, see also (Pang et al. 2009; Bjoernerfeldt et al. 2006)] were most closely related to ancient canids originating from Europe and hence supported a European ancestry of modern dogs. Notably, none of the sequences generated from modern wolves that were sampled from the putative locations of dogs' origin (China or Middle East) showed a closer affinity to modern dogs than the ancient specimens. If the modern wolf population ancestral to all dogs would still exist, this wolf population would be associated closest with modern dogs (phylogenetic argument described in Sect. 3.4). Since the data failed to show such pattern, the authors concluded that the wolf population ancestral to dogs is now extinct (Freedman et al. 2014). Thalmann et al. (2013) further used the inferred phylogeny to assess when dogs and wolves split and derived to estimates between approximately 32,000 and 19,000 years ago, putting the potential onset of dog domestication into context with hunter-gatherers rather than farmer societies (Axelsson et al. 2013). Another result of the analysis confirmed previous findings of an Old World ancestry of New World dogs (Leonard et al. 2002). The study is furthermore an example of how a more extended dataset might help to revise previous results. In 2013, Druzhkova et al. used fragments of the mitochondrial control region (~400 nucleotides) and suggested that the Razboinichya Cave canid (Ovodov et al. 2011) was an early dog (Druzhkova et al. 2013). The same specimen was included in the study by Thalmann et al. (2013), and the analysis of the complete mitochondrial genome led the authors to revise their earlier conclusions and assign the canid a rather basal position in the phylogenetic tree (Fig. 6).

One critical aspect concerned the limited geographical sampling with a bias toward ancient canids from Europe. Thalmann et al. (2013) acknowledged this shortcoming but argued that in order to disprove their conclusions, additional fossils have to first be found in the potential regions of interest, such as Southeast Asia or the Middle East [see Fig. 2 in (Larson et al. 2012)]. Furthermore, the derived mitochondrial sequences would also need to cluster closer to the three dog clades than those of European origin to contradict the interpretations of the study. Altogether, the authors considered this rather unlikely and concluded that based on the evidence at hand, Europe had played a significant role in the domestication of dogs, and for the first time, the study found the paleontological and genetic evidence to finally coincide.

4.2 *Whole-Genome Analyses of Ancient Canids*

The first draft genome sequence of a Pleistocene canid (the Taimyr wolf) was published in 2015 (Skoglund et al. 2015) and revealed that this 35,000-year-old wolf belonged to a population that diverged from the ancestors of both modern dogs and wolves. By employing this radiocarbon-dated individual to re-calibrate the mutation rate, the authors proposed a split between dogs and wolves occurring before the Last Glacial Maximum. This date is at odds with divergence estimates derived from modern canid genomes (Freedman et al. 2014), but if the mutation rate was adjusted accordingly, the estimates of Freedman et al. (2014) correspond to a split of approximately 40,000–27,000 years ago. Finally, Skoglund et al. (2015) found that Siberian huskies and other northern breeds trace part of their ancestry (1.4–27.3%) back to the Taimyr wolf, a pattern in line with multiple domestication scenarios.

This is an intriguing hypothesis and found additional support in a recent study by Frantz et al. (2016). In an attempt to shed further light on the controversy regarding the temporal and geographic origins of modern dogs, the authors sequenced 58 mitochondrial genomes from European dogs and a single high-coverage (28×) nuclear genome of a 4800-year-old dog from Newgrange, Ireland (Frantz et al. 2016). A first assessment of the genetic composition of the Newgrange dog revealed that this specimen did not possess genetic variants that define modern breed traits. In phylogenetic reconstructions based on 170,000 SNPs, the Newgrange dog clustered together with Western Eurasian dogs, and more in-depth analyses uncovered a close affinity of this dog to European breeds. The results in Frantz et al. (2016) indicated that the separation of Western Eurasian and Eastern Asian clades is older than the actual age of the Newgrange dog (NGD; 4800 years). Further evaluation of a scenario leading to the separation of the Western Eurasian and Eastern Asian genomes led the authors to suggest that Western Eurasian dogs underwent a population size reduction after the initial split from Eastern Asian dogs, which was estimated to have occurred between 14,000 and 6000 years ago. When putting these dates into context with the fossil record, it shows reliable evidence of dogs in Europe and East Asia 15,000 and 12,500 years ago, respectively. These results imply that indigenous dogs were already present before the genetic divergence of Western Eurasian and Eastern Asian dogs and further that Paleolithic dogs in Europe were replaced by dogs arriving from the East. This notion gained additional support from the analysis of the mitochondrial DNA, which showed a clear signal of a partial shift in sequences through time in European dogs congruent with a replacement scenario (but see below, Botigué et al. 2017). Finally, the authors developed an evolutionary model explaining their intriguing dual-origin hypothesis (Fig. 7). In short, it describes independent domestication of potentially extinct wolf populations in Eastern Asia and Western Eurasia during the Paleolithic, followed by a westward dispersal of eastern dogs accompanying humans on their migration routes and partially replacing indigenous Western Eurasian dogs around 14,000–6400 years ago. This model describes by far the most complex and potentially the most accurate scenario to date, but future genetic studies will certainly help to fill in the missing pieces of the complex mosaic of dog domestication.

As time progresses and new methods, analytical tools, and samples become available, we often need to revise previous findings and adjust our knowledge. Progress in science opens new avenues and overturns traditional believes and thereby brings us closer to a comprehensive understanding of evolutionary history.

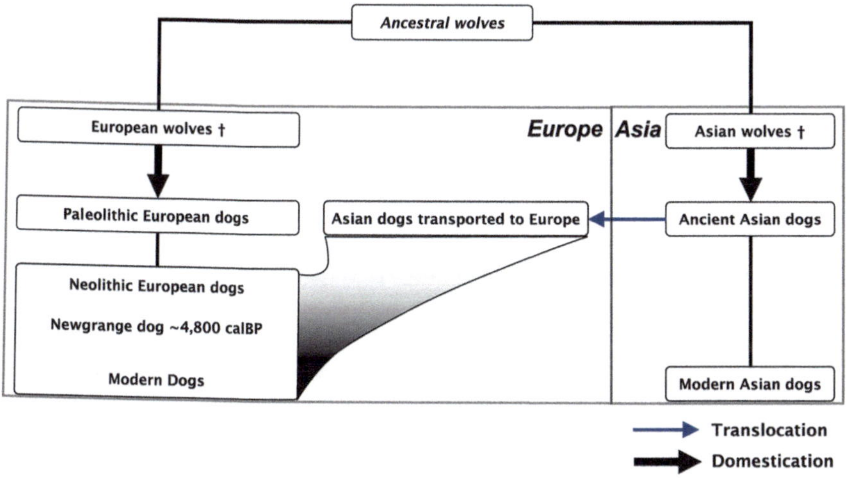

Fig. 7 Demographic model illustrating the histories of dogs and wolves and suggesting a dual domestication process in Europe and Asia, respectively (Frantz et al. 2016). Modified with permission from L.A.F. Frantz

For instance, a recent study has now challenged the evolutionary model developed by Frantz et al. (2016) and suggested a continuum of genetic variation in dogs since the Neolithic in favor of a replacement (Botigué et al. 2017). The authors based their conclusion on two newly generated genomes of dogs from Neolithic Germany, which they put into context with previously published canid genomes, including the Newgrange dog (NGD) analyzed in Frantz et al. (2016). The ages of these two German dogs span the Neolithic era (HXH 7000 and CTC 4700 years old), and their respective genetic makeup showed a degree of continuity rather than a temporal replacement. The two dogs and a reanalyzed version of the Newgrange dog (Frantz et al. 2016) were genetically most similar to modern European breeds in all performed analyses (Fig. 8). Intriguingly, the younger German specimen (CTC) revealed some distant position in the analyses (Fig. 8), and such an arrangement requires a more complex ancestry. While all three Neolithic European dogs (HXH, CTC, and NGD) showed ancestry patterns prevalent in modern Southeast Asian dogs and thus supporting a model that involves admixture of the ancestors of modern European and Southeast Asian dogs, CTC showed an additional component. This was found in modern Indian village dogs and ultimately supports a scenario in which this specimen descended from a population that harbored the older HXH and thus indicating genetic continuity throughout the Neolithic Europe and an outside source most likely of Indian, Asian, or Middle Eastern origin. Such genetic components might have been introduced into Europe by migrating human populations from the East (Haak et al. 2015) and thus once more highlighting the intertwined evolution of man and its best friend.

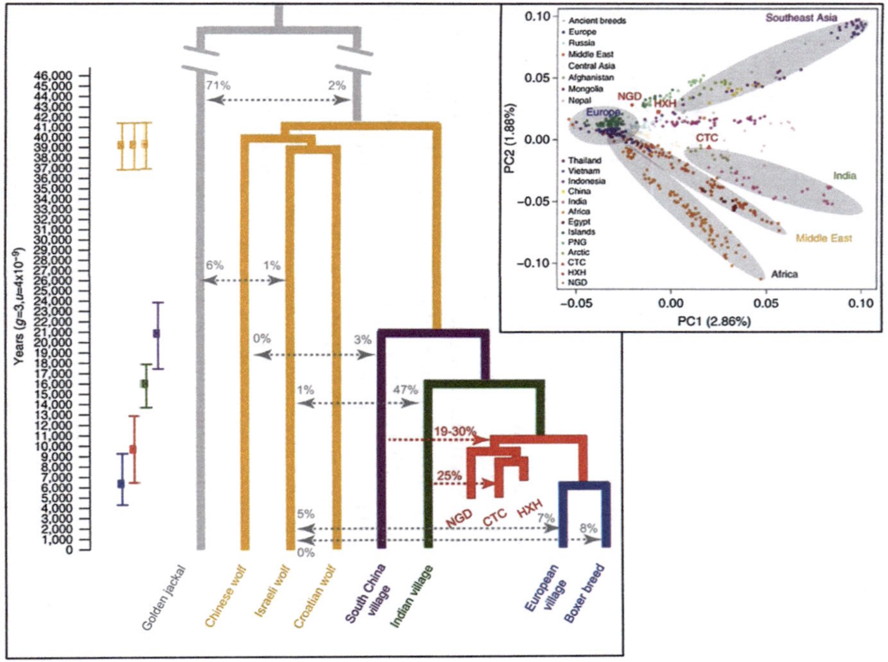

Fig. 8 Model of the demographic history of three ancient dogs from the Neolithic Europe and its genomic relation to contemporary dogs and wolves after Botigué et al. (2017). Inlay depicts the results of the principal component analyses. *NGD* Newgrange dog (Frantz et al. 2016), *CTC* cherry tree cave, *HXH* Herxheim (Botigué et al. 2017). The directionality and amount of gene flow are highlighted by arrows

5 Conclusions and Future Perspectives

Domestication involves both biological and cultural processes (Clutton-Brock 1992, 1995) and has always been difficult to define (Galton 1865; Dyson 1953; Clutton-Brock 1981; Price 1999; Zeder 2006; Ucko and Dimbleby 2007). The history of modern dogs has been the topic of scientific debates for centuries. While dogs are now firmly established as the earliest domesticate, among animals or plants, most questions about their domestication remain largely unresolved, including the timing and location; and we only begin to comprehend the drivers of domestication. One aspect that received support from genetic (Vilà et al. 1997; Lindblad-Toh et al. 2005), morphological (Wayne 1986), and archaeological research (Perri 2016) and is now widely accepted is the ancestry of modern dogs from a gray wolf. The geographic origin of sustained dog domestication is, however, highly debated with suggested locations including Eastern Asia (Savolainen et al. 2002; Pang et al. 2009; Ding et al. 2012), Central Asia (Shannon et al. 2015), the Middle East (von Holdt et al. 2010), and Europe (Thalmann et al. 2013). Recently, Frantz et al. (2016) proposed that dogs were domesticated at least twice, in Europe and East Asia, perhaps reconciling some of the previous divergent conclusions. By providing new evidence, scientific progress will always lead to novel hypotheses and arguments against established views, and so it is not surprising that also the latest model by Frantz et al. (2016) has already been challenged (Botigué et al. 2017).

Inferences about the timing of dog domestication are equally controversial, with a number of researchers reaching conflicting hypotheses. A date in the latest Upper Paleolithic, between 15,000 and 12,000 years ago, has long been the accepted timing of domestication due to clear archaeological evidence of morphologically distinct domesticated dogs by this time [e.g., (Turnbull and Reed 1974; Davis and Valla 1978; Lawrence and Reed 1983; Clutton-Brock 1995; Morey 1994)] and more recent genetic findings [e.g., (Larson et al. 2012; Axelsson et al. 2013; Freedman et al. 2014)]. However, these recent dates are at odds with estimates derived from studies utilizing ancient canids (Thalmann et al. 2013; Druzhkova et al. 2013; Skoglund et al. 2015; Botigué et al. 2017) and also modern specimens (after correcting for mutational biases), which suggest an older time frame for dog domestication some 20,000–40,000 years ago.

Notwithstanding these discrepancies in timing, a consensus exists with regard to the appearance of the first dogs in hunter-gatherer societies (Freedman and Wayne 2016; Larson et al. 2012), but how and why this transition from a wild carnivore into todays' most widespread domesticate happened remains unresolved (Perri 2016). Future paleogenomic research involving genetic studies of ancient materials from dogs and wolves will help to further unravel the precise details of dog domestication and develop a comprehensive evolutionary model that explains the *when*, *where*, and *how* gray wolves became man's best friend.

Acknowledgments O.T. is grateful to M. Arandjelovic and M. Katarzyńska for helpful comments on the manuscript. O.T. further credits M. Katarzyńska and D. Schwochow for providing pictures.

O.T. was supported by the National Science Centre, Poland (2015/19/P/NZ7/03971), with funding from EU's Horizon 2020 program under the Marie Skłodowska-Curie grant agreement (665778). A.R.P. was supported by the Max Planck Society. The authors thank L.A.F. Frantz for providing a figure.

References

Alkorta-Aranburu G, Beall CM, Witonsky DB, Gebremedhin A, Pritchard JK, Di Rienzo A. The genetic architecture of adaptations to high altitude in Ethiopia. PLoS Genet. 2012;8(12): e1003110.

AnzideI AP, Bulgarelli GM, Catalano P, Cerilli E, Gallotti R, Lemorini C, Milli S, Palombo MR, Pantano W, Santucci E. Ongoing research at the late middle Pleistocene site of La Polledrara di Cecanibbio (central Italy), with emphasis on human-elephant relationships. Quat Int. 2012;255:171–87.

Ardalan A, Oskarsson MC, Van Asch B, Rabakonandriania E, Savolainen P. African origin for Madagascan dogs revealed by mtDNA analysis. R Soc Open Sci. 2015;2(5):140552.

Axelsson E, Ratnakumar A, Arendt M-L, Maqbool K, Webster MT, Perloski M, Liberg O, Arnemo JM, Hedhammar ÃK, Lindblad-Toh K. The genomic signature of dog domestication reveals adaptation to a starch-rich diet. Nature. 2013;495:360–4.

Baryshnikov GF, Mol D, Tikhonov AN. Finding of the late Pleistocene carnivores in Taimyr Peninsula (Russia, Siberia) with paleoecological context. Russ J Theriol. 2009;8(2):107–13.

Bell T. A history of British quadrupeds, including the Cetacea. London: John van Voorst; 1837.

Benecke N. Studies on early dog remains from northern Europe. J Archaeol Sci. 1987;14(1):31–49.

Binder WJ, Van Valkenburgh B. A comparison of tooth wear and breakage in Rancho La Brea sabertooth cats and dire wolves across time. J Vertebr Paleontol. 2010;30(1):255–61.

Binder WJ, Thompson EN, Van Valkenburgh B. Temporal variation in tooth fracture among Rancho La Brea dire wolves. J Vertebr Paleontol. 2002;22(2):423–8.

Bjoernerfeldt S, Webster MT, Vilà C. Relaxation of selective constraint on dog mitochondrial DNA following domestication. Genome Res. 2006;16(8):990–4.

Boeskorov G, Baryshnikov G. Late quaternary Carnivora of Yakutia. Saint-Petersburg: Nauka; 2013.

Bonifay M-F. Carnivores quaternaires du Sud-Est de la France. 1971.

Botigué LR, Song S, Scheu A, Gopalan S, Pendleton AL, Oetjens M, Taravella AM, Seregély T, Zeeb-Lanz A, Arbogast R-M. Ancient European dog genomes reveal continuity since the early Neolithic. Nat Commun. 2017;8:16082.

Boudadi-Maligne M, Mallye J-B, Langlais M, Barshay-Szmidt C. Magdalenian dog remains from Le Morin rock-shelter (Gironde, France). Socio-economic implications of a zootechnical innovation. PALEO Revue d'archéologie préhistorique. 2012;(23):39–54.

Boyko AR, Boyko RH, Boyko CM, Parker HG, Castelhano M, Corey L, Degenhardt JD, Auton A, Hedimbi M, Kityo R. Complex population structure in African village dogs and its implications for inferring dog domestication history. Proc Natl Acad Sci. 2009;106(33):13903–8.

Brown SK, Pedersen NC, Jafarishorijeh S, Bannasch DL, Ahrens KD, Wu J-T, Okon M, Sacks BN. Phylogenetic distinctiveness of Middle Eastern and Southeast Asian village dog Y chromosomes illuminates dog origins. PLoS One. 2011;6(12):e28496.

Brugal J-P, Boudadi-Maligne M. Quaternary small to large canids in Europe: taxonomic status and biochronological contribution. Quat Int. 2011;243(1):171–82.

Camarós E, Münzel SC, Mn C, Rivals F, Conard NJ. The evolution of Paleolithic hominin-carnivore interaction written in teeth: stories from the Swabian Jura (Germany). J Archaeol Sci Rep. 2016;6:798–809.

Clutton-Brock J. Domesticated animals from early times. In: Domesticated animals from early times. London: British Museum (Natural History) and William Heinemann Ltd; 1981.
Clutton-Brock J. The process of domestication. Mammal Rev. 1992;22:79–85.
Clutton-Brock J. Origins of the dog: domestication and early history. In: Serpell J, editor. The domestic dog, its evolution, behaviour and interactions with people. Cambridge: Cambridge University Press; 1995. p. 7–20.
Coppinger R, Coppinger L. Dogs: a startling new understanding of canine origin, behavior & evolution. New York: Simon and Schuster; 2001.
Coppinger R, Coppinger L. Dogs: a new understanding of canine origin, behavior and evolution. Chicago: University of Chicago Press; 2002.
Coppinger R, Schneider R. Evolution of working dogs. In: The domestic dog: its evolution, behaviour and interactions with people. 1995:21–47
Crockford SJ, Kuzmin YV. Comments on Germonpré et al., Journal of Archaeological Science 36, 2009 "Fossil dogs and wolves from Palaeolithic sites in Belgium, the Ukraine and Russia: osteometry, ancient DNA and stable isotopes", and Germonpré, Lázničková-Galetová, and Sablin, Journal of Archaeological Science 39, 2012 "Palaeolithic dog skulls at the Gravettian Předmostí site, the Czech Republic". J Archaeol Sci. 2012;39(8):2797–801.
Cuvier G. Le Regne Animal Distribue d'apres son Organisation, pour servir de base a l'histoire naturelle des Animaux et d'introduction a l'Anatomie Comparée. Paris; 1817.
Darwin C. On the origin of species by means of natural selection. London: Murray; 1859.
Darwin C. The variation of animals and plants under domestication, vol. 2. New York: Orange Judd; 1868.
Davis SJ, Valla FR. Evidence for domestication of the dog 12,000 years ago in the Natufian of Israel. Nature. 1978;276(5688):608.
Dikov NN. The Ushki sites, Kamchatka Peninsula. In: West FH, editor. American beginnings, the prehistory and palaeoecology of Beringia. Chicago: University of Chicago Press; 1996. p. 244–50.
Ding Z, Oskarsson M, Ardalan A, Angleby H, L-Gr D, Tepeli C, Kirkness E, Savolainen P, Zhang Y. Origins of domestic dog in southern East Asia is supported by analysis of Y-chromosome DNA. Heredity. 2012;108(5):507–14.
Downs JF. Domestication: an examination of the changing social relationships between man and animals. Kroeber Anthr Soc Paps. 1960;22:18–67.
Drake AG, Coquerelle M, Colombeau G. 3D morphometric analysis of fossil canid skulls contradicts the suggested domestication of dogs during the late Paleolithic. Sci Rep. 2015;5:8299
Druzhkova AS, Thalmann O, Trifonov VA, Leonard JA, Vorobieva NV, Ovodov ND, Graphodatsky AS, Wayne RK. Ancient DNA analysis affirms the Canid from Altai as a primitive dog. PLoS One. 2013;8(3):e57754.
Dyson RH. Archeology and the domestication of animals in the old world. Am Anthropol. 1953;55 (5):661–73.
Ellegren H. Microsatellites: simple sequences with complex evolution. Nat Rev Genet. 2004;5 (6):435–45.
Fan Z, Silva P, Gronau I, Wang S, Armero AS, Schweizer RM, Ramirez O, Pollinger J, Galaverni M, Del-Vecchyo DO. Worldwide patterns of genomic variation and admixture in gray wolves. Genome Res. 2016;26(2):163–73.
Flower LO, Schreve DC. An investigation of palaeodietary variability in European Pleistocene canids. Quat Sci Rev. 2014;96:188–203.
Frantz LAF, Mullin VE, Pionnier-Capitan M, Ol L, Ollivier M, Perri A, Linderholm A, Mattiangeli V, Teasdale MD, Dimopoulos EA. Genomic and archaeological evidence suggest a dual origin of domestic dogs. Science. 2016;352(6290):1228–31.
Freedman AH, Wayne RK. Deciphering the origin of dogs: from fossils to genomes. Annu Rev Anim Biosci. 2016.
Freedman AH, Gronau I, Schweizer RM, Ortega-Del Vecchyo D, Han E, Silva PM, Galaverni M, Fan Z, Marx P, Lorente-Galdos B. Genome sequencing highlights the dynamic early history of dogs. PLoS Genet. 2014;10(1):e1004016.

Freedman AH, Lohmueller KE, Wayne RK. Evolutionary history, selective sweeps, and deleterious variation in the dog. Annu Rev Ecol Evol Syst. 2016;47:73–96.

Galton F. The first steps towards the domestication of animals. Trans Ethnol Soc Lond. 1865;3:122–38.

Galton F. Inquiries into the human faculty & its development. London: JM Dent and Company; 1883.

Garcia M-A. Ichnologie générale de la grotte Chauvet. Bulletin de la Société Préhistorique Française. 2005:103–8.

Garcia-Moreno J, Matocq MD, Roy MS, Geffen E, Wayne RK. Relationships and genetic purity of the endangered Mexican wolf based on analysis of microsatellite loci. Conserv Biol. 1996;10 (2):376–89.

Geffen E, Anderson MJ, Wayne RK. Climate and habitat barriers to dispersal in the highly mobile grey wolf. Mol Ecol. 2004;13(8):2481–90.

Germonpré M, Sablin MV, Stevens RE, Hedges RE, Hofreiter M, Stiller M, Després VR. Fossil dogs and wolves from Palaeolithic sites in Belgium, the Ukraine and Russia: osteometry, ancient DNA and stable isotopes. J Archaeol Sci. 2009;36(2):473–90.

Germonpré M, Lázničková-Galetová M, Sablin MV. Palaeolithic dog skulls at the Gravettian Předmostí site, the Czech Republic. J Archaeol Sci. 2012;39(1):184–202.

Germonpré M, Lázničková-Galetová M, Losey RJ, Räikkönen J, Sablin MV. Large canids at the Gravettian Předmostí site, the Czech Republic: the mandible. Quat Int. 2015a;359:261–79.

Germonpré M, Sablin MV, Lázničková-Galetová M, Després V, Stevens RE, Stiller M, Hofreiter M. Palaeolithic dogs and Pleistocene wolves revisited: a reply to Morey (2014). J Archaeol Sci. 2015b;54:210–6.

Germonpré M, Fedorov S, Danilov P, Galeta P, Jimenez E-L, Sablin M, Losey RJ. Palaeolithic and prehistoric dogs and Pleistocene wolves from Yakutia: identification of isolated skulls. J Archaeol Sci. 2017;78:1–19.

Gou X, Wang Z, Li N, Qiu F, Xu Z, Yan D, Yang S, Jia J, Kong X, Wei Z. Whole-genome sequencing of six dog breeds from continuous altitudes reveals adaptation to high-altitude hypoxia. Genome Res. 2014;24(8):1308–15.

Green RE, Krause J, Briggs AW, Maricic T, Stenzel U, Kircher M, Patterson N, Li H, Zhai W, Fritz MH-Y, Hansen NF, Durand EY, Malaspinas A-S, Jensen JD, Marques-Bonet T, Alkan C, Prufer K, Meyer M, Burbano HA, Good JM, Schultz R, Aximu-Petri A, Butthof A, Hober B, Hoffner B, Siegemund M, Weihmann A, Nusbaum C, Lander ES, Russ C, Novod N, Affourtit J, Egholm M, Verna C, Rudan P, Brajkovic D, Kucan Z, Gusic I, Doronichev VB, Golovanova LV, Lalueza-Fox C, de la Rasilla M, Fortea J, Rosas A, Schmitz RW, Johnson PLF, Eichler EE, Falush D, Birney E, Mullikin JC, Slatkin M, Nielsen R, Kelso J, Lachmann M, Reich D, Paabo S. A draft sequence of the Neandertal genome. Science. 2010;328(5979):710–22.

Haak W, Lazaridis I, Patterson N, Rohland N, Mallick S, Llamas B, Brandt G, Nordenfelt S, Harney E, Stewardson K, Fu Q, Mittnik A, Banffy E, Economou C, Francken M, Friederich S, Pena RG, Hallgren F, Khartanovich V, Khokhlov A, Kunst M, Kuznetsov P, Meller H, Mochalov O, Moiseyev V, Nicklisch N, Pichler SL, Risch R, Rojo Guerra MA, Roth C, Szecsenyi-Nagy A, Wahl J, Meyer M, Krause J, Brown D, Anthony D, Cooper A, Alt KW, Reich D. Massive migration from the steppe was a source for Indo-European languages in Europe. Nature. 2015;522(7555):207–11.

Housley RA, Gamble CS, Street M, Pettitt P. Radiocarbon evidence for the Lateglacial human recolonisation of northern Europe. Proc Prehist Soc. 1997;63:25–54.

Hudson RR. Gene trees, species trees and the segregation of ancestral alleles. Genetics. 1992;131 (2):509–12.

Hunter J, Mears W. Observations tending to Shew that the wolf, jackal, and dog, are all of the same species. Philos Trans R Soc Lond B Biol Sci. 1787;77:253–66.

Hurles ME, Sykes BC, Jobling MA, Forster P. The dual origin of the Malagasy in Island Southeast Asia and East Africa: evidence from maternal and paternal lineages. Am J Hum Genet. 2005;76 (5):894–901.

Kahlke R-D. Taubach, Gebiet der ehemaligen Travertinbrueche. Altenburger Naturwissenschaftliche Forschungen. 1994;7:366–7.

Kirkness EF, Bafna V, Halpern AL, Levy S, Remington K, Rusch DB, Delcher AL, Pop M, Wang W, Fraser CM, Venter JC. The dog genome: survey sequencing and comparative analysis. Science. 2003;301(5641):1898–903.

Koler-Matznick J. The origin of the dog revisited. Anthrozoös. 2002;15(2):98–118.

Kuzmina I, Sablin M. Pozdnepleistotsenovyi pesets verhnei Desny. Materiali po mezozoickoi i kainozoickoi istorii nazemnykh pozvonochnykh Trudy Zoologicheskogo Instituta RAN, vol 249; 1993. p. 93–104.

Larson G, Burger J. A population genetics view of animal domestication. Trends Genet. 2013;29 (4):197–205.

Larson G, Karlsson EK, Perri A, Webster MT, Ho SYW, Peters J, Stahl PW, Piper PJ, Lingaas F, Fredholm M, Comstock KE, Modiano JF, Schelling C, Agoulnik AI, Leegwater PA, Dobney K, Vigne J-D, Vilà C, Andersson L, Lindblad-Toh K. Rethinking dog domestication by integrating genetics, archeology, and biogeography. Proc Natl Acad Sci U S A. 2012;109(23):8878–83.

Lawrence B, Reed CA. The dogs from Jarmo. In: Braidwood LS, Braidwood RJ, Howe B, Reed CA, Watson PJ, editors. Prehistoric archaeology along the Zagros Flanks. Chicago: Oriental Institute, University of Chicago; 1983. p. 485–9.

Ledoux L, Boudadi-Maligne M. The contribution of geometric morphometric analysis to prehistoric ichnology: the example of large canid tracks and their implication for the debate concerning wolf domestication. J Archaeol Sci. 2015;61:25–35.

Leonard JA, Wayne RK, Wheeler J, Valadez R, Guillen S, Vilà C. Ancient DNA evidence for old world origin of new world dogs. Science. 2002;298(5598):1613–6.

Leonard JA, Vilà C, Fox-Dobbs K, Koch PL, Wayne RK, Van Valkenburgh B. Megafaunal extinctions and the disappearance of a specialized wolf Ecomorph. Curr Biol. 2007;17 (13):1146–50.

Lindblad-Toh K, Wade CM, Mikkelsen TS, Karlsson EK, Jaffe DB, Kamal M, Clamp M, Chang JL, Kulbokas EJ, Zody MC, Mauceli E, Xie X, Breen M, Wayne RK, Ostrander EA, Ponting CP, Galibert F, Smith DR, deJong PJ, Kirkness E, Alvarez P, Biagi T, Brockman W, Butler J, Chin C-W, Cook A, Cuff J, Daly MJ, DeCaprio D, Gnerre S, Grabherr M, Kellis M, Kleber M, Bardeleben C, Goodstadt L, Heger A, Hitte C, Kim L, Koepfli K-P, Parker HG, Pollinger JP, Searle SMJ, Sutter NB, Thomas R, Webber C, Lander ES. Genome sequence, comparative analysis and haplotype structure of the domestic dog. Nature. 2005;438(7069):803.

Linnaeus C. Systema Naturae, Editio decima. Holmiae: Laur. Salvius; 1758.

Lorenz K. Man meets dog. London: Methuen; 1954.

Lorenz K. Foreword. In: Fox MW, editor. The wild canids: their systematics, behavioral ecology and evolution. New York: Van Nostrand Reinhold; 1975.

Losey RJ, Garvie-Lok S, Leonard JA, Katzenberg MA, Germonpré M, Nomokonova T, Sablin MV, Goriunova OI, Berdnikova NE, Savel'ev NA. Burying dogs in ancient Cis-Baikal, Siberia: temporal trends and relationships with human diet and subsistence practices. PLoS One. 2013;8 (5):e63740.

Malmström H, Vilà C, Gilbert M, Storå J, Willerslev E, Holmlund G, Götherström A. Barking up the wrong tree: modern northern European dogs fail to explain their origin. BMC Evol Biol. 2008;8(1):71.

Meachen JA, Brannick AL, Fry TJ. Extinct Beringian wolf morphotype found in the continental US has implications for wolf migration and evolution. Ecol Evol. 2016;6(10):3430–8.

Morey DF. The early evolution of the domestic dog. Am Sci. 1994;82(4):336–47.

Morey DF. In search of Paleolithic dogs: a quest with mixed results. J Archaeol Sci. 2014;52:300–7.

Morey DF, Jeger R. Paleolithic dogs: why sustained domestication then? J Archaeol Sci Rep. 2015;3:420–8.

Müller W. Les vestiges osseux. In: Leesch D, Cattin M-I, Müller W (eds) Hauterive-Champréveyres et Neuchâtel-Monruz: Témoins d'implantations magdaléniennes et aziliennes sur la rive nord du lac de Neuchâtel, vol 31. Archéologie neuchâteloise; 2004. p. 96–109.

Müller W. Les témoins animaux. In: Bullinger J, Leesch D, Plumettaz N (eds) Le site magdalénien de Monruz: Premiers éléments pour l'analyse d'un habitat de plein air. Service et musée cantonal d'archéologie de Neuchâtel; 2006. p. 123–37.
Müller W. Le site magdalénien de Monruz 3. Aquisition, traitement et consommation des ressources animales, vol 49. Archéologie neuchâteloise. Neuchâtel: Musée et Service cantonal d'archéologie; 2013.
Musiani M, Leonard JA, Cluff HD, Gat CC, Mariani S, Paquet PC, Vilà C, Wayne RK. Differentiation of tundra/taiga and boreal coniferous forest wolves: genetics, coat colour and association with migratory caribou. Mol Ecol. 2007;16(19):4149–70.
Musil R. Evidence for the domestication of wolves in Central European Magdalenian sites in dogs through time: an archaeological perspective. In: Crockford SJ (ed) Proceedings of the 1st ICAZ symposium on the history of the domestic dog. British Archaeological Reports International Series; 2000.
Napierala H, Uerpmann H-P. A 'new' palaeolithic dog from central Europe. Int J Osteoarchaeol. 2012;22(2):127–37.
Nichols R. Gene trees and species trees are not the same. Trends Ecol Evol. 2001;16(7):358–64.
Nielsen R, Akey JM, Jakobsson M, Pritchard JK, Tishkoff S, Willerslev E. Tracing the peopling of the world through genomics. Nature. 2017;541(7637):302–10.
Nobis G. Der älteste Haushund lebte vor 14000 Jahren. Umschau. 1979;79:610.
Oskarsson MC, Kluetsch CF, Boonyaprakob U, Wilton A, Tanabe Y, Savolainen P. Mitochondrial DNA data indicate an introduction through Mainland Southeast Asia for Australian dingoes and Polynesian domestic dogs. Proc R Soc Lond B Biol Sci. 2011:rspb20111395. http://rspb.royalsocietypublishing.org/content/early/2011/09/06/rspb.2011.1395?version=meter+at+null&module=meter-Links&pgtype=article&contentId=&mediaId=&referrer=&priority=true&action=click&contentCollection=meter-links-click.
Ovodov ND, Crockford SJ, Kuzmin YV, Higham TF, Hodgins GW, van der Plicht J. A 33,000-year-old incipient dog from the Altai Mountains of Siberia: evidence of the earliest domestication disrupted by the last glacial maximum. PLoS One. 2011;6(7):e22821.
Pang J-F, Kluetsch C, Zou X-J, A-b Z, Luo L-Y, Angleby H, Ardalan A, Ekström C, Sköllermo A, Lundeberg J, Matsumura S, Leitner T, Zhang Y-P, Savolainen P. mtDNA data indicate a single origin for dogs south of Yangtze River, less than 16,300 years ago, from numerous wolves. Mol Biol Evol. 2009;26(12):2849–64.
Parker HG, Kim LV, Sutter NB, Carlson S, Lorentzen TD, Malek TB, Johnson GS, DeFrance HB, Ostrander EA, Kruglyak L. Genetic structure of the purebred domestic dog. Science. 2004;304(5674):1160–4.
Perri A. A wolf in dog's clothing: initial dog domestication and Pleistocene wolf variation. J Archaeol Sci. 2016;68:1–4.
Pidoplichko IH. Upper Palaeolithic Dwellings of Mammoth Bones in the Ukraine: Kiev-Kirillovskii, Gontsy, Dobranichevka, Mezin, and Mezhirich, vol. 712. Oxford: British Archaeological Reports International Series; 1998.
Pionnier-Capitan M, Bemilli C, Bodu P, Célérier G, Ferrié J-G, Fosse P, Garcià M, Vigne J-D. New evidence for Upper Palaeolithic small domestic dogs in South-Western Europe. J Archaeol Sci. 2011;38(9):2123–40.
Price EO. Behavioral development in animals undergoing domestication. Appl Anim Behav Sci. 1999;65(3):245–71.
Raisor MJ. Determining the antiquity of dog origins: canine domestication as a model for the consilience between molecular genetics and archaeology, vol. 1367. Oxford, UK: British Archaeological Reports Ltd; 2005.
Reed CA. Animal domestication in the prehistoric near east. Science. 1959;130(3389):1629–39.
Repenning CA, Brouwers EM. Late Pliocene-early Pleistocene ecologic changes in the Arctic Ocean borderland. US Government Printing Office; 1992.

Roy MS, Geffen E, Smith D, Ostrander EA, Wayne RK. Patterns of differentiation and hybridization in north American wolflike canids, revealed by analysis of microsatellite loci. Mol Biol Evol. 1994;11(4):553–70.

Ruvolo M, Pan D, Zehr S, Goldberg T, Disotell TR, von Dornum M. Gene trees and hominoid phylogeny. Proc Natl Acad Sci U S A. 1994;91(19):8900–4.

Sablin M, Khlopachev G. The earliest ice age dogs: evidence from Eliseevichi 11. Curr Anthropol. 2002;43(5):795–9.

Sacks BN, Brown SK, Stephens D, Pedersen NC, Wu J-T, Berry O. Y chromosome analysis of dingoes and Southeast Asian village dogs suggests a Neolithic continental expansion from Southeast Asia followed by multiple Austronesian dispersals. Mol Biol Evol. 2013;30 (5):1103–18.

Sardella R, Bertè D, Iurino DA, Cherin M, Tagliacozzo A. The wolf from Grotta Romanelli (Apulia, Italy) and its implications in the evolutionary history of Canis lupus in the late Pleistocene of southern Italy. Quat Int. 2014;328:179–95.

Savolainen P, Zhang Y-P, Luo J, Lundeberg J, Leitner T. Genetic evidence for an East Asian origin of domestic dogs. Science. 2002;298(5598):1610–3.

Savolainen P, Leitner T, Wilton AN, Matisoo-Smith E, Lundeberg J. A detailed picture of the origin of the Australian dingo, obtained from the study of mitochondrial DNA. Proc Natl Acad Sci U S A. 2004;101(33):12387–90.

Scott J, Fuller J. Dog behavior: the genetic basis. Chicago: University of Chicago Press; 1965.

Serpell J. Pet-keeping and animal domestication: a reappraisal. In: The walking larder: patterns of domestication, pastoralism, and predation. 1989. p. 10–21.

Shannon LM, Boyko RH, Castelhano M, Corey E, Hayward JJ, McLean C, White ME, Said MA, Anita BA, Bondjengo NI. Genetic structure in village dogs reveals a Central Asian domestication origin. Proc Natl Acad Sci. 2015;112(44):13639–44.

Shipman P. How do you kill 86 mammoths? Taphonomic investigations of mammoth megasites. Quat Int. 2015;359:38–46.

Skoglund P, Ersmark E, Palkopoulou E, Dalen L. Ancient wolf genome reveals an early divergence of domestic dog ancestors and admixture into high-latitude breeds. Curr Biol. 2015;25 (11):1515–9.

Sotnikova M, Rook L. Dispersal of the Canini (Mammalia, Canidae: Caninae) across Eurasia during the late Miocene to early Pleistocene. Quat Int. 2010;212(2):86–97.

Street M. Ein Wiedersehen mit dem Hund von Bonn-Oberkassel. In: Hutterer R, editor. Animals in history: Archaeozoological papers in honour of Günter Nobis (1921–2002), Bonner Zoologische Beiträge, vol. 50. Bonn: Zoologisches Forschungsinstitut und Museum Alexander Koenig; 2002.

Street M, Napierala H, Janssens L. The late Paleolithic dog from Bonn Oberkassel in context. The late glacial burial from Oberkassel revisited. Darmstadt: Verlag Phillip von Zabern; 2015. p. 253–74.

Tchernov E, Valla FF. Two new dogs, and other Natufian dogs, from the southern Levant. J Archaeol Sci. 1997;24(1):65–95.

Tedford RH, Wang X, Taylor BE. Phylogenetic systematics of the North American fossil caninae (Carnivora: Canidae). Bull Am Mus Nat Hist. 2009.

Thalmann O, Shapiro B, Cui P, Schuenemann VJ, Sawyer SK, Greenfield DL, Germonpré MB, Sablin MV, López-Giráldez F, Domingo-Roura X, Napierala H, Uerpmann H-P, Loponte DM, Acosta AA, Giemsch L, Schmitz RW, Worthington B, Buikstra JE, Druzhkova A, Graphodatsky AS, Ovodov ND, Wahlberg N, Freedman AH, Schweizer RM, Koepfli K-P, Leonard JA, Meyer M, Krause J, Pääbo S, Green RE, Wayne RK. Complete mitochondrial genomes of ancient Canids suggest a European origin of domestic dogs. Science. 2013;342 (6160):871–4.

Tishkoff SA, Reed FA, Friedlaender FR, Ehret C, Ranciaro A, Froment A, Hirbo JB, Awomoyi AA, Bodo J-M, Doumbo O, Ibrahim M, Juma AT, Kotze MJ, Lema G, Moore JH, Mortensen H, Nyambo TB, Omar SA, Powell K, Pretorius GS, Smith MW, Thera MA, Wambebe C, Weber

JL, Williams SM. The genetic structure and history of Africans and African Americans. Science. 2009;324(5930):1035–44.

Turnbull PF, Reed CA. The fauna from the terminal Pleistocene of Palegawra cave, a Zarzian occupation in northeastern Iraq. Fieldiana Anthropol. 1974;63:81–146.

Ucko PJ, Dimbleby GW. The domestication and exploitation of plants and animals. New Brunswick, NJ: Transaction Publishers; 2007

van Asch B, Zhang A-B, Oskarsson MCR, Klütsch CFC, Amorim A, Savolainen P. Pre-Columbian origins of native American dog breeds, with only limited replacement by European dogs, confirmed by mtDNA analysis. Proc R Soc Lond B Biol Sci. 2013;280(1766):20131142.

Vigne J-D. L'humérus de chien magdalénien de Erralla (Gipuzkoa, Espagne) et la domestication tardiglaciaire du loup en Europe. Munibe. 2005;51:279–87.

Vilà C, Savolainen P, Maldonado JE, Amorim IR, Rice JE, Honeycutt RL, Crandall KA, Lundeberg J, Wayne RK. Multiple and ancient origins of the domestic dog. Science. 1997;276:1687–9.

von Holdt BM, Pollinger JP, Lohmueller KE, Han E, Parker HG, Quignon P, Degenhardt JD, Boyko AR, Earl DA, Auton A, Reynolds A, Bryc K, Brisbin A, Knowles JC, Mosher DS, Spady TC, Elkahloun A, Geffen E, Pilot M, Jedrzejewski W, Greco C, Randi E, Bannasch D, Wilton A, Shearman J, Musiani M, Cargill M, Jones PG, Qian Z, Huang W, Ding Z-L, Y-p Z, Bustamante CD, Ostrander EA, Novembre J, Wayne RK. Genome-wide SNP and haplotype analyses reveal a rich history underlying dog domestication. Nature. 2010;464 (7290):898–902.

Wang G-D, Zhai W, Yang H-C, Fan R-X, Cao X, Zhong L, Wang L, Liu F, Wu H, Cheng L-G, Poyarkov AD, Poyarkov NA Jr, Tang S-S, Zhao W-M, Gao Y, Lv X-M, Irwin DM, Savolainen P, Wu C-I, Zhang Y-P. The genomics of selection in dogs and the parallel evolution between dogs and humans. Nat Commun. 2013;4:1860.

Wayne RK. Limb morphology of domestic and wild canids: the influence of development on morphologic change. J Morphol. 1986;187(3):301–19.

Wayne RK, O'Brien SJ. Allozyme divergence within the Canidae. Syst Biol. 1987;36(4):339–55.

Wurster-Hill D, Centerwall W. The interrelationships of chromosome banding patterns in canids, mustelids, hyena, and felids. Cytogenet Genome Res. 1982;34(1–2):178–92.

Xu S, Li S, Yang Y, Tan J, Lou H, Jin W, Yang L, Pan X, Wang J, Shen Y. A genome-wide search for signals of high-altitude adaptation in Tibetans. Mol Biol Evol. 2011;28(2):1003–11.

Zeder MA. Archaeological approaches to documenting animal domestication. In: Documenting domestication: new genetic and archaeological paradigms. 2006. p. 171–80

Zeder MA. Core questions in domestication research. Proc Natl Acad Sci. 2015;112(11):3191–8.

Zeuner FE. A history of domesticated animals. A history of domesticated animals. London: Hutchinson & Co. (Publishers) Ltd.; 1963.

Zimen E. The wolf: his place in the natural world. London: Souvenir Press; 1981.

Of Cats and Men: Ancient DNA Reveals How the Cat Conquered the Ancient World

Eva-Maria Geigl and Thierry Grange

Abstract Neither genetics and genomics of modern cats nor archeology could so far reconstruct the domestication and dispersal process of the cat. It was only known that all domestic cats belong to the subspecies *Felis silvestris lybica*, that their genomes are close to the ones of wildcats, and that they were translocated to Cyprus by the Neolithic farmers who colonized this island roughly 9,500 years ago. The results of our large-scale paleogenetic study of the mitochondrial DNA of archeological cat remains fill the existing gaps in that they allowed us to reconstruct the history of the dispersal of the cat starting in Southwest Asia during the Neolithic and achieving a new quality in Egypt during the 1st millennium BCE. Together with those from Southwest Asia, these mitochondrial lineages from Egypt showed up in samples from the following centuries all over Southwest Asia, North Africa, and Europe, testifying of the cat's conquest of the ancient world. The dispersal pattern that we reconstructed from our data tells us that cats accompanied seafarers throughout history on their trading and raiding routes.

Keywords Ancient DNA · Cat dispersal · Domestication · Genetics

1 Introduction

Hundreds of millions of cats inhabit the world: They live as cherished companions and pets in urban households where they make families happy; as stray cats in cities; as village and barn cats in villages and farms, where they restrict rodent plagues; but also as feral cats in nature where they become a threat to wild birds and other vertebrates and, in some areas, an invasive species that exterminates the endemic fauna. Despite their popularity and widespread occurrence, however, little is known about their domestication.

E.-M. Geigl (✉) · T. Grange
Institut Jacques Monod, CNRS, UMR 7592, University Paris Diderot, Paris, France
e-mail: eva-maria.geigl@ijm.fr

Charlotte Lindqvist and Om P. Rajora (eds.), *Paleogenomics*, Population Genomics [Om P. Rajora (Editor-in-Chief)], https://doi.org/10.1007/13836_2018_26,

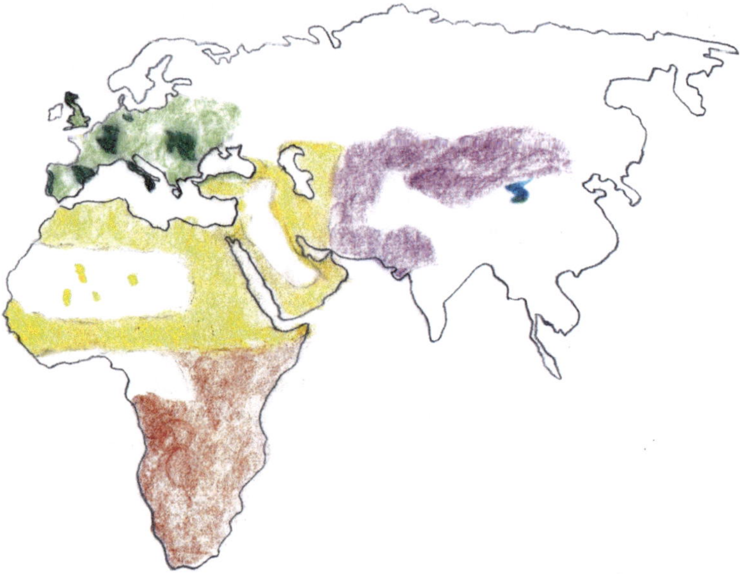

Fig. 1 Distribution of wildcats based on IUCN (Yamaguchi et al. 2015) and data from Ottoni et al. (2017). Drawing: E.-M. Geigl

1.1 Genetics of Present-Day Domestic Cats Based on Mitochondrial and Microsatellite DNA Data

Genetic studies showed that all domestic cats, including feral cats, are descendants from one of the five subspecies of the wildcat *Felis silvestris lybica*, the wildcat from North Africa (NA) and Southwest Asia (SWA), while the other subspecies of *Felis silvestris*, such as the European wildcat *Felis silvestris silvestris*, did not make a major contribution to the gene pool of the domestic cat although hybridization between the two subspecies is clearly detectable (Driscoll et al. 2007; Oliviera et al. 2008; Mattucci et al. 2015) (Fig. 1). Apart from revealing monophyly for the domestic cat, mitochondrial DNA sequences and nuclear microsatellite DNA did not disclose the course of its domestication process since no population structure in the domestic cat population emerged.

1.2 Archeozoology of Cat Domestication

There are two other sources of information about this process. The first one is the archeological record, which is regrettably scarce. The cat has never been a subsistence species and is, therefore, rarely found as refuse in archeological sites. The wildcat being a solitary animal of elusive nature, it was probably never hunted and only occasionally killed for its fur in prehistoric times. This changed only during the Middle Ages in Europe when items made of the pelts and skins of domestic cats became popular among common people (Ewing 1981), as well as their flesh and other body parts (Von den Driesch 1992). There are, however, two archeological finds that stand out and hold a clue to the timing and location of the taming process of cats. The first one is a human burial in Cyprus dated to ca. 7500 BCE in which a cat skeleton was found (Vigne et al. 2004). The most parsimonious although disputed (Rothwell 2004) explanation of this find on a formerly cat-free island is that the cat had been transported by Neolithic farmers to the island on a raft or ship as it would not have been able to swim there on its own. This situation is suggestive of the beginning of the taming process (Vigne et al. 2004). The second important find are six cat skeletons, including those from four kittens, found in a pit in an elite cemetery of Predynastic Egypt around 3700 BCE (Van Neer et al. 2014). Based on the age of the animals, the authors believe that these cats had been held in captivity, at least for some time (Van Neer et al. 2014). As in the case of the cat in Cyprus, this situation testifies to a change in the status of the animal with respect to humans, i.e., a tightening of their relationship. Thus, these early finds hint to the Fertile Crescent and Egypt as the areas where the cat first may have changed its behavior so that a coexistence with humans became possible. Since genetic data of extant cats clearly show that all domestic cats descend from *F. s. lybica*, the wildcat from NA and SWA, archeological and genetic data are in agreement albeit without proposing a refined view of the sequence of events taking place during the domestication process.

1.3 The Cat in Ancient Egyptian Iconography

The second and richest source of information about the important place that cats occupied in ancient societies is Egyptian iconography (Málek 1993–2006; Engels 2001): cats were depicted in Egypt as early as 2200 BCE, first on ivory knives and as outlines carved in walls. Later, during the Middle Kingdom, they were painted as bird hunters in marshes in company of Egyptian hunters ("cat in the marshes") (Fig. 2). The presence of other wild animals, such as the genet and the ichneumon (the Egyptian mongoose), suggests that these cats were wildcats (Von den Driesch 1992). Some of the cats that hunted together with humans might also have been trained jungle cats, *F. chaus* (Morrison-Scott 1952), but these are slightly bigger, lack the typical striped fur pattern of wildcats, and have tufted ears, discriminative features that would have been faithfully transcribed in Egyptian paintings. During the New Kingdom, in the second half of the 2nd millennium BCE, another theme became more and more recurrent, i.e., the cat sitting under the chair of a noble person, mostly a woman, translating the growing bonds between cats and humans, in particular women. Indeed, cats were a symbol of fertility and motherhood (Fig. 3a). By 1450 BCE, cats were common in paintings of domestic scenes. Cats also assumed great importance in Egyptian religion from about 2000 BCE onward

Fig. 2 Detail of cat from the hunting scene (fowling scene) from the tomb of Nebamun, Thebes, Egypt, 18th dynasty, ca. 1400–1350 BCE (British Museum, London, UK). Photo: T. Grange

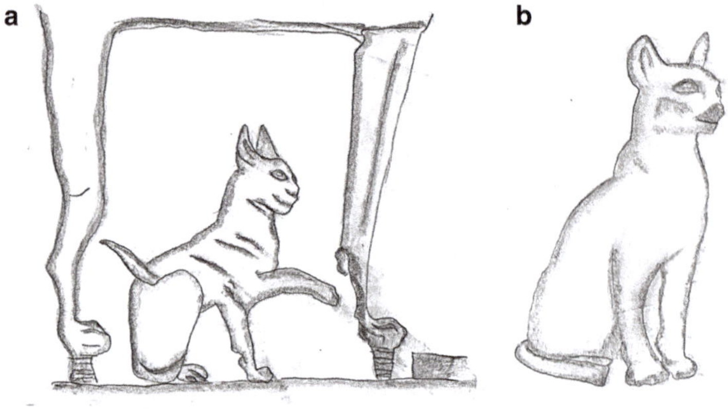

Fig. 3 (**a**) Bas relief of a "cat under the chair" from the tomb of Mery-mery, 1391–1353 BCE, Sakkara, 18th dynasty (Rijksmuseum van Oudheden, Leiden, Netherlands). (**b**) Statue of Bastet (Naturmuseum Senckenberg, Frankfurt, Germany). Drawing: E.-M. Geigl

(Málek 1993–2006; Engels 2001). From about 1500 BCE, it was believed that the sun god Ra could manifest himself in the form of a cat, the "Great Tomcat." Each night Ra would travel to the underworld, confront his enemy, the snake demon Apophis, kill the snake with a knife, and thus ensure the return of the sun the following morning. Many ancient Egyptian paintings depict Ra in the form of a spotted cat slaying Apophis and thus defeating the forces of chaos. By 945 BCE, the cat had become associated with a female divinity, the goddess Bastet, and with the funeral rituals. Indeed, sacred cats kept in temple catteries were worshipped as living embodiments or epiphanies of the goddess (Málek 1993–2006; Engels 2001; Machon 2015; Warmenbol 2015). The popularity of this cult of Bastet continued for over 1,500 years into the Roman era (to 330 CE). Many beautiful bronze sculptures of cats survive from this period (Fig. 3b). Cat mummification became more and more important during the 1st millennium BCE and in particular during the Ptolemaic period, and millions of mummified cats had survived until the nineteenth century CE when large amounts of them were shipped to England where they were ground up and used as fertilizer (Málek 1993–2006).

Thus, Egyptian iconography is the richest archive testifying to the domestication of the cat: without changing appearance, it entered the domestic context. The dichotomy of the situations in which the cats are depicted, on one hand as skilled hunters that killed dangerous snakes and scorpions, as well as intelligent rats and mobile birds, and on the other hand sitting quietly under the chairs of noble people and as guardians of the deceased, suggests that the cat was appreciated as both a most useful hunter and a companion animal. This dual role was maintained in Greek, Etruscan, and Roman iconography (Engels 2001; Luce 2015).

From the carvings, frescos, and statues in Ancient Egypt, it is not possible to distinguish a domestic from a wildcat except for the context in which it is placed. To a certain degree, this holds still true since the shape and appearance of wild-colored domestic cats are not radically different from those of wildcats (Krüger et al. 2009; Müller 2011). Most of the differences that can be noticed today are of very recent origin, such as the fancy breeds that were created from random-bred cat populations in a frenzy of innovation and experimental cross-breeding during the nineteenth century in the Western World (Kurushima et al. 2013). The analysis of genomes of modern wild and domestic cats indeed showed a few differences, most of which are attributable to the genes involved in the development of the neural crest suggesting mainly behavioral changes (Montague et al. 2014). The resemblance of size and shape between the domestic cat and the wild ancestor is also seen in the skeletal remains of cats: wild and domestic cat bones and teeth cannot be distinguished on the basis of their metrics, unless one has complete skeletons or at least skulls (Wim van Neer, personal communication). It is only the archeological context that gives a hint about a change in the relationship between humans and cats, such as in the two examples described above.

Thus, from the three lines of evidence, i.e., mitochondrial DNA sequences and nuclear microsatellite DNA data from extant cats, the archeological record, as well as Egyptian iconography, it was only possible to conclude that the cat must have been domesticated in SWA and/or NA from the subspecies *F. s. lybica* starting in the Early Neolithic, when hunter-gatherers became sedentary farmers. It was clear that in order to better characterize the domestication process of cats in time and space, DNA preserved in archeological cat remains had to be analyzed. Thus, we aimed at establishing the phylogeographic structure of the wildcat in North Africa, Southwest Asia, and Europe and following its evolution across time.

2 Paleogenetic Analysis of Cat Remains

Our team collected 352 ancient and 28 modern cat samples from all around the Mediterranean Basin and from Africa dated from 9000 BP to the beginning of the twentieth century (see supplementary tables of Ottoni et al. (2017)). An example for such archeological remains is shown in Fig. 4. Knowing that we had to analyze archeological specimens from Egypt, the Levant, and Syria, i.e., areas with a very hot climate where DNA is poorly preserved based on our previous experience analyzing other species, we developed a metabarcoding approach tailored to highly degraded, small DNA molecules that combines the sensitivity of PCR and the power of NGS for the analysis of many ancient samples at reasonable costs (Guimaraes et al. 2016). The approach uses highly optimized multiplex PCR to increase sample throughput without losing sensitivity and specificity of the PCR and proved to be powerful in various experimental systems (Cote et al. 2016; Guimaraes et al. 2016; Librado et al. 2017). The experimental design was such that through the analysis of the ND5, ND6, and CYTB region of the mitochondrial DNA (Ottoni et al. 2017), we achieved a phylogenetic resolution that was comparable to the one that had been obtained previously in extant cats (Driscoll et al. 2007). DNA was extracted and purified in the high containment laboratories of the University of Leuven and the Jacques Monod Institute using high stringency and DNA contamination prevention procedures (Champlot et al. 2010; Ottoni et al. 2017). Multiplex-PCR setups were performed in the high containment laboratory of the Jacques Monod Institute in Paris dedicated to ancient DNA research, while the construction of DNA libraries from the PCR products as well as Ion Torrent sequencing was carried out there in various modern DNA laboratories (Guimaraes et al. 2016; Ottoni et al. 2017).

It was the aim of our study to establish the phylogeographic structure of the wildcat populations in Europe, the Middle East, and North Africa and to see whether we can detect changes that would disclose the domestication process. To achieve this aim, we analyzed many samples from these regions in order to compensate for samples that would not yield results and to increase the chance of detecting the emergence of a change in the geographical and temporal distribution pattern of mitotypes. This rationale of our experimental design turned out to be justified since DNA in most samples from Egypt, the Levant, and Syria was so poorly preserved

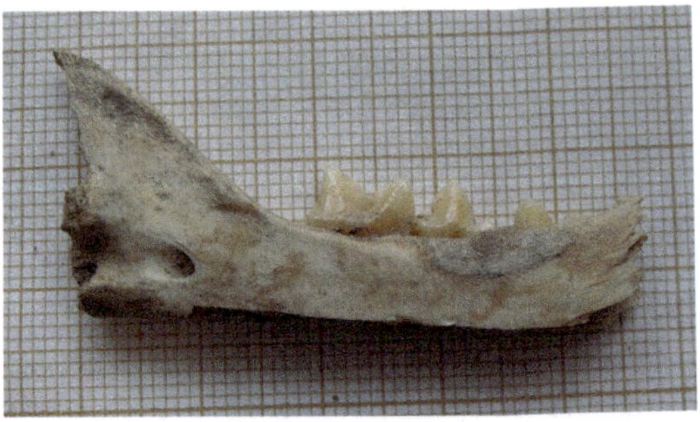

Fig. 4 Cat mandible from the archeological site of Entzheim-Geispolsheim, France, dated to around 2,400 years ago (archeozoological determination: Olivier Putelat)

that they did not yield results. Moreover, it was the sheer number of samples that allowed us to draw conclusions on the domestication history of the cat.

3 Knowledge Gained from Our Paleogenetic Analysis

3.1 *Phylogeography of the Wildcat*

Our oldest samples originated from Western Europe dating to the Mesolithic period prior to the arrival of the Neolithic farmers that colonized Europe coming from Anatolia (e.g., Lazaridis et al. 2016) and introduced agriculture and domestic animals (e.g., Özdogan 2011). These cats as well as more recent ones belonged to the mitotype of the European wildcat *F. s. silvestris* (Fig. 5a). In Southeast Europe, however, we detected in samples dating to 7700 BCE, i.e., before the arrival of the Neolithic farmers, a mitochondrial lineage of *F. s. lybica*, IV-A1, that differed from the one we found in Early Neolithic Anatolia named IV-A* (Fig. 5a). The split date of lineages IV-A1 and IV-A* has been estimated to roughly 20,000 years ago, i.e., the last glacial maximum (LGM), based on the calibration of the maximum likelihood tree obtained from about one third of the mitogenome (Fig. 6). At this time, the Bosphorus was still a land bridge, and it is conceivable that the aridity of the climate in Southeast Europe during the LGM would have allowed *F. s. lybica*, which is adapted to open bushland and deserts, to colonize the Balkans. At the end of the Pleistocene and beginning of the Holocene, forests would have extended again into the area and with it the forest-dwelling *F. s. silvestris*. A mosaic of open bushland and forest environments would have allowed the two different ecotypes, *F. s. silvestris* and *F. s. lybica*, to coexist in this region. And indeed, we found both mitotypes in later periods, and they still coexist there to date although in two independent subspecies (Wozencraft 2005) (Fig. 5b, c).

In Anatolia, not a single specimen analyzed carried the mitotype of *F. s. silvestris*. This finding was unexpected since Anatolia was so far considered to be inhabited by the European wildcat (Yamaguchi et al. 2015). Although we cannot exclude that these latter ones lived in the northern forests of Anatolia lining the Black Sea and had simply not been sampled, our data rather suggest that Anatolia was inhabited by *F. s. lybica*, at least up to the thirteenth century CE (Fig. 5a, b).

3.2 *The Cat During the Neolithic*

Cat remains from the early Neolithic site of Asikli Höyük on the Konya plain in Anatolia that carry mitotype IV-A* testify to both the mitotype of the native Anatolian wildcat population, at least on the Anatolian highlands, and to cats in a human settlement suggestive of some change in the human-cat relationship (Fig. 5b).

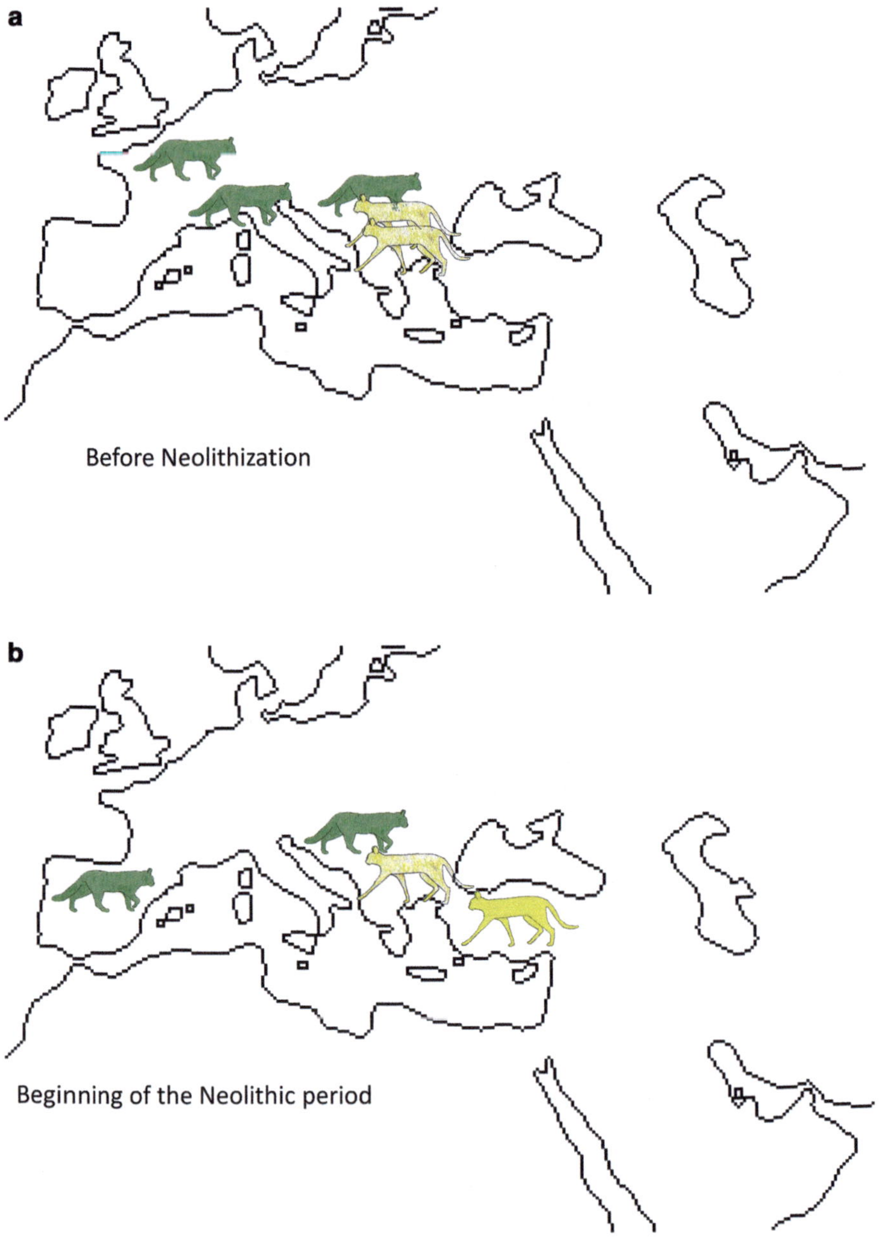

Fig. 5 Occurrence of the various mitotypes (*F. s. silvestris*, green; *F. s. lybica* mitotype IV-A1, fade yellow; *F. s. lybica* mitotype IV-A*, yellow; *F. s. lybica* mitotype IV-C1 and C*, orange; *F. s. lybica* mitotype IV-B, brown; *F. s. lybica* mitotype IV-E, light blue; *F. s. ornate* mitotype, purple) in our data over time. (**a**) Before Neolithization. (**b**) At the beginning of the Neolithic period. (**c**) Neolithic period and Bronze Age. (**d**) Classical Antiquity. (**e**) Byzantine Empire and Middle Ages. (**f**) Ottoman Empire. (**g**) Migration routes of cats deduced from the distribution through time and space of the different mitotypes. Drawings: E.-M. Geigl

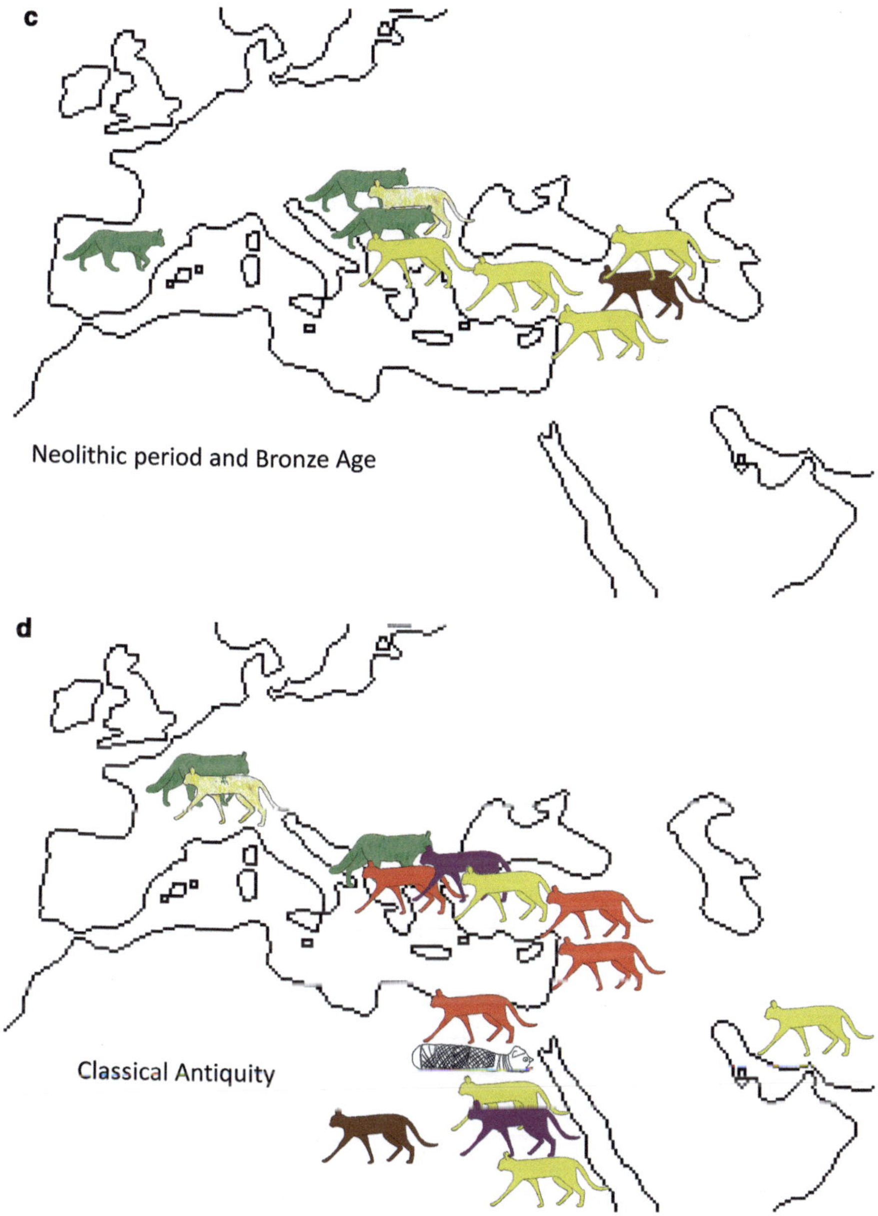

Fig. 5 (continued)

This view is reinforced through cat figurines found in Neolithic sites in Turkey, Syria, and Israel (Vigne et al. 2004). In particular, the catlike animals depicted on ceramic figurines of women from the Anatolian site Haçilar dated to ca. 8000 BP could well be the first iconographic evidence of tamed cats, as proposed already in 1965 (Brentjes 1965) but later questioned (e.g., Gautier 1999). The Anatolian

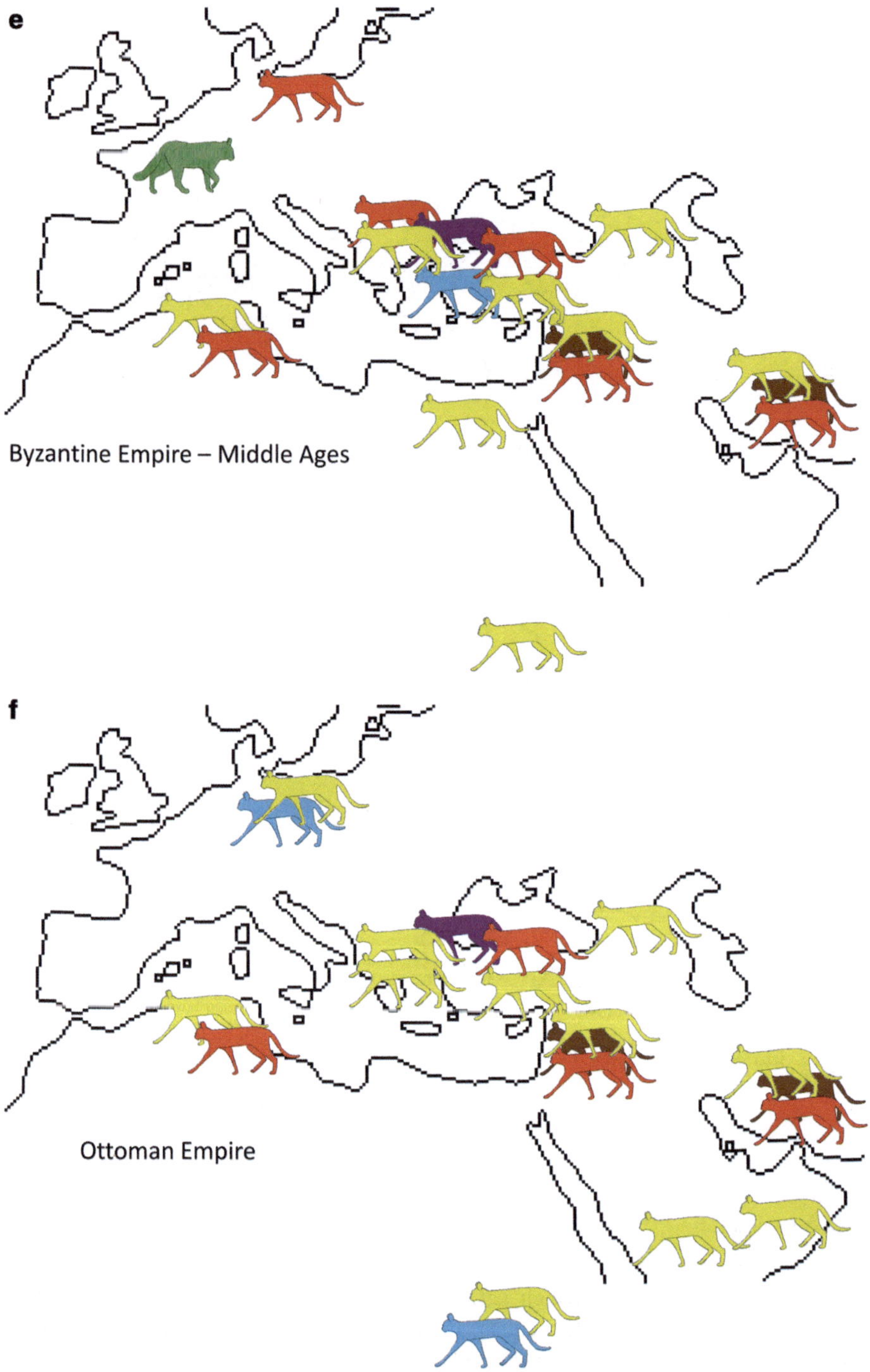

Fig. 5 (continued)

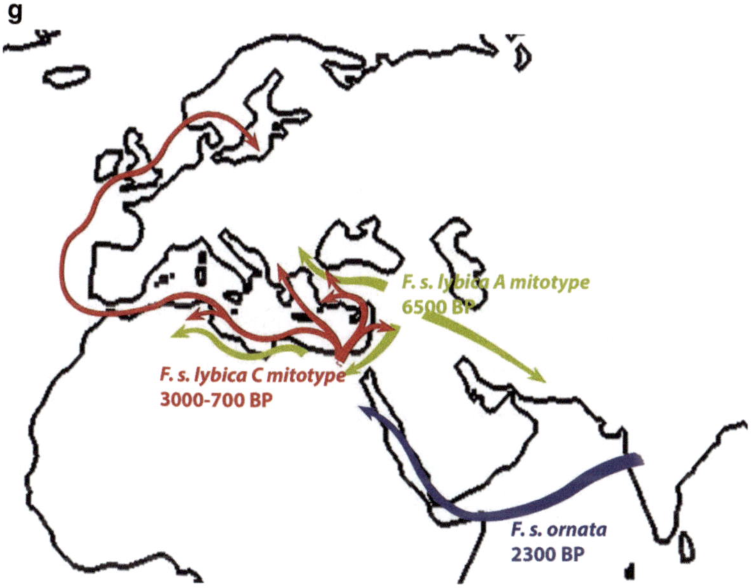

Fig. 5 (continued)

mitotype IV-A* appeared in those of our samples from Southeast Europe that were dated to the time when the Neolithic farmers migrated into Europe (e.g., Lazaridis et al. 2016), but not in earlier European specimens (Fig. 5c). Once again, the most parsimonious interpretation of this finding is that the cats had been translocated to Europe by the early farmers when they crossed the Bosphorus by boat suggesting that the cats were tamed or at least familiar with the presence of people. We also found the Anatolian mitotype in more recent samples (2nd–1st millennium BCE) from the Caucasus, the Levant, Iran, and Egypt (Fig. 5c). Since cats are territorial animals, the most parsimonious conclusion is that these cats have spread following humans on their migrations, in contrast to those types carrying lineage IV-B that were and remained endemic to the Levant. Taken together our genetic, but also archeological and ethological data, it comes naturally to mind that the first steps of the domestication of cats must have happened in the Fertile Crescent. Indeed, it was there where hunter-gatherers established the first permanent settlements, cultivated, collected, and stored large amounts of wild grain as early as 13,000 years ago (Snir et al. 2015) and developed agriculture more than 10,000 years ago (e.g., Bar-Yosef 1998; Belfer-Cohen and Bar-Yosef 2000; Asouti and Fuller 2012). As a consequence, cereals in the field as well as grain accumulations must have attracted rodents that for their parts attracted the local wildcats. Wildcats that overcame their fear of humans and tolerated the presence of other cats would have found much better living conditions in or close to the settlements of the early farmers than in the wild, a situation still encountered nowadays in NA and Arabia (Faure and Kitchener 2009). In this way, the cat would have become a commensal, just as the rodents it preyed on, and its tameness would have been the result of a higher reward

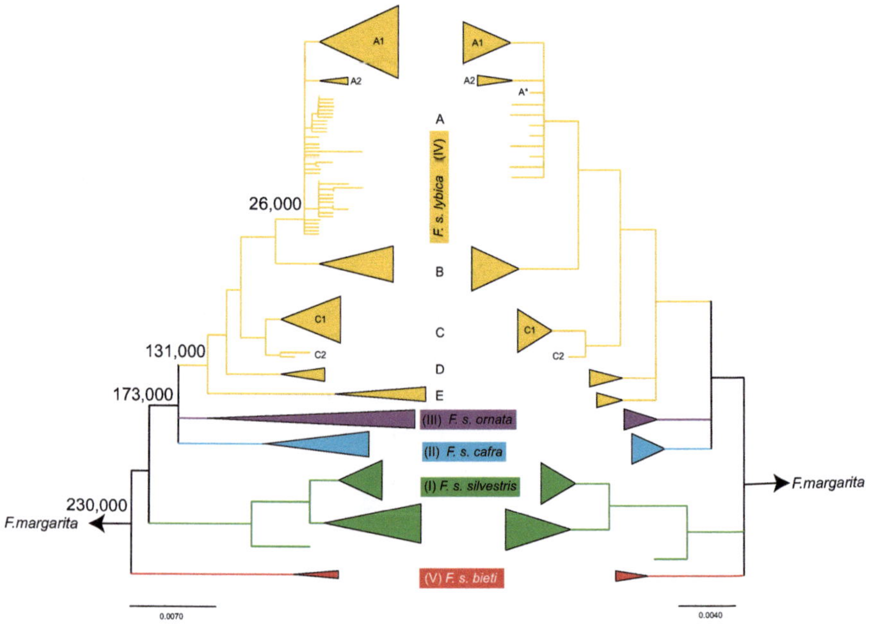

Fig. 6 Maximum-likelihood trees based on haplotype sequence data generated by Driscoll et al. (2007) (left) and of the same haplotype data reduced to the minimum sequence length – 286 bp – generated by our assay (Ottoni et al. 2017) (right). Subspecies and clade names as described by Driscoll et al. (2007) are reported in the rectangular shape between the trees. Names of subclades within clade IV (*F. s. lybica*), as defined by Ottoni et al. (2017), are reported in each tree. Some of the subclades and lineages observed in the 2007 study of Driscoll et al. were collapsed in a single haplotype (A*). Dates of the nodes were as estimated by Driscoll et al. (2007) except for the date of the node at the root of the IV-A haplogroup that we estimated using the Driscoll et al. dates

rate in human company compared to the wild. The human-cat relationship must have satisfied both sides from the very beginning of its establishment. Humans must have benefitted from cats that lived in and around the settlements since cats were the "bulwark" against rodent pests that depleted and spoiled the grain in the fields and in the stores causing starvation, famines, and economic loss. Moreover, the spread of dangerous diseases carried by rodents, and in particular rats, would have been reduced. Egyptian cats were described as very big (Boessneck and von den Driesch 1983) compared to the cats in SWA (Davis 1987; Benecke 1994), as if they had been selected for as ratters (Faure and Kitchener 2009), while present-day domestic cats are poor killers of rats selecting only juveniles (Childs 1986). The find of six rat skeletons in the skeleton of an Egyptian cat (Boessneck and von den Driesch 1983), supports this hypothesis.

Cats would also have fought venomous animals such as scorpions and snakes. Since the latter predate on rodents, they must have come closer to human settlements where they became pests themselves. The depiction of cats as snake killers in Egyptian iconography might therefore be linked directly to the role of cats as adjuncts in early agriculture. The gradual nature of the move from the use of wild stands of cereal to cultivation of domesticated plants that took place at the transition from the hunter-gatherer to the agricultural lifestyle could well constitute the context for an equally gradual development of the human-cat association.

3.3 The Egyptian Cat

The cat mummies from the Ptolemaic period in Egypt that we analyzed carried two different mitochondrial lineages, the Anatolian lineage IV-A* and two lineages, IV-C1 and IV-C*, that we had not found in more ancient cats from other areas (Fig. 5d). These latter lineages were present also in Anatolian cats from the Roman-Byzantine period, where about half of the samples carried these IV-C lineages and a fourth carried the native lineage IV-A* and the rest lineages IV-E and *F. s. ornata* (Fig. 5e). We found them also in samples from the Levant and Iran dated to the Roman period and later even in a Viking port at the Baltic Sea (Fig. 5e).

This fast and massive spread of the "Egyptian cat" over large distances hints to sea vessels as dissemination means. Indeed, cats must have accompanied seafarers, from the Phoenicians (called "cat thieves" by the ancient Egyptians (Faure and Kitchener 2009)) over Greeks, Etruscans, to Romans, on their ships keeping the vessels rodent-free and taking advantage of fish as a food source that naturally is not readily available for cats. Rodents on ships are a threat not only to the food reserves but also to all organic material such as ropes that are being gnawed on. The usefulness of ship's cats can be deduced from the fact that until 1975 cats were compulsory on all the vessels of the British Navy (Beadle 1977) and some of them have been immortalized in novels, poems, and god statues erected to celebrate their memory. That this was a widespread custom at the latest during Roman times can be deduced from a cat sample from the Egyptian-Roman port of Berenike at the Red Sea that carried the mitotype of the Asian wildcat, *F. s. ornata*. This port is known for its lively trading connections with India, and the sailors spent several months on either end of their voyage giving their ship's cats the opportunity to mix with the local Asian wildcat the offspring of which would sometimes make the voyage back to the area where their ancestors came from. Alternatively, or in addition, Asian wildcats boarded the sea vessels and were shipped to Egypt and to the Eastern Mediterranean such as Anatolia, where we also found this mitotype in a few samples from coastal sites. Thus, our ancient mitochondrial data suggest that the spread of the "domestic" cat over the Ancient World occurred through seafaring merchants and soldiers, although the Roman army must have played an important role for the spread of the cat within the European continent as evidenced from archeological finds (e.g., Bökönyi 1974, 1984; Clutton-Brock 1999). During the eighth and ninth century CE, the Vikings, traders as well as raiders, continued the dispersal of cats as evidenced by our finding of the Egyptian mitotype in the Baltic sea port of Ralswiek from the eighth century CE and so did the returning crusaders during the eleventh and twelfth century CE (Lepetz 1996; Sunquist and Sunquist 2002).

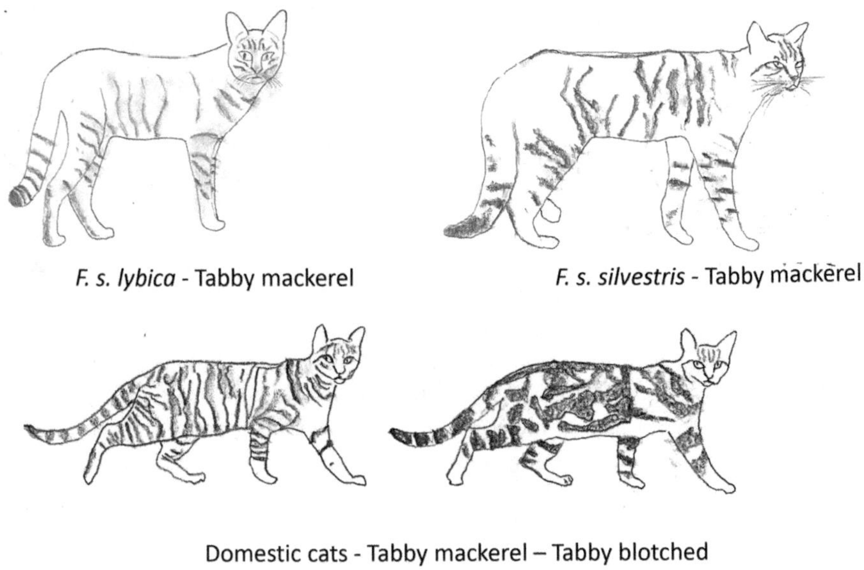

Fig. 7 Tabby mackerel coat and tabby blotched coat pattern in wildcats and domestic cats. Drawings: E.-M. Geigl

3.4 Selection of Coat Patterns in the Domestic Cat

Ancient DNA analyses have shown for some animals such as horses that domestication led to an increase in the variety of coat colors as an early signature of the domestication process (Ludwig et al. 2009). In order to analyze whether this was also the case in cats and thereby obtain some information about the physical changes that the domestic cat potentially underwent through the domestication process, we analyzed the *transmembrane aminopeptidase A* (*Taqpep*) gene coding for the tabby coat pattern, the striped coat pattern of wildcats. A nonsense mutation (W841X) in exon 17 of the *Taqpep* gene causes the blotched pattern present in a large percentage of domestic cats (Kaelin et al. 2012) (Fig. 7). Indeed, the coat marking variations are used to discriminate wildcats and hybrids (e.g., Beaumont et al. 2002; Pierpaoli et al. 2003; Oliviera et al. 2008), while other coat color variations are described mainly in domestic cat breeds (for review, see Lyons 2015).

We investigated in all of our samples three single nucleotide polymorphisms (SNPs) in this recessive allele and found it not earlier than just before the thirteenth century in SWA. It then increased in frequency in domestic cats in Europe, SWA, and Africa suggesting that it was selected for. If we take this marker of appearance as a witness of the timing of selection for physical traits, this would represent a very late selection process in cats compared to other animals. This is not in contradiction to the genomic, archeological, and iconographic evidence but only underlines the hypothesis that the cat has not been actively selected through breeding. Indeed, 85% of the 40–50 extant cat breeds arose only in the past 75 years (Kurushima et al. 2013).

We would like to hypothesize that it was not necessary to "change" the cat since it was the perfect animal for human societies living in farms, villages, and towns. From the very beginning, it naturally performed the tasks that it was expected to do, i.e., hunting vermin and venomous animals, without being dangerous or trying to escape.

4 Conclusions and Future Perspectives

To conclude, we showed that two different clades of cats colonized subsequently the Ancient World, an Anatolian and an Egyptian one, confirming that both locations played an important role in the domestication history of cats. The dates and areas where these clades appeared in our extensive data set speak for translocations as the mechanism of spread: first the somehow tamed cat followed humans on their migrations into new territories (Fig. 5g), and later it was transported as ship's cats with seafarers on routes of trading and warfare confirming a previous hypothesis (Kirk 1977).

Since the underlying cause of cat domestication, grain accumulations in human settlements attracting rodents and thereby attracting cats preying on them, was generalized when the Neolithic began to spread, it is likely that the domestication process had several foci. Once cats interacted more closely with humans, they were translocated over much larger distances than they would have roamed on their own. Thus, any new genetic change affecting cats in one of the areas that would have been interesting to humans could then have been rapidly propagated to other locations. We believe that this is what happened with the Egyptian cats. To detect this process, as we did in our data set, we need to assume that, apart from Egypt, the bonds between cats and humans were probably still loose at the time of Classical Antiquity, and these cats were not abundant enough to dilute the genetic contribution of the newcomers. Later on, the process of spread of novel genetic variations must have continued but becomes increasingly more difficult to detect with the cat-human relationship growing closer everywhere. We think that one should consider the domestication process of cats as a gradual spread over several thousands of years rather than a limited number of key founding domestication events. The simplified view of a limited number of key founding events is prone to be challenged by new archeological finds.

Further insights from the study of whole genomes of ancient cats at archeological key sites are expected to refine our knowledge of the domestication process of the cat. Indeed, additional information could be obtained if it was possible to sequence the genomes of cat remains predating the period when cats started to be translocated in order to establish baseline genomes, to which genomes from later stages of the cat domestication process can be compared. Such baselines are important since it is clear that hybridization between wild and tamed cats associated with the human niche was recurrent with the consequence of blurring the picture of the process that modified the genome of the house cats. The methodologies for this enterprise exist, but the task is not easy since DNA is poorly preserved in most skeletal remains from the presumed centers of cat domestication due to the high temperature in these regions that is detrimental to DNA preservation. Moreover, it is likely that many samples need to be analyzed to detect gradual changes in the frequency of a limited number of relevant alleles. Nevertheless, this analysis is necessary if one wants to confirm the hypothesis deduced from the diachronic analysis of the tabby coat color allele (Ottoni et al. 2017) that selection of physical traits has been a late event in the domestication process, which in turn suggests that the domestication process of the cat followed the commensal pathway (Zeder 2012) and was relaxed and long.

Acknowledgments We would like to thank an anonymous reviewer for his/her interesting propositions.

References

Asouti E, Fuller DQ. From foraging to farming in the southern Levant: the development of Epipaleolithic and pre-pottery Neolithic plant management strategies. Veg Hist Archaeobotany. 2012;21:149–62.

Bar-Yosef O. The Natufian culture in the Levant, threshold to the origins of agriculture. Evol Anthropol. 1998;6(5):159–77.

Beadle M. The cat: history, biology, behavior. New York: Simon and Schuster; 1977.

Beaumont M, Barratt EM, Gottelli D, Kitchener AC, Daniels MJ, Pritchard JK. Genetic diversity and introgression in the Scottish wildcat. Mol Ecol. 2002;10:319–36.

Belfer-Cohen A, Bar-Yosef O. Early sedentism in the near east: a bumpy ride to village life. In: Kuijt I, editor. Life in Neolithic farming communities: social organization, identity, and differentiation. New York: Kluwer/Plenum Press; 2000. p. 19–37.

Benecke N. Der Mensch und seine Haustiere - Die Geschichte einer jahrtausendealten Beziehung. Stuttgart: Konrad Theiss Verlag; 1994.

Boessneck J, von den Driesch A. Ein Katzenskelett der Römerzeit aus Quscir (Koser) am Roten Meer. Spixiana. 1983;6:285–9.

Bökönyi S. History of domestic animals. Mammals in central and eastern Europe. Publicationes Instituti Archaeologici Academiae Scientiarum Hungaricae, Studia Archaeologica. Budapest: Akadémiai Kiado; 1974.

Bökönyi S. Animal husbandry and hunting in Tac-Gorsium: the vertebrate fauna of a Roman town in Pannonia. Publicationes Instituti Archaeologici Academiae Scientiarum Hungaricae, Studia archaeologica. Budapest: Akadémiai Kiado; 1984.

Brentjes B. Die Haustierwerdung im Orient. Der Neue Brehm. Wittenberg: Ziemsen Verlag; 1965.

Champlot S, Berthelot C, Pruvost M, Bennett EA, Grange T, Geigl EM. An efficient multistrategy DNA decontamination procedure of PCR reagents for hypersensitive PCR applications. PLoS One. 2010;5(9):e13042. https://doi.org/10.1371/journal.pone.0013042.

Childs JE. Size-dependent predation on rats (*Rattus norvegicus*) by house cats (*Felis catus*) in an urban setting. J Mammology. 1986;67:196–9.

Clutton-Brock J. A natural history of domesticated animals. London: Cambridge University Press; 1999.

Cote NM, Daligault J, Pruvost M, Bennett EA, Gorge O, Guimaraes S, Capelli N, Le Bailly M, Geigl EM, Grange T. A new high-throughput approach to genotype ancient human gastrointestinal parasites. PLoS One. 2016;11(1):e0146230. https://doi.org/10.1371/journal.pone.0146230.

Davis SJM. The archaeology of animals. London: B.T. Batsford; 1987.

Driscoll CA, Menotti-Raymond M, Roca AL, Hupe K, Johnson WE, Geffen E, Harley EH, Delibes M, Pontier D, Kitchener AC, Yamaguchi N, O'Brien SJ, Macdonald DW. The near eastern origin of cat domestication. Science. 2007;317(5837):519–23. https://doi.org/10.1126/science.1139518.

Engels DW. Classical cats: the rise and fall of the sacred cat. Abingdon: Routledge; 2001.

Ewing E. Fur in dress. London: Batsford; 1981.

Faure E, Kitchener AC. An archaeological and historical review of the relationships between felids and people. Anthrozoös. 2009;22(3):221–38. https://doi.org/10.2752/175303709X457577.

Gautier A. Fauna, domesticated. Encyclopedia of the archaeology of ancient Egypt. London: Routledge; 1999.

Guimaraes S, Pruvost M, Daligault J, Stoetzel E, Bennett EA, Cote NM, Nicolas V, Lalis A, Denys C, Geigl EM, Grange T. A cost-effective high-throughput metabarcoding approach powerful enough to genotype ~44 000 year-old rodent remains from northern Africa. Mol Ecol Resour. 2016;17(3):405–17. https://doi.org/10.1111/1755-0998.12565.

Kaelin CB, Xu X, Hong LZ, David VA, McGowan KA, Schmidt-Kuntzel A, Roelke ME, Pino J, Pontius J, Cooper GM, Manuel H, Swanson WF, Marker L, Harper CK, van Dyk A, Yue B, Mullikin JC, Warren WC, Eizirik E, Kos L, O'Brien SJ, Barsh GS, Menotti-Raymond M. Specifying and sustaining pigmentation patterns in domestic and wild cats. Science. 2012;337(6101):1536–41. https://doi.org/10.1126/science.1220893.

Kirk M. The everlasting cat. New York: Galahad Books; 1977.

Krüger M, Hertwig ST, Jetschke G, Fischer MS. Evaluation of anatomical characters and the question of hybridization with domestic cats in the wildcat population of Thuringia, Germany. J Zool Syst Evol Res. 2009;47:268–82. https://doi.org/10.1111/j.1439-0469.

Kurushima JD, Lipinski MJ, Gandolfi B, Froenicke L, Grahn JC, Grahn RA, Lyons LA. Variation of cats under domestication: genetic assignment of domestic cats to breeds and worldwide random bred populations. Anim Genet. 2013;44(3):311–24. https://doi.org/10.1111/age.12008.

Lazaridis I, Nadel D, Rollefson G, Merrett DC, Rohland N, Mallick S, Fernandes D, Novak M, Gamarra B, Sirak K, Connell S, Stewardson K, Harney E, Fu Q, Gonzalez-Fortes G, Jones ER, Roodenberg SA, Lengyel G, Bocquentin F, Gasparian B, Monge JM, Gregg M, Eshed V, Mizrahi AS, Meiklejohn C, Gerritsen F, Bejenaru L, Bluher M, Campbell A, Cavalleri G, Comas D, Froguel P, Gilbert E, Kerr SM, Kovacs P, Krause J, McGettigan D, Merrigan M, Merriwether DA, O'Reilly S, Richards MB, Semino O, Shamoon-Pour M, Stefanescu G, Stumvoll M, Tonjes A, Torroni A, Wilson JF, Yengo L, Hovhannisyan NA, Patterson N, Pinhasi R, Reich D. Genomic insights into the origin of farming in the ancient near east. Nature. 2016;536(7617):419–24. https://doi.org/10.1038/nature19310.

Lepetz S. L'animal dans la société gallo-romaine de la France du Nord. Revue Archéologique de Picardie (Amiens). 1996;S12.

Librado P, Gamba C, Gaunitz C, Der Sarkissian C, Pruvost M, Albrechtsen A, Fages A, Khan N, Schubert M, Jagannathan V, Serres-Armero A, Kuderna LFK, Povolotskaya IS, Seguin-Orlando A, Lepetz S, Neuditschko M, Theves C, Alquraishi S, Alfarhan AH, Al-Rasheid K, Rieder S, Samashev Z, Francfort HP, Benecke N, Hofreiter M, Ludwig A, Keyser C, Marques-Bonet T, Ludes B, Crubezy E, Leeb T, Willerslev E, Orlando L. Ancient genomic changes associated with domestication of the horse. Science. 2017;356(6336):442–5. https://doi.org/10.1126/science.aam5298.

Luce J-M. Les chats dans l'Antiquité grecque. In: Bellier C, Cattelain L, Cattelain P, editors. Chiens et chats dans la Préhistoire et l'Antiquité. Bruxelles: Editions de Cedarc; 2015.

Ludwig A, Pruvost M, Reissmann M, Benecke N, Brockmann GA, Castanos P, Cieslak M, Lippold S, Llorente L, Malaspinas AS, Slatkin M, Hofreiter M. Coat color variation at the beginning of horse domestication. Science. 2009;324(5926):485. https://doi.org/10.1126/science.1172750.

Lyons LA. DNA mutations of the cat. J Feline Med Surg. 2015;17:203–19.

Machon C. Le culte du chat en Egypte ancienne. In: Bellier C, Cattelain L, Cattelain P, editors. Chiens et chats dans la Préhistoire et l'Antiquité. Bruxelles: Editions de Cedarc; 2015.

Málek J. The cat in ancient Egypt. Rev ed. London: The British Museum Press; 1993–2006.

Mattucci F, Oliveira R, Lyons LA, Alves PC, Randi E. European wildcat populations are subdivided into five main biogeographic groups: consequences of Pleistocene climate changes or recent anthropogenic fragmentation? Ecol Evol. 2015;6(1):3–22. https://doi.org/10.1002/ece3.1815.

Montague MJ, Li G, Gandolfi B, Khan R, Aken BL, Searle SMJ, Minx P, Hillier LW, Koboldt DC, Davis BW, Driscoll CA, Barr CS, Blackistone K, Quilez J, Lorente-Galdos B, Marques-Bonet T, Alkan C, Thomas GWC, Hahn MW, Menotti-Raymond M, O'Brien SJ, Wilson RK, Lyons LA, Murphy WJ, Warren WC. Comparative analysis of the domestic cat genome reveals genetic signatures underlying feline biology and domestication. Proc Natl Acad Sci U S A. 2014;111(48):17230–5. https://doi.org/10.1073/pnas.1410083111.

Morrison-Scott TCS. The mummified cats of ancient Egypt. Proc Zool Soc London. 1952;121:861–7.

Müller F. Körpermerkmale als Unterscheidungskriterien zwischen wildfarbenen Hauskatzen (*Felis s. catus*) und Wildkatzen (*Felis silvestris silvestris, Felidae*) aus Mitteleuropa. Beiträge zur Jagd- und Wildforschung. 2011;36:359–68.

Oliviera R, Godinho R, Randi E, Alves PC. Hybridization vs conservation: are domestic cats threatening the genetic integrity of European wildcat (*Felis silvestris silvestris*) populations in Iberian peninsula? Philos Trans R Soc Lond B Biol Sci. 2008;363:2953–61.

Ottoni C, Van Neer W, De Cupere B, Daligault J, Guimaraes S, Peters J, Spassov N, Prendergast ME, Boivin NL, Morales-Muñiz A, Balasescu A, Benecke N, Boroneanţ A, Buitenhuis H, Chahoud J, Crowther A, Llorente L, Manaseryan N, Monchot H, Onar V, Osypińska M, Putelat O, Quintana Morales EM, Studer J, Wierer U, Decorte R, Grange T, Geigl EM. The palaeogenetics of cat dispersal in the ancient world. Nat Ecol Evol. 2017;1:0139. https://doi.org/10.1038/s41559-017-0139.

Özdogan M. Archaeological evidence on the westward expansion of farming communities from eastern Anatolia to the Aegean and the Balkans. Curr Anthropol. 2011;52:S415–30.

Pierpaoli M, Biro ZS, Herrmann M, Hupe K, Fernandes M, Ragni B, Szemethy L, Randi E. Genetic distinction of wildcat (Felis silvestris) populations in Europe, and hybridization with domestic cats in Hungary. Mol Ecol. 2003;12(10):2585–98.

Rothwell T. Evidence for taming of cats – response to J.-D. Vigne and J. Guilaine. Science. 2004;305:1714.

Snir A, Nadel D, Groman-Yaroslavski I, Melamed Y, Sternberg M, Bar-Yosef O, Weiss E. The origin of cultivation and proto-weeds, long before Neolithic farming. PLoS One. 2015;10(7):e0131422. https://doi.org/10.1371/journal.pone.0131422.

Sunquist MN, Sunquist F. Wild cats of the world. Chicago: University of Chicago Press; 2002.

Van Neer W, Linseele V, Friedman R, De Cupere B. More evidence for cat taming at the Predynastic elite cemetery of Hierakonpolis (Upper Egypt). J Arch Sci. 2014;45:103–11. https://doi.org/10.1016/j.jas.2014.02.014.

Vigne J-D, Guilaine J, Debue K, Haye L, Gérard P. Early taming of the cat in Cyprus. Science. 2004;304(5668):259. https://doi.org/10.1126/science.1095335.

Von den Driesch A. Kulturgeschichte der Hauskatze. In: Schmidt V, Horzinek MC, editors. Krankheiten der Katze, vol. 1. Jena: Gustav Fischer Verlag; 1992. p. 17–40.

Warmenbol E. Le chien et le chat en Egypte pharaonique: à la vie, à la mort. In: Chiens et chats dans la Préhistoire et l'Antiquité. Bruxelles: Editions du Cedarc; 2015.

Wozencraft C. Felis silvestris. In: Wilson DE, Reeder DM, editors. Mammal species of the world. 3rd ed. Baltimore: Johns Hopkins University Press; 2005. p. 2142.

Yamaguchi N, Kitchener AC, Driscoll C, Nussberger B. Felis silvestris. In: The IUCN Red List of Threatened Species 2015. 2015.

Zeder MA. Pathways to animal domestication. In: Gepts P, Famula TR, Bettinger RL, et al., editors. Biodiversity in agriculture. Cambridge: Cambridge University Press; 2012. p. 227–59.

An Ancient DNA Perspective on Horse Evolution

Ludovic Orlando

Abstract With the development of fast transportation and cavalry, the horse represents the domestic animal that most influenced human history. Yet, the evolutionary history of the horse was not limited to the last 5,500 years since it was first domesticated. It is rooted within a 55 million-year-long time span, where a large number of lineages radiated and became extinct. Together with zebras, hemiones, and donkeys, the horse belongs to the genus *Equus*, the only remaining equine lineage living in the planet. Even though the survival of exploitable ancient DNA molecules is at best limited to the last million years, the sequencing of short mitochondrial and nuclear DNA fragments, as well as of complete genome sequence from archaeological and paleontological material, has illuminated our understanding of the evolutionary history of the horse family. Such work has not only revisited the evolutionary tempo of *Equus* and the phylogenetic relationships within and outside the genus but also revealed how past climates and human activities have shaped the genetic makeup of the horse species.

Keywords Ancient DNA · Climate change · Conservation · Domestication · *Equus* · Horses · Speciation

1 Introduction

The history of ancient DNA (aDNA) is intimately linked to the horse family as the first aDNA sequence ever characterized was obtained from the quagga zebra (*Equus quagga quagga*), an extinct member of this family, closely related to the plains zebra (Higuchi et al. 1984). With a short stretch of the mitochondrial DNA (mtDNA) sequence, a new

L. Orlando (✉)
Centre for GeoGenetics, Natural History Museum of Denmark, Copenhagen, Denmark

Laboratoire AMIS, CNRS UMR, Université de Toulouse, Université Paul Sabatier (UPS), Toulouse, France
e-mail: ludovic.orlando@univ-tlse3.fr

Charlotte Lindqvist and Om P. Rajora (eds.), *Paleogenomics*, Population Genomics [Om P. Rajora (Editor-in-Chief)], https://doi.org/10.1007/13836_2018_23,

research area was born, which makes use of the tiny pieces of aDNA preserved in museum specimens and subfossil material to recover direct genetic information from the past (see Ermini et al. (2015), Hofreiter et al. (2015), Pedersen et al. (2015), and Llamas et al. (2017) for reviews). Within no more than three decades and thanks to the developments of high-throughput DNA sequencing, aDNA research has moved from the sequencing of mainly minute amounts of maternally transmitted mitochondrial markers to complete genome sequencing [see Stoneking and Krause (2011) and Orlando et al. (2015) for reviews], first of the woolly mammoth (Miller et al. 2008), ancient human individuals (Rasmussen et al. 2010, 2011), and archaic hominins (Green et al. 2010; Reich et al. 2010; Meyer et al. 2012) but, soon enough, of members of the horse family. The complete genome of the extinct quagga was released in December 2014 and was obtained from only ~50 mg of hair shaft (Jónsson et al. 2014). A year earlier, a bone preserved in the oldest permafrost reported in the planet delivered enough DNA molecules to reconstruct a first draft of the nuclear genome of a Middle Pleistocene horse (Orlando et al. 2011, 2013). It was dated to 560,000–780,000 years (kyrs) ago (Froese et al. 2008) and demonstrated the last million years as a credible temporal limit for genome sequencing in frozen conditions. Still in 2014, the genome sequence of animals that lived prior to domestication times was characterized, enabling the first direct comparison with the genomes of present-day domesticates. Perhaps not surprisingly given their importance for human history, horses provided the first such comparisons (Schubert et al. 2014) and have been followed by others since, including aurochs/cattle (Park et al. 2015; Braud et al. 2017), wolves/dogs (Skoglund et al. 2015; Frantz et al. 2016), and maize (Da Fonseca et al. 2015; Ramos-Madrigal et al. 2016) (see MacHugh et al. 2017; Scheu 2017 for reviews).

Ancient DNA studies applied to equine subfossil material have greatly improved our understanding of the evolutionary history of the horse family. These have addressed a broad range of topics, including phylogenetics, extinction, and population dynamics, as well as domestication, conservation, and human management. The present chapter highlights a broad, complementary panel of the literature within these research areas.

2 Phylogenetics and Taxonomy

2.1 *The Rise of Equus*

The horse belongs to the genus *Equus*, a genus defined by a series of cranial and postcranial morphological features (Eisenmann and Baylac 2000; Franzen 2010). *Equus* comprises three extant species of zebras (the Grévy's zebra, *E. grevyi*; the plains zebra, *E. quagga*; and the mountains zebra, *E. zebra*) and three extant species of asses (the hemione, *E. hemionus*; the Tibetan kiang, *E. kiang*; and the African ass, *E. africanus*) (Fig. 1; see Orlando 2015 for a review), the latter of which has been domesticated some 5,500 years ago to form the domestic donkey (*E. asinus*, Rossel et al. 2008). Molecular phylogenies based on partial mtDNA (Weinstock et al. 2005; Orlando et al. 2009) and/or nuclear genes (Steiner et al. 2012), complete

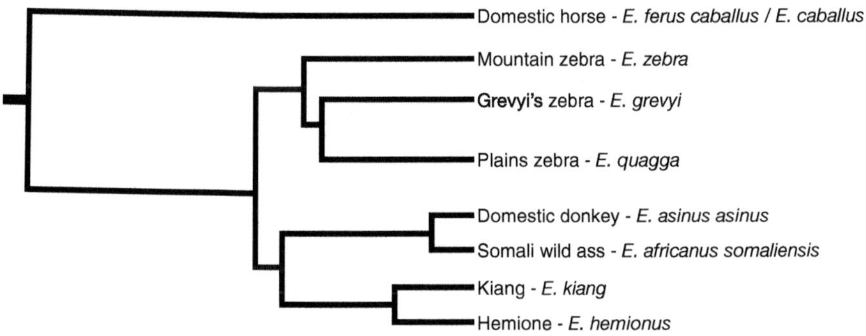

Fig. 1 Phylogenetic relationships within living members of the equine family [adapted from Jónsson et al. (2014)]

mitogenomes (Vilstrup et al. 2013), and whole genome sequences (Jónsson et al. 2014) support a deep split within *Equus*, where the horse (*E. caballus*) and other extinct lineages showing horse-like morphological features, the so-called caballines, form a first clade and where both zebras and asses form a second group, the so-called stenonines.

The time to the most recent common ancestor (tMRCA) of *Equus* remained controversial, with estimates ranging between ~2 and over 6 million years (myrs) depending on the molecular clock calibration methods considered (see Steiner et al. 2012; Orlando et al. 2013, references therein). Sequencing a draft genome of a Middle Pleistocene horse preserved in the permafrost of the Yukon territory, Orlando and colleagues found that the MRCA of caballines and stenonines lived at least some ~4–4.5 myrs ago (Orlando et al. 2013). The authors used coalescence simulations under a divergence model to predict, in the absence of gene flow, the credible distribution range of the F summary statistics between the ancient and present-day horse genomes. This statistics was originally introduced by the Neanderthal Genome Consortium (Green et al. 2010; Reich et al. 2010) and represents the probability to sample a derived allele in the ancient genome at a site where present-day genomes are heterozygous. The F-statistics increases for younger population divergence times as the derived neutral allele is then highly likely to not be lost in any descending population. Conversely, for older population divergences, the persistence of the neutral derived allele in both descending populations is highly unlikely, and the F-statistics decreases. Carefully accounting for demographic changes within the population leading to present-day horses, Orlando and colleagues found that coalescence simulations, where the genome-wide mutation rate was calibrated assuming ~4–4.5 myrs for the tMRCA of caballines and stenonines and where the divergence between ancient and present-day populations was constrained by the age of the Middle Pleistocene fossil, could reproduce the F-statistics observed between the ancient and present-day horse genomes. This suggested that the MRCA of all living *Equus* lived ~4–4.5 myrs ago, in line with the oldest dates of the monodactyle *Plesippus simplicidens*, one candidate for the earliest fossil of *Equus* (MacFadden and Carranza-Castaneda 2002). Further work based on the diploid genome sequence of all living members of the genus and exploiting the CoalHMM statistical model (Mailund et al. 2012) revealed that the caballine and stenonine lineages coexisted in sympatry in North America and remained genetically connected until 2.9–3.8 myrs ago, before they finally became reproductively fully isolated (Jónsson et al. 2014).

2.2 Paleontological Over-splitting

The relatively limited number of equine species living today is in sharp contrast to the large number of Pleistocene species described in the fossil record. Over 50 equine species have been named in the Pleistocene of the Americas, and genetic analyses of short DNA fragments of the mitochondrial hypervariable region have revealed this as a typical case of paleontological over-splitting (Weinstock et al. 2005). In fact, most of the morphological diversity found in the Pleistocene of the Americas can be lumped into three main genetic clusters. The first is the caballine lineage leading to the horse, which only survived in the Old World before it was reintroduced in the Americas following the Spanish conquest. The second is referred to as New World stilt-legged (NWSL) horses as they were endemic to North America and shared hemione-like gracile limbs. They disappeared from the fossil record of Alaska and the Yukon territory approximately ~31 kyrs ago (Guthrie 2003, 2006). Their mtDNA sequences, based on partial fragments (Weinstock et al. 2005), or on the whole mitogenome (Vilstrup et al. 2013), were clearly divergent to those of hemiones (and any extant stenonine), contradicting scenarios positing their origin following the expansion of the hemione distribution range through Beringia during past glacial periods. Finally, a third group includes species endemic to South America, the so-called hippidiforms for their characteristic morphological features, including robust limbs and a prominent nasal notch (Orlando et al. 2003, 2009; Der Sarkissian et al. 2015a).

The cranial and postcranial morphologies of hippidiforms and *Equus* are so different that these lineages have been modeled as two independent branches in the equine tree, possibly diverging some ~10–12 myrs ago (Alberdi and Prado 1993, 1998; MacFadden 1997). The first partial mitochondrial sequences obtained for three Patagonian remains yet indicated hippidiforms nested within *Equus*, as a sister group to caballine horses (Orlando et al. 2003). This first suggested that these specimens belonged to *Equus* that were misclassified within *Hippidion*, a typical hippidiform genus. Further morphological analysis of the specimens rejected the possibility of sample misassignment (Alberdi et al. 2005), and additional partial sequences of the mitochondrial hypervariable region, including specimens from the montane Peruvian range, confirmed the phylogenetic placement within *Equus* (Weinstock et al. 2005). However, the closest living relative of *Equus* – the rhinos and tapirs – diverged some ~55 myrs ago from the equine lineage, which makes the location of the equine root particularly difficult based on partial mitochondrial sequences only. Despite supporting a root placement nesting hippidiforms within *Equus*, as a sister group to caballine horses, a further study indicated possible alternative roots more in line with classical evolutionary models based on paleontological evidence, resulting in hippidiforms and *Equus* in two reciprocally monophyletic separate lineages (Orlando et al. 2009). The latter was finally confirmed using whole mitogenome sequences of nine specimens from two *Hippidion* species (Der Sarkissian et al. 2015a). The divergence time between hippidiforms and *Equus* was, however, estimated to be ~5.6–6.5 myrs ago, which likely roots the deep origins of hippidiforms in North America as the South American *Hippidion* is not known prior to ~3 myrs ago.

Another example of marked morphological plasticity within a single genetic group is provided by the genetic analyses of specimens revealed to be conspecific to South African plains zebras. These included the extinct quagga (Leonard et al. 2005; Jónsson et al. 2014), which showed a range of coat-color morphotypes along its South African range, and the extinct giant Cape zebra, which was both larger and bigger than present-day plains zebras and became extinct by the end of the late Pleistocene (Orlando et al. 2009).

2.3 *Identifying Species and Hybrids*

Complete skulls are classically considered to provide the best morphological characters to discriminate extant equine species and reconstruct their phylogenetic relationships to fossil specimens [Groves and Willoughby 1981; Eisenman 1998; see, however, Cucchi et al. 2017 for recent methodological developments applying geometric morphometrics, GMM (see Lawling and Polly 2010 for a review), approaches to occlusal enamel folding patterns of cheek teeth]. Complete skulls are, however, relatively rare in the fossil record, which often consists of fragmentary remains, limiting the identification of species and possible hybrid forms. In contrast, limited amounts of genetic data can supply archaeologists with such information. For example, a short minibarcode (<90 bp) within the mitochondrial hypervariable region has been found to show limited variation within species but large differences between species (Orlando et al. 2009). Amplifying and sequencing such fragments can, thus, provide a first candidate for the taxonomy of the remains analyzed. The sequencing of the whole mitogenome, instead of the sole minibarcode, can offer a complementary approach (although more work-intensive and less cost-effective), especially now that simple, robust target enrichment methods for the mitogenome are available (Maricic et al. 2010). Applying this methodology, Cardoso and colleagues could identify donkeys within the bone assemblages of a Chalcolithic fortified site from Portugal (Cardoso et al. 2013). This site predated by more than one millennium the supposed introduction of the donkey in the region by the Phoenicians, revealing that even minute amounts of genetic information can significantly change our understanding of the past species dynamics.

However, as mtDNA is maternally inherited, the approach above cannot address whether the specimen analyzed was a purebred individual or a hybrid. Yet, mules – the offspring of a jack and a mare – have been extensively used in the Roman Army for transportation (Johnstone 2004), and their identification requires the analysis of markers that are both maternally and paternally inherited. Leveraging the whole genome sequence data now available for all living members of the horse family (Jónsson et al. 2014), the Zonkey pipeline provides a fast and cost-effective method for a molecular identification of the species, the sex, and the hybrid status of equine remains in the archaeological record. The methodology is based on shotgun sequencing of raw extracts and requires no more than 10,000 endogenous reads (Schubert et al. 2017). F1 hybrids are detected based on both ADMIXTURE profiles (Alexander et al. 2009) and TreeMix phylogenetic reconstruction (Pickrell and Pritchard 2012), allowing one migration edge, whereas the molecular sex is estimated from the proportion of high-quality reads aligned against the X chromosome and autosomes. Applied to 18 archaeological remains, the approach revealed the presence of mules in situations where morphological evidence remained inconclusive (Schubert et al. 2017). This was, for example, the case of six genetically identified mules from the Byzantine site of Yenikapi (Turkey) and one genetically identified mule from the Roman site of Dangstetten (Germany). The approach proved also useful for the identification of species in Southwest Iran, where no less than four equine species coexisted in sympatry and where identifying which among the horse, the donkey, the hemione, and the now extinct European ass, *Equus hydruntinus*, was present in the Chalcolithic faunal assemblage of Mehr Ali based on tooth morphology alone was impossible.

The difficulty in assigning bone and tooth archaeological remains to a particular equine species is perhaps best illustrated in the case of the hydruntine, also known as the European wild ass, *Equus hydruntinus*. This species shows a mosaic of morphological characters, some of which are reminiscent of a stenonine species that lived prior to ~2 myrs ago, but others are present in donkeys, zebras, and hemiones (see references in Orlando et al. 2006), and others again are specific to this taxon. Depending on the material available, the species can thus be missed, or misidentified, and the discordance in the species' assignment among archaeologists can become highly discordant. Geigl and Grange (2012) reported the results of a study where four archaeozoologists were asked to re-identify 23 bones and teeth originally attributed to *E. hydruntinus*. The archaeologists were unanimous for only one single sample, and for approximately a third of the samples investigated, *E. hydruntinus* could be confirmed by only a single archaeozoologist, while all the others disagreed. Luckily, early mitochondrial studies have established the close genetic proximity between hydruntines and hemiones (Orlando et al. 2006), which share a 28 bp deletion in the region covered by the mtDNA minibarcode (Orlando et al. 2009). The identification of horse specimens that were morphologically misassigned to hydruntines is thus genetically straightforward (Geigl and Grange 2012). In what is so far the largest genetic analysis of hydruntine specimens, Bennett and colleagues successfully recovered DNA information encompassing the mitochondrial barcode from 64 alleged hydruntine samples (Bennett et al. 2017). Approximately 40% of those turned out to be horses. Limiting their analyses to the remaining subset of specimens also showing reliable morphological identification, these authors revealed that the hydruntine range was structured into two clades. The first was represented by ~5–8-kyr-old material from Anatolia and the Balkans, while the second was present in France some ~100 kyrs ago and in Iran until the beginning of the twentieth century. The genetic distance observed between these clades and other groups of hemiones is somewhat comparable, suggesting the hydruntine as a conspecific member of hemiones. Hemiones were thus present in the Upper Paleolithic of Europe and were likely the inspirational source of some parietal paintings and engravings showing hemione-like equine silhouettes with particularly long ears.

The genetic analysis of other paleontological remains originally attributed to *E. hydruntinus* revealed the existence of a mitochondrial sequence both divergent to hemiones and all other present-day equine species (Orlando et al. 2009; Vilstrup et al. 2013). Its phylogenetic placement within stenonines still requires further clarification, but a careful morphological reanalysis of the specimens dismissed the original assignment to hydruntines (Eisenman 2010) and showed instead strong affinities with an equine group described in the Middle Pleistocene of Germany. These so-called Sussemiones were originally thought to have become extinct hundreds of thousand years earlier (Eisenman 2006, 2010) but likely survived in the Altai Siberian caves from Proskuriakov, Okladnikov, and Denisova until at least ~45 kyrs ago, as indicated by the age of the youngest sample genetically analyzed (Orlando et al. 2009). It thus seems that the strong morphological plasticity present in the equine paleontological record has not just resulted in situations similar to NWSL horses and hydruntines, where a single genetic species exhibited an entire range of morphotypes. It also caused situations where different genetic species (here, the Sussemione and the hydruntine) were lumped together within a single morphological species.

3 Population Dynamics and Conservation

In addition to revisit the evolutionary tree of the horse family and second archaeozoologists in the identification of bone assemblages, aDNA studies have investigated the population dynamics of several equine species in the past, mostly horses. The two main questions addressed are: first, how did major climatic crises, in particular the Last Glacial Maximum (LGM), impact the horse demography (Lorenzen et al. 2011; Orlando et al. 2013; Schubert et al. 2014)? Second, did climate or human activities, including overhunting, drive the extinction of all equine species in the Americas at the end of the Late Pleistocene (Haile et al. 2009; Lorenzen et al. 2011; Willerslev et al. 2014)? The application of similar methodologies to both present-day individuals and museum specimens has also helped evaluate the impact of decades of captivity in a population generally considered as the last remaining truly wild horse on the planet (Der Sarkissian et al. 2015b). The main findings of these studies are presented below.

3.1 Extinction and Climate Change

Within the last 50 kyrs, a large fraction of the equine biodiversity became extinct. The exact extinction time of the Sussemiones is unknown but took place within the last ~45 kyrs (Orlando et al. 2009). The last NSWL horse remains were found in the Alaskan permafrost and were radiocarbon dated to ~31 kyrs cal. BP (Guthrie 2003). Hippidiform and horse populations vanished in the Americas around ~11–12 kyrs ago (Guthrie 2006), a time that also experienced the extinction of up to two thirds of the megafauna genera and four fifths of their species in this continent [see Stuart (2015) for a review]. The Giant Cape zebra from South Africa became extinct almost at the same time. The trace of the main mitochondrial lineage of the hydruntine is lost after ~3 kyrs ago (Orlando et al. 2006; Bennett et al. 2017), probably due to human-driven fragmentation of their habitat [claims of later survival, possibly up until the Middle Ages, have been dismissed as donkeys by genetic evidence (Orlando et al. 2009)]. The Atlas wild ass died out in Algeria probably due to overhunting by the Romans, some ~2 kyrs ago. By the late 1890s, no quagga zebras were roaming the South African savannahs, and the last captive specimen died in the early 1900s (Leonard et al. 2005).

While there is little doubt that human activities have driven most of these recent extinctions, several hypotheses have been proposed to account for those taking place around the Late Pleistocene–Holocene transition, some ~11 kyrs ago. In the Americas, this period overlaps not only with major climate changes, associated to the massive contraction of grasslands and tundra steppes, which are rich in nutrients, but also with human expansion. At one extreme, models posit climate as the main extinction driver, while at the other extreme, humans overkilled megafauna populations, including equids, possibly almost overnight (the Blietzkrieg hypothesis). Even though the latest horse macrofossils discovered north of the Cordilleran and Laurentide ice sheets dated to ~13–15 kyrs ago, traces of horse DNA have, however, been detected several millennia later in Alaskan permafrozen sediments (Haile et al. 2009). This demonstrated that horses survived at least until ~10.5 kyrs ago in the Americas. They thus overlapped with humans during almost three millennia before becoming extinct (Rasmussen et al. 2014), ruling out the Blitzkrieg extinction model. This does not rule out, however, humans as possible extinction drivers because a massive American potential range is predicted by climate niche modeling in the mid-Holocene, a time when horses already went extinct there (Lorenzen et al. 2011). Since climate niche modeling is purely trained on climatic variables, climate change in itself does not seem sufficient to have led horses to extinction. In Europe and Siberia, where extensive information on the presence of herbivores present in archaeological sites can be compiled, horses also represent the dominant species and show a massive overlap with humans from after the LGM (Lorenzen et al. 2011). Human activities might thus have contributed to shape the horse demographic trajectory within the last 10 kyrs, possibly including their extinction in the Americas.

Additionally, current evidence leaves no doubt that climate changes have also been a major driver of the horse demography during the last 50 kyrs. Bayesian skyline reconstructions based on the mitogenome diversity within present-day horses show a steady demography until their domestication where an exponential demographic increase is recovered (Lippold et al. 2011a, b; Achilli et al. 2012). This profile contrasts with those obtained when including ancient horse mitogenomes in the dataset (Orlando et al. 2013; Schubert et al. 2014) (Fig. 2), suggesting that the present-day mitogenome diversity only captures limited demographic information prior to domestication times. The demographic profile suggests a first expansion phase from ~100 kyrs ago, peaking right before the LGM, and followed by a massive decline. Interestingly, demographic reconstructions based on the pairwise sequential Markov chain (PSMC) model, which exploits patterns of heterozygosity variation along single diploid genomes (Li and Durbin 2011), are consistent with such Bayesian skyline profiles prior to the LGM (Orlando et al. 2013; Schubert et al. 2014; Librado et al. 2016) (Fig. 2). Leveraging partial hypervariable mtDNA sequences from over a hundred of radiocarbon dated horses, and the climate niche envelopes predicted at four time periods (42, 30, 21, and 6 kyrs ago), Lorenzen and colleagues found that the effective population size was significantly correlated to the predicted range size for the

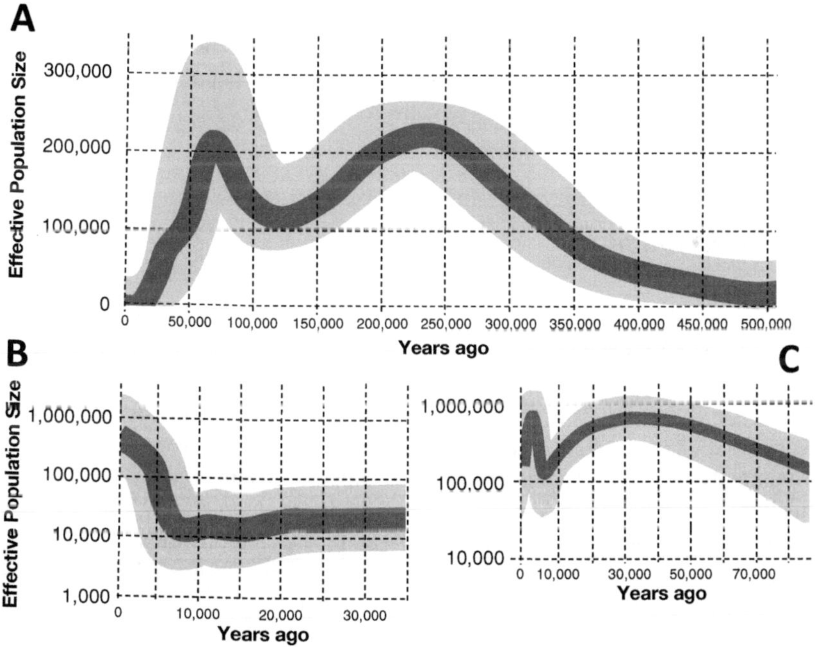

Fig. 2 Horse demographic trajectories within the last 30–500 kyrs [adapted from Orlando et al. (2013), Schubert et al. (2014), Der Sarkissian et al. (2015b), and Librado et al. (2015, 2016, 2017)]. Complementary temporal resolution was obtained using Bayesian skyline modeling on patterns of mitochondrial DNA variation, using ancient DNA data (panel **a**; Orlando et al. 2013; Librado et al. 2017) or not (panel **b**; Lippold et al. 2011a), and PSMC reconstruction applied to the autosomal data underlying single diploid genomes of domestic horses (panel **c**; Orlando et al. 2013; Der Sarkissian et al. 2015b; Librado et al. 2015)

species, confirming climate, and in particular the transition to the cold, arid, and fully glacial marine isotope stage 2 some ~30 kyrs ago, as a major driver of the horse demography (Lorenzen et al. 2011). It is noteworthy that the PSMC profiles obtained from the single diploid genomes of present-day domesticates and wild individuals and Late Pleistocene animals all show similar trajectories (although with changes of a different magnitude), with two additional phases of expansion/collapse within the last 2 myrs (Orlando et al. 2013; Schubert et al. 2014; Librado et al. 2015). In particular, the Eemian period (~130 kyrs ago) also appeared associated with reduced effective sizes, suggesting an important role for glacial/interglacial transitions.

The climate changes underlying the global warming following the Late Pleistocene/Holocene transition might also have played a significant role in the horse extinction in North America. The sequencing of minibarcodes from permafrozen sediments spread across the whole Holarctic region indeed showed a complete ecological turnover in the post-LGM plant communities, with forbs no longer dominating the ecosystem but graminoids (Willerslev et al. 2014). The genetic analysis of pre-LGM coprolites from horses as well as other megafauna species, such as bisons, woolly rhinos, and woolly mammoths, yet suggested that these animals were predominantly forb feeders. Therefore, it is possible that the rise of graminoids associated with the post-LGM climate changes might not have provided conditions compatible with sustaining large horse (megafauna) populations. Whether the nutrient content of forbs helped sustain large herbivore populations, and/or the nitrogen input from such populations favored forbs in their competition with graminoids remains to be investigated.

3.2 Conservation

Except for the plains zebra and the Tibetan kiang, formally a species but recently proposed to represent a subspecies of hemiones (Bennett et al. 2017), all other wild equine species are presently considered vulnerable or (critically) endangered by the International Union for Conservation of Nature (IUCN) (Orlando 2015; see http://www.iucnredlist.org/). The situation is also concerning for horse domesticates, with approximately a quarter of breeds being considered endangered by the Food and Agriculture Organization (FAO; see http://www.fao.org/home/en/). It is thus essential that sound conservation programs are established before most of the equine diversity, be wild or domesticated, disappears. Ancient DNA studies of museum specimens can help toward the objective, as recently demonstrated for Przewalski's horses (Der Sarkissian et al. 2015b).

Since the last tarpan died out in the Hellabrunn Zoo from Munich in 1887, the Przewalski's horses have long been considered to represent the last surviving wild horses on the planet. Recent genomic work, however, demonstrated that these were the direct descendants of the earliest domestic horses known in the archaeological record (Gaunitz et al. 2018). They, thus, belong to a lineage that was once successfully domesticated but further returned feral.

Przewalski's horses were first described as their own species but are now considered to be a subspecies of horses, despite their additional pair of chromosomes (Boyd and Houpt 1994). Following their discovery by the Western world in the second half of the nineteenth century, this population has experienced a massive demographic collapse until the last individual was caught in the wild in 1947, seen in 1969, and soon after declared extinct (Wakefield et al. 2012). The almost ~2,000 individuals currently living are the progeny of captive individuals kept in zoos and provide one of the too few success stories of conservation biology, with thriving reintroduction reserves in Mongolia, Russia, and China. However, the pedigree of all living Przewalski's horses was founded by only 12–16 animals, potentially limiting their evolutionary potential and chances of survival in the future. By using museum specimens to sequence the complete genomes of living Przewalski's horses representing all founding lineages, as well as those from animals that lived in the late nineteenth and early twentieth century, Der Sarkissian and colleagues found a number of features relevant to the conservation of the population (Der Sarkissian et al. 2015b).

First, for two of the horses analyzed, the genetic data were not compatible with the expected pedigree as described within the conservation studbook. The latter is key to predicting what the expected inbreeding levels would be for any parental pair of males and females. The identification and correction of mistakes in the conservation studbook have, thus, an overall positive impact on the ex situ conservation program by enabling more accurate estimates of individual relatedness and limiting the inbreeding levels in future generations.

Second, the Przewalski's horses and domestic horses were found to represent the descending populations of two lineages that split some ~45 kyrs ago, in agreement with earlier estimates based on more limited genome-scale data (Orlando et al. 2013). Those separate populations continued to admix with each other even after the horse was domesticated some ~5.5 kyrs ago and virtually up until their discovery by the modern world. The contribution of both populations to gene flow was, however, asymmetrical, with the lineage leading to Przewalski's horses predominantly contributing to that leading to domesticated horses. Estimates indicate that this contribution into the latter lineage was moderate and altogether corresponded to approximately 2–3% of the genome (Gaunitz et al. 2018).

The situation changed during captivity as Der Sarkissian and colleagues could identify one historical Przewalski's horse as the F1 hybrid of one Przewalski's stallion and one Mongolian domestic mare. Additionally, recent introgression from domesticates could be detected in a fraction of the living Przewalski's horses analyzed. Therefore, while some branches in the Przewalski's horse pedigree are devoid of such introgression introduced during captivity, some individuals show as much as ~30% of domestic genome ancestry. Explicit modeling based on f4-statistics and including the genome data from Botai horses, the direct ancestors of Przewalski's horses, revealed that a minimal ~6–7% of the genome of Przewalski's horses pertained to this recent introgression from domesticates (Gaunitz et al. 2018). Nonetheless, altogether, those studies suggest that the vast majority of the genome of Przewalski's horses has no equivalent in modern domesticates. All possible efforts must thus be done to preserve this genetically unique population of horses, even if they do not represent the last truly wild horses in the planet.

Improving the chances of success of future conservation plans will, however, not necessarily equate to designing mating schemes aimed at diluting the impact of this domestic ancestry in future generations. Other genomic features, such as the levels of inbreeding and deleterious mutations (i.e., the so-called genomic load), as well as the taxonomic diversity of the gut microbiome, which is now known to be impacted by captivity in horses (McKenzie et al. 2017; Metcalf et al. 2017), will also have to be considered.

4 Domestication

Within the last six millennia, two equine species have been domesticated almost at the same time but in different regions of the world. Archaeological evidence from Egypt suggests that the donkey was already domesticated ~5 kyrs ago (Rossel et al. 2008), probably to help the transportation of people and goods across the Sahara in response to drier climatic conditions. Since then, donkeys have remained important pack animals and facilitated the development of mobile pastoralism and major overland trade routes in both Africa and Eurasia. Similarly, the domestication of the horse also had a far-reaching impact on human history (Kelekna 2009). With horses, humans could travel for the first time well above their own speed and carry their germs, culture, and genes across vast geographic areas. The development of horse-drawn chariots and cavalry also radically changed the history of warfare and was instrumental to the emergence and stability of empires until the mechanization of weaponry during the second half of the twentieth century. Beyond the battlefield, farm horses have massively impacted agricultural productivity, especially in the late Middle Ages (Langdon 2006). Although earlier changes in the human/horse relationship have been suggested (Anthony 2007; Anthony and Brown 2011), the bite wear patterns present on the animal teeth indicate that horses were harnessed during the ~5.5-kyr-old Eneolithic culture of Botai from the North central Kazakh steppes (Outram et al. 2009), where the animal represent >99% of the bone assemblages. Additionally, the isotopic signatures of equine milk found in fatty acid traces preserved on ceramics support horse milking. This suggests that horses were already domesticated in the region some ~5.5 kyrs ago, despite the fact that the important size shifts that are generally taken as a domestication marker in other animals (Vigne et al. 2005) are only observed for horses at least a millennium later (Benecke and von den Driesch 2003). Discoveries of corral structures associated to the Botai culture, including at the eponymous site, confirmed domestication (Gaunitz et al. 2018).

The biological changes that accompanied the domestication processes of horses and donkeys are difficult to reconstruct from current patterns of genetic diversity both due to the development of intensively selected and extremely influential breeds during the last two centuries and the almost extinction and, thus, the massive loss of diversity experienced by their wild relatives (Der Sarkissian et al. 2015b; Kimura et al. 2011). By traveling the past, aDNA offers the possibility to catch evolution red-handed and chart through space and time the important genetic changes underlying horse and donkey domestication. An overview of the main findings related to the existence of one single or multiple domestication center(s), the importance of wild restocking of the livestock, the traits selected from early domestication stages to the emergence of particular breeds, is presented in the following section.

4.1 Domestication Center(s) and Wild Restocking

Extensive surveys of the mitochondrial diversity present in extant domestic donkeys across the world have indicated the presence of two main clusters, genetically distinct from hemiones (Beja-Pereira et al. 2004). This ruled out hemiones as possible progenitors for the domestic donkey and suggested that two independent domestication processes gave rise to each genetic cluster. The Nubian wild ass, which historically spread across North Africa, from the Red Sea shores of Sudan to the Moroccan Atlantic coast, shows haplotypes typical of the first cluster and thus represents a likely progenitor for this cluster. The critically endangered Somali wild ass, which only survives today in small pockets within Sudan and Erithrea, represented a possible candidate for the progenitor of the second cluster. However, the Somali wild ass was found to form a cluster of its own, distinct and almost equidistant to the other two (Beja-Pereira et al. 2004), leaving the origins of the second cluster unknown. Perhaps the recent demographic collapse of African wild ass populations resulted in the loss of haplotypes in either wild ass population, precluding any identification of direct genetic affinities within extant populations. By extending the genetic analyses to field-collected feces, museum specimens, and subfossil material, Kimura and colleagues have confirmed the limited genetic diversity present in the Somali wild ass (Kimura et al. 2011). This is in line with the signature of a drastic demographic bottleneck found in the sequence data underlying the first diploid genome sequence of the species, which likely took place within the last ~25 kyrs (Jónsson et al. 2014). Despite the large sampling effort in the study from Kimura and colleagues, the Somali wild ass remained genetically distinct. Furthermore, whole genome sequence data confirmed a population time split at ~350 kyrs ago for the lineages of the Somali wild ass and donkey domesticates, which predates the onset of domestication (Rossel et al. 2008). Therefore, the Somali wild ass has not given rise to any of the two main domestic donkey clusters. The sequence data obtained by Kimura and colleagues confirmed instead the relationship between the Nubian wild ass and the first domestic cluster, as a ~3-kyr-old domestic specimen from Central Sahara showed a mitochondrial haplotype identical to museum specimens of Nubian wild asses (Kimura et al. 2011). However, while eight of nine such museum specimens nested with the first mitochondrial cluster, the haplotype of the ninth specimen was typical of the second cluster, suggesting that the Nubian wild ass, or any extinct population also sharing this haplotype – possibly the Atlas wild ass from the Maghreb and the coast of Yemen – was in fact also the progenitor of the second domestication. Testing this hypothesis will require additional sequence data from ancient African wild ass. This might prove difficult but not impossible given the encouraging success in the characterization of the mitogenome or nuclear genome variation present in ancient remains from the Middle East (Almathen et al. 2016; Mohandesan et al. 2017) and Eastern Africa (Gallego Llorente et al. 2015).

The mitochondrial variation found in present-day and ancient domestic donkeys and African wild asses has revealed something additional, which proved also valid for horses, namely, their domestication process involved substantial restocking of the livestock from wild progenitors (Kimura et al. 2011; Vilà et al. 2001; Jansen et al. 2002; Lippold et al. 2011a; Achilli et al. 2012). In none of these species analyzed does the mitochondrial diversity found in domestic (ancient) animals form a single, monophyletic cluster encompassing only a small fraction of the (ancient) wild genetic diversity. Instead, the haplogroups found in wild (ancient) animals and (ancient) domesticates overlap to a very large extent, suggesting either that lots of such haplogroups were present in the founder group of domesticates or that wild females were mated with male domesticates throughout the domestication process. Analyses based on nuclear data support the latter scenario (Warmuth et al. 2012; Der Sarkissian et al. 2015b). Recent estimates based on whole mitogenome sequences suggest that possibly up to three quarters of the mtDNA diversity once present in the wild has been effectively incorporated into the horse domestic gene pool (Lippold et al. 2011a). Such estimates are not available for donkeys, due to the still relatively limited characterization of the genetic variation in (ancient) wild animals.

Interestingly, in horses, the Y chromosome shows a pattern opposite to mtDNA. There, present-day domesticates show extremely limited variation, with one single haplotype dominating most breeds and populations (Lindgren et al. 2004; Wallner et al. 2013). This has supported the contention that the process of horse domestication involved restocking from the wild mostly through mares, rather than stallions, possibly as the aggressive reproductive behavior of the latter makes them more difficult to manage. However, sequence data from ancient domestic horses, especially from Scythian stallions (Lippold et al. 2011b; Librado et al. 2017), have revealed quite diverse Y-chromosome haplotypes in the Iron Age. Therefore, the number of stallion founders used in early domestication stages was not limited, and the participation of a diversity of stallions has only reduced after the Iron Age. Further studies, charting the Y-chromosome diversity of stallions through time, will reveal when and in which archaeological and/or historical contexts this reduction took place. Although presently unknown in absence of extensive surveys of the Y-chromosome diversity in this species, the wild restocking in the domestic donkey could be expected to show limited sex bias, in contrast to the horse as African pastoralists recruit both jacks and jennets to their herds and trapping of wild animals irrespective of sex is documented historically.

The first complete genome sequences obtained from Late Pleistocene wild horses have revealed that a third population of horses existed some ~16–42 kyrs ago, in addition to the populations underlying present-day domestic and Przewalski's horses (Schubert et al. 2014). Little is known about this population of wild horses as it is currently known from the genome sequences of three specimens only. It was spread across the Taymir peninsula and Yakutia from ~42 kyrs ago to ~5 kyrs ago and split from the population ancestral to both present-day domestic and Przewalski's horses by the time of the Eemian interglacial, some ~130 kyrs ago (Schubert et al. 2014; Librado et al. 2015). As none of the genomes from present-day horses clusters together with any of the three known genomes for this population, it is considered to have become extinct within the last ~5 kyrs. This population, however, was proposed to have contributed to the genetic makeup of present-day horses, possibly representing a minimum of ~13% of their genome ancestry using calculations based on D-statistics (Durand et al. 2011) and f4-statistics (Patterson et al. 2012). However, once the earliest domestic horses from Botai were sequenced (Gaunitz et al. 2018), D-statistics showed that Botai and modern domesticates have similar amounts of shared derived polymorphisms with the archaic population (in contrast to Przewalski's horses, which show a deficit of such variation). This pattern is compatible with two scenarios. Firstly, the admixture from the archaic lineage took place prior to the divergence between the lineages of Botai and modern domesticates but was further eliminated post-Botai in Przewalski's horses (this component was, however, maintained in modern domesticates). Alternatively, an admixture from a yet unidentified ghost lineage into Przewalski's horses may be the reason why they look more distant to the archaic lineage identified. Determining which of these scenarios prevailed requires further work. Either way, it nonetheless illustrates that wild, archaic populations of horses have significantly contributed to the genetic makeup of modern horses, be domesticated, or feral.

Given the extent of wild introgression and the dispersal capacity of the horse, it is perhaps not surprising that the horse mitochondrial variation shows a lack of phylogeographic structure, with haplotypes spreading over vast regions (Leonard et al. 2005; Cieslak et al. 2010), and no major haplogroups specific for a single breed (Jansen et al. 2002). Haplogroup D1, which is quite frequent in Iberian breeds and North-African barb horses (Jansen et al. 2002), was yet proposed to reflect local domestication in Iberia, but aDNA failed to reveal its presence in the Iberian Neolithic and Bronze Age (Lira et al. 2010). Thus, it likely reflects the descent of horses introduced after the Bronze Age. However, some mtDNA haplotypes that populated the Iberian Bronze Age are still occasionally found in Iberian breeds (Lira et al. 2010; Jansen et al. 2002) and could represent signs of local domestication. This, and the discovery of a hotspot of STR diversity in Iberian breeds (Warmuth et al. 2011), has been proposed to reflect the existence of a second domestication center for horses in Iberia, independent from the Kazakh/Pontic-Caspian steppes (Outram et al. 2009; Warmuth et al. 2012). However, Iberia also represents a well-known glacial refugium (Hewitt 2000), where more diverse populations of indigenous wild horses might have survived until exogenous domestic horses finally arrived in the region. A simple local wild restocking would then equally well explain the hotspot of diversity observed. Therefore, pending additional data testing the genetic continuity of Iberian horse population pre- and post-domestication, the question of a possible Iberian domestication center is open. In Europe, horses, however, represent a rare fraction of bone assemblages in the archaeological record prior to the Chalcolithic and probably survived as fragmented populations islets around open patches in the canopy forest (Sommer et al. 2011). The limited nature of the archaeological record in Europe prior to the Chalcolithic might limit future attempts at addressing whether Iberian populations were fully replaced or assimilated into the domestic pool. The recent sequencing of the genome of Botai horses has revived the debate on the possible multiple origins of the domestic horses since none of the domestic horses sequenced within the last ~4,100 years directly relates to Botai. Instead, they form a second, distinct monophyletic group with no more than ~2–3% of Botai ancestry. Together with the signal of a massive demographic expansion at approximately ~4,500 years ago found in patterns of mitochondrial variation, this suggests that another horse group, not related to Botai horses, could have been domesticated around that time and fueled massive expansion of human groups across Eurasia (Gaunitz et al. 2018). Alternatively, the Botai genetic component could have been diluted to almost disappear during such expansion by means of introgressive capture (Larson and Fuller 2014). Further data, particular from animals that lived in the third millennium BCE, are necessary to tease both scenarios apart and to identify the true temporal and geographic locus of horse domestication.

4.2 *Trait Selection and Genetic Load*

Given the importance of the horse industry, representing a yearly €100 billion impact for the EU economy alone, genetic investigations of present-day domestic horses have not been limited to neutral markers, such as mtDNA and the Y chromosome. Many studies have successfully identified genes underlying major phenotypic traits and disorders (see Chowdary 2013 and references therein). Besides coat color (e.g., Brooks et al. 2002; Brooks and Bailey 2005; Brunberg et al. 2006; Reissmann et al. 2007; Bellone et al. 2013; Imsland et al. 2016), variants statistically associated with racing performance have been discovered (Hill et al. 2010; McGivney et al. 2010; Tozaki et al. 2010; Petersen et al. 2013). Perhaps the most spectacular example involves a single mutation and a SINE insertion at the MSTN gene, with homozygous mutants showing hypertrophic muscles and top performance at short-distance sprint races (see Rivero and Hill 2016 for a review). One single C → A mutation at the DMRT3 gene is also known to be permissive for alternate gaits complementary to walk, trot, and gallop (Andersson et al. 2012) and is distributed throughout the world (Promerová et al. 2014), mainly in gaited breeds. Additionally, the GYS1 H allele, which can be found at non-negligible frequencies despite causing severe myopathies, has been proposed to represent an example of "thrifty gene" that conferred a selective advantage in early domestication (it increases muscular glycogen storage) but would be maladapted to modern starch-rich diets (McCue et al. 2008). Besides metabolic disorders, the genes underlying size variation have received much attention (Signer-Hasler et al. 2012; Makvandi-Nejad et al. 2012; Metzger et al. 2013), and four loci (LCORL, HMGA2, ZFAT, and LASP1) seem to explain a majority (as much as 83% in Makvandi-Nejad et al. 2012) of the size differences among breeds, which illustrates how humans have successfully manipulated a fraction of the horse genetic potential to create a whole diversity of breed sizes.

This brief overview is certainly not exhaustive as many other genotype/phenotype associations have been unveiled since the horse reference genome, and high-throughput genotyping tools have been developed (Wade et al. 2009; McCue et al. 2012; Petersen et al. 2013). Ancient DNA has offered a unique opportunity to chart these variants through space and time and track the origins and the geographical, cultural, and historical context into which the underlying characters later expanded. For instance, aDNA from museum samples suggested that the "speed" mutation at MSTN probably entered the pedigree of racing Thoroughbred horses from local British mares, as it was absent from all founding Arab stallions (Bower et al. 2012). In addition to such candidate gene approaches, targeting specific loci of known genetic variation, the advent of whole genome sequencing has enabled the first genome scans in horses, providing a first glimpse at the full suite of genetic changes that have been introduced and/or selected in the course of their domestication. The main findings resulting from these two complementary approaches are presented below.

4.3 Candidate Gene Approaches

Most analyses of the genetic variation present at nuclear loci in ancient horses have focused on loci involved in coat coloration (Ludwig et al. 2009; Pruvost et al. 2011; Ludwig et al. 2015; Wutke et al. 2016a). The Dun phenotype is characterized by pigment dilution and typical so-called primitive markings, corresponding to undiluted black stripes, mostly along the spine and the legs. It represents the wild-type coat coloration phenotype in horses and is common to all Przewalski's horses. The causative variant underlying Dun coloration in horses, a ~1.6 kb-long insertion downstream of the TBX3 gene (Imsland et al. 2016), has only been described recently. The presence/absence of the corresponding 1.6 kb block could not be investigated in earlier studies opting for candidate gene approaches, which can thus not distinguish Dun and non-Dun horses. Deleted alleles have yet been identified in the sequence data underlying two ancient genomes dated to ~43 and ~5 kyrs ago (Schubert et al. 2014; Librado et al. 2015), suggesting that a diversity of color morphs existed prior to domestication (Imsland et al. 2016). The diversity was nonetheless much limited compared to what is seen from the early Bronze Age, as only the black allele at ASIP (Ludwig et al. 2009) and the leopard spotted allele at TRPM1 (Pruvost et al. 2011) could be detected in Late Pleistocene and early Holocene horses. Incidentally, the genetic identification of spotted horses in the Late Pleistocene suggests that the famous cave paintings of the *dappled horses of Pech-Merle* could well be simple representations of living animals, not symbolic expressions (Pruvost et al. 2011).

From the Bronze Age onward, a diversity of variants including chestnut horses, cream and silver dilutions, and sabino and tobiano spotting were present at detectable frequencies and were found to increase in frequency much faster than expected by chance (Ludwig et al. 2009). They were thus likely selection targets to horse herders. This work demonstrated that changes in coat coloration patterns have represented one of the early targets in the process of horse domestication, with different color morphs perhaps providing early herders with a means to differentiate their herds (Wutke et al. 2016a). Importantly, selection patterns were not homogeneous through space and time as the genetic variant underlying leopard spotting was quite common in Western Europe during the early Bronze Age but remained undetectable until it reappeared in Siberia during the Iron Age (Ludwig et al. 2015). Horse herders from different regions and cultures had thus fluctuating selection targets through time.

Importantly, leopard spotting is associated with congenital night blindness in horses (Bellone et al. 2013). Some herders might have had different opinions as to whether the spotted phenotype typical of present-day Appaloosa horses outweighed a diminished vision capacity at night and the related increased risks in terms of predation and thievery. For instance, no spotted (and dilution) alleles were found within Celtic Swiss horses (Elsner et al. 2016), but spotted horses were not uncommon among Iron Age Scythian horses (Ludwig et al. 2015; Librado et al. 2017). Leopard spotting alleles, and more generally other spotting alleles, however dropped in frequency following 400 AD in Europe (Wutke et al. 2016a). While these alleles were selected against, chestnut and black alleles were positively selected and rose to high frequencies, suggesting that strong changes in coat color phenotypes took place in the course of the Middle Ages. Surprisingly, no spotted alleles were detected in early Norse (Viking) horses from Iceland. As these traits are relatively frequent in present-day Icelandic horses, it is likely that the prohibition of horse import in the island was not as strict as commonly believed. Early Norse Icelandic horses were, however, frequent carriers of the DMRT3 allele associated with alternate gaits, such as ambling, which are more comfortable to riders than classical canters and gallops (Wutke et al. 2016b). The mutation was left undetected in mainland Europe, including Scandinavia (although Norway, an important Norse region, was not formally tested), prior to 850–900 AD, where it appeared in Britain. This suggested that Norse first acquired ambling horses on the British Isles before they transported them to Iceland, and possibly further, throughout their whole distribution range.

4.4 Genome Scans

Recent methodological developments in aDNA research have opened access to genome-scale information from past populations. Target enrichment offers the most economical approach for characterizing up to several millions of loci (Haak et al. 2015; Mathieson et al. 2015), whole chromosomes (Fu et al. 2013), and even the non-repeated fraction of nuclear genomes (Carpenter et al. 2013). For DNA extracts showing relatively high levels of endogenous DNA, shotgun sequencing also provides a fast- and cost-effective approach to whole genome sequencing (Orlando et al. 2015). So far, target enrichment has been limited to a handful of loci in ancient equids, mostly the whole mitogenome (Vilstrup et al. 2013), but also several thousands of nuclear loci showing functional and neutral variation (Cruz-Dávalos et al. 2016). Shotgun sequencing has, however, delivered no less than 22 ancient horse genomes (Orlando et al. 2013; Schubert et al. 2014; Der Sarkissian et al. 2015b; Librado et al. 2015, 2017), which, compared to present-day genomes, provided important insights into the selection process underlying horse domestication.

First, by comparing the genomes of two Late Pleistocene horses to the genomes representing a broad range of present-day domestic breeds, Schubert and colleagues could identify 50 kb-long regions showing selection signatures in modern horses (Schubert et al. 2014). The authors implemented four independent tests of positive selection and considered as good candidates those regions supported by a minimum of two independent tests. This provided a total number of 125 regions, which carried genes enriched for four main functional categories of genes. A number of genes were involved in horse locomotion, with genes participating to the organization of skeletal muscles, myotendons, and articular junctions but also to balance and motor coordination (Schubert et al. 2014). Additionally, the candidates included many genes associated with the cardiovascular system and represented possible selection targets to adapt the equine physiology to the energetic demands related to sustained efforts. A third category of candidates included genes associated with skeletal and facial development, possibly echoing the diversity of sizes and faces seen in modern horses. Finally, the last category of selection candidates was indicative of changes potentially associated with behavior and learning capacity, two traits that are central to the development of the human/horse relationship.

As it contrasted to the genomic variation in present-day horses and horses that lived prior to domestication, this work could only detect signatures associated to the domestication process. It could, however, not determine when the detected changes were exactly introduced in the course of horse domestication. The genome sequences of 11 Scythian horses that lived ~2.3 kyrs ago provided a first attempt toward this objective, enabling to look at the changes introduced prior to 2.3 kyrs ago during early domestication stages and those introduced during the last 2.3 kyrs (Librado et al. 2017). Interestingly, the list of candidate genes showing selection signatures in early domestication stages was enriched in genes involved in the neural crest development. Yet, the "neural crest" hypothesis (Wilkins et al. 2014) posits that the full suite of traits that are common to many domestic animals (floppy ears, depigmentation, juvenile behavior, docility, etc., all grouped into the so called domestication syndrome) was first selected by imposing a selective pressure on central pathways affecting the development of the underlying organs and structures. Genes involved in the formation, migration, and differentiation of neural crest cells were natural good candidates for such pathways, given the large number of cell types and tissues deriving from neural crest cells and their general overlap with traits associated with the domestication syndrome. The variation present in the genomes of Scythian horses lends supports to this hypothesis. Interestingly, the authors could also detect selection signatures within the Scythian horses themselves, mostly of genes involved in the development of forelimbs, which confirmed morphometric measurements of metacarpals showing that Scythian horses were more robust than present-day horses living in the same region of Kazakhstan and Mongolia (Librado et al. 2017). Additionally, Scythian horse riders seem to have selected for genes involved in the posterior pituitary, a key production center of vasopressin and oxytocin. The latter neurohypophysal hormone is essential to both lactation and is known to be involved in the development of strong bonds between humans and dogs (Nagasawa et al. 2015). This suggested that Scythian herders might have selected variants facilitating both the milking of horses and their managing.

Additional work applying similar methodology focused on reconstructing the genomic history of one of the most iconic horse breeds on the planet, namely, the Yakutian horses (Librado et al. 2015). Yakutia represents the coldest country in the Northern hemisphere, with winter temperature records dropping below −70°C. Sequencing the genome of both ancient and present-day Yakutian horses has helped reveal the genetic basis for the suite of morphological and physiological adaptations that Yakutian horses develop to survive in this extreme environment. Interestingly, in addition to identifying candidate genes that also show adaptive signatures among other cold mammals, such as the woolly mammoth and Siberian humans, the study revealed that the population of Yakutian horses developed within an extremely short timeframe, probably within the last thousand years, following the migration of the first Yakut settlers in the region (Keyser et al. 2015). The authors proposed that selection at regulatory regions, as supported by the genomic evidence, is key to the success of such fast episodes of adaptation (Librado et al. 2015).

Finally, the whole genome data available for horses have revealed that the domestication process was accompanied with an increase in the genetic load, as present-day domesticates show a higher fraction of potentially deleterious mutations in protein-coding genes (Schubert et al. 2014). The demographic collapse associated with horse domestication, especially within the last 2.3 kyrs (Librado et al. 2017), is proposed to have reduced the effect of negative selection in purging out (slightly) deleterious mutations from the domestic horse gene pool. Therefore, the domestication of the horse did not only come with an improvement of functions of key interest to humans but also resulted in an overall inflation in deleterious mutations, potentially enhancing the odds that horses develop important genetic disorders.

5 Conclusions and Future Perspectives

Members of the horse family have accompanied more than three decades of aDNA research. Even though a wealth of taxonomic groups has been studied the horse has focused most of the attention, especially since the advent of high-throughput DNA sequencing, most logically given the impact that horses have had on human history. The genome-scale methodology currently applied to horses will likely be used for other members of the family in the near future, particularly the donkey, where de novo reference genomes and genome-scale data have now become available (Orlando et al. 2013; Huang et al. 2015; Bertolini et al. 2015; Renaud et al. 2018), but also for endangered close relatives. There is also a lot to be learnt from the extinct members of the family, especially for helping implement sound conservation programs. This chapter strictly focused on how the changes in the genetic variation observed through space and time could help better understand the biology and evolutionary history of equine species. However, with recent developments in aDNA research, it is now possible to recover information of the host itself but also of its pathogens and the whole community of microbes living in the gut and/or in the mouth (see Warinner et al. 2015 for a review). Additionally, epigenetic information can be gathered from aDNA extracts, potentially revealing how environmental cues are integrated at the genomic level to regulate gene expression (Orlando et al. 2015; Gokhman et al. 2016). The application of such methodology to the horse, as well as all other members of its family, will undoubtedly open new exciting avenues for equine research.

References

Achilli A, Olivieri A, Soares P, et al. Mitochondrial genomes from modern horses reveal the major haplogroups that underwent domestication. Proc Natl Acad Sci U S A. 2012;109:2449–54.

Alberdi MT, Prado JL. Review of the genus *Hippidion* Owen, 1869 (Mammalia: Perissofactyla) from the Pleistocene of South America. Zool J Linn Soc. 1993;108:1–22.

Alberdi MT, Prado JL. Comments on Pleistocene horses from Tarija, Bolivia, and the validity of the genus *Onohippidium* (Mammalia: Equidae), by B.J. MacFadden. J Vert Paleontol. 1998;18:669–72.

Alberdi MT, Prado JL, Prieto A. Considerations on the paper "morphological convergence in Hippidion and Equus (Amerhippus) South American equids elucidated by ancient DNA analysis", by Ludovic Orlando, Véra Eisenmann, Frédéric Reynier, Paul Sondaar, Catherine Hänni. J Mol Evol. 2005;61:145–7.

Almathen F, Charruau P, Mohandesan E, et al. Ancient and modern DNA reveal dynamics of domestication and cross-continental dispersal of the dromedary. Proc Natl Acad Sci U S A. 2016;113:6707–12.

Andersson LS, Larhammar M, Memic F, et al. Mutations in DMRT3 affect locomotion in horses and spinal circuit function in mice. Nature. 2012;488:642–6.

Anthony DW. The horse, the wheel and language. Oxford: Princeton University Press; 2007.

Anthony DW, Brown DE. The secondary products revolution, horse-riding, and mounted warfare. J World Prehist. 2011;24:131.

Alexander DH, Novembre J, Lange K. Fast model-based estimation of ancestry in unrelated individuals. Genome Res. 2009;19:1655–64.

Beja-Pereira A, England PR, Ferrand N, et al. African origins of the domestic donkey. Science. 2004;304:1781.

Benecke N, von den Driesch A. Horse exploitation in the Kazakh steppes during the Eneolithic and Bronze Age. In: Levine M, Renfrew C, Boyle K, editors. Prehistoric steppe adaptation and the horse. Cambridge: McDonald Institute for Archaeological Research; 2003. p. 69–82.

Bennett EA, Champlot S, Peters J, et al. Taming the late Quaternary phylogeography of the Eurasiatic wild ass through ancient and modern DNA. BioArXiv. 2017. https://doi.org/10.1101/090928.

Bertolini F, Scimone C, Geraci C, Schiavo G, Utzeri VJ, Chiofalo V, Fontanesi L. Next generation semiconductor based sequencing of the donkey (Equus asinus) genome provided comparative sequence data against the horse genome and a few millions of single nucleotide polymorphisms. PLoS One. 2015;10:e0131925.

Bellone RR, Holl H, Setaluri V, et al. Evidence for a retroviral insertion in TRPM1 as the cause of congenital stationary night blindness and leopard complex spotting in the horse. PLoS One. 2013;8:e78280.

Bower MA, McGivney BA, Campana MG, et al. The genetic origin and history of speed in the Thoroughbred racehorse. Nat Commun. 2012;3:643.
Boyd L, Houpt KA. Przewalski's horse: the history and biology of an endangered species. Albany, New York: State University of New York Press; 1994. isbn:10-ISBN 0-791-41889-8; 13-ISBN 978-0-791-41889-5; OCLC 28256312.
Braud M, Magee DA, Park SD, et al. Genome-wide microRNA binding site variation between extinct wild aurochs and modern cattle identifies candidate microRNA-regulated domestication genes. Front Genet. 2017;8:3.
Brooks SA, Bailey E. Exon skipping in the KIT gene causes a Sabino spotting pattern in horses. Mamm Genome. 2005;16:893–902.
Brooks SA, Terry RB, Bailey E. A PCR-RFLP for KIT associated with tobiano spotting pattern in horses. Anim Genet. 2002;33:301–3.
Brunberg E, Andersson L, Cothran G, et al. A missense mutation in PMEL17 is associated with the silver coat color in the horse. BMC Genet. 2006;7:46.
Cardoso JL, Vilstrup JT, Eisenman V, et al. First evidence of Equus asinus L. in the chalcolithic disputes the Phoenicians as the first to introduce donkeys into the Iberian Peninsula. J Archaeol Sci. 2013;40:4483–90.
Carpenter ML, Buenrostro JD, Valdiosera C, et al. Pulling out the 1%: whole-genome capture for the targeted enrichment of ancient DNA sequencing libraries. Am J Hum Genet. 2013;93:852–64.
Chowdary BP. Equine genomics. Oxford: Wiley-Blackwell; 2013.
Cieslak M, Pruvost M, Benecke N, et al. Origin and history of mitochondrial DNA lineages in domestic horses. PLoS One. 2010;5:e15311.
Cruz-Dávalos DI, Llamas B, Gaunitz C, et al. Experimental conditions improving in-solution target enrichment for ancient DNA. Mol Ecol Resour. 2016. https://doi.org/10.1111/1755-0998.
Cucchi T, Mohaseb A, Debue K, et al. Detecting taxonomic and phylogenetic signals in equids cheek teeth with geometric morphometrics: towards new paleontological and archaeological proxies. R Soc Open Sci. 2017;4:160997. https://doi.org/10.1098/rsos.160997.
Da Fonseca RA, Smith BD, Wales N, et al. The origin and evolution of maize in the Southwestern United States. Nat Plants. 2015;1:14003.
Der Sarkissian C, Vilstrup JT, Schubert M, et al. Mitochondrial genomes reveal the extinct Hippidion as an outgroup to all living equids. Biol Lett. 2015a;11.
Der Sarkissian C, Ermini L, Schubert M, et al. Evolutionary genomics and conservation of the endangered Przewalski's horse. Curr Biol. 2015b;25:2577–83.
Durand EY, Patterson N, Reich D, et al. Testing for ancient admixture between closely related populations. Mol Biol Evol. 2011;28:2239–52.
Eisenman V. Folivores et tondeurs d'herbe: forme de la symphyse mandibulaire des Equidés et des Tapiridés (Perissodactyla, Mammalia). Geobios. 1998;31:113–23.
Eisenman V. Pliocene and Pleistocene Equids: palaeontology versus molecular biology. Cour Forsch Inst Senckenberg. 2006;256:71–89.
Eisenman V. *Sussemionus*, a new subgenus of *Equus* (Perissodactyla, Mammalia). C R Biol. 2010;333:235–40.
Eisenmann V, Baylac M. Extant and fossil Equus (Mammalia, Perissodactyla) skulls: a morphometric definition of the subgenus *Equus*. Zool Scr. 2000;29:89–100.
Elsner J, Deschler-Erb S, Stopp B, et al. Mitochondrial d-loop variation, coat colour and sex identification of Late Iron Age horses in Switzerland. J Archaeol Sci. 2016;6:386–96.
Ermini L, Der Sarkissian C, Willerslev E, et al. Major transitions in human evolution revisited: a tribute to ancient DNA. J Hum Evol. 2015;79:4–20.
Frantz LA, Mullin VE, Pionnier-Capitan M, et al. Genomic and archaeological evidence suggest a dual origin of domestic dogs. Science. 2016;352:1228–31.
Franzen JL. The rise of the horse family. Baltimore, MD: Johns Hopkins University Press; 2010.
Froese DG, Westgate JA, Reyes AV, et al. Ancient permafrost and a future, warmer Arctic. Science. 2008;321:1648.

Fu Q, Meyer M, Gao X, et al. DNA analysis of an early modern human from Tianyuan Cave, China. Proc Natl Acad Sci U S A. 2013;110:2223–7.

Gaunitz C, Fages A, Hanghøj K, et al. Ancient genomes revisit the ancestry of domestic and Przewalski's horses. Science. 2018. https://doi.org/10.1126/science.aao3297.

Geigl EM, Grange T. Eurasian wild asses in time and space: morphological versus genetic diversity. Ann Anat. 2012;194:88–102.

Gallego Llorente M, Jones ER, Eriksson A, et al. Ancient Ethiopian genome reveals extensive Eurasian admixture throughout the African continent. Science. 2015;350:820–2.

Gokhman D, Meshorer E, Carmel L. Epigenetics: it's getting old. Past meets future in Paleoepigenetics. Trends Ecol Evol. 2016;31:290–300.

Green RE, Krause J, Briggs AW, et al. A draft sequence of the Neandertal genome. Science. 2010;328:710–22.

Groves CP, Willoughby DP. Studies on the taxonomy and phylogeny of the genus *Equus*-1. Subgeneric classification of the recent species. Mammalia. 1981;45:321–54.

Guthrie RD. Rapid body size decline in Alaskan Pleistocene horses before extinction. Nature. 2003;426:169–71.

Guthrie RD. New carbon dates link climatic change with human colonization and Pleistocene extinctions. Nature. 2006;441:207–9.

Haak W, Lazaridis I, Patterson N, et al. Massive migration from the steppe was a source for Indo-European languages in Europe. Nature. 2015;522:207–11.

Haile J, Froese DG, Macphee RD, et al. Ancient DNA reveals late survival of mammoth and horse in interior Alaska. Proc Natl Acad Sci U S A. 2009;106:22352–7.

Hewitt G. The genetic legacy of the Quaternary ice ages. Nature. 2000;405:907–13.

Higuchi R, Bowman B, Freiberger M, et al. DNA sequences from the quagga, an extinct member of the horse family. Nature. 1984;312:282–4.

Hill EW, Gu J, Eivers SS, et al. A sequence polymorphism in MSTN predicts sprinting ability and racing stamina in Thoroughbred horses. PLoS One. 2010;5:e8645.

Hofreiter M, Paijmans JL, Goodchild H, et al. The future of ancient DNA: technical advances and conceptual shifts. BioEssays. 2015;37:284–93.

Huang J, Zhao Y, Bai D, et al. Donkey genome and insight into the imprinting of fast karyotype evolution. Sci Rep. 2015;5:14106.

Imsland F, McGowan K, Rubin CJ, et al. Regulatory mutations in TBX3 disrupt asymmetric hair pigmentation that underlies Dun camouflage color in horses. Nat Genet. 2016;48:152–8.

Jansen T, Forster P, Levine MA, et al. Mitochondrial DNA and the origins of the domestic horse. Proc Natl Acad Sci U S A. 2002;99:10905–10.

Johnstone C (2004) A biometric study of equids in the Roman world. PhD. Department of Archaeology, University of York.

Jónsson H, Schubert M, Seguin-Orlando A, et al. Speciation with gene flow in equids despite extensive chromosomal plasticity. Proc Natl Acad Sci U S A. 2014;111:18655–60.

Kelekna P. The horse in human history. Cambridge: Cambridge University Press; 2009.

Keyser C, Hollard C, Gonzalez A, et al. The ancient Yakuts: a population genetic enigma. Philos Trans R Soc Lond Ser B Biol Sci. 2015;370:20130385.

Kimura B, Marshall FB, Chen S, et al. Ancient DNA from Nubian and Somali wild ass provides insights into donkey ancestry and domestication. Proc Biol Sci. 2011;278:50–7.

Langdon J. Horses, oxen and technological innovation: the use of draught animals in English farming from 1066-1500. Cambridge: Cambridge University Press; 2006.

Larson G, Fuller DQ. The evolution of animal domestication. Annu Rev Ecol Evol Syst. 2014;45:115–36.

Lawling AM, Polly PD. Geometric morphometrics: recent applications to the study of evolution and development. J Zool. 2010;280:1–7.

Lindgren G, Backström N, Swinburne J, et al. Limited number of patrilines in horse domestication. Nat Genet. 2004;36:335–6.

Leonard JA, Rohland N, Glaberman S, et al. A rapid loss of stripes: the evolutionary history of the extinct quagga. Biol Lett. 2005;1:291–5.

Li H, Durbin R. Inference of human population history from individual whole-genome sequences. Nature. 2011;475:493–6.

Librado P, Der Sarkissian C, Ermini L, et al. Tracking the origins of Yakutian horses and the genetic basis for their fast adaptation to subarctic environments. Proc Natl Acad Sci U S A. 2015;112:E6889–97.

Librado P, Fages A, Gaunitz C, Leonardi M, Wagner S, Khan N, Hanghøj K, Alquraishi SA, Alfarhan AH, Al-Rasheid KA, Der Sarkissian C, Schubert M, Orlando L. The evolutionary origin and genetic makeup of domestic horses. Genetics. 2016;204:423–34.

Librado P, Gamba C, Gaunitz C, et al. Ancient genomic changes associated with domestication of the horse. Science. 2017;356:442–5.

Lippold S, Matzke NJ, Reissmann M, et al. Whole mitochondrial genome sequencing of domestic horses reveals incorporation of extensive wild horse diversity during domestication. BMC Evol Biol. 2011a;11:328.

Lippold S, Knapp M, Kuznetsova T, et al. Discovery of lost diversity of paternal horse lineages using ancient DNA. Nat Commun. 2011b;2:450.

Lira J, Linderholm A, Olaria C, et al. Ancient DNA reveals traces of Iberian Neolithic and Bronze Age lineages in modern Iberian horses. Mol Ecol. 2010;19:64–78.

Llamas B, Willerslev E, Orlando L. Human evolution: a tale from ancient genomes. Philos Trans R Soc Lond Ser B Biol Sci. 2017;372.

Lorenzen ED, Nogués-Bravo D, Orlando L, et al. Species-specific responses of Late Quaternary megafauna to climate and humans. Nature. 2011;479:359–64.

Ludwig A, Pruvost M, Reissmann M, et al. Coat color variation at the beginning of horse domestication. Science. 2009;324:485.

Ludwig A, Reissmann M, Benecke N, et al. Twenty-five thousand years of fluctuating selection on leopard complex spotting and congenital night blindness in horses. Philos Trans R Soc Lond Ser B Biol Sci. 2015;370:20130386.

MacFadden BJ. Pleistocene horses from Tarija, Bolivia, and the validity of the genus *Onohippidium* (Mammalia: Equidae). J Vert Paleontol. 1997;17:199–218.

MacFadden BJ, Carranza-Castaneda O. Cranium of Dinohippus mexicanus (Mammalia Equidae) from the early Pliocene (latest Hemphillian) of central Mexico and the origin of Equus. Bull Florida Mus Nat Hist. 2002;43:163–85.

MacHugh DE, Larson G, Orlando L, et al. Taming the past: ancient DNA and the study of animal domestication. Annu Rev Anim Biosci. 2017;5:329–51.

Mailund T, Halager AE, Westergaard M, et al. A new isolation with migration model along complete genomes infers very different divergence processes among closely related great ape species. PLoS Genet. 2012;8:e1003125.

Makvandi-Nejad S, Hoffman GE, Allen JJ, et al. Four loci explain 83% of size variation in the horse. PLoS One. 2012;7:e39929.

Maricic T, Whitten M, Pääbo S. Multiplexed DNA sequence capture of mitochondrial genomes using PCR products. PLoS One. 2010;5:e14004.

Mathieson I, Lazaridis I, Rohland N, et al. Genome-wide patterns of selection in 230 ancient Eurasians. Nature. 2015;528:499–503.

McCue ME, Valberg SJ, Miller MB, et al. Glycogen synthase (GYS1) mutation causes a novel skeletal muscle glycogenosis. Genomics. 2008;91:458–66.

McCue ME, Bannasch DL, Petersen JL, et al. A high density SNP array for the domestic horse and extant Perissodactyla: utility for association mapping, genetic diversity, and phylogeny studies. PLoS Genet. 2012;8:e1002451.

McGivney BA, McGettigan PA, Browne JA, et al. Characterization of the equine skeletal muscle transcriptome identifies novel functional responses to exercise training. BMC Genomics. 2010;11:398.

McKenzie VJ, Song SJ, Delsuc F, et al. The effects of captivity on the mammalian gut microbiome. Integr Comp Biol. 2017;57:690–704.

Metcalf JL, Song SJ, Morton JT, et al. Evaluating the impact of domestication and captivity on the horse gut microbiome. Sci Rep. 2017;7:15497.

Metzger J, Philipp U, Lopes MS, et al. Analysis of copy number variants by three detection algorithms and their association with body size in horses. BMC Genomics. 2013;14:487.

Meyer M, Kircher M, Gansauge MT, et al. A high-coverage genome sequence from an archaic Denisovan individual. Science. 2012;338:222–6.

Miller W, Drautz DI, Ratan A, et al. Sequencing the nuclear genome of the extinct woolly mammoth. Nature. 2008;456:387–90.

Mohandesan E, Speller CF, Peters J, et al. Combined hybridization capture and shotgun sequencing for ancient DNA analysis of extinct wild and domestic dromedary camel. Mol Ecol Resour. 2017;17:300–13.

Nagasawa M, Mitsui S, En S, et al. Oxytocin-gaze positive loop and the coevolution of human-dog bonds. Science. 2015;348:333–6.

Orlando L. Equids. Curr Biol. 2015;25:R973–8.

Orlando L, Eisenmann V, Reynier F, et al. Morphological convergence in Hippidion and Equus (Amerhippus) South American equids elucidated by ancient DNA analysis. J Mol Evol. 2003;57(suppl 1):S29–40.

Orlando L, Mashkour M, Burke A, et al. Geographic distribution of an extinct equid (Equus hydruntinus: Mammalia, Equidae) revealed by morphological and genetical analyses of fossils. Mol Ecol. 2006;15:2083–93.

Orlando L, Metcalf JL, Alberdi MT, et al. Revising the recent evolutionary history of equids using ancient DNA. Proc Natl Acad Sci U S A. 2009;106:21754–9.

Orlando L, Ginolhac A, Raghavan M, et al. True single-molecule DNA sequencing of a pleistocene horse bone. Genome Res. 2011;21:1705–19.

Orlando L, Ginolhac A, Zhang G, et al. Recalibrating Equus evolution using the genome sequence of an early Middle Pleistocene horse. Nature. 2013;499:74–8.

Orlando L, Gilbert MT, Willerslev E. Reconstructing ancient genomes and epigenomes. Nat Rev Genet. 2015;16:395–408.

Outram AK, Stear NA, Bendrey R, et al. The earliest horse harnessing and milking. Science. 2009;323:1332–5.

Park SD, Magee DA, McGettigan PA, et al. Genome sequencing of the extinct Eurasian wild aurochs, Bos primigenius, illuminates the phylogeography and evolution of cattle. Genome Biol. 2015;16:234.

Patterson N, Moorjani P, Luo Y, et al. Ancient admixture in human history. Genetics. 2012;192:1065–93.

Petersen JL, Mickelson JR, Rendahl AK, et al. Genome-wide analysis reveals selection for important traits in domestic horse breeds. PLoS Genet. 2013;9:e1003211.

Pedersen MW, Overballe-Petersen S, Ermini L, et al. Ancient and modern environmental DNA. Philos Trans R Soc Lond Ser B Biol Sci. 2015;370:20130383.

Pickrell JK, Pritchard JK. Inference of population splits and mixtures from genome-wide allele frequency data. PLoS Genet. 2012;8:e1002967.

Promerová M, Andersson LS, Juras R, et al. Worldwide frequency distribution of the 'Gait keeper' mutation in the DMRT3 gene. Anim Genet. 2014;45:274–82.

Pruvost M, Bellone R, Benecke N, et al. Genotypes of predomestic horses match phenotypes painted in Paleolithic works of cave art. Proc Natl Acad Sci U S A. 2011;108:18626–30.

Ramos-Madrigal J, Smith BD, Moreno-Mayar JV, et al. Genome sequence of a 5,310-year-old maize cob provides insights into the early stages of maize domestication. Curr Biol. 2016;26:3195–201.

Rasmussen M, Li Y, Lindgreen S, et al. Ancient human genome sequence of an extinct Palaeo-Eskimo. Nature. 2010;463:757–62.

Rasmussen M, Guo X, Wang Y, et al. An Aboriginal Australian genome reveals separate human dispersals into Asia. Science. 2011;334:94–8.

Rasmussen M, Anzick SL, Waters MR, et al. The genome of a Late Pleistocene human from a Clovis burial site in western Montana. Nature. 2014;506:225–9.
Reich D, Green RE, Kircher M, et al. Genetic history of an archaic hominin group from Denisova Cave in Siberia. Nature. 2010;468:1053–60.
Reissmann M, Bierwolf J, Brockmann GA. Two SNPs in the SILV gene are associated with silver coat colour in ponies. Anim Genet. 2007;38:1–6.
Renaud G, Petersen B, Seguin-Orlando A, Bertelsen MF, Waller A, Newton R, Paillot R, Bryant N, Vaudin M, Librado P, Orlando L. Improved de novo genomic assembly for the domestic donkey. Sci Adv. 2018;4:eaaq0392. https://doi.org/10.1126/sciadv.aaq0392.
Rivero JL, Hill EW. Skeletal muscle adaptations and muscle genomics of performance horses. Vet J. 2016;209:5–13.
Rossel S, Marshall F, Peters J, et al. Domestication of the donkey: timing, processes, and indicators. Proc Natl Acad Sci U S A. 2008;105:3715–20.
Scheu A (2017) Neolithic animal domestication as seen from ancient DNA. Quat Int. https://doi.org/10.1016/j.quaint.2017.02.009.
Schubert M, Jónsson H, Chang D, et al. Prehistoric genomes reveal the genetic foundation and cost of horse domestication. Proc Natl Acad Sci U S A. 2014;111:E5661–9.
Schubert M, Mashkour M, Gaunitz C, et al. Fast, accurate and sensitive pipeline to genetically identify equine F1-hybrids in archaeological assemblages. J Archaeol Sci. 2017;78:147–57.
Steiner CC, Mitelberg A, Tursi R, et al. Molecular phylogeny of extant equids and effects of ancestral polymorphism in resolving species-level phylogenies. Mol Phylogenet Evol. 2012;65:573–81.
Signer-Hasler H, Flury C, Haase B, et al. A genome-wide association study reveals loci influencing height and other conformation traits in horses. PLoS One. 2012;7:e37282.
Skoglund P, Ersmark E, Palkopoulou E. Ancient wolf genome reveals an early divergence of domestic dog ancestors and admixture into high-latitude breeds. Curr Biol. 2015;25:1515–9.
Sommer RS, Benecke L, Lougas O, et al. Holocene survival of the wild horse in Europe: a matter of open landscape? J Quat Sci. 2011;26:1099–417.
Stoneking M, Krause J. Learning about human population history from ancient and modern genomes. Nat Rev Genet. 2011;12:603–14.
Stuart AJ. Late quaternary megafaunal extinctions on the continents. Geol J. 2015;50:338–63.
Tozaki T, Miyake T, Kakoi H, et al. A genome-wide association study for racing performances in Thoroughbreds clarifies a candidate region near the MSTN gene. Anim Genet. 2010;41(suppl 2):28–35.
Vilà C, Leonard JA, Gotherstrom A, et al. Widespread origins of domestic horse lineages. Science. 2001;291:474–7.
Vigne FD, Helmer D, Peters J. First steps of animal domestication: new archaeozoological approaches. Oxford: Oxbow Books; 2005.
Vilstrup JT, Seguin-Orlando A, Stiller M, et al. Mitochondrial phylogenomics of modern and ancient equids. PLoS One. 2013;8:e55950.
Wade CM, Giulotto E, Sigurdsson S, et al. Genome sequence, comparative analysis, and population genetics of the domestic horse. Science. 2009;326:865–7.
Wakefield S, Knowles J, Zimmermann W, et al. Status and action plan for the Przewalski's horse (Equus ferus przewalskii). In: Moehlman P, editor. Equids: zebras, asses and horses, vol. 2002. Cambridge: IUNC/SSC Equid Specialist Group, IUCN Publications Services Unit; 2012. p. 82–92.
Warmuth V, Eriksson A, Bower MA, et al. European domestic horses originated in two holocene refugia. PLoS One. 2011;6:e18194.
Warmuth V, Eriksson A, Bower MA, et al. Reconstructing the origin and spread of horse domestication in the Eurasian steppe. Proc Natl Acad Sci U S A. 2012;109:8202–6.
Warinner C, Speller C, Collins M. A new era in palaeomicrobiology: prospects for ancient dental calculus as a long-term record of the human oral microbiome. Philos Trans R Soc Lond Ser B Biol Sci. 2015;370:20130376.
Weinstock J, Willerslev E, Sher A, et al. Evolution, systematics, and phylogeography of pleistocene horses in the new world: a molecular perspective. PLoS Biol. 2005;3:e241.

Wilkins AS, Wrangham RW, Fitch WT. The "domestication syndrome" in mammals: a unified explanation based on neural crest cell behavior and genetics. Genetics. 2014;197:795–808.
Willerslev E, Davison J, Moora M, et al. Fifty thousand years of Arctic vegetation and megafaunal diet. Nature. 2014;506:47–51.
Wallner B, Vogl C, Shukla P, et al. Identification of genetic variation on the horse y chromosome and the tracing of male founder lineages in modern breeds. PLoS One. 2013;8:e60015.
Wutke S, Benecke N, Sandoval-Castellanos E, et al. Spotted phenotypes in horses lost attractiveness in the Middle Ages. Sci Rep. 2016a;6:38548.
Wutke S, Andersson L, Benecke N, et al. The origin of ambling horses. Curr Biol. 2016b;26:R697–9.

Primate Paleogenomics

Krishna R. Veeramah

Abstract The field of paleogenomics is revolutionizing our understanding of a variety of species, including humans, dogs, horses, and even extinct mammoths. Yet, despite sequencing over 1,000 anatomically modern and archaic human ancient genomes, there has yet to be a single paleogenome for any nonhuman primate. In this review I outline the problems facing the application of paleogenomics to nonhuman primates. The major issue is that primates are predominantly found in regions of the world that are hot and humid and have acidic soil conditions, such as tropical rainforests, where DNA preservation is poor. I then identify multiple possible directions for future research that focus on questions that could be addressed based on the existing paleontological record from the Late Pleistocene and Holocene. One of these, the study of extinct lemurs, has already produced results using ancient mitogenomes that have challenged existing ideas of the lemur phylogeny based on morphology. A similar process of anthropogenic-mediated extinction could potentially be studied in the case of monkeys that once resided in Caribbean. In addition, there are also possibilities to learn more about the past history and subsequent local extinction of macaques in Europe and apes on mainland Asia during the Late Pleistocene.

Keywords Ancient DNA · Extinction · Gibbons · Lemurs · Macaques · New World monkeys · Orangutans · Paleogenomics · Primates

1 Introduction

Primates have long been a major area of study for both paleontology and genetics. The rapid development of paleogenomics over the last decade should have provided an ideal platform for these two fields to interact through the sequencing of DNA

K. R. Veeramah (✉)
Department of Ecology and Evolution, Stony Brook University, Stony Brook, NY, USA
e-mail: krishna.veeramah@stonybrook.edu

Charlotte Lindqvist and Om P. Rajora (eds.), *Paleogenomics*,
Population Genomics [Om P. Rajora (Editor-in-Chief)],
https://doi.org/10.1007/13836_2018_47,

from primate fossils. However, to date no non-hominid primate paleogenome exists, and only a handful of studies have obtained any ancient DNA. This is in stark contrast to humans, where the field is now approaching 2,000 ancient genomes, while multiple ancient genomes exist for horses (Orlando et al. 2013; Librado et al. 2017) and dogs (Botigué et al. 2017; Frantz et al. 2016; Ní Leathlobhair et al. 2018), which, while clearly a crucial part of our past, are much more evolutionarily distant than primates. In this review I outline why progress in this area has been slow thus far, and identify some potential avenues of research where paleogenomics could contribute substantially to our knowledge of primate evolution.

2 The Rise of *Homo* Paleogenomics

Progress in the sequencing of ancient hominin genomes over the past decade has been staggering, in large part due to the development of next-generation sequencing (NGS), which is ideally suited to sequencing small ancient DNA (aDNA) fragments (Stoneking and Krause 2011), and the later finding that the petrous bone often contains magnitudes more endogenous aDNA than teeth and long bones, which were previously thought to be the best material to study (Pinhasi et al. 2015). Technical innovations such as single-stranded library preparation (Gansauge and Meyer 2013), ancient DNA capture enrichment techniques (Fu et al. 2013; Maricic et al. 2010), and DNA treatment strategies (Rohland et al. 2015) have also been vital in increasing the time depth and throughput of ancient genome production.

The primary focus of paleogenomic studies for hominin primates to date has been on (a) the possibility and subsequent amount of introgression between anatomically modern humans and archaic hominins and (b) the degree to which European Paleolithic/Mesolithic hunter-gatherers were replaced by incoming Neolithic farmers from the Near East, followed by the surprising finding of an additional contribution from eastern Bronze Age herders from the steppe (Haak et al. 2015; Lazaridis et al. 2014; Bramanti et al. 2009; Skoglund et al. 2012; Hofmanová et al. 2016; Allentoft et al. 2015). The sequencing of the first Neanderthal genome in 2010 (Green et al. 2010) found evidence of gene flow between Neanderthals and non-African anatomically modern humans (AMHs), and the subsequent sequencing of nine more Neanderthal genomes ranging from 70 to 39 kyo (thousand years old) has further refined our understanding of the migration and admixture events involved to an unprecedented level of resolution (Hajdinjak et al. 2018; Prüfer et al. 2014, 2017). In addition, a new archaic hominin species named Denisova has been discovered based solely on sequencing aDNA from a phalanx found in a cave in the Altai Mountains. Denisovans appear to have admixed with AMHs from Southeast Asia (particularly Melanesians) (Reich et al. 2010), and four genomes now exist from this previously unknown hominin (Slon et al. 2017a). Nuclear DNA (albeit limited amounts) has even been obtained from two 430 kyo hominins from fossils found in the Sima de los Huesos cave (Meyer et al. 2016).

3 What Have Been the Barriers to Primate Paleogenomics?

Given these successes, why has paleogenomics not found success in nonhuman primates? The most obvious reason is where primates are geographically distributed in the world today. Primates are diverse; there are 504 species across 79 genera and 16 families. However, two thirds of the species are found in just four countries, Brazil, Indonesia, the Democratic Republic of the Congo (DRC), and Madagascar, the first three of which are dominated by tropical rainforests (this was probably the case for Madagascar at some point as well). Primates more generally are predominantly found in tropical and subtropical parts of the world (Fig. 1), where conditions (high temperature, high humidity, acidic soil) are not conducive to long-term DNA preservation (Hofreiter et al. 2015). There are only a handful of cases where aDNA has been successfully extracted from such regions, the vast majority of which involve mitochondrial DNA (mtDNA) sequences from fairly young samples (Kehlmaier et al. 2017; Gutiérrez-García et al. 2014; Brace et al. 2015, 2016; Mohandesan et al. 2017). To appreciate the scale of the problem, a recent paper by David Reich's group at Harvard, who are very much leading the way in terms of human paleogenomic data, attempted to sequence ancient DNA from human individuals from Vietnam and Thailand from 4.1 to 1.7 thousand years ago (kya). Despite generating 350 NGS libraries from the petrous bones of 146 individuals, only 18 yielded usable DNA (12% success rate), and meaningful analysis was only possible after generating ~5 NGS libraries for each of these 18 samples (Lipson et al. 2018).

Even if the technical issues regarding working in tropical and subtropical regions can eventually be resolved (Mohandesan et al. 2017), researchers may find actual available fossils to be lacking. The paleontological record is extremely sparse were most primates are found today and where they have likely resided for millions of years, the rainforests of the Amazon, Congo, and Southeast Asia (there is an absence of exposed rock, and logistically and politically it can be difficult to work in some of these regions). Methods to obtain mammalian aDNA directly from soil in caves have recently emerged and proved successful for obtaining mtDNA from even Middle Pleistocene fossils (Slon et al. 2017b). However, the acidity of rainforest soil may prove a difficult barrier even for this method.

A third problem that may stall progress in the area of primate paleogenomics is a lack of appropriate reference genomes. Due to the short DNA fragments that result from postmortem degradation, any kind of paleogenomic analysis of autosomal, X and Y chromosome regions will require mapping onto an appropriate reference genome; de novo assembly from aDNA, with fragments typically less than 100 bp in length, is likely to be impossible for mammalian species. Currently high-quality reference genomes (genomes with high N50 values where much of the sequence is represented in long continuous contigs and scaffolds) exist for all the apes (all have a scaffold N50 $>$ 50 Mb, except Bonobos with an N50 of 8 Mb (Chimpanzee Sequencing and Analysis Consortium 2005; Scally et al. 2012; Locke et al. 2011; Carbone et al. 2014; Prüfer et al. 2012; Gordon et al. 2016; Kronenberg et al. 2018).

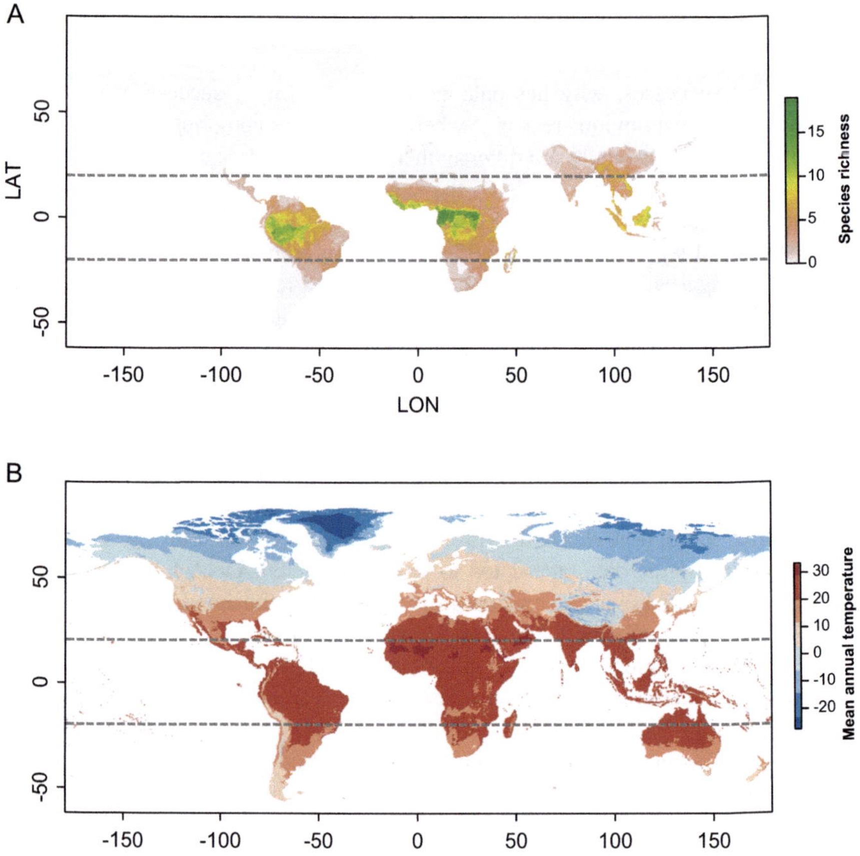

Fig. 1 Global distribution of (**a**) primate species richness and (**b**) mean annual temperature. Plots created in R using the raster, sp., rgdal, and RColorBrewer packages. Primate richness data obtained from the International Union for Conservation of Nature Red List Terrestrial Mammals spatial dataset, downloaded on 24 July 2018. Climate data obtained by averaging monthly figures from the WorldClim Version 2 dataset, release 1, June 2016 (Fick and Hijmans 2017). World borders used to frame plots were obtained using the World Borders Dataset (http://thematicmapping.org/downloads/world_borders.php). Dashed gray lines indicate interval for the tropics

However, for other primates there are published high-quality reference genomes for only three Old World monkey species (*Macaca mulatta* [macaques, scaffold N50 ~ 4 Mb] (Rhesus Macaque Genome Sequencing and Analysis Consortium et al. 2007), *Cercocebus atys* [sooty mangabeys, scaffold N50 ~ 13 Mb] (Palesch et al. 2018), and *Chlorocebus aethiops sabaeus* [vervets, scaffold N50 ~ 81 Mb] (Warren et al. 2015)), one New World monkey species (*Callithrix jacchus* [common marmosets, scaffold N50 ~ 7 Mb] (Marmoset Genome Sequencing and Analysis Consortium 2014)), and one lemur species (*Microcebus murinus* [mouse lemurs, scaffold N50 ~ 93 Mb] (Larsen et al. 2017)), in addition to draft genomes (fairly low to medium N50 genome builds) for *Daubentonia madagascariensis* [the aye-aye, scaffold N50 = 0.01 Mb] (Perry et al. 2012), *Rhinopithecus roxellana* [snub-nosed monkey, scaffold N50 = 1.5 Mb] (Zhou et al. 2014), and *Tarsius syrichta* [tarsiers, scaffold N50 = 0.4 Mb] (Schmitz et al. 2016). This is five high-quality reference genomes to represent almost 500 distinct species across 71 genera. Such a paucity of close reference genomes for the majority of species would likely lead to biases in downstream population genetic analysis of distantly related primates, and these biases would be further amplified for aDNA due to lower mapping confidence introduced by short DNA fragments and postmortem damage. Fortunately, the development of "third-generation" sequencing technologies, such as PacBio sequencing, with the production of long reads in excess of 10 kb, promises to overcome this issue by democratizing the production of high-quality de novo assembly of primate genomes (Gordon et al. 2016; Kronenberg et al. 2018). By using long-read techniques, high-quality reference genomes have the potential to be constructed via simple shotgun sequencing, rather than, for example, via extensive Sanger-based sequencing of laboriously constructed and overlapping BAC clones (Lander et al. 2001).

4 What Kind of Questions Can We Ask Using Primate Paleogenomics?

It is clear that obtaining paleogenomes from nonhuman primates will be extremely challenging and will require substantial resources (in terms of money, time, labor, and sequencing technology). If such resources are to be devoted to this task in the future, it is prudent to ask what kind of questions can be addressed from the application of this technology. Clearly a major constraint is the age of the fossils that can be sequenced. Although there are some notable exceptions [the 430 kyo archaic *Homo* species from the Sima de los Huesos (Meyer et al. 2016) and an almost 800 kyo horse from permafrost (Orlando et al. 2013)], the vast majority of samples that have successfully yielded paleogenomes from temperate climates have been ~50 kyo and younger (Hajdinjak et al. 2018; Reich et al. 2010; Fu et al. 2014, 2015; Raghavan et al. 2014), a large proportion of which are <10 kyo (Haak et al. 2015; Allentoft et al. 2015; Olalde et al. 2018; Mathieson et al. 2018; Damgaard et al. 2018; Lipson et al. 2017; Lazaridis et al. 2016). Clearly, success rates will be even lower in subtropical and tropical regions. Hofreiter et al. (2015) recently developed a model of DNA degradation as a function of mean global temperature that showed that, even in cave sites, there is only a 0–10% chance for DNA fragments of at least 25 bp to survive after 10,000 years in places where primates are abundant today.

Thus, realistically, even with major technical improvements for obtaining aDNA from tropical/subtropical regions and major financial investment, it is unlikely that we would be able to obtain usable aDNA from primate fossils much older than

100 kyo, essentially spanning the Late Pleistocene and Holocene. This is a period of time that is clearly of considerable interest for researchers (archaeologists and geneticists) working on human history and evolution given that AMHs only emerge ~200 kya (Veeramah and Hammer 2014). However, many of the major questions addressed by paleontologists with regard to primates operate over time spans that cover millions to tens of millions of years, including who were the earliest primates (Silcox and López-Torres 2017), how and when did terrestrial bipedal motion emerge in apes (Richmond et al. 2001), and how and when did some of the deep splits in the primate phylogeny occur such as *Platyrrhini* (New World Monkeys) diverging from other *Simiiformes* or *Lemuriformes* (lemurs) branching from other *Strepsirrhini* (Ali and Huber 2010; Perelman et al. 2011; Fleagle 2013; Bond et al. 2015; Ni et al. 2016). The study of primate paleogenomes that we could realistically retrieve is not likely to help answer such questions.

However, paleontology is a broad field, and there are still many opportunities where primate paleogenomics could potentially be of use. In particular there are four classes of questions that could be addressed from obtaining primate aDNA from the last 100 ky: (a) studying the impact of human colonization on primate extinctions, (b) inferring the demographic history of extant primates, (c) understanding how novel extinct primate species identified from paleogenomes fit into the existing primate phylogeny, and (d) using ancient genomes to refine inference of evolutionary processes.

5 Paleogenomics of Large Extinct Lemurs

One of the major areas of aDNA research involves studying to what extent human activity or changes in climate were the primary cause of megafauna extinctions during the Late Pleistocene and into the Holocene (Campos et al. 2010; Lorenzen et al. 2011; Stiller et al. 2010; Palkopoulou et al. 2013, 2015; Barnes et al. 2002; Shapiro et al. 2004; Kuhn et al. 2010), with approximately two thirds of all mammals >44 kg having gone extinct by 10 kya (Barnosky et al. 2004). The high rate of extinction events experienced by large mammals in the Americas and Australia did not greatly affect Africa [in particular primates (Faith 2014)], except for one exception, lemur primates residing in Madagascar.

The best accepted theory of lemur origins in Madagascar involves early/ancestral *Strepsirrhini* primates rafting on Indian Ocean currents from northeast Mozambique/Tanzania 50–60 million years ago (Mya) (Ali and Huber 2010). Since arriving on the fourth largest island in the world (and one that contains no large carnivores), lemurs appear to have undergone an adaptive radiation, and more than 100 individual species now exist spread across five families (*cheirogaleids*, *lemurids*, *lepilemurids*, *indriids*, and *daubentoniids*). They demonstrate considerable phenotypic (behavioral, dietary, locomotor, and morphological) diversity, ranging from the tiny mouse lemur (~30 g) to species such as the sifaka that approaches 10 kg. However, prior to the arrival of humans ~2,000 years ago, there were at least additional 17 species, many of which were considerably larger than any extant lemurs, including one that was as big as a modern male gorilla (*Archaeoindris fontoynontii*) (Fig. 2). A combination of human-mediated habitat fragmentation/modification and hunting appears to have been the primary contributor to this mass extinction event, though this process likely took centuries (Fleagle 2013; Crowley 2010). These extinct lemurs have been placed into three new families, *Archaeolemuridae* (monkey lemurs), *Megaladapis* (koala lemurs), and *Palaeopropithecidae* (sloth lemurs), while there was also once a much larger relative of the aye-aye, *Daubentonia robusta*, and some extinct species of *Lemuridae*.

These phylogenetic placements have primarily been based on morphological criteria. However, some of the large lemur fossils are ~1,000 years old, which would place them within the realms of current paleogenomic NGS techniques. Indeed, the only published aDNA work performed on primates has involved lemurs. Two studies were able to use classical PCR-based aDNA techniques to obtain partial mitochondrial fragments from members of each of the extinct families, demonstrating that *Archaeolemuridae* and *Palaeopropithecidae* were both closely related to *Indridae*, consistent with prior morphological hypotheses, though *Megaladapis* was

Fig. 2 Body mass distribution of extinct and extant lemurs. Note *y*-axis for mass is $\log_{10}$ scaled. Extant data from Barnosky et al. (2004), extinct data from Fleagle (2013). *Microcebus berthae* figure is modified (cropped) from https://commons.wikimedia.org/wiki/File:Madame_Berthe%27s_Mouse_Lemur.jpg by FC Casuario and is licensed under the Creative Commons Attribution-Share Alike 4.0 International license. *Indri Indri* figure is modified (cropped) from https://commons.wikimedia.org/wiki/File:Indri_Andasibe.JPG by Karen Coppock and is licensed under the Creative Commons Attribution 3.0 Unported license. *Archaeoindris* figure is modified (cropped) from https://commons.wikimedia.org/wiki/File:Archaeoindris_fontoynonti.jpg by Smokeybjb and is licensed under the Creative Commons Attribution-Share Alike 3.0 Unported license

found to be sister taxa with *lemurids* rather than the expected *lepilemurids* (Karanth et al. 2005; Orlando et al. 2008). However, it is a study by Kistler et al. (2015) that is currently the most comprehensive paleogenomic primate study to date. The authors used NGS and a combination of shotgun and capture techniques to construct almost full mtDNA genomes from four of the five families with extinct species (they were only lacking the giant Aye-Aye). This provided a high-resolution mitochondrial phylogeny that also included divergence dates for the different families. The phylogenetic result supported the previous hypothesis of Orlando et al. (2008) that *Archaeolemuridae* was an outgroup to *Palaeopropithecidae* and *Indridae* based on a proposed model of locomotion evolution, with the latter two species sharing a slow sloth-like hanging skill. By obtaining additional mtDNA data from multiple individuals per species, they were even able to demonstrate that the large extinct lemurs had smaller long-term effective population sizes compared to their smaller extant relatives, providing new information regarding the dynamics of such population genetic parameters during extinction events and how they relate to phenotype (large body size appears to be associated with both small population sizes and higher anthropogenic extinction risks).

Though significant challenges still remain, the next natural step will clearly be to obtain whole genomes from some of these extinct species. Lemurs are one of the premier examples of an adaptive radiation, with John Fleagle referring to them as a "a natural experiment in evolution" (Fleagle 2013). Obtaining aDNA from the current extant taxa as well as the larger extinct taxa is important for fully characterizing the evolutionary and ecological consequences of such a radiation. There is in fact very little fossil record for Madagascar prior to the Holocene (Crowley 2010). The sequencing of the full range of late Holocene lemurs would not only provide a refined phylogeny with more precise node dating but also allow us to better infer species demography than is possible with mtDNA (Chang and Shapiro 2016), for example, by the use of analytical methods that model changes in effective population size hundreds of thousands and even millions of years into the past (Schiffels and Durbin 2014; Terhorst et al. 2017; Li and Durbin 2011; Hobolth et al. 2011) or identifying complex patterns of historical between species gene flow (Patterson et al. 2012). A high-resolution definition of the long-term population dynamics of small, medium, and large lemurs, especially if related to data on phenotype evolution and climate change, would add greatly to our understanding of this classical adaptive radiation.

6 Deciphering Caribbean Primate Extinctions

There is a general scarcity of New World monkey fossils, which suggests that much of their evolution occurred in the Amazon Basin where they are usually found today (Fleagle 2013). Pleistocene primate fossils from at least two distinct genera have been found in two cave deposits in Brazil that appear to be double the size of any extant platyrrhine. The relationship of these extinct species to modern New World monkeys is unclear (*Protopithecus brasiliensis* has features that would link it to *Ateles*, *Brachyteles*, and *Alouatta* genera, though *Caipora bambuiorum* is more clearly related to *Alouatta*) (Halenar and Rosenberger 2013), and aDNA would be informative with regard to their phylogeny and the proposed hypothesis of some kind of evolutionary size constraint on New World monkeys (Ford and Davis 1992), though the limited fossil material coupled with the tropical conditions makes future aDNA retrieval unlikely.

The other major region containing Late Quaternary primate fossils is the Greater Antilles portion of the Caribbean. Similar to Madagascar, the introduction of humans appears to have led to a major loss of endemic mammals in the region. However, in this case a much wider variety of fauna were lost, with overall mammalian extinction rates the highest observed anywhere in the world during the Holocene (Cooke et al. 2017a). No primates exist on the islands today, but there is evidence for their presence beginning in the early Miocene (16.1–21.5 Mya) (MacPhee et al. 2003) up to ~1,000 years ago. There are four Late Pleistocene to Holocene species of extinct primate on these islands (Fig. 3): *Xenothrix mcgregori* (Jamaica), *Antillothrix bernensis* (Dominican Republic), *Insulacebus toussaintiana* (Haiti), and *Paraloutta varonai* (Cuba), the last of which appears to be very similar to the earliest Miocene fossil found on the same island, *Paraloutta marianae*. The most recent New World monkey occurrence in the Caribbean was *mcgregori* ~900 years ago (Cooke et al. 2017b), demonstrating clear overlap of primates with the earliest human occupation in the region [there is evidence of a Lithic culture entering the region at least 6–7 kya, as well as additional colonization waves 5 kya and 2 kya (Cooke et al. 2017a)]. These four extinct species were quite morphologically different to existing mainland monkeys, and generally larger in body size as a group (perhaps indicative of island gigantism, which would suggest they had been isolated from the mainland for a significant period of time).

Three general models have been proposed for the origins of extinct Caribbean monkeys: (a) they are the result of a single radiation within an existing *Platyrrhini* family such as *Pitheciidae* (Macphee and Horovitz 2004), (b) they are distinct from any extant family and are instead a stem *Platyrrhini* family (Kay 2015), or (c) they are related to multiple different extant taxa from diverse families on the mainland (Rosenberger et al. 2011) (there are abundant nearby extant *Platyrrhini* species from different families and genera on the surrounding mainland in the south and west, in particular, the Yucatan State of Mexico next to Cuba and in Venezuela). Clearly the sequencing of paleogenomes, even if only mtDNA, would provide answers to these questions in similar fashion to the phylogenetic placement of giant lemurs in Madagascar. Ancient mtDNA has already been sequenced from Caribbean animal fossils to ask similar questions (Kehlmaier et al. 2017; Brace et al. 2015), and a 12× coverage whole genome has even been obtained from a 1,000-year-old human (Schroeder et al. 2018), suggesting obtaining aDNA from the extinct primates of these islands is a realistic possibility in the near future.

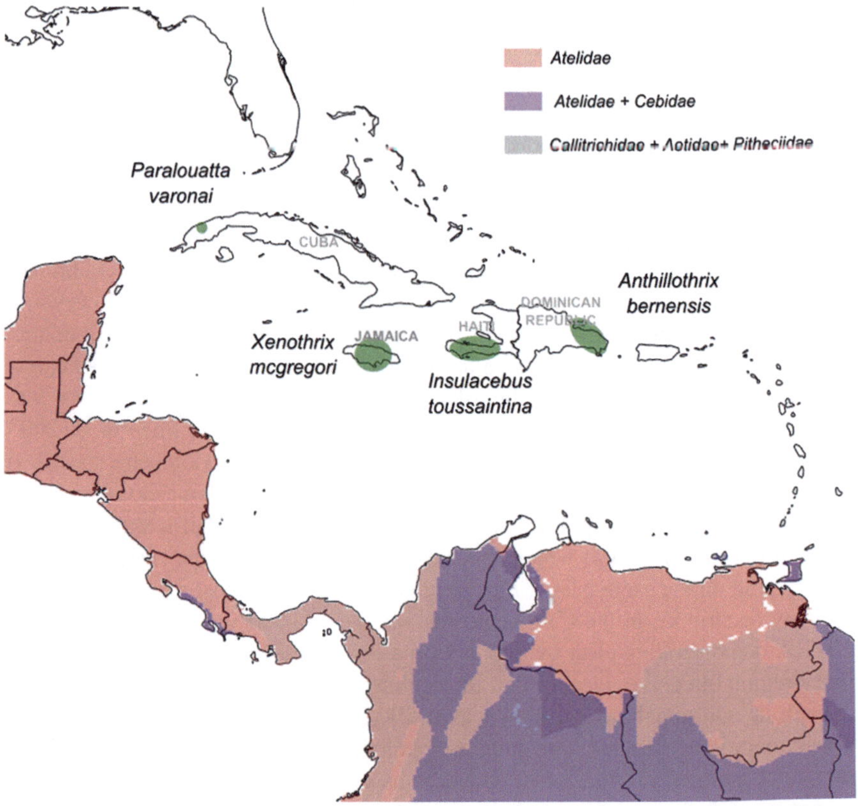

Fig. 3 Locations of extinct monkeys in the Caribbean along with distribution of extant *Platyrrhini* families. Base maps created in R using the raster, sp., rgdal, and RColorBrewer package. New World monkey distribution data obtained from the International Union for Conservation of Nature Red List Terrestrial Mammals spatial dataset, downloaded on 24 July 2018. World borders used to frame plots were obtained using the World Borders Dataset (http://thematicmapping.org/downloads/world_borders.php). Green circles represent broad range of fossil finds for an extinct species. *Cebidae* always overlaps with the much broadly distributed *Atelidae*

7 The Demography of Macaques

The majority of living primates are threatened by extinction (Estrada et al. 2017). The foremost exception to this are members of the genus *Macaca*, or macaques, who are found through much of South, Southeast, and East Asia as well as parts of north Africa. There are 23 different species, and they are highly adaptable, in some circumstances even being considered invasive and reaching their highest densities in areas occupied by humans (Fleagle 2013). The best known species is the Rhesus macaque (*Macaca mulatta*), which has become a major model organism for biomedical purposes (Vallender and Miller 2013). The most southern extant nonhuman primate is the crab-eating macaque (*Macaca fascicularis*), and the most northern extant nonhuman primate is the Japanese macaque (*Macaca fuscata*). The latter actually spends substantial parts of its life in snow, and it has likely lived in these extreme conditions for the last 300–800 ky (Marmi et al. 2004). Fossils from this species (Iwamoto 1975; Fooden and Aimi 2005; Iwamoto and Hasegawa 1972) might make an excellent candidate for aDNA extraction and sequencing in order to explore both the species' demography and the molecular dynamics that underlie how it has adapted to such extreme conditions (Enari 2014; Enari and Sakamaki-Enari 2013).

There is a small colony of ~200 Barbary macaques (*Macaca sylvanus*) in Gibraltar, Spain, that was introduced from North Africa, but otherwise Europe is absent of this and other Old World primates [the Gibraltar colony are considered to be recent imports from Morocco and Algeria (Modolo et al. 2005)]. However, there is ample evidence for a wide distribution of this species across much of Europe in the past starting in the Late Miocene (Alba et al. 2014; Köhler et al. 2000), and in the Middle Pleistocene they even appear as far north as Great Britain (Elton and O'Regan 2014). These fossil macaques do not appear to greatly differ from modern Barbary macaques, suggesting a very stable morphology for the last 5 million years. While other non-ape primate genera had resided in Europe up to the Middle Pleistocene, macaques are the only fossils found in the Late Pleistocene (Elton and O'Regan 2014). Though macaque fossils are not common during this period compared to the Middle Pleistocene, for which almost four times as many fossils have been found, they are widely distributed across a 1 million km^2 area. Fossil/climate modelling suggest there was an excellent chance of finding macaques across mainland Europe during the Last Glacial Maximum (Elton and O'Regan 2014), which is within the timeframe for obtaining human European paleogenomes (Fu et al. 2016). However, the youngest fossils actually found to date are ~70–50 kyo from southern Europe and Germany (Mazza et al. 2005; Castaños et al. 2011; Rosendahl et al. 2011), which may make any aDNA extraction difficult using existing material. The cause of the extinction has been primarily attributed to climatic effects on vegetation, but unlike Asian species of macaques, Barbary macaques are negatively impacted by human occupation (Ménard 2003), so the impact of some kind of resident *Homo* species such as *Neanderthals* may have also played some role (Rosendahl et al. 2011) (AMH likely did not enter Europe before Barbary macaques went extinct in the region). As the only known recent primate in Europe, obtaining aDNA from some of these ancient European macaques would be of interest for examining their long-term population dynamics and how they may have interacted with archaic humans, as well as help advise conservation efforts for extant Barbary macaques in North Africa that are currently in serious decline (IUCN lists them as endangered).

Despite their wide distribution today, macaques appear to have entered Asia later than Europe. There are abundant *Macaca* Pleistocene fossils throughout China and Southeast Asia, though their range contracts in a southerly direction approaching the Late Pleistocene, presumably to avoid more temperate climates that emerged in the north (Jablonski 1998) (the exception of course being the Japanese macaque). There are an array of Late Pleistocene fossils from different macaque species, with many of them showing high similarity to those residing in the same location, suggesting long-term population continuity (Fleagle 2013). In particular, because of its status as a key model organism, there has been substantial high-profile work using modern genomic data to model Rhesus Macaque demographic history (Hernandez et al. 2007; Xue et al. 2016). The current model suggests a divergence between Indian and Chinese rhesus macaques ~100–150 kya followed by a population decrease in the former and increase in the latter. There are also suggestions of hybridization with *Macaca cynomolgus* (Stevison and Kohn 2009). The sequencing of ancient macaque sequences would help refine these demographic inferences, which have wide confidence intervals and depend on uncertain assumptions for key parameters such as the mutation rate and generation time (Scally and Durbin 2012), by acting as calibration points. Having genomes from both the present day and ancient macaques from the same species would also help better understand the molecular and evolutionary basis of any recent adaptations (Mathieson et al. 2015), which may have implications for downstream biomedical research using these organisms.

8 Ancient Asian Apes

Probably because of their long-term presence in dense rainforest, almost no ancient ape fossils exist that would be considered specific to the chimpanzee or gorilla lineages [the notable exception being some Middle Pleistocene *Pan* teeth from the Rift Valley in Kenya that extends beyond the geographic range of modern chimpanzees (McBrearty and Jablonski 2005)]. Such material would be of value for illuminating African ape history [e.g., subspecies divergence (Prado-Martinez et al. 2013; McManus et al. 2015; de Manuel et al. 2016; Wegmann and Excoffier 2010)] and for studying molecular evolutionary processes such as mutation rates across the ape family (Scally and Durbin 2012; Besenbacher et al. 2018; Thomas et al. 2018; Moorjani et al. 2016; Venn et al. 2014), but it does not look likely that such paleogenomes will be generated in the future. However, Asian ape paleogenomics may prove a more fruitful avenue of research. Though now highly endangered and restricted to three species [the third of which was only recently identified based on genome sequencing (Nater et al. 2017)] in the Southeast Asian islands of Borneo and Sumatra, there is ample fossil evidence for orangutans on mainland South and Southeast Asia (Ibrahim et al. 2013) [including full skeletons (Bacon and The Long V 2001)], as well as a diverse range of other now extinct apes from the Miocene and beyond that appear to be related to the *Pongo* genus (Fleagle 2013). *Pongo* fossils appear in the fossil record throughout the Pleistocene and even into the Holocene (Fig. 4), consistent with a mainland extinction that may have been a result of a failure to adapt to a more seasonal and drier environment during the last glacial maximum (Jablonski 1998). Some fossils appear to be as young as 50–20 ky old and thus potentially amenable to aDNA analysis (Ibrahim et al. 2013).

The modeling of demographic history using modern orangutan genomes has produced some curious patterns. While possessing a census population size that is

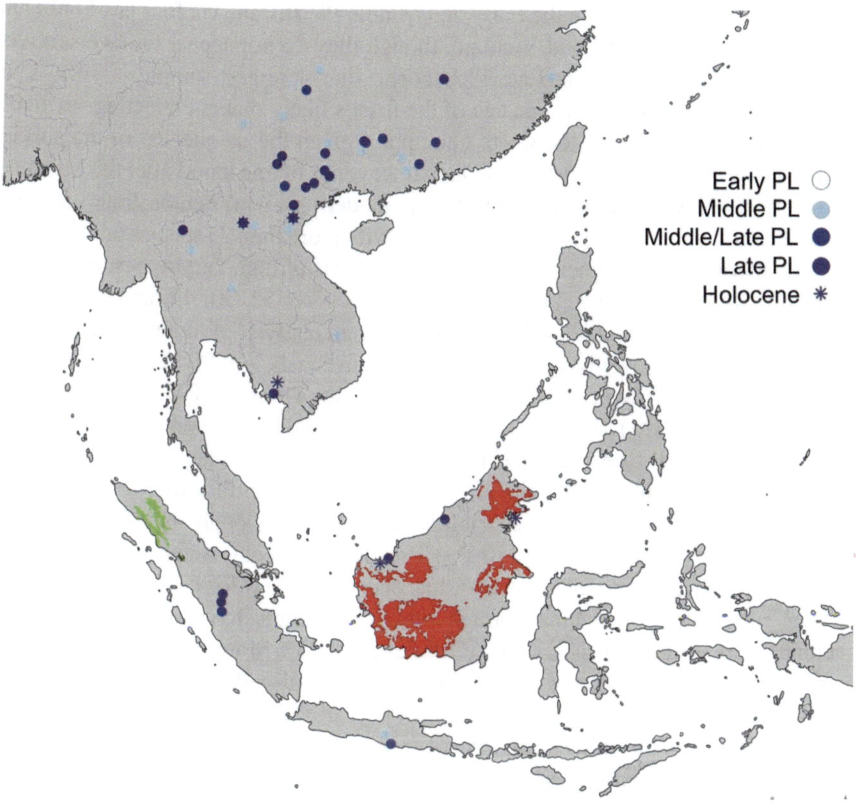

Fig. 4 Distribution of Pleistocene and Holocene *Pongo* fossils on mainland Asia. Data from Table 1 of Ibrahim et al. (2013). Red, green, and dark green areas are the modern distributions of *Pongo pygmaeus*, *Pongo abelii*, and *Pongo tapanuliensis*, respectively. *Pongo* distribution data obtained from the International Union for Conservation of Nature Red List Terrestrial Mammals spatial dataset, downloaded on 24 July 2018. *PL* Pleistocene

~7 times larger than Sumatran orangutans (*Pongo abelii*), Bornean orangutans (*Pongo pygmaeus*) have maintained a much smaller effective population size since diverging ~1–0.4 Mya (Locke et al. 2011; Prado-Martinez et al. 2013). In addition, despite being found on the same island as *Pongo abelii* and only being separated by Lake Toba, the newly identified third species, *Pongo tapanuliensis*, appears to have diverged from the ancestor of the other two species 3.4 Mya (Nater et al. 2017). It has been suggested that *Pongo tapanuliensis* may be the descendents of the original population that migrated from mainland Asia. Paleogenomes from mainland *Pongo* fossils would be extremely useful for disentangling the population processes that have since occurred for the island species.

Another related ape that would be fascinating to examine using paleogenomic methods would be *Gigantopithecus blacki*. This is the largest known primate to have ever lived, weighing as much as 330 kg and standing almost 10 ft tall. There is fossil evidence for its presence in the Early and Middle Pleistocene via four jaw bones and ~2,000 teeth from China and Vietnam, though they do not appear to have survived the climatic changes of the Late Pleistocene (the youngest sample is ~300 kyo) (Zhang and Harrison 2017). The age of the fossils likely makes extracting verifiable aDNA difficult but would be worth attempting given the uniqueness of the species and the rare occurrence of fossils for an extinct genus of ape found after the Miocene.

Despite being highly endangered today, modern gibbons demonstrate high species diversity compared to other extant apes, with more than 20 species or subspecies found among 4 major genera identified on the basis of their karyotypes. *Nomascus*, *Symphalangus*, *Hylobates*, and *Hoolock* each possess 52, 50, 44, and 38 chromosomes, respectively, and it has been proposed from analysis of modern genomes that these 4 genera likely diverged almost instantaneously ~5 Mya, possibly due to climatic shifts and fragmentation of the Sunda Shelf forests (Carbone et al. 2014; Veeramah et al. 2015). Gibbons are highly endangered but are spread across tropical rainforests close to the coasts of mainland South and Southeast Asia and island Southeast Asia. Gibbons are also represented in the Asian Pleistocene fossil record and were once more widely distributed than today, though assignment to any particular genus is difficult as most fossil remains are teeth. A separate genus from the extant four, *Bunopithecus sericus*, has been suggested for a Middle Pleistocene partial mandible from China, though other researchers have proposed that it is actually part of the *Hoolock* genus (Ortiz et al. 2015). More interestingly from the perspective of paleogenomic applications, Turvey and colleagues recently reported that some fossil remains from a 2,200–2,300-year-old tomb in China was that of a new genus of gibbon, *Junzi imperialis* (Turvey et al. 2018). However, before the use of karyotypes, the correct assignment of modern gibbon species based only on morphological criteria was fraught with issues (Mootnick 2006), and therefore basing this on only partial skeletal information is likely to be even more problematic. The idea that genera distinct from those living today may have existed more widely in mainland Asia has important consequences for the evolution and radiation of gibbons and small apes in Asia and the process that underlies their somewhat unique karyotype evolution. Obtaining paleogenomic data is likely the only way the question of genera membership will be truly clarified, though in this case ancient mtDNA is unlikely to have resolution to do this based on previous work, which has shown extensive incomplete lineage sorting across all four existing genera (Veeramah et al. 2015; Thinh et al. 2010; Matsudaira and Ishida 2010; Wall et al. 2013).

9 Conclusion and Future Perspectives

Obtaining paleogenomes (even just mtDNA) from primates in the future is likely to continue to be a difficult proposition. However, if sufficient resources are devoted to the task, there are possibilities to advance many areas of ecology and evolution across the primate kingdom, as well as help answer larger questions about extinctions and adaptive radiations. The work of George Perry and colleagues on lemurs

(Kistler et al. 2015) is leading the way in this regard. Clearly, some technical innovations will be required to enhance the probability of obtaining DNA from regions of high temperatures. The scarcity of valuable primate fossils, often from single elements such as teeth, means that the chance of success needs a priori to be high, as the downstream laboratory processing will almost certainly result in the destruction of much of the available material, a factor that has likely prevented aDNA work on many primate fossils already [e.g., see the recent case of *Junzi imperialis* (Turvey et al. 2018)]. CT scanning coupled with 3D printing has the potential to mitigate this loss somewhat, but the decision to attempt aDNA extraction should not be taken lightly and be done in a framework that involves close collaboration between paleontologists and genetics at all stages.

References

Alba DM, Delson E, Carnevale G, Colombero S, Delfino M, Giuntelli P, et al. First joint record of Mesopithecus and cf. Macaca in the Miocene of Europe. J Hum Evol. 2014;67:1–18.

Ali JR, Huber M. Mammalian biodiversity on Madagascar controlled by ocean currents. Nature. 2010;463:653–6.

Allentoft ME, Sikora M, Sjögren K-G, Rasmussen S, Rasmussen M, Stenderup J, et al. Population genomics of Bronze Age Eurasia. Nature. 2015;522:167–72.

Bacon AM, The Long V. The first discovery of a complete skeleton of a fossil orang-utan in a cave of the Hoa Binh Province, Vietnam. J Hum Evol. 2001;41:227–41.

Barnes I, Matheus P, Shapiro B, Jensen D, Cooper A. Dynamics of Pleistocene population extinctions in Beringian brown bears. Science. 2002;295:2267–70.

Barnosky AD, Koch PL, Feranec RS, Wing SL, Shabel AB. Assessing the causes of late Pleistocene extinctions on the continents. Science. 2004;306:70–5.

Besenbacher S, Hvilsom C, Marques-Bonet T. Direct estimation of mutations in great apes reveals significant recent human slowdown in the yearly mutation rate. bioRxiv. 2018. https://www.biorxiv.org/content/early/2018/03/23/287821.abstract.

Bond M, Tejedor MF, Campbell KE Jr, Chornogubsky L, Novo N, Goin F. Eocene primates of South America and the African origins of New World monkeys. Nature. 2015;520:538–41.

Botigué LR, Song S, Scheu A, Gopalan S, Pendleton AL, Oetjens M, et al. Ancient European dog genomes reveal continuity since the early Neolithic. Nat Commun. 2017;8:16082.

Brace S, Turvey ST, Weksler M, Hoogland MLP, Barnes I. Unexpected evolutionary diversity in a recently extinct Caribbean mammal radiation. Proc Biol Sci. 2015;282:20142371.

Brace S, Thomas JA, Dalén L, Burger J, MacPhee RDE, Barnes I, et al. Evolutionary history of the Nesophontidae, the last unplaced recent mammal family. Mol Biol Evol. 2016;33:3095–103.

Bramanti B, Thomas MG, Haak W, Unterlaender M, Jores P, Tambets K, et al. Genetic discontinuity between local hunter-gatherers and Central Europe's first farmers. Science. 2009;326:137–40.

Campos PF, Willerslev E, Sher A, Orlando L, Axelsson E, Tikhonov A, et al. Ancient DNA analyses exclude humans as the driving force behind late Pleistocene musk ox (Ovibos moschatus) population dynamics. Proc Natl Acad Sci U S A. 2010;107:5675–80.

Carbone L, Harris RA, Gnerre S, Veeramah KR, Lorente-Galdos B, Huddleston J, et al. Gibbon genome and the fast karyotype evolution of small apes. Nature. 2014;513:195–201.

Castaños P, Murelaga X, Arrizabalaga A, Iriarte M-J. First evidence of Macaca sylvanus (Primates, Cercopithecidae) from the late Pleistocene of Lezetxiki II cave (Basque Country, Spain). J Hum Evol. 2011;60:816–20.

Chang D, Shapiro B. Using ancient DNA and coalescent-based methods to infer extinction. Biol Lett. 2016;12:20150822.

Chimpanzee Sequencing and Analysis Consortium. Initial sequence of the chimpanzee genome and comparison with the human genome. Nature. 2005;437:69–87.

Cooke SB, Dávalos LM, Mychajliw AM, Turvey ST, Upham NS. Anthropogenic extinction dominates Holocene declines of West Indian mammals. Annu Rev Ecol Evol Syst. 2017a;48:301–27.

Cooke SB, Mychajliw AM, Southon J, MacPhee RDE. The extinction of Xenothrix mcgregori, Jamaica's last monkey. J Mammal. 2017b;98:937–49.

Crowley BE. A refined chronology of prehistoric Madagascar and the demise of the megafauna. Quat Sci Rev. 2010;29:2591–603.

Damgaard PB, Marchi N, Rasmussen S, Peyrot M, Renaud G, Korneliussen T, et al. 137 ancient human genomes from across the Eurasian steppes. Nature. 2018;557:369–74.

de Manuel M, Kuhlwilm M, Frandsen P, Sousa VC, Desai T, Prado-Martinez J, et al. Chimpanzee genomic diversity reveals ancient admixture with bonobos. Science. 2016;354:477–81.

Elton S, O'Regan HJ. Macaques at the margins: the biogeography and extinction of Macaca sylvanus in Europe. Quat Sci Rev. 2014;96:117–30.

Enari H. Snow tolerance of Japanese macaques inhabiting high-latitude mountainous forests of Japan. In: Grow NB, Gursky-Doyen S, Krzton A, editors. High altitude primates. New York: Springer; 2014. p. 133–51.

Enari H, Sakamaki-Enari H. Influence of heavy snow on the feeding behavior of Japanese macaques (macaca fuscata) in northern Japan. Am J Primatol. 2013;75:534–44.

Estrada A, Garber PA, Rylands AB, Roos C, Fernandez-Duque E, Di Fiore A, et al. Impending extinction crisis of the world's primates: why primates matter. Sci Adv. 2017;3:e1600946.

Faith JT. Late Pleistocene and Holocene mammal extinctions on continental Africa. Earth-Sci Rev. 2014;128:105–21.

Fick SE, Hijmans RJ. WorldClim 2: new 1-km spatial resolution climate surfaces for global land areas: new climate surfaces for global land areas. Int J Climatol. 2017;37:4302–15.

Fleagle JG. Primate adaptation and evolution. Cambridge: Academic; 2013.

Fooden J, Aimi M. Systematic review of Japanese macaques, Macaca fuscata. Fieldiana. 2005;104:1–200.

Ford SM, Davis LC. Systematics and body size: implications for feeding adaptations in New World monkeys. Am J Phys Anthropol. 1992;88:415–68.

Frantz LAF, Mullin VE, Pionnier-Capitan M, Lebrasseur O, Ollivier M, Perri A, et al. Genomic and archaeological evidence suggest a dual origin of domestic dogs. Science. 2016;352:1228–31.

Fu Q, Meyer M, Gao X, Stenzel U, Burbano HA, Kelso J, et al. DNA analysis of an early modern human from Tianyuan Cave, China. Proc Natl Acad Sci U S A. 2013;110:2223–7.

Fu Q, Li H, Moorjani P, Jay F, Slepchenko SM, Bondarev AA, et al. Genome sequence of a 45,000-year-old modern human from western Siberia. Nature. 2014;514:445–9.

Fu Q, Hajdinjak M, Moldovan OT, Constantin S, Mallick S, Skoglund P, et al. An early modern human from Romania with a recent Neanderthal ancestor. Nature. 2015;524:216–9.

Fu Q, Posth C, Hajdinjak M, Petr M, Mallick S, Fernandes D, et al. The genetic history of ice age Europe. Nature. 2016;534:200–5.

Gansauge M-T, Meyer M. Single-stranded DNA library preparation for the sequencing of ancient or damaged DNA. Nat Protoc. 2013;8:737–48.

Gordon D, Huddleston J, Chaisson MJP, Hill CM, Kronenberg ZN, Munson KM, et al. Long-read sequence assembly of the gorilla genome. Science. 2016;352:aae0344.

Green RE, Krause J, Briggs AW, Maricic T, Stenzel U, Kircher M, et al. A draft sequence of the Neandertal genome. Science. 2010;328:710–22.

Gutiérrez-García TA, Vázquez-Domínguez E, Arroyo-Cabrales J, Kuch M, Enk J, King C, et al. Ancient DNA and the tropics: a rodent's tale. Biol Lett. 2014;10 https://doi.org/10.1098/rsbl.2014.0224.

Haak W, Lazaridis I, Patterson N, Rohland N, Mallick S, Llamas B, et al. Massive migration from the steppe was a source for Indo-European languages in Europe. Nature. 2015;522:207–11.

Hajdinjak M, Fu Q, Hübner A, Petr M, Mafessoni F, Grote S, et al. Reconstructing the genetic history of late Neanderthals. Nature. 2018;555:652–6.

Halenar LB, Rosenberger AL. A closer look at the "Protopithecus" fossil assemblages: new genus and species from Bahia, Brazil. J Hum Evol. 2013;65:374–90.

Hernandez RD, Hubisz MJ, Wheeler DA, Smith DG, Ferguson B, Rogers J, et al. Demographic histories and patterns of linkage disequilibrium in Chinese and Indian rhesus macaques. Science. 2007;316:240–3.

Hobolth A, Dutheil JY, Hawks J, Schierup MH, Mailund T. Incomplete lineage sorting patterns among human, chimpanzee, and orangutan suggest recent orangutan speciation and widespread selection. Genome Res. 2011;21:349–56.

Hofmanová Z, Kreutzer S, Hellenthal G, Sell C, Diekmann Y, Díez-Del-Molino D, et al. Early farmers from across Europe directly descended from Neolithic Aegeans. Proc Natl Acad Sci U S A. 2016; https://doi.org/10.1073/pnas.1523951113.

Hofreiter M, Paijmans JLA, Goodchild H, Speller CF, Barlow A, Fortes GG, et al. The future of ancient DNA: technical advances and conceptual shifts. BioEssays. 2015;37:284–93.

Ibrahim YK, Tshen LT, Westaway KE, Cranbrook EO, Humphrey L, Muhammad RF, et al. First discovery of Pleistocene orangutan (Pongo sp.) fossils in Peninsular Malaysia: biogeographic and paleoenvironmental implications. J Hum Evol. 2013;65:770–97.

Iwamoto M. On a skull of a fossil macaque from the Shikimizu limestone quarry in the Shikoku District, Japan. Primates. 1975;16:83–94.

Iwamoto M, Hasegawa Y. Two macaque fossil teeth from the Japanese Pleistocene. Primates. 1972;13:77–81.

Jablonski NG. The response of catarrhine primates to Pleistocene environmental fluctuations in East Asia. Primates. 1998;39:29–37.

Karanth KP, Delefosse T, Rakotosamimanana B, Parsons TJ, Yoder AD. Ancient DNA from giant extinct lemurs confirms single origin of Malagasy primates. Proc Natl Acad Sci U S A. 2005;102:5090–5.

Kay RF. Biogeography in deep time - what do phylogenetics, geology, and paleoclimate tell us about early platyrrhine evolution? Mol Phylogenet Evol. 2015;82(Pt B):358–74.

Kehlmaier C, Barlow A, Hastings AK, Vamberger M, Paijmans JLA, Steadman DW, et al. Tropical ancient DNA reveals relationships of the extinct Bahamian giant tortoise Chelonoidis alburyorum. Proc Biol Sci. 2017;284 https://doi.org/10.1098/rspb.2016.2235.

Kistler L, Ratan A, Godfrey LR, Crowley BE, Hughes CE, Lei R, et al. Comparative and population mitogenomic analyses of Madagascar's extinct, giant "subfossil" lemurs. J Hum Evol. 2015;79:45–54.

Köhler M, Moyà-Solà S, Alba DM. Macaca (Primates, Cercopithecidae) from the late Miocene of Spain. J Hum Evol. 2000;38:447–52.

Kronenberg ZN, Fiddes IT, Gordon D, Murali S, Cantsilieris S, Meyerson OS, et al. High-resolution comparative analysis of great ape genomes. Science. 2018;360:eaar6343. https://doi.org/10.1126/science.aar6343.

Kuhn TS, McFarlane KA, Groves P, Mooers AØ, Shapiro B. Modern and ancient DNA reveal recent partial replacement of caribou in the Southwest Yukon. Mol Ecol. 2010;19:1312–23.

Lander ES, Linton LM, Birren B, Nusbaum C, Zody MC, Baldwin J, et al. Initial sequencing and analysis of the human genome. Nature. 2001;409:860–921.

Larsen PA, Harris RA, Liu Y, Murali SC, Campbell CR, Brown AD, et al. Hybrid de novo genome assembly and centromere characterization of the gray mouse lemur (Microcebus murinus). BMC Biol. 2017;15:110.

Lazaridis I, Patterson N, Mittnik A, Renaud G, Mallick S, Kirsanow K, et al. Ancient human genomes suggest three ancestral populations for present-day Europeans. Nature. 2014;513:409–13.

Lazaridis I, Nadel D, Rollefson G, Merrett DC, Rohland N, Mallick S, et al. Genomic insights into the origin of farming in the ancient near east. Nature. 2016;536:419–24.

Li H, Durbin R. Inference of human population history from individual whole-genome sequences. Nature. 2011;475:493–6.

Librado P, Gamba C, Gaunitz C, Der Sarkissian C, Pruvost M, Albrechtsen A, et al. Ancient genomic changes associated with domestication of the horse. Science. 2017;356:442–5.

Lipson M, Szécsényi-Nagy A, Mallick S, Pósa A, Stégmár B, Keerl V, et al. Parallel palaeogenomic transects reveal complex genetic history of early European farmers. Nature. 2017;551:368–72.

Lipson M, Cheronet O, Mallick S, Rohland N, Oxenham M, Pietrusewsky M, et al. Ancient genomes document multiple waves of migration in southeast Asian prehistory. Science. 2018;361:92–5. https://doi.org/10.1126/science.aat3188.

Locke DP, Hillier LW, Warren WC, Worley KC, Nazareth LV, Muzny DM, et al. Comparative and demographic analysis of orang-utan genomes. Nature. 2011;469:529–33.

Lorenzen ED, Nogués-Bravo D, Orlando L, Weinstock J, Binladen J, Marske KA, et al. Species-specific responses of late quaternary megafauna to climate and humans. Nature. 2011;479:359–64.

Macphee RDE, Horovitz I. New craniodental remains of the Quaternary Jamaican monkey Xenothrix mcgregori (Xenotrichini, Callicebinae, Pitheciidae), with a reconsideration of the Aotus hypothesis. Am Mus Novit. 2004;3434:1–51.

MacPhee RDE, Iturralde-Vinent MA, Gaffney ES. Domo de Zaza, an early Miocene vertebrate locality in South-Central Cuba, with notes on the tectonic evolution of Puerto Rico and the Mona Passage. Am Mus Novit. 2003;3394:1–42.

Maricic T, Whitten M, Pääbo S. Multiplexed DNA sequence capture of mitochondrial genomes using PCR products. PLoS One. 2010;5:e14004.

Marmi J, Bertranpetit J, Terradas J, Takenaka O, Domingo-Roura X. Radiation and phylogeography in the Japanese macaque, Macaca fuscata. Mol Phylogenet Evol. 2004;30:676–85.

Marmoset Genome Sequencing and Analysis Consortium. The common marmoset genome provides insight into primate biology and evolution. Nat Genet. 2014;46:850–7.

Mathieson I, Lazaridis I, Rohland N, Mallick S, Patterson N, Roodenberg SA, et al. Genome-wide patterns of selection in 230 ancient Eurasians. Nature. 2015;528:499–503.

Mathieson I, Alpaslan-Roodenberg S, Posth C, Szécsényi-Nagy A, Rohland N, Mallick S, et al. The genomic history of southeastern Europe. Nature. 2018;555:197–203.

Matsudaira K, Ishida T. Phylogenetic relationships and divergence dates of the whole mitochondrial genome sequences among three gibbon genera. Mol Phylogenet Evol. 2010;55:454–9.

Mazza P, Rustioni M, Agostini S, Rossi A. An unexpected Late Pleistocene macaque remain from Grotta degli Orsi Volanti (Rapino, Chieti, central Italy). Geobios Mem Spec. 2005;38:211–7.

McBrearty S, Jablonski NG. First fossil chimpanzee. Nature. 2005;437:105–8.

McManus KF, Kelley JL, Song S, Veeramah KR, Woerner AE, Stevison LS, et al. Inference of gorilla demographic and selective history from whole-genome sequence data. Mol Biol Evol. 2015;32:600–12.

Ménard N. Ecological plasticity of Barbary macaques (Macaca sylvanus). Evol Anthropol. 2003;11:95–100.

Meyer M, Arsuaga J-L, de Filippo C, Nagel S, Aximu-Petri A, Nickel B, et al. Nuclear DNA sequences from the middle Pleistocene Sima de los Huesos hominins. Nature. 2016;531:504–7.

Modolo L, Salzburger W, Martin RD. Phylogeography of Barbary macaques (Macaca sylvanus) and the origin of the Gibraltar colony. Proc Natl Acad Sci U S A. 2005;102:7392–7.

Mohandesan E, Speller CF, Peters J, Uerpmann H-P, Uerpmann M, De Cupere B, et al. Combined hybridization capture and shotgun sequencing for ancient DNA analysis of extinct wild and domestic dromedary camel. Mol Ecol Resour. 2017;17:300–13.

Moorjani P, Amorim CEG, Arndt PF, Przeworski M. Variation in the molecular clock of primates. Proc Natl Acad Sci U S A. 2016;113:10607–12.

Mootnick AR. Gibbon (Hylobatidae) species identification recommended for rescue or breeding centers. Primate Conserv. 2006;21:103–38.

Nater A, Mattle-Greminger MP, Nurcahyo A, Nowak MG, de Manuel M, Desai T, et al. Morphometric, behavioral, and genomic evidence for a new orangutan species. Curr Biol. 2017;27:3576–7.

Ní Leathlobhair M, Perri AR, Irving-Pease EK, Witt KE, Linderholm A, Haile J, et al. The evolutionary history of dogs in the Americas. Science. 2018;361:81–5.

Ni X, Li Q, Li L, Beard KC. Oligocene primates from China reveal divergence between African and Asian primate evolution. Science. 2016;352:673–7.

Olalde I, Brace S, Allentoft ME, Armit I, Kristiansen K, Booth T, et al. The beaker phenomenon and the genomic transformation of Northwest Europe. Nature. 2018;555:190–6.

Orlando L, Calvignac S, Schnebelen C, Douady CJ, Godfrey LR, Hänni C. DNA from extinct giant lemurs links archaeolemurids to extant indriids. BMC Evol Biol. 2008;8:121.

Orlando L, Ginolhac A, Zhang G, Froese D, Albrechtsen A, Stiller M, et al. Recalibrating Equus evolution using the genome sequence of an early middle Pleistocene horse. Nature. 2013;499:74–8.

Ortiz A, Pilbrow V, Villamil CI, Korsgaard JG, Bailey SE, Harrison T. The taxonomic and phylogenetic affinities of Bunopithecus sericus, a fossil hylobatid from the Pleistocene of China. PLoS One. 2015;10:e0131206.

Palesch D, Bosinger SE, Tharp GK, Vanderford TH, Paiardini M, Chahroudi A, et al. Sooty mangabey genome sequence provides insight into AIDS resistance in a natural SIV host. Nature. 2018;553:77–81.

Palkopoulou E, Dalén L, Lister AM, Vartanyan S, Sablin M, Sher A, et al. Holarctic genetic structure and range dynamics in the woolly mammoth. Proc Biol Sci. 2013;280:20131910.

Palkopoulou E, Mallick S, Skoglund P, Enk J, Rohland N, Li H, et al. Complete genomes reveal signatures of demographic and genetic declines in the woolly mammoth. Curr Biol. 2015;25:1395–400.

Patterson N, Moorjani P, Luo Y, Mallick S, Rohland N, Zhan Y, et al. Ancient admixture in human history. Genetics. 2012;192:1065–93.

Perelman P, Johnson WE, Roos C, Seuánez HN, Horvath JE, Moreira MAM, et al. A molecular phylogeny of living primates. PLoS Genet. 2011;7:e1001342.

Perry GH, Reeves D, Melsted P, Ratan A, Miller W, Michelini K, et al. A genome sequence resource for the aye-aye (Daubentonia madagascariensis), a nocturnal lemur from Madagascar. Genome Biol Evol. 2012;4:126–35.

Pinhasi R, Fernandes D, Sirak K, Novak M, Connell S, Alpaslan-Roodenberg S, et al. Optimal ancient DNA yields from the inner ear part of the human petrous bone. PLoS One. 2015;10: e0129102.

Prado-Martinez J, Sudmant PH, Kidd JM, Li H, Kelley JL, Lorente-Galdos B, et al. Great ape genetic diversity and population history. Nature. 2013;499:471–5.

Prüfer K, Munch K, Hellmann I, Akagi K, Miller JR, Walenz B, et al. The bonobo genome compared with the chimpanzee and human genomes. NatureNature. 2012;486:527–31.

Prüfer K, Racimo F, Patterson N, Jay F, Sankararaman S, Sawyer S, et al. The complete genome sequence of a Neanderthal from the Altai Mountains. Nature. 2014;505:43–9.

Prüfer K, de Filippo C, Grote S, Mafessoni F, Korlević P, Hajdinjak M, et al. A high-coverage Neandertal genome from Vindija Cave in Croatia. Science. 2017;358:655–8.

Raghavan M, Skoglund P, Graf KE, Metspalu M, Albrechtsen A, Moltke I, et al. Upper Palaeolithic Siberian genome reveals dual ancestry of native Americans. Nature. 2014;505:87–91.

Reich D, Green RE, Kircher M, Krause J, Patterson N, Durand EY, et al. Genetic history of an archaic hominin group from Denisova Cave in Siberia. Nature. 2010;468:1053–60.

Rhesus Macaque Genome Sequencing and Analysis Consortium, Gibbs RA, Rogers J, Katze MG, Bumgarner R, Weinstock GM, et al. Evolutionary and biomedical insights from the rhesus macaque genome. Science. 2007;316:222–34.

Richmond BG, Begun DR, Strait DS. Origin of human bipedalism: the knuckle-walking hypothesis revisited. Am J Phys Anthropol. 2001;Suppl 33:70–105.

Rohland N, Harney E, Mallick S, Nordenfelt S, Reich D. Partial uracil–DNA–glycosylase treatment for screening of ancient DNA. Philos Trans R Soc Lond Ser B Biol Sci. 2015;370:20130624.

Rosenberger AL, Cooke SB, Rímoli R, Ni X, Cardoso L. First skull of Antillothrix bernensis, an extinct relict monkey from the Dominican Republic. Proc Biol Sci. 2011;278:67–74.

Rosendahl W, Ambros D, Hilpert B, Hambach U, Alt KW, Knipping M, et al. Neanderthals and monkeys in the Würmian of Central Europe: the middle Paleolithic site of Hunas, southern Germany. In: Conard NJ, Richter J, editors. Neanderthal lifeways, subsistence and technology: one hundred fifty years of Neanderthal study. Dordrecht: Springer; 2011. p. 15–23.

Scally A, Durbin R. Revising the human mutation rate: implications for understanding human evolution. Nat Rev Genet. 2012;13:745–53.

Scally A, Dutheil JY, Hillier LW, Jordan GE, Goodhead I, Herrero J, et al. Insights into hominid evolution from the gorilla genome sequence. Nature. 2012;483:169–75.

Schiffels S, Durbin R. Inferring human population size and separation history from multiple genome sequences. Nat Genet. 2014;46:919–25.

Schmitz J, Noll A, Raabe CA, Churakov G, Voss R, Kiefmann M, et al. Genome sequence of the basal haplorrhine primate Tarsius syrichta reveals unusual insertions. Nat Commun. 2016;7:12997.

Schroeder H, Sikora M, Gopalakrishnan S, Cassidy LM, Maisano Delser P, Sandoval Velasco M, et al. Origins and genetic legacies of the Caribbean Taino. Proc Natl Acad Sci U S A. 2018;115:2341–6.

Shapiro B, Drummond AJ, Rambaut A, Wilson MC, Matheus PE, Sher AV, et al. Rise and fall of the Beringian steppe bison. Science. 2004;306:1561–5.

Silcox MT, López-Torres S. Major questions in the study of primate origins. Ann Rev. 2017;45:113–37. https://doi.org/10.1146/annurev-earth-063016-015637.

Skoglund P, Malmström H, Raghavan M, Storå J, Hall P, Willerslev E, et al. Origins and genetic legacy of Neolithic farmers and hunter-gatherers in Europe. Science. 2012;336:466–9.

Slon V, Viola B, Renaud G, Gansauge M-T, Benazzi S, Sawyer S, et al. A fourth Denisovan individual. Sci Adv. 2017a;3:e1700186.

Slon V, Hopfe C, Weiß CL, Mafessoni F, de la Rasilla M, Lalueza-Fox C, et al. Neandertal and Denisovan DNA from Pleistocene sediments. Science. 2017b;356:605–8.

Stevison LS, Kohn MH. Divergence population genetic analysis of hybridization between rhesus and cynomolgus macaques. Mol Ecol. 2009;18:2457–75.

Stiller M, Baryshnikov G, Bocherens H, Grandal d'Anglade A, Hilpert B, Münzel SC, et al. Withering away--25,000 years of genetic decline preceded cave bear extinction. Mol Biol Evol. 2010;27:975–8.

Stoneking M, Krause J. Learning about human population history from ancient and modern genomes. Nat Rev Genet. 2011;12:603–14.

Terhorst J, Kamm JA, Song YS. Robust and scalable inference of population history from hundreds of unphased whole genomes. Nat Genet. 2017;49:303–9.

Thinh VN, Mootnick AR, Geissmann T, Li M, Ziegler T, Agil M, et al. Mitochondrial evidence for multiple radiations in the evolutionary history of small apes. BMC Evol Biol. 2010;10:74.

Thomas GWC, Wang RJ, Puri A, Alan Harris R, Raveendran M, Hughes D, et al. Reproductive longevity predicts mutation rates in primates [internet]. bioRxiv. 2018:327627. https://doi.org/10.1101/327627.

Turvey ST, Bruun K, Ortiz A, Hansford J, Hu S, Ding Y, et al. New genus of extinct Holocene gibbon associated with humans in Imperial China. Science. 2018;360:1346–9.

Vallender EJ, Miller GM. Nonhuman primate models in the genomic era: a paradigm shift. ILAR J. 2013;54:154–65.

Veeramah KR, Hammer MF. The impact of whole-genome sequencing on the reconstruction of human population history. Nat Rev Genet. 2014;15:149–62.

Veeramah KR, Woerner AE, Johnstone L, Gut I, Gut M, Marques-Bonet T, et al. Examining phylogenetic relationships among gibbon genera using whole genome sequence data using an approximate Bayesian computation approach. Genetics. 2015;200:295–308.

Venn O, Turner I, Mathieson I, de Groot N, Bontrop R, McVean G. Strong male bias drives germline mutation in chimpanzees. Science. 2014;344:1272–5.

Wall JD, Kim SK, Luca F, Carbone L, Mootnick AR, de Jong PJ, et al. Incomplete lineage sorting is common in extant gibbon genera. PLoS One. 2013;8:e53682.

Warren WC, Jasinska AJ, García-Pérez R, Svardal H, Tomlinson C, Rocchi M, et al. The genome of the vervet (Chlorocebus aethiops sabaeus). Genome Res. 2015;25:1921–33.

Wegmann D, Excoffier L. Bayesian inference of the demographic history of chimpanzees. Mol Biol Evol. 2010;27:1425–35.

Xue C, Raveendran M, Harris RA, Fawcett GL, Liu X, White S, et al. The population genomics of rhesus macaques (Macaca mulatta) based on whole-genome sequences. Genome Res. 2016;26:1651–62.

Zhang Y, Harrison T. Gigantopithecus blacki: a giant ape from the Pleistocene of Asia revisited. Am J Phys Anthropol. 2017;162(Suppl 63):153–77.

Zhou X, Wang B, Pan Q, Zhang J, Kumar S, Sun X, et al. Whole-genome sequencing of the snub-nosed monkey provides insights into folivory and evolutionary history. Nat Genet. 2014;46:1303–10.

Structural Variants in Ancient Genomes

Skyler D. Resendez, Justin R. Bradley, Duo Xu, and Omer Gokcumen

Abstract The last decade has witnessed a myriad of advancements in the field of genomics, drastically changing our understanding of how genomes evolve; how genetic variation is maintained, gained, and lost; and how this variation affects gene function. In our opinion, the most relevant conceptual development has to be the renewed appreciation of the impact of genomic structural variation within species and across different species. In parallel, our newly gained ability to sequence the genomes collected from ancient populations has revolutionized how we conduct population and evolutionary genetics analyses. Combining these two exciting developments, we argue that studying the structural variation in ancient genomes will open new doors to previously unexplored areas of mammalian genome evolution. In this review, we summarize some of the recent developments in this field, most of which comes from studies in humans, and give an example where we determined the Neanderthal origins of a polymorphic gene deletion in humans combining information from modern and ancient genomes.

Keywords Ancient DNA · Genetic mapping · Genome analysis · Structural variants

1 Introduction

1.1 Genomic Structural Variants Are Major Drivers of Mammalian Phenotypic Variation

Structural variants (SVs) refer to genomic differences between individuals with regards to copy number, orientation, or genomic location of large segments of DNA (Fig. 1a). It is our working hypothesis that SVs are the major drivers of

S. D. Resendez · J. R. Bradley · D. Xu · O. Gokcumen (✉)
Department of Biological Sciences, University at Buffalo, The State University of New York (SUNY), Buffalo, NY, USA
e-mail: omergokc@buffalo.edu

Charlotte Lindqvist and Om P. Rajora (eds.), *Paleogenomics*, Population Genomics [Om P. Rajora (Editor-in-Chief)], https://doi.org/10.1007/13836_2018_34,

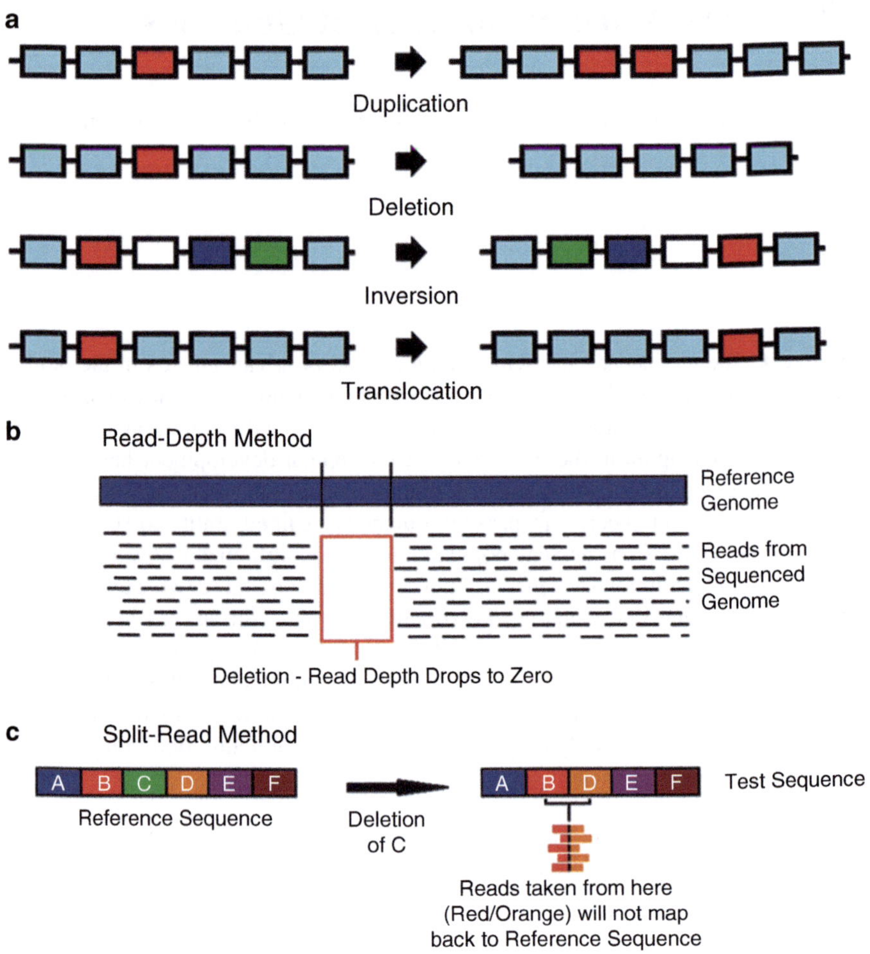

Fig. 1 Graphic depiction of the events that create different structural variants within the genome (**a**) and of the read-depth and split-read methodologies of structural variant discovery. Different colors represent different genetic segments. (**b** and **c**) show how read-depth and split-read methodologies (respectively) would visually depict a deletion in the test sequence compared to the reference sequence

mammalian adaptive evolution. This is a bold claim but has merit based on two observations. First, SVs constitute the majority of variable base pairs when two mammalian genomes are compared to each other (Conrad et al. 2009; Sudmant et al. 2015b). For example, SVs constitute two to seven times more base pairs that are variable among primate species when compared to single-nucleotide variants (Sudmant et al. 2013; Gokcumen et al. 2013a). The implication is that SVs simply have more of an effect on phenotype due to their sheer size. In fact, the world of plant genomics already has a great recognition of the impact of structural variants. The majority of phenotypic variations among plant species are thought of at the level of gene family expansions and contractions, as well as large chromosome-level variations (Ibarra-Laclette et al. 2013; Denoeud et al. 2014). However, we are just beginning to appreciate the phenotypic impact of structural variants in mammalian genomes, where most of the data comes from studies done on the human genome (Weischenfeldt et al. 2013). We have several examples in which SVs contribute to many common and rare diseases (Stefansson et al. 2008; Cooper et al. 2011; Sekar et al. 2016; Boettger et al. 2016), as well as nonpathogenic phenotypic variations in humans (Perry et al. 2007; Traherne et al. 2010; Lou et al. 2015; Xu et al. 2016). It is likely that these SVs have been evolving under adaptive constraints.

The second reason why we think that SVs may be major drivers of mammalian evolution is because SVs can contribute to phenotypic evolution in a myriad of ways. More specifically, SVs can alter gene function in several, often unpredictable ways. This allows for a wide range of phenotypic modifications, from subtle to drastic (Stranger et al. 2007; Hurles et al. 2008; Gokcumen et al. 2013a). It is now well established that the evolution of new genes is primarily due to gene duplication, followed by one of the duplicates gaining new function (Neofunctionalization) (Ohno 1970). This phenomenon has been studied extensively in multiple organisms (Rodríguez-Trelles et al. 2003; Loppin et al. 2005; Cui et al. 2006; Duarte et al. 2006; Byrne and Wolfe 2007). Innan and Kondrashov provide a comprehensive review of different fates of gene duplications (Innan and Kondrashov 2010).

One of the most fascinating examples of this phenomenon is the extensive duplication and diversification of a particular gene family in snakes, leading to an "arsenal" of venom (Lynch 2007). In addition to neofunctionalization, gene duplications can affect function in other ways. Of course, deletions, inversions, and translocations can also have surprising, adaptively relevant functional effects. For example, in humans, the functional effects of SVs can encompass a range of outcomes. These include, but are not limited to, changes in expression due to gene duplications causing an increased dosage effect (Perry et al. 2007), the inhibition of the activity of a gene due to a truncated gene duplication (Dennis et al. 2012), the deletion or translocation of regulatory sequences (McLean et al. 2011; Gokcumen et al. 2013b), changes in the copy number of exonic tandem repeats (Xu et al. 2016, 2017b; Eaaswarkhanth et al. 2016), and the inversion of megabases worth of gene-rich sequences (Stefansson et al. 2005).

The plethora of ways in which SVs affect function contribute to the overall difficulty of studying these variants. Thus, most of our knowledge of the phenotypic impact of SVs comes from studies of locus-specific analyses in human genomes. Regardless, our argument here is that SVs are common and have a high, underappreciated, and wide-ranging genomic impact that may underlie the majority of phenotypic variations in mammalian species.

1.2 Why Do We Know So Little About Genomic Structural Variants?

This question actually extends beyond mammals as it is clear that the scientific community still knows little about SVs in general. The reason for this is simple: SVs are difficult to study. More precisely, it is much more challenging to discover and genotype SVs than it is to discover and genotype single-nucleotide variants.

There are two interrelated reasons for the difficulty associated with studying SVs. The first reason is that most SVs are located in repetitive and thus complex regions of the genomes (Conrad et al. 2009). For example, recent gene duplicates are often formed by segmental duplications, which create large (>1 kb), nearly identical repeats (Marques-Bonet et al. 2009). Even with relatively long Sanger sequences, accurate assembly of these regions is difficult. Consequently, there are several errors and inaccuracies in such segmental duplication-rich regions of the reference genomes. Even when the reference genomes are accurate, mapping the short reads from resequencing experiments can be problematic and lead to false positives and, maybe more importantly, false negatives (Mills et al. 2011). Long-read sequencing technologies, such as the PacBio platform, produce individual sequence reads that average 15,000 bases in length. Because of their size, these reads can be mapped accurately, even in complex, repeat rich regions (Khost et al. 2017). They can also span entire deletions, duplications, inversions, and translocations, allowing us to accurately and comprehensively detect these structural variants. In fact, recent SV maps created by utilizing emerging long-read sequencing technologies have shown that tens of thousands of SVs per genome are missed by more traditional, short-read-based methods (Chaisson et al. 2014; Gordon et al. 2016; Huddleston et al. 2017).

Second, over one fourth of human SVs do not show strong linkage disequilibrium ($R^2 < 0.6$) with neighboring variants (Saitou et al. 2018). High recurrence rate, as well as frequent gene conversion events may explain the lack of linkage disequilibrium (Saitou et al. 2018). Regardless of the mechanistic basis of the loss of linkage disequilibrium between SVs and neighboring single-nucleotide variants, it is a fact that it is often difficult to use "tag" variants to genotype a significant portion of SVs: an essential method for most genome-wide association studies, as well as for gaining phylogenetic and evolutionary insights.

The combined effect of these two challenges, in our opinion, leads to a general underestimation of the functional and adaptive relevance of SVs. There once was a general excitement with regards to SVs carried by the scientific community as it was thought that SVs might be able to fill in the "heritability gap" (Eichler et al. 2010). This excitement was diminished by a series of negative results. We argue that these negative results were often caused by a lack of methodological power needed to comprehensively detect SVs. For example, one extensive study concluded that copy number variants (i.e., duplications and deletions) that "can be typed on existing platforms are unlikely to contribute greatly to the genetic basis of common human diseases" (Wellcome Trust Case Control Consortium et al. 2010). This conclusion loses its significance if one considers that less than 40% of the targeted SVs could actually be integrated into the study. The rest either could not be genotyped properly or could not be tagged by single-nucleotide variants. Now, more sophisticated tools for resolving the haplotype structure of SVs allow us to clearly show that multiple SVs that were previously overlooked have important biomedical and evolutionary impacts (Sekar et al. 2016; Boettger et al. 2016; Xu et al. 2017b). With the advent of population-level discovery and the genotyping of SVs, along with the increasing availability of long-read sequencing platforms, we argue that the recognition of SVs as a major driver of mammalian evolution and phenotypic variation will re-emerge.

1.3 Why Do We Care About Ancient Genomes?

Ancient genomes have transformed our understanding of speciation, population movements, and adaptation. The combination of novel experimental and bioinformatics approaches has enabled the sequencing of hundreds of ancient genomes in dozens of extant and extinct mammalian species (Orlando et al. 2015). The data from these studies were used, often with transformative results, to shed light on three broad, interrelated areas of inquiry.

First, it is now possible to look at the genetic variation of ancient populations to gain direct insight into past demographic events. For example, ancient genomic data have shown that the genetic makeup of human populations can change substantially within a few thousand years (Somel et al. 2016; Lazaridis et al. 2016; Kılınç et al. 2016). Similarly, ancient genomes show that the genetic variation of domesticated animals has changed drastically, showing the effect of rapid changes (and sometimes continuity) of selective pressures caused by domestication over short periods of time (Orlando et al. 2013; Skoglund et al. 2015; Botigué et al. 2017). Finally, ancient genomes have helped scientists understand the process of extinction to a greater degree, allowing empirically robust timelines and demographic histories to be created using extinct species from previous eras (Kistler et al. 2015; Palkopoulou et al. 2015).

Second, it is now possible to resolve the impact of introgression from past species and populations to extant groups (Allentoft et al. 2015; Taskent et al. 2017). In fact, some of the most groundbreaking and surprising findings of the last 5 years have involved evidence of ancient introgression events from now extinct lineages to modern populations, as is exemplified in humans (Slatkin and Racimo 2016) and bears (Miller et al. 2012).

Third, ancient genomics studies allow researchers to trace back the presence or absence, as well as the frequencies of functional alleles across time, allowing the development of novel and powerful methods to detect selection. For example, Mathieson et al. (2015) used an allele frequency-based approach to discover soft sweeps that occurred over the last 5,000 years. This is especially important for studying the adaptive significance of SVs since it is emerging that soft sweeps may be a leading force in shaping the distribution of structural variation within species (Salojärvi et al. 2017). In addition, other scientists have documented functional haplotypes that were introgressed from ancient species into extant ones, affecting phenotypes that we still observe today (Nédélec et al. 2016; Gittelman et al. 2016). Overall, ancient genomes have been utilized to develop multiple innovative methodologies through which we can study mammalian evolution. However, there have been almost no efforts to date to study genomic structural variants in ancient genomes, with a few notable exceptions (Prüfer et al. 2014; Lin et al. 2015; Lou et al. 2015).

2 Methods of Discovery and Genotyping of SVs in Ancient Genomes

The first and most direct approach for SV discovery in ancient genomes measures deviations from read-depth when comparing the depth of coverage between a sequenced genome and a reference genome (Fig. 1b). A deletion with respect to the reference genome sequence leads to a reduction in read-depth within a given genomic segment, while a duplication leads to an increase. However, there are multiple limitations of the read-depth method: (1) it can only detect copy number variations and not inversions or translocations; (2) its power depends on the overall read-depth of the data, making direct comparisons between samples problematic; (3) it does not detect small events sufficiently due to a decrease in statistical power associated with a decrease in size; and (4) it cannot predict the chromosomal location of duplicates. Despite these shortcomings, it remains a reliable genotyping and discovery tool, especially for large duplications and deletions of nonrepetitive sequences. For example, such read-depth methods have been used to investigate copy number variation in thousands of modern humans, as well as the Altai Neanderthal (Sudmant et al. 2015a).

Split-read methods use the sequence reads that contain the breakpoints of a structural variant within their sequence. In such cases, the read will map to multiple locations, but only partially at each region (Fig. 1c). This signature can be used to detect structural variants, especially deletions. For example, a deletion in the sample sequence will potentially lead to reads that harbor the breakpoints of that deletion. In these cases, the sequence reads will not be mapped properly to the reference genome. However, using appropriate algorithms, it is possible to bioinformatically split each of these reads into two and then map these pieces to their corresponding reference genome locations. Doing so will help indicate the locations of the borders of the deletion. This can be a powerful method, especially for experiments that utilize long sequence lengths where the probability of having reads that span breakpoints with enough remaining flanking sequence is higher. The major limitation of this method is that it only works when one of the structural variant's breakpoints is located in the middle of a sequence read. If this does not occur in enough reads, the algorithm can overlook these instances, causing it to miss the majority of SVs, especially when read-depth is low. However, if these reads are accurately identified, this method can discover the actual start and end points of a structural variant down to an individual base pair. Breakpoint resolution allows elucidating the mutational mechanisms of SV formation (Lam et al. 2010; Sudmant et al. 2015b).

An emerging approach for genotyping SVs is to use "tag" single-nucleotide variants to impute the allelic state of a SV. This requires a large dataset of modern genomes with accurate SV and single-nucleotide variation genotypes [e.g., 1000 Genomes Project phase 3 (1000 Genomes Project Consortium et al. 2015)]. With such datasets, it is plausible to resolve the haplotypic architecture of SVs, even when they are recurrent. Once the haplotypic variations harboring different SV alleles are found, the neighboring single-nucleotide variants with high linkage disequilibrium to specific SV alleles can be used to estimate the allele frequencies in ancient populations. This can be done using a combination of these neighboring single-nucleotide variants, even when they are not individually in perfect linkage disequilibrium with the structural variant in question, as is often the case. Now that phased genomes are more readily available, especially for human populations, "tag" methods are used more frequently to study structural variation (Usher et al. 2015; Pajic et al. 2016; Sekar et al. 2016; Eaaswarkhanth et al. 2016). As mentioned above, this method is limited to subset of SVs that reside in well-defined haplotype blocks.

It is possible that these methods will largely be replaced by more straightforward assembly and/or direct comparison-type methods as the use of long-read sequencing platforms is becoming more widespread in modern genomic studies. However, these traditional methods based on short-read sequencing will most likely remain relevant for analysis of structural variants in ancient genomes due to the nature of DNA degradation and consequent limitations in sequence length in ancient genomes.

3 Complications in SV Discovery and Genotyping Specific to Ancient DNA

Two major challenges associated with sequencing ancient genomes are the miniscule proportion of ancient DNA (aDNA) to modern DNA and the fragmented nature of aDNA caused by post-mortem DNA degradation (Prüfer et al. 2010). This degradation is accelerated in hot, humid climates, explaining the variation of DNA quality from ancient samples depending on the geographic region (Schwarz et al. 2009). Because of aDNA fragmentation, the DNA molecules often only extend to an average of approximately 60 to 150 base pairs in length (Miller et al. 2008; Briggs et al. 2009). This is shorter than the current read length in Illumina sequencing and far shorter than what PacBio long-read technologies are capable of capturing. Therefore, the emerging long-read technologies may have limited benefits in ancient genomics.

This natural fragmentation of DNA across time also prevents aDNA studies from benefiting from paired-end mapping-based approaches, which are now mainstream for SV discovery in modern genomes (Korbel et al. 2007). As the name implies, these methodologies utilize short-read sequences that are each "paired" with one other read. This is because paired-end mapping usually involves sequencing the initial and final segments of a given length of base pairs. For example, libraries can be constructed by sequencing the first and last 125 base pairs (250 base pairs in total) of a 750 nucleotide region. This leaves the central 500 base pairs unsequenced among these two reads, but it also provides an expected distance between the two linked reads. Once these reads are mapped to a reference genome, if this distance deviates from the expected by a significant amount, then there may be a structural variant between the two reads. A deletion would decrease the distance between the paired reads, while an insertion would increase it. With this being said, DNA fragments that are shorter in length than what is required to make paired-end reads cannot benefit from this approach.

Furthermore, when in the presence of water molecules, DNA goes through a process known as deamination (Hofreiter et al. 2001). This is a hydrolysis reaction, meaning that the affected region is cleaved in two while incorporating a water molecule. This occurs specifically to cytosine bases, causing them to transform into uracil as they release ammonia (hence the term deamination). These uracils will then be read as thymines during the sequencing process. Interestingly, deamination occurs at accelerated rates at the ends of the DNA molecules. With this in mind, it is possible to look for reads with a greater than expected amount of thymines near the ends of the sequences in order to differentiate ancient DNA from modern DNA. This has been an extremely useful tool in many studies of ancient DNA. However, it also underlines the error-prone nature of sequencing studies in ancient genomes.

The properties of ancient sequences create four limitations in studying SVs in ancient genomes. First is the reduced power of split-read analyses due to the fragmentation of ancient DNA. This power is reduced exponentially as the sequence fragments get shorter. Split-read methods work by mapping a single read to two different locations in the genome, suggesting the breakpoints of an SV. Split-read analyses may not be possible in most cases if the reads are shorter than 100 bp, even for noncomplicated, repeat-free breakpoints.

The second issue with SV discovery and genotyping in ancient genomes is also related to DNA fragment length. When it comes to SVs in segmental duplication regions or other repetitive sequences, longer read lengths have a better chance of mapping to the correct location due to the recognition of a unique nucleotide in a given repeat (Chaisson et al. 2014). If this does not happen, the repetitive reads can be theoretically mapped to more than one region of the genome. Such inaccuracies in mapping are an issue, even for relatively long Illumina sequences of modern DNA (e.g., 150 bp $\times$ 2 in paired-end sequences). Mapping errors currently prohibit genotyping and discovery in most duplicated regions in ancient genomes. However, read-depth methods that use promiscuous mapping, such as Mrs. Fast (Hach et al. 2010), may be used to estimate the copy number of these complex regions. This is exemplified in a study by Sudmant et al. (2015b).

The third issue is the accuracy and sensitivity of single-nucleotide variant calling in ancient genomes. As mentioned previously, one of the emerging ways in which to genotype SVs in genomes is to impute the SVs using nearby single-nucleotide variants. However, due to DNA degradation and the deamination process described above, single-nucleotide variant calling in ancient sequences can be erroneous (Orlando et al. 2015). In our experience, most researchers correctly err on the side of being conservative with regards to their calls, especially for ancient samples. This translates into a larger proportion of false-negative single-nucleotide variant calls. Hence, the power for imputation likely remains lower in ancient genomes [even in high-quality assemblies, such as the Altai Neanderthal genome (Prüfer et al. 2014)], especially for regions that are rich in segmental duplications and heterozygous SVs.

The fourth limitation in studying SVs in ancient genomes is that ancient reads are often mapped back to whichever reference genome sequence is available from the most closely related organism. For example, the Altai Neanderthal and Denisovan sequences were generated by mapping their reads back to the modern human genome reference sequence. This was necessary given that a de novo assembly formed by ancient reads would be either impossible to generate or highly inaccurate given the short sequence lengths. Depending on the evolutionary closeness of the species and completeness of the reference genome, this practice may create ascertainment biases. The overall effect and extent of these biases are still unknown. Specifically, we fear that particular sequences in ancient genomes may have no homologous counterparts in the reference genome. This would cause the reads mapping to these regions to be discarded since ancient genome sequences are filtered of highly divergent reads to reduce bacterial sequence contamination. However, it is plausible to capture some of these unique, ancient species-specific sequences using some of the methodologies described for modern human insertions (Kidd et al. 2010).

4 Tracing Back the Evolutionarily and Phenotypically Relevant SVs Among Human Genomes

Recently, ancient hominin genomes have been utilized to better understand the evolution of functionally relevant SVs by tracing allele frequencies back in time. This gives a direct estimation of the emergence time of these SVs and gives us a better understanding of their change in frequency across time. For example, we genotyped more than 10,000 deletion polymorphisms observed in the Denisovan and Neanderthal genomes using the read-depth and split-read methods described above (Lin et al. 2015). In this genome-wide study, we showed that one of the most common gene deletions among modern human populations is also deleted in the Denisovan genome, but not in the Altai Neanderthal genome. This indicates that the deletion evolved before the modern human-Denisovan split. This deleted region includes two late cornified envelope genes (*LCE3B* and *LCE3C*), which code for structurally important proteins expressed in the skin. Furthermore, this common deletion is strongly associated with susceptibility to psoriasis. We were then able to further resolve the haplotypic variations that harbor this deletion and showed that the allele frequency of this deletion has remained near 50% over the last 10,000 years in European populations (Pajic et al. 2016). Our results were concordant with the notion that this deletion has evolved under balancing selection – a conclusion that was made possible largely due to the observations made using ancient genomes.

Another related application of the direct genotyping of SVs in ancient hominin genomes is to pinpoint the causal variants of putatively adaptive-introgressed haplotypes. The best example of this is the fascinating case of a 3.4 kb deletion found in the *EPAS1* haplotype (Lou et al. 2015). This haplotype was shown to be introgressed into modern Asian populations from the Denisovan genome and swiftly increased to a high allele frequency in Tibetan populations, now reaching approximately 90% (Huerta-Sánchez et al. 2014). It was further suggested that the allele frequency difference observed between Tibetans and other lower altitude Asian populations was so high that it could not be explained by neutral evolution alone (Beall et al. 2010). The most parsimonious explanation put forward was that this Denisovan haplotype reached near fixation in Tibetan populations after its introgression because it provides an adaptive advantage to high altitude environments (Huerta-Sánchez et al. 2014). However, it was not clear what the exact causal variant was within this haplotype. To answer this question, another study (Lou et al. 2015) investigated the evolution of the previously mentioned 3.4 kb deletion found in humans within the Denisovan-introgressed haplotype. Since this deletion is so large, it is highly possible that it has a powerful phenotypic effect, making it a potential causal variant within the haplotype. However, direct genotyping of this deletion showed that it does not exist in the Denisovan genome. As such, it is plausible that the deletion evolved on one of the introgressed haplotypes after the introgression event occurred. The selection for this deletion would have then facilitated the selective sweep observed in contemporary populations, rather than the sweep being directed by anything that came from the original introgression. Further studies are needed to resolve this issue. However, it is important to note that direct genotyping of this deletion in the Denisovan genome enabled a better investigation of the adaptive variation in this locus.

5 A Case Example: Neanderthal Introgression of the Spermatogenesis-Associated SPATA45 Gene Deletion

Another common deletion found within the human genome is the partial deletion of the *SPATA45* gene. This deletion spans ~11.7 kb on chromosome 1 (Hg19: chr1: 213,002,038 to 213,013,756) (Sudmant et al. 2015b). In a previous study, we showed that this deletion is shared with Neanderthals (Lin et al. 2015). We argued that it was likely introgressed from Neanderthals based on the observation that the deletion has not been detected anywhere in Sub-Saharan African genomes. The alternative hypothesis would be that this haplotype is a relic of the ancestral structure observed in human populations (Plagnol and Wall 2006). Here, we will use this deletion to provide a methodological example as to how one can study structural variation in ancient genomes.

The *SPATA45* gene is believed to be a spermatogenesis gene. This means that it is part of the process of sperm cell development (Sejian et al. 2012). It was previously hypothesized that the haplotypic variation that overlaps the members of the *SPATA* gene family was introgressed from Neanderthals and then adaptively increased in allele frequency within modern populations (Vernot and Akey 2014). This matches our proposed evolutionary history of the *SPATA45* gene deletion.

First, we directly genotyped and visualized this deletion in the Denisovan, Altai Neanderthal, and some modern human genomes (Fig. 2a). As is clear from the figure, it is possible to manually determine whether a large deletion is present in a genome by comparing it to a reference sequence (in this case, the human reference sequence) if the read-depth of the data is high enough. We found that the deletion observed in these ancient genomes shares precise breakpoints with the modern deleted variant, which strongly suggests that they are identical by descent. We then looked into the allelic distribution of this deletion in contemporary populations using a linked single-nucleotide variant, rs77760940 (Fig. 2b), and confirmed that it is not found in Sub-Saharan Africans. This supported our hypothesis that the deletion was introgressed from Neanderthals. However, it was still possible that the current allelic distribution of the deletion was caused by genetic drift removing the deleted variant from ancestral Sub-Saharan African populations.

To independently investigate whether this deletion was indeed introgressed, we determined which flanking single-nucleotide variants are in high linkage disequilibrium with the deletion (Fig. 2c). We found that there is an ~9.3 kb haplotype flanking the deletion with 31 single-nucleotide variants showing at least near perfect ($R^2 = 0.95$) linkage disequilibrium with the deletion allele, with the vast majority of them being in perfect ($R^2 = 1.0$) linkage disequilibrium. This haplotype block is an ideal proxy to conduct population genetic analyses in order to predict the evolutionary history of this deletion.

We then used the software entitled VCFtoTree (Xu et al. 2017a) to conduct a phylogenetic analysis of the haplotype block that we described above. Specifically, we aligned sequences from thousands of modern human haplotypes, as well as those from the Neanderthal and Denisovan genomes, and constructed a maximum likelihood tree using RAxML (Stamatakis 2014). Then, we manually superimposed the deletion alleles onto this tree (Fig. 2d). Concordant with the introgression model, the haplotype that carries the deletion clusters tightly with the Neanderthal haplotype with no exceptions. To quantify this observation, we used the program R (R Development Core Team 2014) to compute mean pairwise distances among modern and ancient haplotypes, using the Wilcoxon rank-sum test (Wilcoxon and Wilcox 1964). The pairwise distances observed between haplotypes show that those that carry the deletion are significantly closer to each other and the Neanderthal haplotype than they are to those that do not carry the deletion (distance by species $W = 186{,}290$, p-value = 2.366e-06, at a 95% confidence interval). The fact that there is little variation among the haplotypes that carry the deletion indicates that the

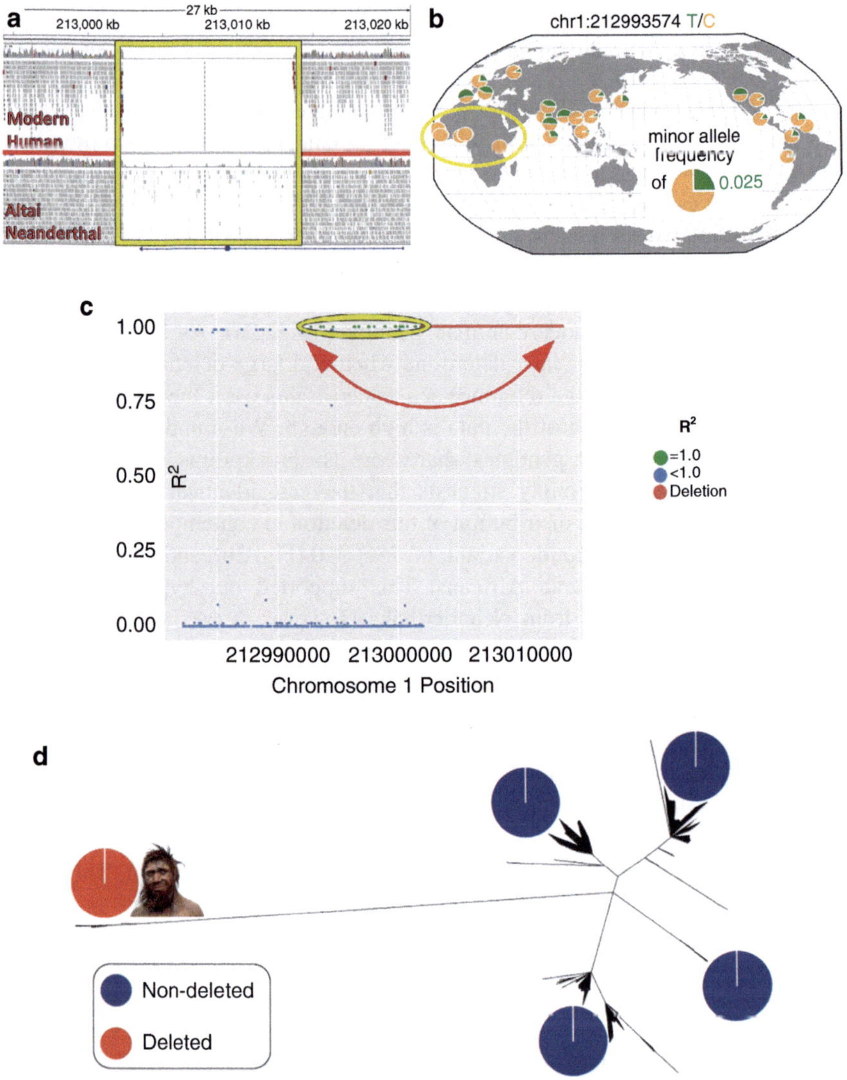

Fig. 2 A hominin structural variation case study involving the partial gene deletion of *SPATA45*. This deletion can be found on chromosome 1 between base pair positions 213,002,038 and 213,013,756. (**a**) Edited Integrated Genomics Viewer (IGV) screencap depicting a deletion found within a modern human (NA12763) and the Altai Neanderthal genomes when compared to the human reference sequence. This depiction of the deletion utilizes read-depth methods. (**b**) Geographic distribution and frequencies of the single-nucleotide polymorphism, rs77760940, located on chromosome 1 at nucleotide position 212993574. The minor allele, the single-nucleotide variant consisting of a thymine at this location, is in perfect linkage disequilibrium with the *SPATA45* gene deletion. Each pie is representative of only 10% of the population at that location. The green portions of the pie graphs specify the proportion of haplotypes within that 10% that carry the minor allele (T) in that population. The blue portions of the pie graphs specify the proportion of haplotypes that carry the major allele, a cytosine at this locus, in that 10% of the population. Note that the other 90% of each population also carry the major allele but are not shown as a method to better visualize the rarer minor allele comparatively across populations. All continental African populations are

origin of this haplotype in humans is relatively recent. All these observations are inconsistent with the ancestral structure scenario. Instead, the most parsimonious explanation that fits the data is that the *SPATA45* deletion was introgressed from Neanderthals at some point over the last 50,000 years.

6 Future Perspectives and Conclusions

In this review, we emphasized the pivotal yet underappreciated role of genomic structural variation within the emerging field of ancient population genomics. We did this by first making a case to underlie the importance of genomic structural variants with regards to phenotypic and adaptive variation within and between species. We then summarized the methodological and biological reasons underlying the well-documented difficulties associated with studying the evolutionary impact of structural variation. Our primary goal was to make the case that now is the time to start assessing the monumental impact of structural variation in ancient genomes, especially to elucidate the evolutionary histories of previously understudied phenotypically relevant variants. To facilitate such endeavors, we have summarized some of the methodologies that have recently been developed to trace back structural variation in ancient populations. Furthermore, we provided examples where ancient genomes have already been leveraged to shed new light on the evolution of structural variants. One specific example detailed here involved our novel insights into the polymorphic deletion of the *SPATA45* gene, which we systematically showed to have been introgressed from Neanderthals into Eurasian genomes.

Fig. 2 (continued) circled for emphasis because they carry no minor allele (deleted) haplotypes, which would be expected in a Neanderthal-introgressed allele. (**c**) A scatterplot showing R^2 values for different chromosomal positions within and around the high linkage disequilibrium (LD) region, located upstream of the *SPATA45* deletion. An R^2 value of 1.0 indicates perfect LD between the single-nucleotide polymorphism (SNP) at that location and the deletion. The green points indicate chromosomal positions that have an R^2 score of 1.0, and the blue points indicate positions that have an R^2 score of less than 1.0. The circled region was chosen for sequence variation tests due to it containing a high number of perfectly linked SNPs. The red bar indicates the location of the deletion (Hg19, chr1: 213,002,038–213,013,756). (**d**) Phylogenetic tree of 5,008 human haplotypes from the 1000 Genomes Project (1000 Genomes Project Consortium et al. 2015), along with the haplotypes from the Altai Neanderthal and the Denisovan genomes. This tree was created using variable SNP data from the high LD region. The red portion of the pie graphs specifies the percentage of haplotypes that carry the deletion, and the blue portions specify the percentage of haplotypes that do not carry the deletion. There is a perfect separation of the deleted and non-deleted haplotypes. The Altai Neandertal clustered tightly with the deleted haplotypes, which indicates that the Altai Neanderthal shares the same haplotype as deleted modern humans. The Denisovan clusters tightly with the non-deleted haplotypes, indicating the Denisovan and Neanderthal do not share the same haplotype. This means that the *SPATA45* deletion may have formed specifically in the Denisovan lineage

As the field of ancient genomics continues to advance, we foresee a future in which population level ancient genome data will be available for multiple species. Such datasets will allow for a more comprehensive, temporal understanding of the evolution of SVs. Specifically, one can trace back allele frequencies across time using these data while also testing for putative adaptive forces acting on SVs. In fact, such an analysis was performed for human single-nucleotide variants by comparing ancient and modern Eurasian populations as we briefly summarized above (Mathieson et al. 2015). Complete maps of variation, including SVs, at the population level will allow us to predict inherited phenotypes in ancient genomes by utilizing the correlations found in modern populations. This has already been an important aspect of ancient genome work in humans (Keller et al. 2012), and we expect this to become increasingly relevant in other species.

In sum, it is clear to us that there are a number of structural variants important to evolutionary history that have yet to be explored. We expect that increased understanding of the genome evolution and more comprehensive ancient genome datasets will lead to better characterization of structural variants among past and modern populations. We hope that our review encourages a better understanding of the role of genomic structural variation in mammalian evolution, as we believe that many secrets still hide within ancient genomes, waiting to be discovered.

Acknowledgements The authors would like to acknowledge members of Gokcumen Laboratory for their input during the development of their review. We would also like to acknowledge National Science Foundation Award (1714867).

References

1000 Genomes Project Consortium, Auton A, Brooks LD, et al. A global reference for human genetic variation. Nature. 2015;526:68–74.

Allentoft ME, Sikora M, Sjögren K-G, et al. Population genomics of bronze age Eurasia. Nature. 2015;522:167–72.

Beall CM, Cavalleri GL, Deng L, et al. Natural selection on EPAS1 (HIF2α) associated with low hemoglobin concentration in Tibetan highlanders. Proc Natl Acad Sci. 2010;107:11459–64.

Boettger LM, Salem RM, Handsaker RE, et al. Recurring exon deletions in the HP (haptoglobin) gene contribute to lower blood cholesterol levels. Nat Genet. 2016;48:359–66.

Botigué LR, Song S, Scheu A, Gopalan S, Pendleton AL, Oetjens M, Taravella AM, Seregély T, Zeeb-Lanz A, Arbogast R-M, Bobo D, Daly K, Unterländer M, Burger J, Kidd JM, Veeramah KR. Ancient European dog genomes reveal continuity since the Early Neolithic. Nat Commun. 2017;8:16082.

Briggs AW, Good JM, Green RE, et al. Targeted retrieval and analysis of five Neandertal mtDNA genomes. Science. 2009;325:318–21.

Byrne KP, Wolfe KH. Consistent patterns of rate asymmetry and gene loss indicate widespread neofunctionalization of yeast genes after whole-genome duplication. Genetics. 2007;175:1341–50.

Chaisson MJP, Huddleston J, Dennis MY, et al. Resolving the complexity of the human genome using single-molecule sequencing. Nature. 2014. https://doi.org/10.1038/nature13907.

Conrad DF, Pinto D, Redon R, et al. Origins and functional impact of copy number variation in the human genome. Nature. 2009;464:704–12.
Cooper GM, Coe BP, Girirajan S, et al. A copy number variation morbidity map of developmental delay. Nat Genet. 2011;43:838–46.
Cui L, Wall PK, Leebens-Mack JH, et al. Widespread genome duplications throughout the history of flowering plants. Genome Res. 2006;16:738–49.
Dennis MY, Nuttle X, Sudmant PH, et al. Evolution of human-specific neural SRGAP2 genes by incomplete segmental duplication. Cell. 2012;149:912–22.
Denoeud F, Carretero-Paulet L, Dereeper A, et al. The coffee genome provides insight into the convergent evolution of caffeine biosynthesis. Science. 2014;345:1181–4.
Duarte JM, Cui L, Wall PK, et al. Expression pattern shifts following duplication indicative of subfunctionalization and neofunctionalization in regulatory genes of Arabidopsis. Mol Biol Evol. 2006;23:469–78.
Eaaswarkhanth M, Xu D, Flanagan C, et al. Atopic dermatitis susceptibility variants in Filaggrin Hitchhike Hornerin selective sweep. Genome Biol Evol. 2016;8:3240–55.
Eichler EE, Flint J, Gibson G, et al. Missing heritability and strategies for finding the underlying causes of complex disease. Nat Rev Genet. 2010;11:446–50.
Gittelman RM, Schraiber JG, Vernot B, et al. Archaic hominin admixture facilitated adaptation to out-of-Africa environments. Curr Biol. 2016. https://doi.org/10.1016/j.cub.2016.10.041.
Gokcumen O, Tischler V, Tica J, et al. Primate genome architecture influences structural variation mechanisms and functional consequences. Proc Natl Acad Sci U S A. 2013a;110:15764–9.
Gokcumen O, Zhu Q, Mulder LCF, et al. Balancing selection on a regulatory region exhibiting ancient variation that predates human-neanderthal divergence. PLoS Genet. 2013b;9:e1003404.
Gordon D, Huddleston J, Chaisson MJP, et al. Long-read sequence assembly of the gorilla genome. Science. 2016;352:aae0344.
Hach F, Hormozdiari F, Alkan C, et al. mrsFAST: a cache-oblivious algorithm for short-read mapping. Nat Methods. 2010;7:576–7.
Hofreiter M, Jaenicke V, Serre D, et al. DNA sequences from multiple amplifications reveal artifacts induced by cytosine deamination in ancient DNA. Nucleic Acids Res. 2001;29:4793–9.
Huddleston J, Chaisson MJP, Steinberg KM, et al. Discovery and genotyping of structural variation from long-read haploid genome sequence data. Genome Res. 2017;27:677–85.
Huerta-Sánchez E, Jin X, Asan, et al. Altitude adaptation in Tibetans caused by introgression of Denisovan-like DNA. Nature. 2014;512:194–7.
Hurles ME, Dermitzakis ET, Tyler-Smith C. The functional impact of structural variation in humans. Trends Genet. 2008;24:238–45.
Ibarra-Laclette E, Lyons E, Hernández-Guzmán G, et al. Architecture and evolution of a minute plant genome. Nature. 2013;498:94–8.
Innan H, Kondrashov F. The evolution of gene duplications: classifying and distinguishing between models. Nat Rev Genet. 2010;11:97–108.
Keller A, Graefen A, Ball M, et al. New insights into the Tyrolean Iceman's origin and phenotype as inferred by whole-genome sequencing. Nat Commun. 2012;3:698.
Khost DE, Eickbush DG, Larracuente AM. Single-molecule sequencing resolves the detailed structure of complex satellite DNA loci in Drosophila melanogaster. Genome Res. 2017;27:709–21.
Kidd JM, Sampas N, Antonacci F, et al. Characterization of missing human genome sequences and copy-number polymorphic insertions. Nat Methods. 2010;7:365–71.
Kılınç GM, Omrak A, Özer F, et al. The demographic development of the first farmers in Anatolia. Curr Biol. 2016;26:2659–66.
Kistler L, Ratan A, Godfrey LR, et al. Comparative and population mitogenomic analyses of Madagascar's extinct, giant "subfossil" lemurs. J Hum Evol. 2015;79:45–54.
Korbel JO, Urban AE, Affourtit JP, et al. Paired-end mapping reveals extensive structural variation in the human genome. Science. 2007;318:420–6.

Lam HYK, Mu XJ, Stütz AM, et al. Nucleotide-resolution analysis of structural variants using BreakSeq and a breakpoint library. Nat Biotechnol. 2010;28:47–55.

Lazaridis I, Nadel D, Rollefson G, et al. Genomic insights into the origin of farming in the ancient Near East. Nature. 2016;536:419–24.

Lin Y-L, Pavlidis P, Karakoc E, et al. The evolution and functional impact of human deletion variants shared with archaic hominin genomes. Mol Biol Evol. 2015;32:1008–19.

Loppin B, Lepetit D, Dorus S, et al. Origin and neofunctionalization of a Drosophila paternal effect gene essential for zygote viability. Curr Biol. 2005;15:87–93.

Lou H, Lu Y, Lu D, et al. A 3.4-kb copy-number deletion near EPAS1 is significantly enriched in high-altitude Tibetans but absent from the Denisovan sequence. Am J Hum Genet. 2015;97:54–66.

Lynch VJ. Inventing an arsenal: adaptive evolution and neofunctionalization of snake venom phospholipase A2 genes. BMC Evol Biol. 2007;7:2.

Marques-Bonet T, Kidd JM, Ventura M, et al. A burst of segmental duplications in the genome of the African great ape ancestor. Nature. 2009;457:877–81.

Mathieson I, Lazaridis I, Rohland N, et al. Genome-wide patterns of selection in 230 ancient Eurasians. Nature. 2015;528:499–503.

McLean CY, Reno PL, Pollen AA, et al. Human-specific loss of regulatory DNA and the evolution of human-specific traits. Nature. 2011;471:216–9.

Miller W, Drautz DI, Ratan A, et al. Sequencing the nuclear genome of the extinct woolly mammoth. Nature. 2008;456:387–90.

Miller W, Schuster SC, Welch AJ, et al. Polar and brown bear genomes reveal ancient admixture and demographic footprints of past climate change. Proc Natl Acad Sci U S A. 2012;109: E2382–90.

Mills RE, Walter K, Stewart C, et al. Mapping copy number variation by population-scale genome sequencing. Nature. 2011;470:59–65.

Nédélec Y, Sanz J, Baharian G, et al. Genetic ancestry and natural selection drive population differences in immune responses to pathogens. Cell. 2016;167:657–669.e21.

Ohno S. Evolution by gene duplication. London/Berlin: George Alien & Unwin/Springer; 1970.

Orlando L, Ginolhac A, Zhang G, et al. Recalibrating Equus evolution using the genome sequence of an early Middle Pleistocene horse. Nature. 2013;499:74–8.

Orlando L, Gilbert MTP, Willerslev E. Reconstructing ancient genomes and epigenomes. Nat Rev Genet. 2015;16:395–408.

Pajic P, Lin Y-L, Xu D, Gokcumen O. The psoriasis-associated deletion of late cornified envelope genes LCE3B and LCE3C has been maintained under balancing selection since human Denisovan divergence. BMC Evol Biol. 2016;16:265.

Palkopoulou E, Mallick S, Skoglund P, et al. Complete genomes reveal signatures of demographic and genetic declines in the woolly mammoth. Curr Biol. 2015;25:1395–400.

Perry GH, Dominy NJ, Claw KG, et al. Diet and the evolution of human amylase gene copy number variation. Nat Genet. 2007;39:1256–60.

Plagnol V, Wall JD. Possible ancestral structure in human populations. PLoS Genet. 2006;2:e105.

Prüfer K, Stenzel U, Hofreiter M, et al. Computational challenges in the analysis of ancient DNA. Genome Biol. 2010;11:R47.

Prüfer K, Racimo F, Patterson N, et al. The complete genome sequence of a Neanderthal from the Altai Mountains. Nature. 2014;505:43–9.

R Development Core Team. R: A language and environment for statistical computing. Vienna: R Foundation for Statistical Computing; 2014.

Rodríguez-Trelles F, Tarrío R, Ayala FJ. Convergent neofunctionalization by positive Darwinian selection after ancient recurrent duplications of the xanthine dehydrogenase gene. Proc Natl Acad Sci U S A. 2003;100:13413–7.

Saitou M, Satta Y, Gokcumen O. Complex haplotypes of metabolizing GSTM1 gene deletion harbors signatures of a selective sweep in East Asian populations. bioRxiv. 2018. https://doi.org/10.1101/287417.

Salojärvi J, Smolander O-P, Nieminen K, et al. Genome sequencing and population genomic analyses provide insights into the adaptive landscape of silver birch. Nat Genet. 2017;49:904–12.

Schwarz C, Debruyne R, Kuch M, et al. New insights from old bones: DNA preservation and degradation in permafrost preserved mammoth remains. Nucleic Acids Res. 2009;37:3215–29.

Sejian V, SMK N, Ezeji T, et al., editors. Environmental stress and amelioration in livestock production. Berlin: Springer; 2012.

Sekar A, Bialas AR, de Rivera H, et al. Schizophrenia risk from complex variation of complement component 4. Nature. 2016;530:177–83.

Skoglund P, Ersmark E, Palkopoulou E, Dalén L. Ancient wolf genome reveals an early divergence of domestic dog ancestors and admixture into high-latitude breeds. Curr Biol. 2015;25:1515–9.

Slatkin M, Racimo F. Ancient DNA and human history. Proc Natl Acad Sci U S A. 2016;113:6380–7.

Somel M, Kilinc GM, Ozer F, et al. Archaeogenomic analysis of ancient Anatolians: first genetic indication for Neolithic cultural diffusion in the Near East. Am J Phys Anthropol. 2016;159:297–8.

Stamatakis A. RAxML version 8: a tool for phylogenetic analysis and post analysis of large phylogenies. Bioinformatics. 2014;30:1312e1313.

Stefansson H, Helgason A, Thorleifsson G, et al. A common inversion under selection in Europeans. Nat Genet. 2005;37:129–37.

Stefansson H, Rujescu D, Cichon S, et al. Large recurrent microdeletions associated with schizophrenia. Nature. 2008;455:232–6.

Stranger BE, Forrest MS, Dunning M, et al. Relative impact of nucleotide and copy number variation on gene expression phenotypes. Science. 2007;315:848–53.

Sudmant PH, Huddleston J, Catacchio CR, et al. Evolution and diversity of copy number variation in the great ape lineage. Genome Res. 2013;23:1373–82.

Sudmant PH, Mallick S, Nelson BJ, et al. Global diversity, population stratification, and selection of human copy-number variation. Science. 2015a;349(6253):aab3761. https://doi.org/10.1126/science.aab3761.

Sudmant PH, Rausch T, Gardner EJ, et al. An integrated map of structural variation in 2,504 human genomes. Nature. 2015b;526:75–81.

Taskent RO, Alioglu D, Fer E, et al. Variation and functional impact of Neanderthal ancestry in Western Asia. Genome Biol Evol. 2017. https://doi.org/10.1093/gbe/evx216.

Traherne JA, Martin M, Ward R, et al. Mechanisms of copy number variation and hybrid gene formation in the KIR immune gene complex. Hum Mol Genet. 2010;19:737–51.

Usher CL, Handsaker RE, Esko T, et al. Structural forms of the human amylase locus and their relationships to SNPs, haplotypes and obesity. Nat Genet. 2015;47:921–5.

Vernot B, Akey JM. Resurrecting surviving Neandertal lineages from modern human genomes. Science. 2014;343:1017–21.

Weischenfeldt J, Symmons O, Spitz F, Korbel JO. Phenotypic impact of genomic structural variation: insights from and for human disease. Nat Rev Genet. 2013;14:125–38.

Wellcome Trust Case Control Consortium, Craddock N, Hurles ME, et al. Genome-wide association study of CNVs in 16,000 cases of eight common diseases and 3,000 shared controls. Nature. 2010;464:713–20.

Wilcoxon F, Wilcox RA. Some rapid approximate statistical procedures. Pearl River, NY: Lederle Laboratories; 1964.

Xu D, Pavlidis P, Thamadilok S, et al. Recent evolution of the salivary mucin MUC7. Sci Rep. 2016;6:31791.

Xu D, Jaber Y, Pavlidis P, Gokcumen O. VCFtoTree: a user-friendly tool to construct locus-specific alignments and phylogenies from thousands of anthropologically relevant genome sequences. BMC Bioinformatics. 2017a;18:426.

Xu D, Pavlidis P, Taskent RO, et al. Archaic hominin introgression in Africa contributes to functional salivary MUC7 genetic variation. Mol Biol Evol. 2017b. https://doi.org/10.1093/molbev/msx206.

Genomics of Extinction

Johanna von Seth, Jonas Niemann, and Love Dalén

Abstract Many species went extinct during the Late Pleistocene, including a large proportion of the Earth's megafauna. Recent research on Pleistocene extinctions has started to reveal that species responded individualistically to environmental fluctuations and human interference. Through paleogenomics, it is now possible to study the extinction process in more detail, which could help disentangle why some species went extinct while others did not. Several species seem to have gone through a sudden decline right before extinction, whereas others reached the point of extinction via a gradual decline. In addition, some species experienced an initial severe bottleneck but survived for thousands of years more at reduced numbers before their final extinction. The use of temporally spaced complete genomes allows for a more direct examination of changes in genomic parameters through time, such as declines in standing genetic variation and accumulation of deleterious mutations, as a consequence of these pre-extinction processes. Additionally, the increasing access to complete ancient genomes will in the future allow researchers to investigate whether species were capable of adapting to environmental changes as well as the small population size that they were subject to prior to the extinction.

Keywords Ancient DNA · Demography · Extinction · Genetic drift · Paleogenomics

J. von Seth
Department of Bioinformatics and Genetics, Swedish Museum of Natural History, Stockholm, Sweden

Division of Systematics and Evolution, Department of Zoology, Stockholm University, Stockholm, Sweden

J. Niemann
Centre for GeoGenetics, Natural History Museum of Denmark, Copenhagen, Denmark

L. Dalén (✉)
Department of Bioinformatics and Genetics, Swedish Museum of Natural History, Stockholm, Sweden
e-mail: love.dalen@nrm.se

Charlotte Lindqvist and Om P. Rajora (eds.), *Paleogenomics*,
Population Genomics [Om P. Rajora (Editor-in-Chief)],
https://doi.org/10.1007/13836_2018_53,

1 Introduction

For any single species, extinction represents the end of evolutionary change. In a short time perspective, extinction represents the disappearance of unique genetic variation and also that an ecological niche is vacated. Put in a wider perspective, however, extinctions are merely fundamental biological processes. Through the history of life on Earth, extinctions have continuously battered millions of species while in parallel having been balanced by a continuous formation of new species.

During the Late Pleistocene (~110–11.7 thousand calendar years before present (cal kyr BP)), extraordinarily many species went extinct, not least a large portion of the megafauna (Cooper et al. 2015). Moreover, several species that survived until present day also went through dramatic population declines in the Late Pleistocene (e.g. Gordon et al. 2016; Johnson et al. 2018). The numerous Late Pleistocene extinctions have often been attributed to climate change or human interference (both directly through hunting and indirectly as the human population expanded and outrivalled other species in the competition for resources) (Barnosky et al. 2004; Lorenzen et al. 2011; Cooper et al. 2015; Saltre et al. 2016). However, recent research on Pleistocene extinctions has started to reveal a more complex story, suggesting that one factor alone cannot explain the high number of extinctions. Rather, the emerging pattern is that species responded individualistically to environmental fluctuations (Lorenzen et al. 2011; Cooper et al. 2015). Finding the causes of extinctions becomes even more challenging with the addition of components such as human interference, interspecific competition and unstable population dynamics.

Using the fossil record to track extinctions, as well as formation of new species, has often been successful. On the other hand, little can be said about the causes behind the extinctions through fossil records alone. Even with the arrival of ancient DNA (aDNA) analyses, it has proved challenging to capture a comprehensive depiction of those last moments before extinction. However, the aDNA research field keeps developing and is no longer dependent on short mitochondrial and nuclear DNA sequences. Instead it is now possible to make use of complete genomes for tracking past biological events in extinct species, and thus the prospects of finding out why some species became extinct while others did not have improved.

In a time of climate change and numerous species facing extinction, understanding the underlying mechanisms that are pushing some species towards extinction is crucial. Through paleogenomics, it is now possible to study past extinction processes in more detail and to fill in gaps that cannot be filled by morphological data and modern genomics alone (Orlando and Cooper 2014). Furthermore, paleogenomics can be used to add a more complete story of currently threatened species by studying their long-term population histories, as a complement to the snapshot of their present-day genetic status provided by modern DNA.

2 Modes of Decline

All species headed towards extinction first go through a substantial demographic population decline. However, the mode of the decline may vary and is related to the life history of the species as well as the external factors causing the decline (Fig. 1) (Purvis et al. 2000). Some species seem to go extinct without any immediately apparent reasons for the extinction. Others reach the point of extinction in a much slower rate, making it easier to track the decline in the fossil records.

Using the coalescent theory as a starting point, it is possible to investigate the demographic history of a population, since past changes in effective population size can be traced within a population's DNA (Fisher 1930; Wright 1931; Kingman 1982a, b; Kimura 1983). Several demographic history modelling methods have been developed over the past decades, such as the Bayesian skyline plot and the Pairwise Sequentially Markovian Coalescent (PSMC) model, which make use of the coalescent theory within a Bayesian statistical framework (Pybus et al. 2000; Strimmer and Pybus 2001; Drummond et al. 2005; Opgen-Rhein et al. 2005; Heled and Drummond 2008; Minin et al. 2008; Li and Durbin 2011). Shared between the methods is the testing of the hypothesis that a population has been of constant size through time by using the relationship between coalescent time and effective population size (N_e) (Kingman 1982a; Emerson et al. 2001). The relationship states that two randomly chosen DNA sequences in a small population have a higher likelihood of sharing a more recent ancestor (corresponding to fewer substitutional differences between the sequences) than two sequences randomly drawn from a large population (Kingman 1982a). Thus, changes in population size over time will leave signatures in terms of differential substitutions between sequences, where simply put a population decline corresponds to fewer substitutions and vice versa. It is also possible to infer from these methods whether a demographic event happened recently or at a more ancient time point (Emerson et al. 2001).

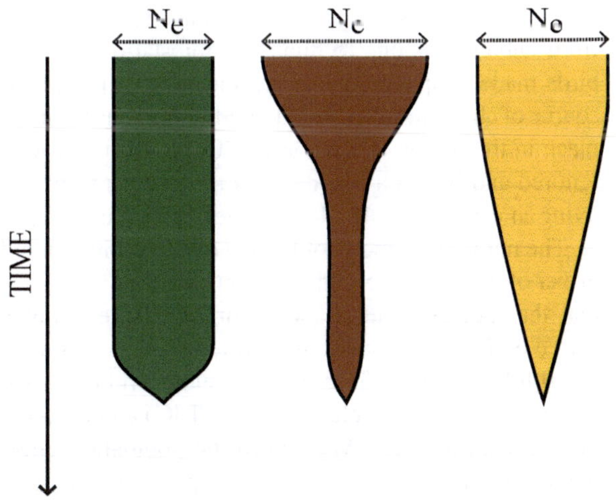

Fig. 1 Conceptual figure depicting three modes of decline before a species goes extinct; sudden decline (green), terminal refugium decline (brown) and gradual decline (yellow)

It has been argued that these models are sensitive to datasets that are too small or sparsely sampled, population structure, as well as choice of genetic loci and that only certain modes of extinction can be detected using these models (Chang and Shapiro 2016). Still, implementing these methods on high-quality ancient genomes of extinct species can enable an approximation of the mode of population decline before their extinction.

2.1 Sudden Decline

Sudden declines occur when a previously stable population collapses within a few generations, typically due to dramatic changes in its environment. Endemic island populations are particularly susceptible to this type of pre-extinction declines due to their relatively small population sizes, low genetic diversity and adaptation to an environment that is often highly distinct from the mainland (Frankham 1997). Changes to the island environment – notably the effects of human colonization – can have catastrophic consequences. Extensive overhunting and the introduction of predators and diseases led to the demise of the moa (Perry et al. 2014), thylacine (Prowse et al. 2013; Feigin et al. 2017) and dodo (Millberg and Tyrberg 1993) among many others. Even though island species only represent a fraction of all species, 75% of the species that have gone extinct in the past 400 years were endemic to islands (Frankham 1998; Sax and Gaines 2008). Extinctions that are preceded by a sudden decline are however not limited to small island populations. Before the passenger pigeon (*Ectopistes migratorius*) rapidly went extinct at the beginning of the twentieth century, it was a highly abundant species endemic to North America, potentially comprising up to 40% of the continent's avian population (Schorger 1955; Bucher 1992). In the early and mid-1800s, the population was reported to consist of billions of individuals, constantly migrating between suitable habitats in the search for food and breeding locations while hugely impacting the ecosystems along their path (Schorger 1955; Bucher 1992). The species was however suffering from habitat loss due to human deforestation. Additionally, the large numbers of birds made people associate the species to a pest, and the seemingly never-ending source of cheap meat triggered overhunting once European settlements started taking place in the region (Fulton et al. 2012). Thus, conservation legislation was largely ignored and in just a few decades the species went extinct, with the last individual dying in captivity in 1914 (Schorger 1955; Fulton et al. 2012).

The numerous reports of human overexploitation indicated that this was the main driver of the passenger pigeon extinction, but a study by Hung et al. (2014) revealed that the species regularly went through large population fluctuations in the past. Based on PSMC analyses using three ancient passenger pigeon genomes with a 13- to 20-fold average coverage, the study reported a significant decrease in N_e that started in the last interglacial period (LIG) and reached its lowest number at the last glacial maximum (LGM), before the population once again recovered. They also noted a surprisingly low N_e of the population in comparison with their large census population size (N_c). The authors thus reasoned that there must have been large fluctuations in N_c that lowered the N_e, following a previously proposed hypothesis, which stated that fluctuations in population size is one of the most important factors explaining variations in the N_e/N_c ratio (Wright 1938; Frankham 1995; Vucetich et al. 1997). To further test the hypothesis of a highly fluctuating N_c, Hung et al. (2014) analysed past fluctuations in various food and habitat resources and argued that these fluctuations would have been large enough to affect the ecosystem's carrying capacity for the passenger pigeon population. Taken together, the researchers suggested that the extinction of the passenger pigeon was a matter of bad timing. The intense hunting coincided with a low N_c during its natural cycle of fluctuations and thereby prevented the population from recovering (Hung et al. 2014).

On the other hand, in a recent study by Murray et al. (2017), analyses of 41 mitochondrial and 4 high-coverage (13- to 51-fold median coverage) nuclear passenger pigeon genomes revealed a stable population size during the approximately 20 kyr prior to the extinction and that the size of the population remained stable even when food and habitat availability was limited. In this study the researchers found indications of strong selection on diversity at linked loci, which could have led to misleading results when estimating population history using PSMC analyses (Murray et al. 2017). Both studies did however agree that human interference may have caused disruptions in the population dynamics that were strong enough to drive the species towards extinction.

2.2 *Gradual Decline*

Other species go through slower, more gradual declines that are often easier to detect than a sudden decline. Stiller et al. (2010) used mitochondrial DNA (mtDNA) to compare the demographic histories of the extinct cave bear (*Ursus spelaeus*) and the extant brown bear (*Ursus arctos*). These two species are especially good to compare since they were closely related, are thought to have had similar life history strategies, and shared habitats when the cave bear was still extant. In the study, by analysing mitochondrial D-loop sequences from 59 temporally spaced cave bears and 40 temporally spaced brown bears, they used the Bayesian coalescent approach to infer the demographic histories of the two species (Drummond et al. 2005; Stiller et al. 2010). From that, they could report that while the extant brown bears appeared to have had a constant and stable demographic history through time, the cave bear population started to decrease some 50 cal kyr BP and then continued to decrease up until their extinction approximately 24 cal kyr BP. Thus, something in the environment appears to have been affecting the cave bears negatively while leaving the brown bears undisturbed. It has been suggested that since cave bears were predominantly herbivorous (Bocherens et al. 1994; Nelson et al. 1998), they were more sensitive to climate changes causing vegetation shifts than were brown bears (Pacher and Stuart 2009). However, the onset of the decline in the cave bear population did not coincide with extreme changes in vegetation, since the population started to decline long before the onset of the cooling of the climate (Stiller et al. 2010). Another potential difference between the two species was their hibernation strategies. Reports on the relative higher amount of cave bear remains in caves in comparison with brown bear remains imply that cave bears were more reliant on caves for hibernation than were brown bears (Kurtén 1976; Stiller et al. 2010). This might have triggered a competition for access to caves between cave bears, anatomically modern humans and Neanderthals upon the latter two species' arrival to the area that forced cave bears out of the caves where they had to search for other, potentially less favourable, hibernation locations (Grayson and Delpech 2003). Furthermore, a study by Fortes et al. (2016) demonstrated that cave bears might have had a higher tendency to return to their hibernation sites year after year while brown bears did not, which would have intensified the competition between cave bears and human species even more (Fortes et al. 2016).

Taken together, these results suggest that competition of resources between the two bear species, or some other unknown environmental factor, affected the cave bears negatively while leaving the brown bears more or less unaffected long before human arrival and the initiation of extreme climate change. If the subsequent human arrival then forced cave bears out of their caves and the cooling of the climate had started, it is not unlikely that the cave bear population was struggling to remain viable (Stiller et al. 2010). In either case, the cave bear population gradually declined until it went locally extirpated and was later on replaced by another cave bear population. However, this population too could not manage to survive, and the species went globally extinct only a few thousands of years later, at approximately 24 cal kyr BP (Pacher and Stuart 2009; Stiller et al. 2010).

2.3 Terminal Refugium Decline

In a third mode of decline, species go through severe population bottlenecks, leaving just a portion of the original population behind, but still survive for thousands of years more. The well-studied woolly mammoth (*Mammuthus primigenius*) seems to have had a quite stable population size during the Late Pleistocene (Palkopoulou et al. 2013, 2015). However, the last surviving mainland mammoth population disappeared approximately 11 cal kyr BP (Nikolskiy et al. 2011). Thereafter, the last remaining populations were situated on the remote St Paul Island and Wrangel Island for another approximately 5 and 6 kyr, respectively (Vartanyan et al. 1993; Veltre et al. 2017). Analyses of the demographic history of the last surviving population, the one located on Wrangel Island, have revealed a dramatic population bottleneck some 8 kyr before their actual extinction (around the same time as the elimination of the mainland population) (Palkopoulou et al. 2015). Although the population subsequently survived for several thousand years more in this terminal refugium, several studies have shown that the population suffered from the bottleneck as well as ensuing small population size, in terms of loss of genetic diversity in both coding and non-coding regions of the genome (Lister and Stuart 2008; Palkopoulou et al. 2015; Pečnerová et al. 2016; Rogers and Slatkin 2017).

Lister and Stuart (2008) pointed out that this time lag between an extreme range contraction into a terminal refugium and the final extinction is similar to what has been termed an 'extinction lag'. This phenomenon has already been described for areas that have gone through fragmentation in modern times. Populations that appear to have remained viable post fragmentation are when further investigated discovered to be at risk of future extirpation, mainly due to gene flow barriers, increased demographic allee effects (positive density dependence), as well as decreased genetic diversity within each fragment of the population and a decreased carrying capacity of the area (Brooks et al. 1999; Dixo et al. 2009). Thus, the 'extinction lag' or 'extinction debt' refers to the future ecological and genetic cost of the fragmentation (Tilman et al. 1994; Lister and Stuart 2008). So while the last woolly mammoth population survived for some thousands years more, the fact that the Wrangel Island population was the last extant population with no possibilities for genetic rescue through gene flow into the population, as well as apparent negative genetic effects of the bottleneck, implies that the population may not have been large enough to be viable.

3 Local Population Turnovers

One of the most significant insights in paleoecology obtained through aDNA analyses is the identification of temporal population discontinuity within specific geographic regions. Such lack of continuity has either been through partial replacement of resident populations (Skoglund et al. 2012) or through extinctions followed by recolonization from genetically different source populations (Barnes et al. 2002). The latter type of population turnovers, extinctions/recolonizations, seem to have been common during the Late Pleistocene and have been described for a wide variety of wild animals as well as humans (e.g. Hofreiter et al. 2007; Leonard et al. 2007; Campos et al. 2010; Posth et al. 2016). The most pronounced example of extinctions/recolonizations comes from the collared lemming (*Dicrostonyx torquatus*), which was a keystone small herbivore that inhabited the Late Pleistocene Eurasian steppe tundra. Analyses of mtDNA sampled across a broad geographical scale and covering the last 50 kyr have indicated that the collared lemming went through a series of population extinctions throughout western Eurasia, with subsequent and repeated recolonizations from further east (Brace et al. 2012; Palkopoulou et al. 2016). These extinctions imply an unexpected instability of the Late Pleistocene ecosystem during the last Ice Age, likely caused by brief warm periods (Dansgaard-Oeschger events).

Most previous paleogenetic studies that have identified local extinctions have been based on analyses of mtDNA. However, mtDNA has limited power since it only provides information on a single gene tree, which may deviate from the species phylogeny due to its maternal inheritance, lineage sorting and introgression.

Moreover, the absence of recombination in mtDNA makes it sensitive to hitchhiking selection (Galtier et al. 2009). Because of this, future paleogenetic studies will likely use genome-wide data to revisit earlier mtDNA-based studies to re-examine the existence and timing of local extinctions. This has recently been done for Neanderthals, where analyses of multiple genomes (Hajdinjak et al. 2018) led to support for an earlier hypothesis that Neanderthals in western Europe went through a population turnover (Dalén et al. 2012). Moreover, a recent study on Paleolithic humans using genome-wide data (Fu et al. 2016) indicated that a previously identified mtDNA replacement (Posth et al. 2016) during the Allerød interstadial likely was caused by migration rather than extinction/recolonization.

4 Genomic Consequences of Demographic Declines

Regardless of the mode of decline, the mere decrease in size of a population increases its risk for extinction simply because small populations are more vulnerable to stochastic events, be they demographic, environmental or genetic (Frankham 2005). This increased risk of extinction related to decreased population size is known as the small population paradigm and was first defined by Caughley (1994). In terms of genetics, loss of genetic diversity and the exposure of recessive deleterious alleles are thought to be the most serious threats for such small populations.

In theory, loss of genetic diversity is inversely proportional to the effective population size (Frankham 2005). This is due to genetic drift, i.e. the random fixation of alleles that occurs within all populations but becomes much stronger in small ones. As a population declines, the fixation of alleles and consequently the loss of all other alleles at the corresponding loci increase (Wright 1950). This loss of standing genetic variation may in turn limit the evolutionary potential of the population (Kohn et al. 2006; Willi et al. 2006), thus reducing its capacity to evolve in response to environmental change, competition or disease. At the same time, inbreeding is likely to increase in a declining population even if mating occurs randomly, simply because the number of non-related potential mating partners decreases. While this does not necessarily result in a loss of genetic variation in the population, other than the loss that can be explained by genetic drift, inbreeding does decrease the within-individual genetic variation as more loci are becoming homozygous when individuals are more often inheriting alleles that are identical by descent (Crow 2010).

Both loss of genetic diversity and inbreeding can cause a lowered individual fitness in the population. This can take place either through an increased homozygosity at loci where heterozygote genotypes have an advantage over homozygote genotypes as, for example, in the major histocompatibility complex (MHC) (Carrington et al. 1999; Bernatchez and Landry 2003; Spurgin and Richardson 2010) or through an increased exposure of recessive deleterious alleles in homozygotes (Charlesworth and Charlesworth 1999). Recessive deleterious alleles are seldom exposed to selection in large populations and can therefore remain fairly unnoticed within a population for a relatively long time. However, since individuals

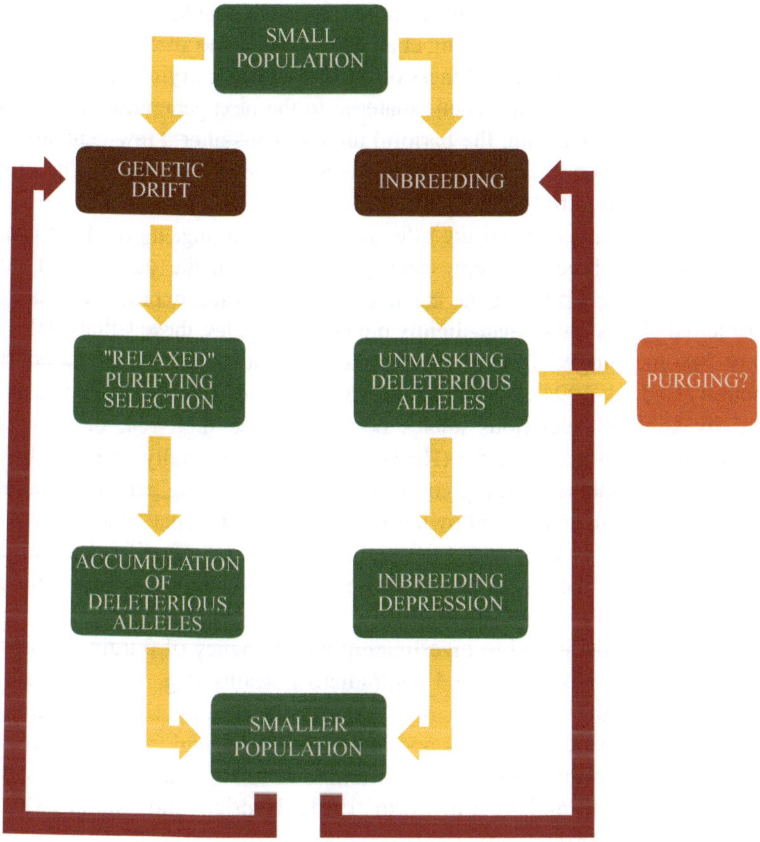

Fig. 2 Genetic processes in small populations

in declining and inbred populations more often become homozygous at loci, including those that that are carrying harmful alleles, the individual fitness in small populations is expected to become reduced (Fig. 2). When a population has reached this stage, i.e. several individuals showing clear signs of lowered fitness due to inbreeding, the population is experiencing an inbreeding depression (Charlesworth and Charlesworth 1999). However, as long as variation remains in the population at loci where recessive deleterious alleles are located, these alleles can potentially be purged from the population through purifying selection.

4.1 *Purifying Selection*

In theory, if a population is maintained at low numbers so that already existing deleterious recessive alleles (and novel ones that originate through mutation) become exposed, these alleles can be purged from the population through purifying selection (Lynch et al. 1995b; Wang et al. 1999). Since the alleles, when expressed, are expected to cause a lowered fitness of the individuals carrying them, they will be less likely to contribute with genetic material to the next generation in comparison with individuals not carrying the harmful alleles. Thus after a few generations, the population can in theory have a higher individual fitness than it had right before the harmful alleles started to become expressed.

In a study that investigated the effectiveness of purging, it was found that the genetic basis of inbreeding depression greatly affected the outcome of purging (Hedrick 1994). Generally, if the cause of the inbreeding depression was genetic load of lethal alleles rather than slightly deleterious alleles, these lethal alleles could quickly become purged from the population without a highly increased risk of extinction (Hedrick 1994). The opposite was true if the inbreeding depression was caused by slightly deleterious alleles because of the high risk of these alleles becoming fixed via genetic drift (Hedrick 1994). Additionally, concern has been raised regarding whether purging could decrease the standing genetic variation of a population by simultaneously allowing for a decrease in genetic variation at other, non-lethal, loci as a consequence of the maintained small effective population size, thereby decreasing the population's evolutionary potential (Hedrick and Miller 1992; Hedrick 1994).

Several studies dedicated to investigating the efficiency of purging in small and inbred populations have presented contradictive results (e.g. Bryant et al. 1990; Kalinowski et al. 2000). To summarize, it seems that the effectiveness of purging is highly relative, dependent on how purging is measured, and the measurements are sensitive to confounding factors such as temporal environmental changes (Bryant et al. 1990; Barrett and Charlesworth 1991; Hedrick and Kalinowski 2000; Kalinowski et al. 2000).

4.2 The Theory of Mutational Meltdown

Genetic drift can become so strong in small populations that instead of purifying selection removing new detrimental mutations that appear in the population, these mutations become fixed (Fig. 2) (Lynch and Gabriel 1990; Hedrick 1994; Lynch et al. 1995a). Once fixed within a reproductively isolated population, they are bound to be carried onto the following generations unless new mutations appear. As more harmful mutations are accumulating for each generation, the population size is likely to decrease even further (Lynch and Gabriel 1990; Wang et al. 1999; Hedrick and Kalinowski 2000). This decline in population size will in turn lead to further increased strength of genetic drift and additional fixation of detrimental mutations, thus resulting in a negative feedback loop for the population (Lynch and Gabriel 1990; Lynch et al. 1995a; Gaggiotti 2003; Charlesworth and Willis 2009). This phenomenon, where the increasing strength of genetic drift causes a negative feedback loop in population size, has been termed the population mutational meltdown by Lynch and Gabriel (1990).

4.3 *Fragmentation of Populations*

In many extinct and endangered species, demographic declines also lead to population fragmentation, which in turn can lead to increased genetic drift and inbreeding within each subpopulation (Brooks et al. 1999; Dixo et al. 2009; Frankham et al. 2017). While fragmentation can have natural causes, e.g. as rising sea levels create isolated islands with small isolated populations, human-caused fragmentation is one of the main anthropogenic threats for species and population survival in modern times (Haddad et al. 2015). The split of one population into several small populations, e.g. due to loss of suitable habitats or newly introduced barriers, at best only limits gene flow and at worst eliminates any possibilities for gene flow between the populations (Goossens et al. 2005). Regardless, the smaller population sizes caused by fragmentation increases the vulnerability to and effects of stochastic events, including genetic drift and inbreeding (Dixo et al. 2009; Pečnerová et al. 2016).

5 Paleogenomics to Study Effects of Decline

5.1 *Genetic Parameters*

One of the most important aspects of assessing the genomic consequences of a demographic decline is to determine the pre-decline status of important genomic erosion parameters, such as genome-wide diversity, inbreeding levels, as well as the amount genetic load within a population (Fig. 3). Several recent studies have indicated that there are some discordances in the theoretically acknowledged correlation between population size and the level of heterozygosity (Leffler et al. 2012; Díez-del-Molino et al. 2018). For example, the Sumatran orangutan (*Pongo abelii*) and the bonobo (*Pan paniscus*) are two currently endangered species, the former critically so (Prado-Martinez et al. 2013; IUCN 2016). Still, even though the current population sizes of the two species are similar, the Sumatran orangutan population has approximately three times higher genome-wide heterozygosity than the bonobo (Leffler et al. 2012; Prado-Martinez et al. 2013). Similarly, the giant panda, classified as vulnerable according to the IUCN red list, has significantly higher heterozygosity than humans (Cho et al. 2013; IUCN 2016). It has therefore been suggested that ancient bottlenecks and different life history strategies among species are likely to give rise to varying pre-decline levels of diversity, inbreeding and genetic load. In order to be able to distinguish the genomic effects of pre-extinction declines from the effects of more ancient events and life history traits, analysing pre-decline genomes

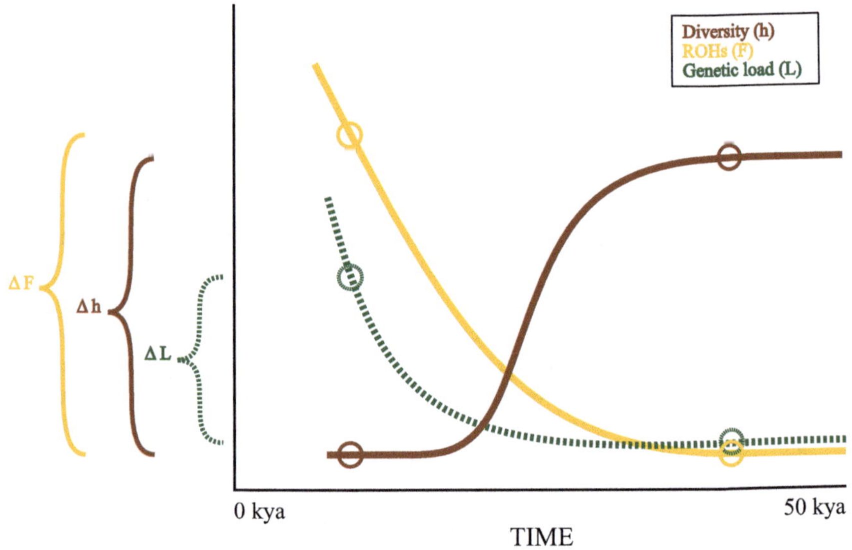

Fig. 3 Conceptual figure showing how pre-decline sampling enables direct estimates of the changes in genomic erosion parameters, such as genome-wide diversity (brown), inbreeding levels measured as amount of runs of homozygosity (ROH) (yellow) and genetic load (green), as a direct consequence of the pre-extinction decline

using, for example, century-old museum specimens can be a valuable approach in conservation genomics (Díez-del-Molino et al. 2018).

5.2 *Genomes*

When inferring past demographic events and conducting population genetic analyses based on ancient data, the most common DNA marker previously used has been mtDNA, such as the D-loop (Hofreiter et al. 2004; Valdiosera et al. 2007; Lorenzen et al. 2011). There are several benefits of using mtDNA, with one important benefit being the much higher copy number of the mtDNA genome in each cell in comparison with the nuclear genome (Clayton 1982). However, while all mtDNA is inherited from a single parent, the nuclear genome comprises several million independently, biparentally inherited loci and will therefore facilitate greater statistical power for the conduction of population genetic analyses than does mtDNA (Shapiro and Hofreiter 2014). For example, by using genome-wide single nucleotide polymorphism (SNP) sites, it is possible to estimate individual levels of heterozygosity in the population, making estimates of changes in genetic diversity more robust (Park et al. 2015).

Using whole genomes is of high importance when evaluating genetic consequences of declines, since such data in parallel enables analyses of effects on standing genetic variation, effects on fitness by analysing functional regions as well as genome-wide scans for runs of homozygosity (ROHs), i.e. long genomic fragments completely depleted from diversity (Broman and Weber 1999; McQuillan et al. 2008; Kardos et al. 2016). While generating high-coverage genome data is always preferential, there are some unneglectable obstacles for this goal when working with ancient material. First of all, as an organism dies, the natural post-mortem degradation of DNA is initiated, through, for example, enzymatic processes occurring shortly post-mortem, hydrolytic strand cleavage, lesions induced by oxygen-free radicals and cytosine deamination (Pääbo 1989; Pääbo et al. 2004; Wandeler et al. 2007; Skoglund et al. 2014). The rate of DNA degradation is to a large extent dependent on the environment in which the remains are preserved (in general, cold and dry environments can facilitate a slower rate of degradation) (Lindahl 1993). Secondly, as a consequence of this, the quality and the amount of endogenous DNA can vary greatly between different samples, and ancient samples are known to be highly sensitive to modern DNA contamination (Pääbo 1989; Pääbo et al. 2004).

With good aDNA preservation, however, high-coverage genome data can be generated. In this scenario, given the large number of independently inherited loci, only a handful of genomes or so are sufficient for inferring the extent of inbreeding and loss of genomic diversity in a population prior to its extinction (Shapiro and Hofreiter 2014). Quantification of genome-wide diversity as well as the inbreeding levels based on ROHs were recently done in two different studies of the extinct Denisovans and woolly mammoths, respectively (Meyer et al. 2012; Palkopoulou et al. 2015). Here, in-depth analyses such as long-term demographic changes, as well as individual genome-wide heterozygosity and inbreeding estimates, were generated (Meyer et al. 2012; Palkopoulou et al. 2015). Using the software mlRho (Haubold et al. 2010), the two studies reported extremely low to low heterozygosity in one 30-fold coverage Denisovan genome and one 17-fold coverage woolly mammoth genome, respectively. In both cases, the low heterozygosity could not be explained by inbreeding of immediate ancestors since no unusually long ROHs could be detected in the Denisovan genome and in the woolly mammoth genome the lengths of the ROHs were relatively short, a pattern typical for when mating between distant relatives has been taking place for several generations rather than close relatives having mated more recently (Broman and Weber 1999; Gibson et al. 2006; Meyer et al. 2012; Palkopoulou et al. 2015).

With poor aDNA quality on the other hand, only low-coverage genome data can be generated. As a consequence, the analyses will be constrained to population-level analyses. Still, a lot of new insights can come from these types of analyses, like in the case of camel evolutionary history. Through high-coverage mitochondrial genomes from two ancient Yukon *Camelops* specimens and low-coverage nuclear genomes from one of these individuals, results contradicting previous morphology-based phylogenetic analyses of the relationships between different camels species could be reported (Heintzman et al. 2015). As another example, by using the high-coverage mitochondrial and low-coverage nuclear genome of a wolf sample dated to 35 cal kyr BP, Skoglund et al. (2015) found support for the divergence between wolves and dogs haven taken place ~27–40 kyr earlier than previously suggested.

Up until recently, few high-quality genomes (>10-fold average genome coverage) of extinct species had successfully been generated. In 2008, the first report of an attempt to sequence an extinct mammalian genome was published along with a partial genome sequence covering roughly 70% of the genome, this by sequencing DNA from woolly mammoth hair (Miller et al. 2008). Subsequently in 2015, two complete woolly mammoth genomes were generated with a 17-fold and 11-fold average coverage, respectively (Palkopoulou et al. 2015). The first ancient human genome was sequenced in 2010, with a 20-fold average coverage across 79% of the genome of a Paleo-Eskimo human (Rasmussen et al. 2010). In 2014, the complete genome sequence of one Neanderthal (52-fold average coverage) (Prufer et al. 2014) and four passenger pigeons (5–20-fold coverage) (Hung et al. 2014) were generated. One year later, Park et al. (2015) managed to sequence the complete genome of the extinct aurochs (*Bos primigenius*) (six-fold average coverage). In 2017, two additional high-coverage passenger pigeon genomes (51- and 41-fold median coverage) (Murray et al. 2017) as well as the complete genome (43-fold average coverage) of the Tasmanian tiger (*Thylacinus cynocephalus*) (Feigin et al. 2017) were published. Thus, the recent advances in sequencing technologies now means that the possibilities to analyse genomes of extinct species have increased immensely and along with it comes the increasing potential for understanding pre-extinction genetic processes.

6 Future Challenges

6.1 Reference Genomes

When working with genomic data generated from extant species, there are either de novo assembled reference genomes already available for mapping the sequencing reads or such de novo genome assemblies can relatively easily be generated for the study species in question (Li et al. 2010). However, since de novo assembly requires high-quality DNA to generate high coverage across the entire genome, this is considered impossible for extinct species. Instead, one has to rely on the most closely related extant species as a reference for mapping sequencing reads. Since the most closely related species can often correspond to a divergence time of millions of years, aligning sequencing reads from an extinct species is not trivial and can often result in gaps in parts of the genome that are non-existing in the genome of the related extant species (Prüfer et al. 2010; Shapiro and Hofreiter 2014; Richmond et al. 2016). It is however still possible to conduct some analyses without a proper reference genome, such as changes in genome-wide diversity, while other important biological questions such as functional genomics may be more difficult to answer.

6.2 *Sequence Analysis*

With whole-genome sequencing comes the generation of massive amounts of data, which has led to the development of bioinformatics softwares that can be applied to such large data sets. Numerous different pipelines and bioinformatics tools are now available for filtering away low-quality sequencing reads (e.g. John 2011; Bolger et al. 2014), mapping high-quality reads to a reference genome (e.g. Li et al. 2009; Li 2013), consensus sequence generation and the conduction of data analyses to statistically test a large range of biological questions (e.g. McKenna et al. 2010; DePristo et al. 2011; Lunter and Goodson 2011). However, analysing whole genomes from ancient samples requires pipelines and bioinformatics tools that can handle data generated from poor-quality DNA and that can distinguish endogenous DNA from contaminant DNA, as well as identify nucleotide changes caused by post-mortem DNA damage. While there are some best practices available (Mourier et al. 2012) and tools applicable to low-quality data are on the rise (e.g. Schubert et al. 2014; Peltzer et al. 2016), a general issue with bioinformatics software development and usage concerns software version updates. As an increasing amount of researchers apply various bioinformatics tools to their specific data sets, new unforeseen issues arise leading to version updates of the tools to correct for these issues. Thus, during the course of a research project, the initial version of a program might have been updated several times making the first analyses irrelevant by the end of the project. Maintenance of previous versions is moreover often abandoned in favour of newer versions, and most research groups include custom-made scripts or programs specifically designed for their data, making it difficult to replicate analyses from previously published studies.

The field of paleogenomics is expanding rapidly, especially due to the increasing possibilities to generate large data sets from degraded DNA, and good practice guidelines for processing and analysing this type of data are desirable.

6.3 *De-extinction*

With the rise of whole-genome sequencing, advanced laboratory techniques that enable in vitro fertilization and cloning, as well as genetic engineering techniques to edit genomes such as CRISPR, the debate about bringing back extinct species has intensified (Jinek et al. 2012). Some argue that it is our moral responsibility to bring back the species we have once driven to extinction. Others suggest that de-extinction could be used to counteract environmental change by bringing back key species to important ecosystems, such as grasslands, where, e.g. woolly mammoths could contribute by haltering releases of carbon from soils (Zimov et al. 2012). The advancement of cloning has yielded several successful cloned animals over the past decade, e.g. the successful generation of an afghan dog puppy clone and the creation of a viable mouse clone from a dead mouse donor that had been frozen at −20°C for 16 years, both of these through somatic cell nuclear transfer (SCNT) (Lee et al. 2005; Wakayama et al. 2008). However, bringing back extinct species is even more complicated. For example, the first trial of bringing back an extinct species through cloning resulted in a Pyrenean ibex clone (*Capra pyrenaica pyrenaica*) that survived for only a few minutes (Folch et al. 2009).

Since it is today possible to sequence more or less complete genomes of extinct species, these genomes could theoretically be synthetically generated and introduced into empty nuclei of egg cells carried by surrogate mothers of closely related species. Since DNA goes through post-mortem degradation, however, the probability of generating high-quality, high-coverage genomes of extinct species decreases with the age of the specimen (Wandeler et al. 2007; Skoglund et al. 2014). In other words, the highest-quality samples are also going to be comparatively close in time to the extinction event and are consequently likely to carry genomes that are depleted of genetic diversity, contain high numbers of ROHs, and most importantly may have an excess of fixed deleterious mutations. With genetic engineering methods, such as CRISPR-Cas9, it is however to some extent possible to circumvent this issue by replacing harmful mutations (Jinek et al. 2012).

The typical read length of degraded DNA poses another challenge, as it is not possible to align fragmented ancient DNA sequences to the regions of the genome that are highly repetitive or duplicated, as this requires much longer sequencing reads than what can be retrieved from ancient samples (Treangen and Salzberg 2012). It is therefore hopeless to retrieve the complete genetic information of historical and ancient individuals, even if the DNA is relatively well preserved. Apart from the previously mentioned technical challenges, it is thus impossible to recreate a perfect clone of long-extinct individuals, as we have no knowledge of a significant part of the genome. Most de-extinction efforts therefore focus on creating hybrids that retain some key phenotypes from the extinct species. In the case of the proposed woolly mammoth hybrid, only 45 genes have been modified so far to carry woolly mammoth alleles in the Asian elephant genome (Campbell and Whittle 2017). It is unclear how mammoth-like such a hybrid would be in appearance and behaviour and whether the outcome justifies the immense efforts.

Besides molecular and genetics issues that need to be addressed before de-extinction can be realized, there are other ecological and behavioural aspects that might affect the outcome. Would the closest living relative be able to teach a newborn the way of life of another species? Are there any remaining suitable habitats for an extinct species in modern times? Are the external factors that originally contributed to the extinction gone?

Instead, perhaps the idea of de-extinction would be best applied to currently threatened species, by using genetic engineering to bring back genetic diversity and ancient, healthier allele variants (Shapiro 2017). The endangered Tasmanian devil (*Sarcophilus harrisii*), for example, is suffering from low genetic diversity not least in the MHC complex and is severely affected by a transmittable cancer (Siddle et al. 2010). Here, the use of genomic data from healthy, long-dead individuals and the CRISPR-Cas9 technique could potentially provide an alternative way to obtain a genetic rescue effect (Tallmon et al. 2004), for example, by adding diversity to the immune system within the population, thereby increasing its resilience towards the cancer and in the long-term extinction (Jinek et al. 2012).

6.4 Adaptation During the Extinction Process

While the majority of studies on extinct species focus on the cause of extinction and population demography, not much is known about whether populations are capable of adapting to the additional challenges of inbreeding depression or accumulation of deleterious alleles prior to extinction. Species such as the cheetah, channel island fox and the wandering albatross went through severe bottlenecks that led to a very low genetic heterozygosity in the present-day population, yet these species have persisted in relatively stable populations for thousands of years (Milot et al. 2007; Dobrynin et al. 2015; Robinson et al. 2016). It is not clear whether these species currently are in terminal refugia or whether they have escaped from the extinction vortex at the time of the bottleneck due to stochastic factors or species-specific behavioural strategies. Alternatively, the survival of these species could be explained by them having been able to genetically adapt to a small population size during the decline.

This question could in the future be addressed with paleogenomics by comparing the adaptive potential of in-decline populations that went extinct with those populations that persisted after the bottleneck. Genes under positive selection that enabled the population to be less vulnerable to the effects of the extinction vortex, as well as possible decreases in genetic load due to purifying selection, might also be detectable by comparing pre-decline with post-decline individuals. As neither population size nor low heterozygosity is a good proxy for the immediate extinction risk of a species (Díez-del-Molino et al. 2018), the additional information of an estimated adaptive potential could be valuable in conservation to prioritize especially vulnerable populations.

The ability of small populations to adapt to changes in the environment is especially relevant today, given the ongoing changes in climate that is likely to put additional stress on endangered species throughout the world. Paleogenomic analyses on species that became extinct in conjunction with the severe changes in climate that took place at the end of the Pleistocene could provide highly valuable information in this context. In particular, knowledge on the extent to which species were able to adapt to prehistoric temperature increases may help conservation biologists to predict how resilient present-day species will be to future climate change.

7 Conclusions and Future Perspectives

Extinction is one of the most fundamental processes in evolution. However, despite its importance to better understand today's biodiversity crisis, little is known about the demographic trajectories that precede extinction as well as how population declines affect genomic parameters. Paleogenomic analyses of taxa that went extinct in the past offer a unique opportunity to investigate how species demographies changed prior to their disappearance. Moreover, serially sampled genomic data can be used to test whether genome erosion in itself can contribute to the extinction process. Although only a few ancient genomes from wild species have been sequenced to date, this is probably going to change in the near future given the continuous decrease in high-throughput DNA sequencing costs and ongoing developments in ancient DNA recovery methods. It therefore seems highly likely that genomes from several additional extinct species will soon be made available. While this will inevitably result in an increased debate about the possibility of resurrecting these species, comparisons of genomes from multiple extinct species with those from their closest living relatives will also help emphasize the importance of having suitable genome assemblies from related extant species to use for reference-based mapping. In the near future, we are also likely to see comprehensive genomic catalogues for several extinct species, comprising multiple genomes sampled through time leading up to the extinction. Such genomic catalogues will enable detailed studies of how changes in the environment and population size have affected microevolutionary processes through time.

References

Barnes I, Matheus P, Shapiro B, Jensen D, Cooper A. Dynamics of Pleistocene population extinctions in Beringian brown bears. Science. 2002;295:2267–70.

Barnosky AD, Koch PL, Feranec RS, Wing SL, Shabel AB. Assessing the causes of Late Pleistocene extinctions on the continents. Science. 2004;306:70–5.

Barrett SCH, Charlesworth D. Effects of a change in the level of inbreeding on the genetic load. Nature. 1991;352:522–4.

Bernatchez L, Landry C. MHC studies in nonmodel vertebrates: what have we learned aboutnatural selection in 15 years? J Evol Biol. 2003;16:363–77.

Bocherens H, Fizet M, Mariotti A. Diet, physiology and ecology of fossil mammals as inferred from stable carbon and nitrogen isotope biogeochemistry: implications for Pleistocene bears. Palaeogeogr Palaeoclimatol Palaeoecol. 1994;107:213–25.

Bolger AM, Lohse M, Usade B. Trimmomatic: a flexible trimmer for Illumina sequence data. Bioinformatics. 2014;30:2114–20.

Brace S, Palkopoulou E, Dalen L, Lister AM, Miller R, Otte M, Germonpré M, Blockley SPE, Stewart JR, Barnes I. Serial population extinctions in a small mammal indicate Late Pleistocene ecosystem instability. Proc Natl Acad Sci U S A. 2012;109:20532–6.

Broman KW, Weber JL. Long homozygous chromosomal segments in reference families from the centre d'Etude du polymorphisme humain. Am J Hum Genet. 1999;65:1493–500.

Brooks TM, Pimm SL, Oyugi JO. Time lag between deforestation and bird extinciton in tropical forest fragments. Conserv Biol. 1999;13:1140–50.
Bryant EH, Meffert LM, McCommas SA. Fitness rebound in serially bottlenecked populations of the house fly. Am Nat. 1990;136:542–9.
Bucher EH. The causes of extinction of the passenger pigeon. Curr Ornithol. 1992;9:1–36.
Campbell DL, Whittle PM. Three case studies: aurochs, mammoths and passenger pigeons. In: Resurrecting extinct species. Cham: Palgrave MacMillan; 2017.
Campos PF, Willerslev E, Sher A, Orlando L, Axelsson E, Tikhonov A, Aaris-Sorensen K, Greenwood AD, Kahlke RD, Kosintsev P, Krakhmalnaya T, Kuznetsova T, Lemey P, MacPhee R, Norris CA, Shepherd K, Suchard MA, Zazula GD, Shapiro B, Gilbert MTP. Ancient DNA analyses exclude humans as the driving force behind late Pleistocene musk ox (Ovibos moschatus) population dynamics. Proc Natl Acad Sci U S A. 2010;107:5675–80.
Carrington M, Nelson GW, Martin MP, Kissner T, Vlahov D, Goedert JJ, Kaslow R, Buchbinder S, Hoots K, O'Brien SJ. HLA and HIV-1: heterozygote advantage and B*35-Cw*04 disadvantage. Science. 1999;283:1748–52.
Caughley G. Directions in conservation biology. J Anim Ecol. 1994;63:215–44.
Chang D, Shapiro B. Using ancient DNA and coalescent-based methods to infer extinction. Biol Lett. 2016;12:20150822.
Charlesworth B, Charlesworth D. The genetic basis of inbreeding depression. Genet Res. 1999;74:329–40.
Charlesworth D, Willis JH. The genetics of inbreeding depression. Nat Rev Genet. 2009;10:783–96.
Cho YS, Hu L, Hou H, Lee H, Xu J, Kwon S, Oh S, Kim HM, Jho S, Kim S, Shin YA, Kim BC, Kim H, Kim CU, Luo SJ, Johnson WE, Koepfli KP, Schmidt-Kuntzel A, Turner JA, Marker L, Harper C, Miller SM, Jacobs W, Bertola LD, Kim TH, Lee S, Zhou Q, Jung HJ, Xu X, Gadhvi P, Xu P, Xiong Y, Luo Y, Pan S, Gou C, Chu X, Zhang J, Liu S, He J, Chen Y, Yang L, Yang Y, He J, Liu S, Wang J, Kim CH, Kwak H, Kim JS, Hwang S, Ko J, Kim CB, Kim S, Bayarlkhagva D, Paek WK, Kim SJ, O'Brien SJ, Wang J, Bhak J. The tiger genome and comparative analysis with lion and snow leopard genomes. Nat Commun. 2013;4:2433.
Clayton DA. Replication of animal mitochondrial DNA. Cell. 1982;28:693–705.
Cooper A, Turney C, Hughen KA, Brook BW, McDonald HG, Bradshaw CJA. Abrupt warming events drove Late Pleistocene Holarctic megafaunal turnover. Science. 2015;349:602–6.
Crow JF. Wright and fisher on inbreeding and random drift. Genetics. 2010;184:609–11.
Dalén L, Orlando L, Shapiro B, Brandström-Durling M, Quam R, Gilbert MTP, Fernández-Lomana JCD, Willerslev E, Arsuaga JL, Götherström A. Partial genetic turnover in neandertals: continuity in the east and population replacement in the west. Mol Biol Evol. 2012;29:1893–7.
DePristo MA, Banks E, Poplin R, Garimella KV, Maguire JR, Hartl C, Philippakis AA, del Angel G, Rivas MA, Hanna M, McKenna A, Fennell TJ, Kernytsky AM, Sivachenko AY, Cibulskis K, Gabriel SB, Altshuler D, Daly MJ. A framework for variation discovery and genotyping using next-generation DNA sequencing data. Nat Genet. 2011;43:491–8.
Díez-del-Molino D, Sánchez-Barreiro F, Barnes I, Gilbert MTP, Dalén L. Quantifying temporal genomic erosion in endangered species. Trends Ecol Evol. 2018;33:176–85.
Dixo M, Metzger JP, Morgante JS, Zamudio KR. Habitat fragmentation reduces genetic diversity and connectivity among toad populations in the Brazilian Atlantic Coastal Forest. Biol Conserv. 2009;142:1560–9.
Dobrynin P, Liu S, Tamazian G, Xiong Z, Yurchenko AA, Krasheninnikova K, Kliver S, Schmidt-Kuntzel A, Koepfli KP, Johnson W, Kuderna LF, Garcia-Perez R, Manuel M, Godinez R, Komissarov A, Makunin A, Brukhin V, Qiu W, Zhou L, Li F, Yi J, Driscoll C, Antunes A, Oleksyk TK, Eizirik E, Perelman P, Roelke M, Wildt D, Diekhans M, Marques-Bonet T, Marker L, Bhak J, Wang J, Zhang G, O'Brien SJ. Genomic legacy of the African cheetah, Acinonyx jubatus. Genome Biol. 2015;16:277.
Drummond AJ, Rambaut A, Shapiro B, Pybus OG. Bayesian coalescent inference of past population dynamics from molecular sequences. Mol Biol Evol. 2005;22:1185–92.

Emerson BC, Paradis E, Thébaud C. Revealing the demographic histories of species using DNA sequences. Trends Ecol Evol. 2001;16:707–16.

Feigin CY, Newton AH, Doronina L, Schmitz J, Hipsley CA, Mitchell KJ, Gower G, Llamas B, Soubrier J, Heider TN, Menzies BR, Cooper A, O'Neill RJ, Pask AJ. Genome of the Tasmanian tiger provides insights into the evolution and demography of an extinct marsupial carnivore. Nat Ecol Evol. 2017;2:182–92.

Fisher RA. The genetical theory of natural selection. Oxford: Clarendon Press; 1930.

Folch J, Cocero MJ, Chesné P, Alabart JL, Dominguez V, Cognie Y, Roche A, Fernandez-Arias A, Marti JI, Sanchez P, Echegoyen E, Beckers JF, Bonastre AS, Vignon X. First birth of an animal from an extinct subspecies (Capra pyrenaica pyrenaica) by cloning. Theriogenology. 2009;71:1026–34.

Fortes GG, Grandal-d'Anglade A, Kolbe B, Fernandes D, Meleg IN, Garcia-Vazquez A, Pinto-Llona AC, Constantin S, de Torres TJ, Ortiz JE, Frischauf C, Rabeder G, Hofreiter M, Barlow A. Ancient DNA reveals differences in behaviour and sociality between brown bears and extinct cave bears. Mol Ecol. 2016;25:4907–18.

Frankham R. Effective population size/adult population size in wildlife: a review. Genet Res. 1995;66:95–107.

Frankham R. Do island populations have less genetic variation than mainland populations? Heredity. 1997;78:311–27.

Frankham R. Inbreeding and extinction: island populations. Conserv Biol. 1998;12:665–75.

Frankham R. Genetics and extinction. Biol Conserv. 2005;126:131–40.

Frankham R, Ballou JD, Ralls K, Eldridge M, Dudash MR, Fenster CB, Lacy RC, Sunnucks P. Genetic management of fragmented animal and plant populations. Oxford: Oxford University Press; 2017.

Fu Q, Posth C, Hajdinjak M, Petr M, Mallick S, Fernandes D, Furtwangler A, Haak W, Meyer M, Mittnik A, Nickel B, Peltzer A, Rohland N, Slon V, Talamo S, Lazaridis I, Lipson M, Mathieson I, Schiffels S, Skoglund P, Derevianko AP, Drozdov N, Slavinsky V, Tsybankov A, Cremonesi RG, Mallegni F, Gely B, Vacca E, Morales MR, Straus LG, Neugebauer-Maresch C, Teschler-Nicola M, Constantin S, Moldovan OT, Benazzi S, Peresani M, Coppola D, Lari M, Ricci S, Ronchitelli A, Valentin F, Thevenet C, Wehrberger K, Grigorescu D, Rougier H, Crevecoeur I, Flas D, Semal P, Mannino MA, Cupillard C, Bocherens H, Conard NJ, Harvati K, Moiseyev V, Drucker DG, Svoboda J, Richards MP, Caramelli D, Pinhasi R, Kelso J, Patterson N, Krause J, Paabo S, Reich D. The genetic history of Ice Age Europe. Nature. 2016;534:200–5.

Fulton TL, Wagner SM, Fisher C, Shapiro B. Nuclear DNA from the extinct Passenger Pigeon (Ectopistes migratorius) confirms a single origin of New World pigeons. Ann Anat. 2012;194:52–7.

Gaggiotti OE. Genetic threats to population persistence. Ann Zool Fenn. 2003;40:155–68.

Galtier N, Nabholz B, Glemin S, Hurst GD. Mitochondrial DNA as a marker of molecular diversity: a reappraisal. Mol Ecol. 2009;18:4541–50.

Gibson J, Morton NE, Collins A. Extended tracts of homozygosity in outbred human populations. Hum Mol Genet. 2006;15:789–95.

Goossens B, Chikhi L, Jalil MF, Ancrenaz M, Lackman-Ancrenaz I, Mohamed M, Andau P, Bruford MW. Patterns of genetic diversity and migration in increasingly fragmented and declining orang-utan (Pongo pygmaeus) populations from Sabah, Malaysia. Mol Ecol. 2005;14:441–56.

Gordon D, Huddleston J, Chaisson MJP, Hill CM, Kronenberg ZN, Munson KM, Malig M, Raja A, Fiddes I, Hillier LW, Dunn C, Baker C, Armstrong J, Diekhans M, Paten B, Shendure J, Wilson RK, Haussler D, Chin C-S, Eichler EE. Long-read sequence assembly of the gorilla genome. Science. 2016;352:aae0344.

Grayson DK, Delpech F. Ungulates and the middle-to-upper Paleolithic transition at Grotte XVI (Dordogne, France). J Archaeol Sci. 2003;30:1633–48.

Haddad NM, Brudvig LA, Clobert J, Davies KF, Gonzalez A, Holt RD, Lovejoy TE, Sexton JO, Austin MP, Collins CD, Cook WM, Damschen EI, Ewers RM, Foster BL, Jenkins CN, King AJ,

Laurance WF, Levey DJ, Margules CR, Melbourne BA, Nicholls AO, Orrock JL, Song D-X, Townshend JR. Habitat fragmentation and its lasting impact on Earth's ecosystems. Sci Adv. 2015;1:e1500052.

Hajdinjak M, Fu Q, Hubner A, Petr M, Mafessoni F, Grote S, Skoglund P, Narasimham V, Rougier H, Crevecoeur I, Semal P, Soressi M, Talamo S, Hublin JJ, Gusic I, Kucan Z, Rudan P, Golovanova LV, Doronichev VB, Posth C, Krause J, Korlevic P, Nagel S, Nickel B, Slatkin M, Patterson N, Reich D, Prufer K, Meyer M, Paabo S, Kelso J. Reconstructing the genetic history of late Neanderthals. Nature. 2018;555:652–6.

Haubold B, Pfaffelhuber P, Lynch M. mlRho – a program for estimating the population mutation and recombination rates from shotgun-sequenced diploid genomes. Mol Ecol. 2010;19(Suppl 1):277–84.

Hedrick PW. Purging inbreeding depression and the probability of extinction: full-sib mating. Heredity. 1994;73:363–72.

Hedrick PW, Kalinowski ST. Inbreeding depression in conservation biology. Annu Rev Ecol Syst. 2000;31:139–62.

Hedrick PW, Miller PS. Conservation genetics – techniques and fundamentals. Ecol Appl. 1992;2:30–46.

Heintzman PD, Zazula GD, Cahill JA, Reyes AV, MacPhee RD, Shapiro B. Genomic data from extinct North American Camelops revise camel evolutionary history. Mol Biol Evol. 2015;32:2433–40.

Heled J, Drummond AJ. Bayesian inference of population size history from multiple loci. BMC Evol Biol. 2008;8:289.

Hofreiter M, Serre D, Rohland N, Rabeder G, Nagel D, Conard N, Munzel S, Paabo S. Lack of phylogeography in European mammals before the last glaciation. Proc Natl Acad Sci U S A. 2004;101:12963–8.

Hofreiter M, Muenzel S, Conard NJ, Pollack J, Slatkin M, Weiss G, Paabo S. Sudden replacement of cave bear mitochondrial DNA in the Late Pleistocene. Curr Biol. 2007;17:R122–3.

Hung CM, Shaner PJL, Zink RM, Liu WC, Chu TC, Huang WS, Li SH. Drastic population fluctuations explain the rapid extinction of the passenger pigeon. Proc Natl Acad Sci U S A. 2014;111:10636–41.

IUCN. The IUCN red list of threatened species 2016. 2016.

Jinek M, Chylinski K, Fonfara I, Hauer M, Doudna JA, Charpentier E. A programmable dual-RNA-guided DNA endonuclease in adaptive bacterial immunity. Science. 2012;337:816–21.

John JS. SeqPrep. 2011. https://github.com/jstjohn/SeqPrep.

Johnson RN, O'Meally D, Chen Z, Etherington GJ, Ho SYW, Nash WJ, Grueber CE, Cheng Y, Whittington CM, Dennison S, Peel E, Haerty W, O'Neill RJ, Colgan D, Russell TL, Alquezar-Planas DE, Attenbrow V, Bragg JG, Brandies PA, Chong AY-Y, Deakin JE, Di Palma F, Duda Z, Eldridge MDB, Ewart KM, Hogg CJ, Frankham GJ, Georges A, Gillett AK, Govendir M, Greenwood AD, Hayakawa T, Helgen KM, Hobbs M, Holleley CE, Heider TN, Jones EA, King A, Madden D, Graves JAM, Morris KM, Neaves LE, Patel HR, Polkinghorne A, Renfree MB, Robin C, Salinas R, Tsangaras K, Waters PD, Waters SA, Wright B, Wilkins MR, Timms P, Belov K. Adaptation and conservation insights from the koala genome. Nat Genet. 2018;50:1102–11.

Kalinowski ST, Hedrick PW, Miller PS. Inbreeding depression in the Speke's gazelle captive breeding program. Conserv Biol. 2000;14:1375–84.

Kardos M, Taylor HR, Ellegren H, Luikart G, Allendorf FW. Genomics advances the study of inbreeding depression in the wild. Evol Appl. 2016;9:1205–18.

Kimura M. The neutral theory of molecular evolution. New York: Cambridge University Press; 1983.

Kingman JFC. On the genealogy of large populations. J Appl Probab. 1982a;19A:27–43.

Kingman JFC. The coalescent. Stoch Process Appl. 1982b;13:235–48.

Kohn MH, Murphy WJ, Ostrander EA, Wayne RK. Genomics and conservation genetics. Trends Ecol Evol. 2006;21:629–37.

Kurtén B. The cave bear story. New York: Columbia University Press; 1976.
Lee BC, Kim MK, Jang G, Oh HJ, Yuda F, Kim HJ, Hossein MS, Kim JJ, Kang SK, Schatten G, Hwang WS. Dogs cloned from adult somatic cells. Nature. 2005;436:641.
Leffler EM, Bullaughey K, Matute DR, Meyer WK, Ségurel L, Venkat A, Andolfatto P, Przeworski M. Revisiting an old riddle: what determines genetic diversity levels within species? PLoS Biol. 2012;10:e1001388.
Leonard JA, Vila C, Fox-Dobbs K, Koch PL, Wayne RK, Van Valkenburgh B. Megafaunal extinctions and the disappearance of a specialized wolf ecomorph. Curr Biol. 2007;17:1146–50.
Li H. Aligning sequence reads, clone sequences and assembly contigs with BWA-MEM. arXiv:1303.3997 [q-bio.GN]. 2013.
Li H, Durbin R. Inference of human population history from individual whole-genome sequences. Nature. 2011;475:493–6.
Li H, Handsaker B, Wysoker A, Fennell T, Ruan J, Homer N, Marth G, Abecasis G, Durbin R, 1000 Genome Project Data Processing Subgroup. The sequence alignment/map format and SAMtools. Bioinformatics. 2009;25:2078–9.
Li Y, Hu Y, Bolund L, Wang J. State of the art de novo assembly of human genomes from massively parallel sequencing data. Hum Genomics. 2010;4:271–7.
Lindahl T. Instability and decay of the primary structure of DNA. Nature. 1993;362:709–15.
Lister AM, Stuart AJ. The impact of climate change on large mammal distribution and extinction: evidence from the last glacial/interglacial transition. Compt Rendus Geosci. 2008;340:615–20.
Lorenzen ED, Nogues-Bravo D, Orlando L, Weinstock J, Binladen J, Marske KA, Ugan A, Borregaard MK, Gilbert MTP, Nielsen R, Ho SYW, Goebel T, Graf KE, Byers D, Stenderup JT, Rasmussen M, Campos PF, Leonard JA, Koepfli KP, Froese D, Zazula G, Stafford TW, Aaris-Sorensen K, Batra P, Haywood AM, Singarayer JS, Valdes PJ, Boeskorov G, Burns JA, Davydov SP, Haile J, Jenkins DL, Kosintsev P, Kuznetsova T, Lai XL, Martin LD, McDonald HG, Mol D, Meldgaard M, Munch K, Stephan E, Sablin M, Sommer RS, Sipko T, Scott E, Suchard MA, Tikhonov A, Willerslev R, Wayne RK, Cooper A, Hofreiter M, Sher A, Shapiro B, Rahbek C, Willerslev E. Species-specific responses of Late Quaternary megafauna to climate and humans. Nature. 2011;479:359–U195.
Lunter G, Goodson M. Stampy: a statistical algorithm for sensitive and fast mapping of Illumina sequence reads. Genome Res. 2011;21:936–9.
Lynch M, Gabriel W. Mutational load and the survival of small populations. Evolution. 1990;44:1725–37.
Lynch M, Conery J, Bürger R. Mutational meltdowns in sexual populations. Evolution. 1995a;49:1067–80.
Lynch M, Conery J, Burger R. Mutation accumulation and the extinction of small populations. Am Nat. 1995b;146:489–518.
McKenna A, Hanna M, Banks E, Sivachenko A, Cibulskis K, Kernytsky A, Garimella K, Altshuler D, Gabriel S, Daly M, DePristo MA. The Genome Analysis Toolkit: a MapReduce framework for analyzing next-generation DNA sequencing data. Genome Res. 2010;20:1297–303.
McQuillan R, Leutenegger AL, Abdel-Rahman R, Franklin CS, Pericic M, Barac-Lauc L, Smolej-Narancic N, Janicijevic B, Polasek O, Tenesa A, Macleod AK, Farrington SM, Rudan P, Hayward C, Vitart V, Rudan I, Wild SH, Dunlop MG, Wright AF, Campbell H, Wilson JF. Runs of homozygosity in European populations. Am J Hum Genet. 2008;83:359–72.
Meyer M, Kircher M, Gansauge M-T, Li H, Racimo F, Mallick S, Schraiber JG, Jay F, Prüfer K, de Filippo C, Sudmant PH, Alkan C, Fu Q, Do R, Rohland N, Tandon A, Siebauer M, Green RE, Bryc K, Briggs AW, Stenzel U, Dabney J, Shendure J, Kitzman J, Hammer MF, Shunkov MV, Derevianko AP, Patterson N, Andrés AM, Eichler EE, Slatkin M, Reich D, Kelso J, Pääbo S. A high-coverage genome sequence from an archaic Denisovan individual. Science. 2012;338:222–6.
Millberg P, Tyrberg T. Naïve birds and noble savages – a review of man-caused prehistoric extinctions of island bird. Ecography. 1993;16:229–50.
Miller W, Drautz DI, Ratan A, Pusey B, Qi J, Lesk AM, Tomsho LP, Packard MD, Zhao F, Sher A, Tikhonov A, Raney B, Patterson N, Lindblad-Toh K, Lander ES, Knight JR, Irzyk GP,

Fredrikson KM, Harkins TT, Sheridan S, Pringle T, Schuster SC. Sequencing the nuclear genome of the extinct woolly mammoth. Nature. 2008;456:387–90.

Milot E, Weimerskirch H, Duchesne P, Bernatchez L. Surviving with low genetic diversity: the case of albatrosses. Proc Biol Sci. 2007;274:779–87.

Minin VN, Bloomquist EW, Suchard MA. Smooth skyride through a rough skyline: Bayesian coalescent-based inference of population dynamics. Mol Biol Evol. 2008;25:1459–71.

Mourier T, Ho SY, Gilbert MT, Willerslev E, Orlando L. Statistical guidelines for detecting past population shifts using ancient DNA. Mol Biol Evol. 2012;29:2241–51.

Murray GGR, Soares AER, Novak BJ, Schaefer NK, Cahill J, Baker AJ, Demboski JR, Doll A, Da Fonseca RR, Fulton TL, Gilbert TP, Heintzman PD, Letts B, McIntosh G, O'Connell BL, Peck M, Pipes M-L, Rice ES, Santos KM, Sohrweide AG, Vohr SH, Corbett-Detig RB, Green RE, Shapiro B. Natural selection shaped the rise and fall of passenger pigeon genomic diversity. Sci Rep. 2017;358:951–4.

Nelson DE, Angerbjorn A, Liden K, Turk I. Stable isotopes and the metabolism of the European cave bear. Oecologia. 1998;116:177–81.

Nikolskiy PA, Sulerzhitsky LD, Pitulko VV. Last straw versus Blitzkrieg overkill: climate-driven changes in the Arctic Siberian mammoth population and the Late Pleistocene extinction problem. Quat Sci Rev. 2011;30:2309–28.

Opgen-Rhein R, Fahrmeir L, Strimmer K. Inference of demographic history from genealogical trees using reversible jump Markov chain Monte Carlo. BMC Evol Biol. 2005;5:6.

Orlando L, Cooper A. Using ancient DNA to understand evolutionary and ecological processes. Annu Rev Ecol Evol Syst. 2014;45(45):573–98.

Pääbo S. Ancient DNA: extraction, characterization, molecular cloning, and enzymatic amplification. Proc Natl Acad Sci U S A. 1989;86:1939–43.

Pääbo S, Poinar H, Serre D, Jaenicke-Despres V, Hebler J, Rohland N, Kuch M, Krause J, Vigilant L, Hofreiter M. Genetic analyses from ancient DNA. Annu Rev Genet. 2004;38:645–79.

Pacher M, Stuart AJ. Extinction chronology and palaeobiology of the cave bear (Ursus spelaeus). Boreas. 2009;38:189–206.

Palkopoulou E, Dalen L, Lister AM, Vartanyan S, Sablin M, Sher A, Edmark VN, Brandstrom MD, Germonpre M, Barnes I, Thomas JA. Holarctic genetic structure and range dynamics in the woolly mammoth. Proc R Soc B Biol Sci. 2013;280:20131910.

Palkopoulou E, Mallick S, Skoglund P, Enk J, Rohland N, Li H, Omrak A, Vartanyan S, Poinar H, Gotherstrom A, Reich D, Dalen L. Complete genomes reveal signatures of demographic and genetic declines in the woolly mammoth. Curr Biol. 2015;25:1395–400.

Palkopoulou E, Baca M, Abramson NI, Sablin M, Socha P, Nadachowski A, Prost S, Germonpre M, Kosintsev P, Smirnov NG, Vartanyan S, Ponomarev D, Nystrom J, Nikolskiy P, Jass CN, Litvinov YN, Kalthoff DC, Grigoriev S, Fadeeva T, Douka A, Higham TFG, Ersmark E, Pitulko V, Pavlova E, Stewart JR, Weglenski P, Stankovic A, Dalen L. Synchronous genetic turnovers across Western Eurasia in Late Pleistocene collared lemmings. Glob Chang Biol. 2016;22:1710–21.

Park SDE, Magee DA, McGettigan PA, Teasdale MD, Edwards CJ, Lohan AJ, Murphy A, Braud M, Donoghue MT, Liu Y, Chamberlain AT, Rue-Albrecht K, Schroeder S, Spillane C, Tai SS, Bradley DG, Sonstegard TS, Loftus BJ, MacHugh DE. Genome sequencing of the extinct Eurasian wild aurochs, Bos primigenius, illuminates the phylogeography and evolution of cattle. Genome Biol. 2015;16:234.

Pečnerová P, Díez-del-Molino D, Vartanyan S, Dalén L. Changes in variation at the MHC class II DQA locus during the final demise of the woolly mammoth. Sci Rep. 2016;6:25274.

Peltzer A, Jäger G, Herbig A, Seitz A, Kniep C, Krause J, Nieselt K. EAGER: efficient ancient genome reconstruction. Genome Biol. 2016;17:60.

Perry GLW, Wheeler AB, Wood JR, Wilmshurst JM. A high-precision chronology for the rapid extinction of New Zealand moa (Aves, Dinornithiformes). Quat Sci Rev. 2014;105:126–35.

Posth C, Renaud G, Mittnik A, Drucker DG, Rougier H, Cupillard C, Valentin F, Thevenet C, Furtwangler A, Wissing C, Francken M, Malina M, Bolus M, Lari M, Gigli E, Capecchi G,

Crevecoeur I, Beauval C, Flas D, Germonpre M, van der Plicht J, Cottiaux R, Gely B, Ronchitelli A, Wehrberger K, Grigorescu D, Svoboda J, Semal P, Caramelli D, Bocherens H, Harvati K, Conard NJ, Haak W, Powell A, Krause J. Pleistocene mitochondrial genomes suggest a single major dispersal of non-Africans and a Late Glacial population turnover in Europe. Curr Biol. 2016;26:827–33.

Prado-Martinez J, Sudmant PH, Kidd JM, Li H, Kelley JL, Lorente-Galdos B, Veeramah KR, Woerner AE, O'Connor TD, Santpere G, Cagan A, Theunert C, Casals F, Laayouni H, Munch K, Hobolth A, Halager AE, Malig M, Hernandez-Rodriguez J, Hernando-Herraez I, Prufer K, Pybus M, Johnstone L, Lachmann M, Alkan C, Twigg D, Petit N, Baker C, Hormozdiari F, Fernandez-Callejo M, Dabad M, Wilson ML, Stevison L, Camprubi C, Carvalho T, Ruiz-Herrera A, Vives L, Mele M, Abello T, Kondova I, Bontrop RE, Pusey A, Lankester F, Kiyang JA, Bergl RA, Lonsdorf E, Myers S, Ventura M, Gagneux P, Comas D, Siegismund H, Blanc J, Agueda-Calpena L, Gut M, Fulton L, Tishkoff SA, Mullikin JC, Wilson RK, Gut IG, Gonder MK, Ryder OA, Hahn BH, Navarro A, Akey JM, Bertranpetit J, Reich D, Mailund T, Schierup MH, Hvilsom C, Andres AM, Wall JD, Bustamante CD, Hammer MF, Eichler EE, Marques-Bonet T. Great ape genetic diversity and population history. Nature. 2013;499:471–5.

Prowse TA, Johnson CN, Lacy RC, Bradshaw CJ, Pollak JP, Watts MJ, Brook BW. No need for disease: testing extinction hypotheses for the thylacine using multi-species metamodels. J Anim Ecol. 2013;82:355–64.

Prüfer K, Stenzel U, Hofreiter M, Pääbo S, Kelso J, Green RE. Computantional challenges in the analysis of ancient DNA. Genome Biol. 2010;11:R47.

Prufer K, Racimo F, Patterson N, Jay F, Sankararaman S, Sawyer S, Heinze A, Renaud G, Sudmant PH, de Filippo C, Li H, Mallick S, Dannemann M, Fu Q, Kircher M, Kuhlwilm M, Lachmann M, Meyer M, Ongyerth M, Siebauer M, Theunert C, Tandon A, Moorjani P, Pickrell J, Mullikin JC, Vohr SH, Green RE, Hellmann I, Johnson PL, Blanche H, Cann H, Kitzman JO, Shendure J, Eichler EE, Lein ES, Bakken TE, Golovanova LV, Doronichev VB, Shunkov MV, Derevianko AP, Viola B, Slatkin M, Reich D, Kelso J, Paabo S. The complete genome sequence of a Neanderthal from the Altai Mountains. Nature. 2014;505:43–9.

Purvis A, Gittleman JL, Cowlishaw G, Mace GM. Predicting extinction risk in declining species. Proc Biol Sci. 2000;267:1947–52.

Pybus OG, Rambaut A, Harvey PH. An integrated framework for the inference of viral population history from reconstructed genealogies. Genetics. 2000;155:1429–37.

Rasmussen M, Li Y, Lindgreen S, Pedersen JS, Albrechtsen A, Moltke I, Metspalu M, Metspalu E, Kivisild T, Gupta R, Bertalan M, Nielsen K, Gilbert MT, Wang Y, Raghavan M, Campos PF, Kamp HM, Wilson AS, Gledhill A, Tridico S, Bunce M, Lorenzen ED, Binladen J, Guo X, Zhao J, Zhang X, Zhang H, Li Z, Chen M, Orlando L, Kristiansen K, Bak M, Tommerup N, Bendixen C, Pierre TL, Gronnow B, Meldgaard M, Andreasen C, Fedorova SA, Osipova LP, Higham TF, Ramsey CB, Hansen TV, Nielsen FC, Crawford MH, Brunak S, Sicheritz-Ponten T, Villems R, Nielsen R, Krogh A, Wang J, Willerslev E. Ancient human genome sequence of an extinct Palaeo-Eskimo. Nature. 2010;463:757–62.

Richmond DJ, Sinding M-HS, Gilbert MTP. The potential and pitfalls of de-extinction. Zool Scr. 2016;45:22–36.

Robinson JA, Ortega-Del Vecchyo D, Fan Z, Kim BY, von Holdt BM, Marsden CD, Lohmueller KE, Wayne RK. Genomic flatlining in the endangered island fox. Curr Biol. 2016;26:1183–9.

Rogers RL, Slatkin M. Excess of genomic defects in a woolly mammoth on Wrangel island. PLoS Genet. 2017;13:e1006601.

Saltre F, Rodriguez-Rey M, Brook BW, Johnson CN, Turney CS, Alroy J, Cooper A, Beeton N, Bird MI, Fordham DA, Gillespie R, Herrando-Perez S, Jacobs Z, Miller GH, Nogues-Bravo D, Prideaux GJ, Roberts RG, Bradshaw CJ. Climate change not to blame for Late Quaternary megafauna extinctions in Australia. Nat Commun. 2016;7:10511.

Sax DF, Gaines SD. Colloquium paper: species invasions and extinction: the future of native biodiversity on islands. Proc Natl Acad Sci U S A. 2008;105(Suppl 1):11490–7.

Schorger AW. The passenger pigeon: its natural history and extinction. Whitefish: Literary Licensing, LLC; 1955.

Schubert M, Ermini L, Der Sarkissian C, Jonsson H, Ginolhac A, Schaefer R, Martin MD, Fernandez R, Kircher M, McCue M, Willerslev E, Orlando L. Characterization of ancient and modern genomes by SNP detection and phylogenomic and metagenomic analysis using PALEOMIX. Nat Protoc. 2014;9:1056–82.

Shapiro B. Pathways to de-extinction: how close can we get to resurrection of an extinct species? Funct Ecol. 2017;31:996–1002.

Shapiro B, Hofreiter M. A paleogenomic perspective on evolution and gene function: new insights from ancient DNA. Science. 2014;343:1236573.

Siddle HV, Marzec J, Cheng Y, Jones M, Belov K. MHC gene copy number variation in Tasmanian devils: implications for the spread of a contagious cancer. Proc Biol Sci. 2010;277:2001–6.

Skoglund P, Malmström H, Raghavan M, Storå J, Hall P, Willerslev E, Gilbert TP, Götherström A, Jakobsson M. Origins and genetic legacy of neolithic farmers and hunter-gatherers in Europe. Science. 2012;336:466–9.

Skoglund P, Northoff BH, Shunkov MV, Derevianko AP, Paabo S, Krause J, Jakobsson M. Separating endogenous ancient DNA from modern day contamination in a Siberian Neandertal. Proc Natl Acad Sci U S A. 2014;111:2229–34.

Skoglund P, Ersmark E, Palkopoulou E, Dalen L. Ancient wolf genome reveals an early divergence of domestic dog ancestors and admixture into high-latitude breeds. Curr Biol. 2015;25:1515–9.

Spurgin LG, Richardson DS. How pathogens drive genetic diversity: MHC, mechanisms and misunderstandings. Proc Biol Sci. 2010;277:979–88.

Stiller M, Baryshnikov G, Bocherens H, Grandal d'Anglade A, Hilpert B, Munzel SC, Pinhasi R, Rabeder G, Rosendahl W, Trinkaus E, Hofreiter M, Knapp M. Withering away – 25,000 years of genetic decline preceded cave bear extinction. Mol Biol Evol. 2010;27:975–8.

Strimmer K, Pybus OG. Exploring the demographic history of DNA sequences using the generalized skyline plot. Mol Biol Evol. 2001;18:2298–305.

Tallmon DA, Luikart G, Waples RS. The alluring simplicity and complex reality of genetic rescue. Trends Ecol Evol. 2004;19:489–96.

Tilman D, May RM, Lehman CL, Nowak MA. Habitat destruction and the extinction debt. Nature. 1994;371:65.

Treangen TJ, Salzberg SL. Erratum: repetitive DNA and next-generation sequencing: computational challenges and solutions. Nat Rev Genet. 2012;13:146.

Valdiosera CE, Garcia N, Anderung C, Dalen L, Cregut-Bonnoure E, Kahlke RD, Stiller M, Brandstrom M, Thomas MG, Arsuaga JL, Gotherstrom A, Barnes I. Staying out in the cold: glacial refugia and mitochondrial DNA phylogeography in ancient European brown bears. Mol Ecol. 2007;16:5140–8.

Vartanyan S, Garutt VE, Sher AV. Holocene dwarf mammoths from Wrangel Island in the Siberian Arctic. Nature. 1993;362:337–40.

Veltre DW, Yesner DR, Crossen KJ, Graham RW, Coltrain JB. Patterns of faunal extinction and paleoclimatic change from mid-Holocene mammoth and polar bear remains, Pribilof Islands, Alaska. Quat Res. 2017;70.40–50.

Vucetich JA, Waite TA, Nunney L. Fluctuating population size and the ratio of effective to census population size. Evolution. 1997;51:2017–21.

Wakayama S, Ohta H, Hikichi T, Mizutani E, Iwaki T, Kanagawa O, Wakayama T. Production of healthy cloned mice from bodies frozen at −20°C for 16 years. Proc Natl Acad Sci U S A. 2008;105:17318–22.

Wandeler P, Hoeck PE, Keller LF. Back to the future: museum specimens in population genetics. Trends Ecol Evol. 2007;22:634–42.

Wang J, Hill WG, Charlesworth D, Charlesworth B. Dynamics of inbreeding depression due to deleterious mutations in small populations: mutation parameters and inbreeding rate. Genet Res. 1999;74:165–78.

Willi Y, van Buskirk J, Hoffmann AA. Limits to the adaptive potential of small populations. Annu Rev Ecol Evol Syst. 2006;37:433–58.
Wright S. Evolution in Mendelian populations. Genetics. 1931;16:139–56.
Wright S. Size of population and breeding structure in relation to evolution. Science. 1938;87:430–1.
Wright S. Genetical structure of populations. Nature. 1950;166:247–9.
Zimov SA, Zimov NS, Tikhonov AN, Chapin FS. Mammoth steppe: a high productivity phenomenon. Quat Sci Rev. 2012;57:26–45.

Index

C. Lindqvist, O. P. Rajora (eds.), *Paleogenomics*, Population Genomics,
https://doi.org/10.1007/978-3-030-04753-5

B

GPSR Compliance

The European Union's (EU) General Product Safety Regulation (GPSR) is a set of rules that requires consumer products to be safe and our obligations to ensure this.

If you have any concerns about our products, you can contact us on ProductSafety@springernature.com

In case Publisher is established outside the EU, the EU authorized representative is:

Springer Nature Customer Service Center GmbH
Europaplatz 3
69115 Heidelberg, Germany

Batch number: 10755329

Printed by Printforce, the Netherlands